G Protein-Coupled Receptors in Drug Discovery

Drug Discovery Series

Series Editor

Andrew Carmen

Johnson & Johnson PRD, LLC
San Diego, California, U.S.A.

G Protein-Coupled Receptors in Drug Discovery

Edited by
Kenneth H. Lundstrom
Mark L. Chiu

CRC Press
Taylor & Francis Group
Boca Raton London New York

CRC Press is an imprint of the
Taylor & Francis Group, an **informa** business
A TAYLOR & FRANCIS BOOK

CRC Press
Taylor & Francis Group
6000 Broken Sound Parkway NW, Suite 300
Boca Raton, FL 33487-2742

First issued in paperback 2019

© 2006 by Taylor & Francis Group, LLC
CRC Press is an imprint of Taylor & Francis Group, an Informa business

No claim to original U.S. Government works

ISBN-13: 978-0-8247-2573-0 (hbk)
ISBN-13: 978-0-367-39251-2 (pbk)

Library of Congress Card Number 2005048556

Library of Congress Cataloging-in-Publication Data

G protein-coupled receptors in drug discovery / [edited by] Kenneth H. Lundstrom and Mark L. Chiu.
    p. cm.
  Includes bibliographical references and index.
  ISBN 0-8247-2573-5
  1. G proteins – receptors. 2. Drug development. I. Lundstrom, Kenneth H. II. Chiu, Mark L.

QP552.G16G1745 2005
612'.01575 – dc 22   2005048556

Library of Congress Card Number 2005048556

Visit the Taylor & Francis Web site at
http://www.taylorandfrancis.com

and the CRC Press Web site at
http://www.crcpress.com

Foreword

The most widely used and safest medicines today are ligands to GPCRs. The first generation of antihypertensives β-blocked the β-adrenergic receptor, the later generations include blockers of the angiotensin receptor, both GPCRs. Antihistamines to fight hay fever and allergy, antipsychotics, and some antidepressant drugs are also ligands to GPCRs. Furthermore, the analgesic actions of morphine and its analogs are exerted at GPCRs and lately antiviral agents blocking HIV entry into cells are also targeting GPCRs. The pharmaceutical industry at every level — biology, chemistry and toxicology — became so familiar with medicines that target GPCRs for chronic drug therapy that this class of proteins is by far the most favored class of new drug targets. There are still orphan GPCRs whose endogenous ligands and function is not known, these GPCRs hold additional promises for drug therapy together with the very large group of GPCRs that bind neuropeptides and peptide hormones. We are going to see rapid exploration of these GPCRs as drug targets in the treatment of diabetes, obesity and major depression. Hence all new knowledge on expression structure and function of GPCRs is of great value for all who work on developing new medicines.

Tamas Bartfai
Chair and Professor of Neuropharamcology
The Scripps Research Institute
La Jolla, CA, USA
formerly Senior VP for CNS Research at Hoffmann-La Roche, Basel, Switzerland

Preface

The impacts of G protein-coupled receptors (GPCRs) from social, therapeutic, and economic points of view are enormous. The broad range of GPCRs covers all areas of modern medicine. The effects of point mutations, overexpression or perturbed regulation of GPCR function, and their overstimulation or inhibition affect such areas as cardiovascular, metabolic, neurodegenerative, psychiatric, and viral diseases and cancers. Currently, more than 50% of drug targets are based on GPCRs and annual worldwide sales reached $47 billion in 2003. The increase is anticipated to continue.

Novel mechanisms of GPCR action through dimerization and the continuous deorphanization of orphan GPCRs may lead to discovery of novel drug interactions and help fuel continued interest by the pharmaceutical industry. Furthermore, the success of structural biology in providing the means to achieve tailor-made structure-based drug design will soon come of age for GPCRs. The first structure of a natively isolated GPCR was solved in 2000 and, based on strong indications, a breakthrough in high-resolution structures for recombinantly expressed GPCRs is at our door step. This will further intensify the drug development programs focused on GPCRs because it will lead to discovery of better medicines with improved efficacy and better selectivity along with fewer side effects.

The physiological roles of GPCRs and their involvements, sometimes with other proteins, in various human diseases are described in this book. The chapters also present current approaches in drug discovery that include target selection, establishment of screening and functional assays, recombinant GPCR expression for drug screening and structural biology, different methods to obtain structures of GPCRs, and the importance of bioinformatics. Despite serious editing, some overlap of certain topics could not be avoided. In fact, some duplication was intentionally permitted so that each chapter could function as an independent unit.

We would like to thank all the chapter authors for their valuable contributions to this book. It has been a pleasure to work with them and inspiring to observe the positive attitudes and enthusiasm for this project. Finally, we are grateful to CRC Press and especially to Anita Lekhwani, senior acquisitions editor, chemistry, Jill Jurgensen, production coordinator, and Robert Sims, project editor, for publishing this book.

Kenneth H. Lundstrom, Ph.D.
Mark L. Chiu, Ph.D.

Editors

Kenneth H. Lundstrom, Ph.D. earned his Ph.D. from the University of Helsinki, Finland on overexpression of viral membrane proteins in *Bacillus subtilis*. He conducted postdoctoral research at Cetus Corporation in California on antisense expression and PCR technologies. Dr. Lundstrom then returned to his native Finland where he was appointed a senior scientist at Orion Pharmaceuticals where he worked on cloning, expression, and structural studies of catechol-o-methyltransferases.

From 1992 through 1995, Dr. Lundstrom developed Semliki Forest virus vectors for overexpression of receptors (GPCRs) and ion channels at the Glaxo Institute of Molecular Biology, in Geneva, Switzerland. He worked as a principal biologist at the Glaxo Medicines Research Centre in Stevenage, United Kingdom in 1995 and 1996. Between 1996 and 2001, he researched receptor expression in the central nervous system for Hoffmann-La Roche in Basel, Switzerland.

Dr. Lundstrom was appointed the scientific coordinator of the MePNet Program in 2001 and became the chief scientific officer of BioXtal in Lausanne, Switzerland in 2002. He is also part of the senior management team (vice president, science and technology) of Regulon Inc., Mountain View, California, a biotech company involved in cancer therapy. Dr. Lundstrom has published more than 100 scientific papers and reviews in international journals, acts as an editor for books in the fields of GPCRs, structural genomics and gene therapy, and is a frequent speaker at international conferences.

Mark L. Chiu, Ph.D. earned a bachelor's in biophysics from the University of California at Berkeley. He then joined Microgenics Corporation as a synthetic organic chemist. He later obtained his Ph.D. in biochemistry from the University of Illinois at Urbana–Champaign, working on the overexpression and biophysical characterizations of heme proteins. He began his postdoctoral work in the field of nuclear magnetic resonance methodology at the Federal Institute of Technology at Zurich, Switzerland and later joined the Biocenter of the University of Basel, working on the mechanisms of bacterial multidrug resistance and the general application of lipidic cubic phases for membrane protein crystallization. After that, he worked in the Department of Chemistry at Seton Hall University, characterizing properties of bacterial membrane proteins. Dr. Chiu is presently working in the department of structural biology at Abbott Laboratories, Abbott Park, Illinois, developing tools for structure-based drug design.

Contributors

Remko A. Bakker, Ph.D.
Division of Medicinal Chemistry
Faculty of Sciences
Vrije Universiteit Amsterdam
Amsterdam, The Netherlands

Marc Baldus, Ph.D.
Department for NMR-Based Structural
  Biology
Max Planck Institute for Biophysical
  Chemistry
Göttingen, Germany

David Burns, Ph.D.
Department of Biological Screening
Global Pharmaceutical Research and
  Development
Abbott Laboratories
Abbott Park, Illinois

Mark L. Chiu, Ph.D.
Department of Structural Biology
Global Pharmaceutical Research and
  Development
Abbott Laboratories
Abbott Park, Illinois

Olivier Civelli, Ph.D.
Department of Developmental and Cell
  Biology and Department of
  Pharmacology
University of California at Irvine
Irvine, California

Christine A. Collins, Ph.D.
Metabolic Disease Research
Global Pharmaceutical Research and
  Development
Abbott Laboratories
Abbott Park, Illinois

Timothy A. Esbenshade, Ph.D.
Neuroscience Research
Global Pharmaceutical Research and
  Development
Abbott Laboratories
Abbott Park, Illinois

Sujatha Gopalakrishnan, M.Sc.
Department of Biological Screening
Global Pharmaceutical Research and
  Development
Abbott Laboratories
Abbott Park, Illinois

Deborah S. Hartman, Ph.D.
Lead Discovery Department
AstraZeneca Pharmaceuticals LP
Wilmington, Delaware

Judith Klein-Seetharaman, Ph.D.
University of Pittsburgh
School of Medicine
Pittsburgh, Pennsylvania

Adam Lange, B.Sc.
Department for NMR-Based Structural
  Biology
Max Planck Institute for Biophysical
  Chemistry
Göttingen, Germany

Michèle C. Loewen, Ph.D.
Plant Biotechnology Institute
National Research Council of Canada
Saskatoon, Saskatchewan, Canada

Kenneth H. Lundstrom, Ph.D.
BioXtal
Epalinges, Switzerland

Maria P. MacWilliams, Ph.D.
Department of Biology
University of Wisconsin Parkside
Kenosha, Wisconsin

Graeme Milligan, Ph.D.
Institute of Biomedical and Life
  Sciences
University of Glasgow
Glasgow, Scotland, United Kingdom

Heike Obermann, Ph.D.
Institute for Hormone & Fertility
  Research
University of Hamburg
Hamburg, Germany

James B. Procter, Ph.D.
School of Life Sciences
University of Dundee
Dundee, Scotland

Rita Raddatz, Ph.D.
Lead Discovery Department
AstraZeneca Pharmaceuticals LP
Wilmington, Delaware

Regina M. Reilly, Ph.D.
Metabolic Disease Research
Global Pharmaceutical Research and
  Development
Abbott Laboratories
Abbott Park, Illinois

Yumiko Saito, Ph.D.
Department of Pharmacology
Saitama Medical School
Saitama, Japan

Mika Scheinin, M.D., Ph.D.
Department of Pharmacology and
  Clinical Pharmacology
University of Turku
Turku, Finland

Martine J. Smit, Ph.D.
Division of Medicinal Chemistry
Faculty of Sciences
Vrije Universiteit Amsterdam
Amsterdam, The Netherlands

Amir Snapir, M.D., Ph.D.
Department of Pharmacology and
  Clinical Pharmacology
University of Turku
Turku, Finland

Jurgen Vanhauwe, Ph.D.
Drug Discovery Department
Commercial Operations
InVitrogen Corporation
Carlsbad, California

Zhiwei Wang, Ph.D.
Department of Pharmacology
University of California at Irvine
Irvine, California

Usha Warrior, Ph.D.
Department of Biological Screening
Global Pharmaceutical Research and
  Development
Abbott Laboratories
Abbott Park, Illinois

Kevan P. Willey, Ph.D.
Centre for Bioinformatics
University of Hamburg
Hamburg, Germany

Contents

1 Introduction

Kenneth H. Lundstrom and Mark L. Chiu

CONTENTS

ABSTRACT

G protein-coupled receptors (GPCRs) represent the largest and most important family of drug targets today. More than 50% of the drugs currently on the market are based on GPCRs and worldwide annual sales reached $47 billion in 2003. The highly remunerative nature of this family of proteins is derived from their importance in many aspects of human physiology. GPCRs have an extremely broad range of mechanisms by which they transduce information through various signaling pathways within cells, but they are also involved in intercellular mechanisms. The activation of GPCRs occurs first through binding of peptides, neurotransmitters, hormones, odors, ions, light, odorants, pheromones, amino acids, amines, nucleotides, nucleosides, prostaglandins, and other small molecular weight compounds. Subsequent G protein activation via GPCRs plays important roles in many types of human maladies such as cardiovascular, metabolic, neurodegenerative, neurological, and viral diseases, as well as cancers.

1.1 SCOPE OF BOOK

This book describes the principal mechanisms for signal transduction through activation of GPCRs and their interactions with G proteins. Fairly recently, it was discovered that GPCRs can also signal through alternative pathways without interactions with G proteins. This finding has cast doubts on the correctness of naming these receptors GPCRs. Rather it has been suggested that the scientific community refer to them as seven transmembrane (7TM) receptors based on their common structural topologies of seven transmembrane-spanning domains. These 7TM receptors can interact with many proteins other than G proteins to modulate signal transduction. Since most of the 7TM receptors mediate signal transduction through G proteins, it is therefore still relevant to call them GPCRs.

Chapter 2 describes the signal transduction mechanisms in detail and includes various examples. The importance of the GPCRs to drug discovery in general and as drug targets is outlined in Chapter 3. Specific areas of medicine such as cardiovascular disease (Chapter 4), cancer (Chapter 5), metabolic disease (Chapter 6), and neuroscience including neurodegeneration and psychiatry indicators (Chapter 7) are also covered. Chapter 9 is dedicated to high throughput screening methods for GPCRs that naturally play important roles in modern drug discovery.

Much attention has been paid to the structural biology of GPCRs. Such efforts can open new avenues for designing drugs with higher levels of potency and engineering specificity to certain GPCRs. Although bovine rhodopsin is the only GPCR for which a high resolution structure has been obtained to date, the strong trend today is to make serious investments in this area. Because structural biology demands milligram quantities of proteins, a need exists to develop robust expression systems for GPCRs.

Since most GPCRs are not overexpressed intrinsically *in vivo* and *in situ*, the facilitation of high-level recombinant heterologous expression systems has been one of the major bottlenecks in investigating the structural biologies of membrane proteins. Obviously, recombinant GPCR expression *also* serves an important function in the drug screening process. All applied and potentially interesting expression systems are described in Chapter 8. *In silico* methods including application of bioinformatics and molecular modeling for GPCRs as tools to support structural biology approaches are covered in Chapter 10. Chapter 11 provides further insight into structural biology identification and deals with the structures and dynamics of GPCRs using rhodopsin as a model protein. Chapter 12 deals with the problems and recent development in crystallization of GPCRs, and Chapter 13 demonstrates how novel solid-state NMR methods can be applied to GPCRs. Chapter 14 is an overview of several recently established national and international networks that bring together expertise from various areas — expression, purification, and crystallization — as a means to study a large number of GPCRs in parallel.

Finally, the dimerization phenomenon, originally considered a curiosity, has been demonstrated to occur much more frequently than originally anticipated through improved analytical methods. The importance of dimerization in relation to drug discovery is described in Chapter 15. Such interactions provide more avenues for G protein activation and potential drug design. Another issue of great interest, especially in relation to the development of novel drugs, is the discovery of a large number of orphan receptors and the extensive subsequent program of their deorphanization, presented in Chapter 16.

1.2 SUMMARY

The identification of the roles of the different GPCRs will shed even more light on the dynamic interplay of signal transduction. During the past 10 to 15 years, the vast amount of information gathered on the functions of GPCRs through studies involving molecular biology, cell biology, pharmacology, and structural biology of these pharmaceutically important receptors has significantly increased. However, the wide range of GPCR functions and signaling, their complicated signaling mechanisms, and interactions with other cellular components give the impression that we have only scratched the surface of the iceberg. We anticipate therefore that continuous committed research will widen our understanding and potentially lead to the future development of greatly improved drugs.

2 Biology of G Protein-Coupled Receptors

Kenneth H. Lundstrom

CONTENTS

ABSTRACT

Many intracellular and intercellular functions are mediated by G protein-coupled receptors (GPCRs). These signal transduction events can occur through the interactions of GPCRs with G proteins and are triggered by a wide variety of inducers such as hormones, neurotransmitters, chemokines, ions, light, and odors. The activation results in a cascade of signal transduction events such as stimulation or inhibition of adenylyl cyclase and activation of phospholipase C. GPCRs must be inactivated or desensitized; this can take place at the receptor or G protein level, generally through phosphorylation mediated by protein kinase A or C. G protein-coupled kinases (GRKs) can also be involved in the desensitization events.

The interference of arrestin with receptor molecules can prevent further GPCR and G protein interactions. Cellular signal transduction pathways can also be established through the interaction of GPCRs with proteins other than G proteins. For example, tyrosine kinase interactions with the third intracellular domains of GPCRs result in the activation of the extracellular signal-regulated kinase (ERK)/mitogen-activated protein kinase (MAPK) cascade. Moreover, the interaction between β-arrestin and c-Src can facilitate GPCR-dependent activation of the ERK/MAPK pathway and may play an important role in the GPCR-mediated glucose transport through the stimulation of the GLUT4 glucose transporter. C-terminal interaction of GPCRs with Janus kinases can lead to the activation of the transcription factor STAT (signal transducer and activator of transcription). Arrestin interactions with GPCRs have been shown to be involved in trafficking of GPCRs, that is, their internalization and recycling to the plasma membrane.

2.1 INTRODUCTION

GPCRs serve as crucial mediators for various cellular signal transduction events that provide the means for cells, tissues, organs and whole organisms to react properly to changing environmental requirements. Their functions are extremely diverse as they regulate many physiological processes related to neurological and neurodegenerative functions, cardiovascular mechanisms, and metabolic control.[1] GPCRs can also act as co-receptors for cellular entry of the human immune deficiency virus (HIV).[2]

Extracellular signaling is triggered through hormones, neurotransmitters, chemokines, calcium ions, light, and odorants and leads to the activation of GPCRs, resulting in a cascade of signaling through various cellular pathways.[3] The GPCR designation relates to the intracellular signaling through guanine nucleotide-binding proteins (G proteins) although alternative mechanisms have been described recently.[4]

The estimated number of GPCRs in the human genome is 800. A large number belong to the subfamily of odorant receptors. Due to their many essential physiological functions, GPCRs play an important role in drug discovery. More than 60% of the current drug targets are focused on GPCRs and a quarter of the 200 top selling drugs are based on GPCRs.[5] The various indications and the more detailed mechanisms of drug and GPCR interactions are described in subsequent chapters.

Common to all GPCRs is their topology of seven transmembrane-spanning domains (7TMs) consisting of α-helical structures and they are therefore also called 7TM receptors. Each GPCR possesses an extracellular N-terminus and an intracellular C-terminus with various intracellular and extracellular loop regions connecting the transmembrane regions. This chapter will describe the different families of GPCRs, their functions, and their couplings to G proteins. Alternative signaling mechanisms for GPCRs are also discussed. Finally, the cellular trafficking of GPCRs is described.

2.2 FAMILIES OF GPCRs

The overall amino acid sequence homology among GPCRs is rather low. Certain homologous regions arise from sequence alignment of the 7TMs within the GPCR families. The largest group of GPCRs is designated family A or class 1 and consists of light receptors (rhodopsin), adrenaline (adrenergic) receptors, and olfactory receptors. Every member shows a highly conserved arginine in the Asp–Arg–Tyr (DRY) motif[6] located at the cytoplasmic side of the third transmembrane domain (TM3).

Approximately 25 peptide hormone and neuropeptide receptors belong to family B or class 2. They include secretin, glucagon and vasoactive intestinal peptides, calcitonin, and corticotrophin-releasing hormone. Family B GPCRs generally couple through the G_s G protein-activating adenylyl cyclase. The family C receptors consist of the metabotropic glutamate receptors (mGluRs), the γ-amino butyric acid (GABA) receptors, the calcium-sensing receptor, and certain taste receptors. Their typical feature is an exceptionally large extracellular N-terminal region that plays a crucial role in ligand binding and receptor activation. Additionally, the yeast pheromone receptors and the cAMP receptors from *Dictyostelium discoideum* form three smaller families designated D, E, and F. The classification of GPCRs (including the novel GRAFS system) is described in more detail in Chapter 10.

2.3 COUPLING TO G PROTEINS

The classical mechanism by which GPCRs are functionally active is defined by their coupling to guanine nucleotide-binding proteins or G proteins.[7] Typically, a GPCR is in its resting or low affinity state in the absence of agonists. The binding of an agonist to a receptor triggers a modification in the receptor conformation and the formation of a complex with the intracellular G proteins (Figure 2.1). The heterotrimeric G protein consists of a membrane-associated G_α subunit and the more

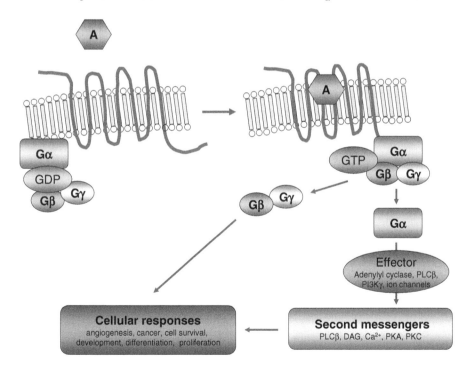

FIGURE 2.1 GPCR signalling through G proteins. The interaction with agonist leads to a conformational change of the receptor and the coupling to G proteins. One consequence is the activation of a signal transduction cascade through the G_α subunit. Also, G_β and G_γ can induce activation of cellular responses. A = agonist or antagonist. DAG = diacylglycerol. PI3Kγ = phosphoinositide 3-kinase-γ. PLCβ = phospholipase Cβ. PKA = protein kinase A. PKC = protein kinase C.

TABLE 2.1
G Proteins and Signal Transduction Events They Mediate

G protein	Signal Transduction	Reference
Gαq, Gα11	Phospholipase C activation; regulation of phosphoinositide-specific phospholipase C activity	44
Gα14, Gα15, Gα16	Rho GEF activation	45
Gαs	Adenylyl cyclase activation; RGS-PX1 acts as a GTPase-activating protein for Gαs	46
Gαolf	Calcium channel interaction; c-Src tyrosine kinase induction	47
Gαi	Adenylyl cyclase inhibition; induction of c-Src tyrosine kinase	48, 49
Gαo	K$^+$ channel activation; increase in K$^+$ leads to hyperpolarization	50
Gαz	Cellular cross-talk through Rap1GAP1	51
Gα12	Rho activation	52
Gα13	E-Cadherin; β-catenin release	53
GαT transducin	cGMP phosphodiesterase increase	54
Gαgust gustiducin	Taste-cell specific G protein, closely related to transducin	55
G$\beta\gamma$	Activation of GIRK K$^+$ channels	8
GIRKs	Adenylyl cyclase activation; phospholipase C activation	56, 57

mobile G_β and G_γ subunits. Modern genomic analysis has made it possible to identify several G protein subunits to reach the current number of 16 G_α, 5 G_β and 12 G_γ subunits that have been cloned and sequenced.[3] The G_α subunit binds GTP and GDP and is responsible for GTP hydrolysis. The G_β and G_γ subunits form a $\beta\gamma$ complex to a stoichiometry of one to one.

After the interaction of GPCR and G protein is established, GDP is released and replaced by GTP, after which the G_α subunit is dissociated from the $G_{\beta\gamma}$ dimer. Both the G_α subunit and the $G_{\beta\gamma}$ dimer can activate cellular effector molecules. For example, G_α can activate adenylyl cyclase, resulting in increased cyclic AMP (cAMP) levels, which in turn can induce protein kinase A (PKA) activation. PKA is a serine–threonine kinase that phosphorylates transcription factors, other kinases, and GPCRs.

The different types of G_α subunits play crucial roles in signal transduction and can be divided into four groups. The $G_{\alpha s}$ subunit is involved in the stimulation of adenylyl cyclase, whereas the $G_{\alpha i}$ subunit functions in the inhibition of adenylyl cyclase and the activation of GIRK (G protein-coupled inwardly rectifying potassium) channels.[8] $G_{\alpha q}$ is responsible for phospholipase C activation and $G_{\alpha 12}$ activates the Rho guanidine nucleotide exchange factors (GEFs). The $G_{\beta\gamma}$ dimers can independently activate GIRK channels, adenylyl cyclase, and phospholipases. Additional functions of G protein subunits are summarized in Table 2.1.

2.4   GPCR DESENSITIZATION

Because GPCR signaling potentiates many important cellular functions, it is essential that the mechanism of activation is initiated and terminated rapidly, specifically, and precisely even when agonist stimulation is continued. This phenomenon of desensitization can function both on the GPCR and G protein levels. Rapid termination of GPCR function can occur through receptor phosphorylation mediated by PKA or PKC (protein kinase C). Receptor phosphorylation is a typical example of negative feedback regulation as the GPCR is directly uncoupled from the G proteins. The desensitization can also target other receptors such as in cases where kinase activation of one GPCR leads to the phosphorylation and desensitization of another GPCR. Receptor phosphorylation can also provide the mechanism to specify coupling to a particular G protein. For example, PKA-mediated phosphorylation of the β2 adrenergic receptor resulted in desensitization in relation to Gs in favor of enhanced coupling to Gi.[9]

G protein-coupled receptor kinases (GRKs) can also mediate desensitization of GPCRs.[4] To date, seven GRK genes have been identified; GRK1 is also called rhodopsin kinase. GRK2 is also known as the β-adrenergic receptor kinase. In the case of GRKs, phosphorylation can occur only on activated or agonist-occupied receptors. Another interacting component is arrestin that binds to the receptor and thereby prevents further GPCR and G protein interactions. There are four arrestins of which two are expressed in the retina as visual and cone arrestins.[10] The other two, β-arrestin-1 (arrestin-2) and β-arrestin-2 (arrestin-3), are expressed in many tissues.

Other processes that contribute to GPCR desensitization are receptor degradation, which generally occurs in lysosomes,[11] and regulation of gene transcription or translation.[12] The termination of GPCR activity is, however, rather slow and can take several hours, whereas the events described for receptor phosphorylation require only minutes. The regulation of GPCRs also occurs at a post-expression level. The family of 25 proteins known as regulators of G protein signaling (RGS) serve as GTPase-activating proteins (GAPs) for G proteins.[13] The RGSs possess characteristic 130 amino acid homologous regions. The various RGSs seem to have selectivity for specific G proteins and even distinct receptors. The mechanism of RGSs is to enhance the hydrolysis rate of GTP, which should aid in termination of receptor activity. Knockout studies on RGS9 clearly demonstrated that light-activated rhodopsin was deactivated.[14]

2.5 OTHER SIGNALING PATHWAYS

In addition to coupling to G proteins, GPCRs have been shown to mediate signal transduction events through interactions with other cellular proteins (Figure 2.2). For example, c-Src tyrosine kinase interacts with the proline-rich SH3 (Src homology-3) domain in the third intracellular loop of the β3-adrenergic receptor, which leads to the activation of the extracellular signal-regulated kinase (ERK)–mitogen-activated protein kinase (MAPK) cascade.[15] It has also been demonstrated that the interaction between β-arrestin-1 and c-Src can facilitate the β2-adrenergic receptor-dependent activation of the ERK–MAPK pathway.[16]

The β-arrestin–c-Src interaction plays an important role in the glucose transport mediated by endothelin receptors through its stimulation of the GLUT4 glucose transporter.[17] In a similar way, the interleukin-8-stimulated granule release is facilitated.[18] Another example of a receptor–protein interaction relates to the activation

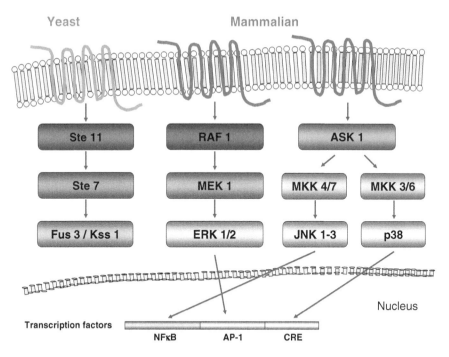

FIGURE 2.2 Other signalling pathways. GPCRs can also trigger the activation of other signalling pathways such as the ERK/MAPK and JNK pathways, which leads to the induction of transcription factors.

of the STAT (signal transducer and activator of transcription) transcription factor.[19] In this case, the angiotensin 1A receptor interacts through its C-terminus with JAK2 (Janus kinase).

Several other examples demonstrate interactions of GPCRs and cellular proteins. Typically, the interaction of β2-adrenergic receptor and the NHERF1 and NHERF2 sodium–hydrogen exchanger regulatory proteins is mediated by PDZ[20,21] Metabotropic glutamate receptors such as mGluR7 also can interact with PDZ-domain-containing proteins.[22] The biological and pharmacological significance of non-G protein signaling is probably to function as a means to enhance the specificity of GPCR signaling. It may also allow distinct physiological functions of two different GPCRs otherwise activated by the same ligand and coupled to the same G protein.

In addition to the interactions of individual receptors and individual cellular proteins, more general interactions can occur between GPCRs and adaptor/scaffolding proteins. The function of β-arrestin as an adaptor/scaffolding protein for the activation of MAPK cascades has been described above. Recently, it was discovered that β-arrestin can interact with the last three kinases in both the ERK/MAPK pathway and the JNK3 (c-Jun amino-terminal kinase 3) cascade.[23,24] The function of β-arrestin seems to be establishment of the contacts of the kinases in both the ERK/MAPK and JNK3 cascades with each other and the GPCRs, which facilitates the signal transduction from receptor to effector. Interestingly, the β-arrestin-mediated scaffolding of MAPK cascades results in kinase retention in the cytoplasm, which promotes the internalization of receptors as described below.[25] In summary, scaffolding proteins can enhance the overall activation of proteins in signal transduction pathways and can increase specificity by interaction with only those proteins that will be activated. In this context, β-arrestin-2 only induces the activation of JNK3, although two other JNK isoforms exist.[24]

An important fairly recent phenomenon observed for the signal transduction of GPCRs is the dimerization of receptors.[26] In case of signaling through receptor tyrosine kinases, for dimeric receptors each monomer can phosphorylate the other monomer during ligand-stimulated autophosphorylation to achieve a fully active state of the kinase. GPCR dimerization is described in more detail in Chapter 15.

2.6 TRAFFICKING OF GPCRs

Although the localization of GPCRs has been associated with the plasma membrane, several studies have demonstrated that they possess the means of being internalized and present various ways of trafficking in different cellular compartments (Figure 2.3). In principle, at least two endocytic GPCR pathways exist; both lead to the formation of specific membrane vesicles but different involvement of β-arrestins.[27] The internalization can be mediated by flask-shaped membrane invaginations called caveolae, which contain large amounts of caveolin proteins and cholesterol. It is

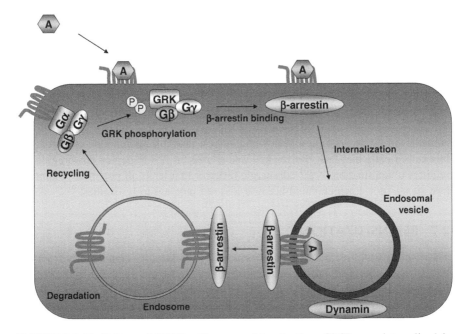

FIGURE 2.3 Trafficking of GPCRs. Upon agonist activation, GPCRs are internalized in endosomes with the aid of β-arrestins. Depending on the receptor, they are either recycled to the plasma membrane or subjected to degradation.

thought that signaling molecules such as receptor-activated G proteins transiently bind to caveolin and this protein–protein interaction decreases the signaling capacity of the GPCR, which results in desensitization of the receptor.[28] In addition, caveolin has been shown to interact with GRKs, which results in a modulated kinase activity.[29] Agonist-dependent endocytosis into caveolae has been described for endothelin-A,[30] bradykinin B2,[31] β2-adrenergic receptors,[32] and adenosine A1 receptors.[28]

GPCRs can also be internalized through specialized plasma membrane structures called clathrin-coated pits.[33] These complexes first invaginate and then detach from the plasma membrane, forming clathrin-coated vesicles. Two components, clathrin and adapter protein complex AP2, are responsible for the assembly of clathrin-coated pits. In the case of GPCR endocytosis, β-arrestin forms a bridge between the activated receptor and the clathrin-coated pit.[34] Studies of agonist-activated thyrotrophin-releasing hormone receptor-1 demonstrated that receptor–arrestin complexes are internalized into pre-existing clathrin-coated pits and do not promote the formation of specific coated pits for endocytosis.[35] The involvement of dynamin has been demonstrated for both clathrin- and caveolin-mediated sequestration of GPCRs.[36,37]

It has also been revealed that β-arrestin can regulate GPCR internalization and degradation through ubiquitination — a process by which ubiquitin is added to lysine residues of the target protein.[38] The process requires three enzymes: ubiquitin-activating enzyme, ubiquitin-carrying enzyme, and ubiquitin–ligase. A direct interaction occurs between β-arrestin and the ubiquitin–ligase mouse double minute 2 (MDM2) that results in ubiquitinated of both β-arrestin and β2-adrenergic receptor. The ubiquitination β-arrestin regulates the internalization of the β2-adrenergic receptor. β-arrestin can also modulate receptor internalization through an interaction with proteins involved in vesicle budding. For instance, β-arrestin interaction occurs with ARF6 (ADP-ribosylation factor 6) and the N-ethylmaleimide-sensitive factor (NSF).[39]

The overexpression of NSF in cells expressing recombinant β2-adrenergic receptor increases the amount of internalized receptor and the recycling of receptor to the plasma membrane is promoted after agonist removal. It has also been suggested that the β-arrestin-dependent internalization aids in receptor downregulation. In this context, the internalization and downregulation of the β2-adrenergic receptors are blocked in cells that lack β-arrestin-1 and β-arrestin 2.[40] Receptor downregulation requires ubiquitination. A β2-adrenergic receptor that lacks all potential ubiquitination sites is not downregulated because it is not targeted to lysosomes.[38]

2.7   RESENSITIZATION

The mechanism of regaining responsiveness to ligands varies significantly from one GPCR to another. Certain internalized receptors are rapidly recycled to the plasma membrane in a fully resensitized state, whereas other GPCRs are slowly recycled or not recycled at all.[41] In the case of the β2-adrenergic receptor, the dephosphorylation of GRK-phosphorylated receptors in early endosomes seems critical for resensitization.[42] The balance between protein kinase and protein phosphatase activities dictates the phosphorylation state and therefore has an effect on resensitization.

For instance, protein phosphatases 2A and 2B have been shown to dephosphorylate the β2-adrenergic receptor.[43] This receptor recruits β-arrestin to the plasma membrane when exposed to agonists, the receptor–arrestin complex dissociates, the β2-adrenergic receptor is internalized into clathrin-coated vesicles, rapidly dephosphorylated, recycled, and resensitized within minutes. In contrast, the activated vasopressin V2 receptor is internalized with β-arrestin into endosomes, slowly dephosphorylated, recycled, and resensitized and the process takes hours.

2.8 CONCLUSIONS

The importance of GPCRs as mediators of various biological functions has become clearer as more information about the roles of their mechanisms on cellular signal transduction events is gathered. In addition to their coupling to G proteins as a means of establishment of signaling cascades, it is evident that GPCRs (that perhaps should be called 7TM receptors in this context) can interact with other cellular proteins. As described above, the direct interaction with β-arrestin plays an important role in receptor internalization and downregulation.

Additionally, the interactions of ERK/MAPK and JNK with GPCRs mediated by β-arrestin can promote the activation of signaling pathways and increase specificity. Further research should reveal additional mechanisms of GPCRs and their interactions with other cellular proteins. Identification of proteins involved with 7TM receptors still leaves many questions open. Additional studies in metabonomics and the questions related to the time of appearance, protein fluxes, and structural mechanisms of protein interaction must be addressed also.

REFERENCES

1. Bockaert J, Pin JP. Molecular tinkering of G protein-coupled receptors: an evolutionary success. EMBO J 1999; 18:1723–1729.
2. Feng Y, Broder CC, Kennedy PE, Berger EA. HIV-1 entry cofactor: functional cDNA cloning of a seven-transmembrane, G protein-coupled receptor. Science 1996; 272:872–877.
3. Pierce KL, Premont RT, Lefkowitz RJ. Seven-transmembrane receptors. Nature Rev 2002; 3:639–650.
4. Pitcher JA, Freedman NJ, Lefkowitz RJ. G protein-coupled receptor kinases. Annu Rev Biochem 1998; 67:653–692.
5. Vanti WB, Swaminathan S, Blevins R, Bonini JA, O'Dowd BF, George SR, Weinshank RL, Smith KE, Bailey WJ. Patent status of the therapeutically important G protein-coupled receptors. Exp Opin Ther Patents 2001; 11:1861–1887.
6. Gether U. Uncovering molecular mechanisms involved in activation of G protein-coupled receptors. Endocr Rev 2000; 21:90–113.
7. Birnbaumer L, Abramowitz J, Brown AM. Receptor-effector coupling of G proteins. Biochim Biophys Acta 1990; 1031:163–224.
8. Mark MD, Herlitze S. G-protein mediated gating of inward-rectifier K^+ channels. Eur J Biochem 2000; 267:5830–5836.
9. Daaka Y, Luttrell LM, Lefkowitz RJ. Switching of the coupling of the beta 2-adrenergic receptor to different G proteins by protein kinase A. Nature 1997; 390:88–91.
10. Krupnick JG, Benovic JL. The role of receptor kinases and arrestins in G protein-coupled receptor regulation. Annu Rev Pharmacol Toxicol 1998; 38:289–319.

11. Tsao P, von Zastrow M. Downregulation of G protein-coupled receptors. Curr Opin Neurobiol 2000; 10:365–369.

12. Collins S, Caron MG, Lefkowitz RJ. Regulation of adrenergic receptor responsiveness through modulation of receptor gene expression. Annu Rev Physiol 1991; 53:497–508.

13. Ross EM, Wilkie TM. GTPase-activating proteins for heterotrimeric G proteins: regulators of G protein signaling (RGS) and RGS-like proteins. Annu Rev Biochem 2000; 69:795–827.

14. Lyubarsky AL, Naarendorp F, Zhang X, Wensel T, Simon MI, Pugh EN Jr. RGS9-1 is required for normal inactivation of mouse cone phototransduction. Mol Vis 2001; 7:71–78.

15. Cao W, Luttrell LM, Medvedev AV, Pierce KL, Daniel KW, Dixon TM, Lefkowitz RJ, Collins S. Direct binding of activated c-Src to the beta 3-adrenergic receptor is required for MAP kinase activation. J Biol Chem 2000; 275:38131–38134.

16. Luttrell LM, Ferguson SS, Daaka Y, Miller WE, Maudsley S, Della Rocca GJ, Lin F, Kawakatsu H, Owada K, Luttrell DK, Caron MG, Lefkowitz RJ. Beta-arrestin-dependent formation of beta 2 adrenergic receptor–Src protein kinase complexes. Science 1999; 283:655–661.

17. Imamura T, Huang J, Dalle S, Ugi S, Usui I, Luttrell LM, Miller WE, Lefkowitz RJ, Olefsky JM. Beta-arrestin-mediated recruitment of the Src family kinase Yes mediates endothelin-1-stimulated glucose transport. J Biol Chem 2001; 276:43663–43667.

18. Barlic J, Andrews JD, Kelvin AA, Bosinger SE, DeVries ME, Xu L, Dobransky T, Feldman RD, Ferguson SS, Kelvin DJ. Regulation of tyrosine kinase activation and granule release through beta-arrestin by CXCRI. Nat Immunol 2000; 1:227–233.

19. Marrero MB, Schieffer B, Paxton WG, Heerdt L, Berk BC, Delafontaine P, Bernstein KE. Direct stimulation of Jak/STAT pathway by the angiotensin II AT1 receptor. Nature 1995; 375:247–250.

20. Hall RA, Premont RT, Chow CW, Blitzer JT, Pitcher JA, Claing A, Stoffel RH, Barak LS, Shenolikar S, Weinman EJ, Grinstein S, Lefkowitz RJ. The beta 2-adrenergic receptor interacts with the Na^+/H^+-exchanger regulatory factor to control Na^+/H^+ exchange. Nature 1998; 392:626–630.

21. Hall RA, Ostedgaard LS, Premont RT, Blitzer JT, Rahman N, Welsh MJ, Lefkowitz RJ. A C-terminal motif found in the beta 2-adrenergic receptor, P2Y1 receptor and cystic fibrosis transmembrane conductance regulator determines binding to the Na^+/H^+ exchanger regulatory factor family of PDZ proteins. Proc Natl Acad Sci USA 1998; 95:8496–8501.

22. Sheng M, Sala C. PDZ domains and the organization of supramolecular complexes. Annu Rev Neurosci 2001; 24:1–29.

23. DeFea KA, Zalevsky J, Thoma MS, Dery O, Mullins RD, Bunnett NW. Beta-arrestin-dependent endocytosis of proteinase-activated receptor 2 is required for intracellular targeting of activated ERK1/2. J Cell Biol 2000; 148:12677–12681.

24. McDonald PH, Chow CW, Miller WE, Laporte SA, Field ME, Lin FT, Davis RJ, Lefkowitz RJ. Beta-arrestin 2: a receptor-regulated MAPK scaffold for the activation of JNK3. Science 2000; 290:1574–1577.

25. Luttrell LM, Roudabush FL, Choy EW, Miller WE, Field ME, Pierce KL, Lefkowitz RJ. Activation and targeting of extracellular signal-regulated kinases by beta-arrestin scaffolds. Proc Natl Acad Sci USA 2001; 98:2449–2454.

26. Heldin CH. Dimerization of cell surface receptors in signal transduction. Cell 1995; 80:213–223.

27. Zhang J, Ferguson SS, Barak LS, Menard L, Caron MG. Dynamin and beta-arrestin reveal distinct mechanisms for G protein-coupled receptor internalization. J Biol Chem 1996; 271:18302–18305.

28. Gines S, Ciruela F, Burgueno J, Casado V, Canela EI, Mallol J, Lluis C, Franco R. Involvement of caveolin in ligand-induced recruitment and internalization of A(1) adenosine receptor and adenosine deaminase in an epithelial cell line. Mol Pharmacol 2001; 59:1314–1323.

29. Carman CV, Lisanti MP, Benovic JL. Regulation of G protein-coupled receptor kinases by caveolin. J Biol Chem 1999; 274:8858–8864.

30. Chun M, Liyanage UK, Lisanti MP, Lodish HF. Signal transduction of a G protein-coupled receptor in caveolae: colocalization of endothelin and its receptor with caveolin. Proc Natl Acad Sci USA 1994; 91:11728–11732.

31. de Weerd WF, Leed-Lundberg LM. Bradykinin sequesters B2 bradykinin receptors and the receptor-coupled G alpha subunits G alpha q and G alpha i in caveolae in DDT1 MF-2 smooth muscle cells. J Biol Chem 1997; 272:17858–17866.

32. Schwencke C, Okumura S, Yamamoto M, Geng YJ, Ishikawa Y. Colocalization of beta-adrenergic receptors and caveolin within the plasma membrane. J Cell Biochem 1999; 75:64–72.

33. Marsh M, McMahon HT. The structural era of endocytosis. Science 1999; 285:215–220.

34. Miller WE, Lefkowitz RJ. Expanding roles for beta-arrestins as scaffolds and adapters in GPCR signaling and trafficking. Curr Opin Cell Biol 2001; 13:139–145.

35. Scott MG, Benmerah A, Muntaner O, Marullo S. Recruitment of activated G protein-coupled receptors to pre-existing clathrin-coated pits in living cells. J Biol Chem 2002; 277:3552–3559.

36. Damke H, Baba T, Warnock DE, Schmid SL. Induction of mutant dynamin specifically blocks endocytic coated vesicle formation. J Cell Biol 1994; 127:915–934.

37. Henley JR, Krueger EW, Oswald BJ, McNiven MA. Dynamin-mediated internalization of caveolae. J Cell Biol 1998; 141:85–99.

38. Shenoy SK, McDonald PH, Kohout TA, Lefkowitz RJ. Regulation of receptor fate by ubiquitination of activated beta 2-adrenergic receptor and beta-arrestin. Science 2001; 294:1307–1313.

39. McDonald PH, Cote NL, Lin FT, Premont RT, Pitcher JA, Lefkowitz RJ. Identification of NSF as a beta-arrestin-1-binding protein: implications for beta 2-adrenergic receptor regulation. J Biol Chem 1999; 274:10677–10680.

40. Kohout TA, Lin FS, Perry SJ, Conner DA, Lefkowitz RJ. Beta-arrestin 1 and 2 differentially regulate heptahelical receptor signaling and trafficking. Proc Natl Acad Sci USA 2001; 98:1601–1606.

41. Anborgh PH, Seachrist JL, Dale LB, Ferguson SS. Receptor/beta-arrestin complex formation and the differential trafficking and desensitization of beta 2-adrenergic and angiotensin II type 1A receptors. Mol Endocrinol 2000; 14:2040–2053.

42. Zhang J, Barak LS, Winkler KE, Caron MG, Ferguson SS. A central role for beta-arrestins and clathrin-coated vesicle-mediated endocytosis in beta 2-adrenergic receptor resensitization: differential regulation of receptor resensitization in two distinct cell types. J Biol Chem 1997; 272:27005–27014.

43. Pitcher JA, Fredericks ZL, Stone WC, Premont RT, Stoffel RH, Koch WJ, Lefkowitz RJ. Phosphatidylinositol 4,5-bisphosphate (PIP2)-enhanced G protein-coupled receptor kinase (GRK) activity: location, structure, and regulation of the PIP2 binding site distinguishes the GRK subfamilies. J Biol Chem 1996; 271:24907–24913.

44. Chikumi H, Vazquez-Prado J, Servitja JM, Miyazaki H, Gutkind JS. Potent activation of Rho A by G alpha q and G q-coupled receptors. J Biol Chem 2002; 277:27130–27134.

45. Smrcka AV, Sternweis PC. Regulation of purified subtypes of phosphatidylinositol-specific phospholipase C beta by G protein alpha and beta gamma subunits. J Biol Chem 1993; 268:9667–9674.

46. Zheng B, Ma YC, Ostrom RS, Lavoie C, Gill GN, Insel PA, Huang XY, Farquhar MG. RGS-PX1, a GAP for G alpha S and sorting nexin in vesicular trafficking. Science 2001; 294:1939–1942.

47. Ma YC, Huang J, Ali S, Lowry W, Huang XY. Src tyrosine kinase is a novel direct effector of G proteins. Cell 2000; 102:635–646.

48. Katada T, Bokoch GM, Northup JK, Ui M, Gilman AG. The inhibitory guanine nucleotide-binding regulatory component of adenylate cyclase: properties and function of the purified protein. J Biol Chem 1984; 259:3568–3577.

49. Bokoch GM, Katada T, Northup JK, Ui M, Gilman AG. Purification and properties of the inhibitory guanine nucleotide-binding regulatory component of adenylate cyclase. J Biol Chem 1984; 259:3560–3567.

50. Saunders C, Limbird LE. Localization and trafficking of alpha 2-adrenergic receptor subtypes in cells and tissues. Pharmacol Ther 1999; 84:193–205.

51. Meng J, Glick JL, Polakis P, Casey PJ. Functional interaction between G alpha(z) and Rap1 GAP suggests a novel form of cellular cross-talk. J Biol Chem 1999; 274:36663–36669.

52. Booden MA, Siderovski DP, Der CJ. Leukemia-associated Rho guanine nucleotide exchange factor promotes G alpha q-coupled activation of RhoA. Mol Cell Biol 2002; 22:4053–4061.

53. Meigs TE, Fields TA, McKee DD, Casey PJ. Interaction of Galpha 12 and Galpha 13 with the cytoplasmic domain of cadherin provides a mechanism for beta-catenin release. Proc Natl Acad Sci USA 2001; 98:519–524.

54. McLaughlin SK, McKinnon PJ, Margolskee RF. Gustducin is a taste-cell-specific G protein closely related to the transducins. Nature 1992; 357:563–569.

55. Kwok-Keung Fung B, Streyr L. Photolyzed rhodopsin catalyzes the exchange of GTP for bound GDP in retinal rod outer segments. Proc Natl Acad Sci USA 1980; 77:2500–2504.

56. Tang WJ, Gilman AG. Type-specific regulation of adenylyl cyclase by G protein beta gamma subunits. Science 1991; 254:1500–1503.

57. Boyer JL, Waldo GL, Harden TK. Beta gamma subunit activation of G-protein-regulated phospholipase C. J Biol Chem 1992; 267:25451–25456.

3 G Protein-Coupled Receptors as Targets for Drug Discovery

Timothy A. Esbenshade

CONTENTS

ABSTRACT

Nearly 25% of total drug sales and prescriptions are directed at GPCRs, making them the most highly desired drug discovery targets by the pharmaceutical industry and resulting in the development of many of the best-selling drugs on the market today including salmeterol, olanzapine, clopidrogel, losartan, and risperidone. These and other drugs directed at GPCRs are used to treat a variety of disease states affecting the cardiovascular and central nervous systems, gastrointestinal, and metabolic functions, as well as cancer, pain, and allergies. This chapter will review current drugs directed at GPCR targets and describe the process of discovering new GPCR therapeutics including the selection and validation of GPCR drug targets and lead optimization. Additionally, the opportunity for the development of potentially useful therapeutics that may be realized from continued investigation of orphan GPCRs and

GPCR oligomerization will be discussed. The past successes of GPCR-targeted drugs and the future potential of GPCR research suggest that these receptors will continue to be the pharmaceutical industry's prime therapeutic drug discovery targets.

3.1 INTRODUCTION

Long before we understood biochemical and molecular activities of GPCRs, humans used the natural products belladonna, ergot and opium from a variety of plant sources to elicit pharmacological effects at then unknown targets to treat a number of maladies. It was not until the beginning of the twentieth century that the idea of receptors was conceived when Paul Ehrlich proposed that drug substances could interact with discrete "chemoreceptors" and produce pharmacological effects. Other prominent pharmacologists including Henry Dale, Otto Loewi, R.P. Ahlquist, and James Black were instrumental in describing the actions of compounds at receptors later identified as GPCRs, laying the foundation for rational drug design.

The advent of molecular biological approaches led to the identification of the GPCR superfamily of approximately 750 genes comprising roughly 1% of the genome.[1] However, the functions and endogenous ligands of a large number of these receptors (orphan GPCRs) remain to be elucidated (see Chapter 16). The cloning of GPCRs has produced a significant impact on drug discovery, leading to the biomolecular identification of both the genes and proteins of those receptors previously described pharmacologically (e.g., those for biogenic amines), additional isoform subtypes of such receptors, and novel, previously uncharacterized GPCRs whose functions are just now being deciphered and for which drugs are being developed. Whereas other chapters in this book deal specifically with a number of therapeutic areas of GPCR drug discovery, this chapter reviews some of the more general and critical aspects of drug discovery research, ranging from target validation to the end result: the synthesis of a drug that interacts selectively with a GPCR target to elicit a therapeutic effect.

3.2 GPCR FAMILY OVERVIEW

A wide variety of extracellular stimuli including photons, ions, neurotransmitters, peptides, and lipids can elicit intracellular signaling events through activation of cell surface GPCRs. These receptors all share a common heptahelical transmembrane-spanning structure that allows the interaction of external stimuli at selective binding sites in extracellular domains or within the transmembrane-spanning domains of GPCRs. This binding in turn conveys conformational changes in the GPCRs, promulgating a cascade of signaling events initiated by the activation of coupled heterotrimeric G proteins from which GPCRs derive their name. GPCRs couple to a variety of heterotrimeric G proteins including those from the $G\alpha_s$, $G\alpha_{i/o}$, $G\alpha_q$, and $G\alpha_{12/13}$ families that in turn activate adenylate cyclase, inhibit adenylate cyclase, activate phospholipase C-β, and activate phospholipase C-β small molecular weight G proteins, respectively. A detailed description of GPCR signal transduction can be found in Chapter 2.

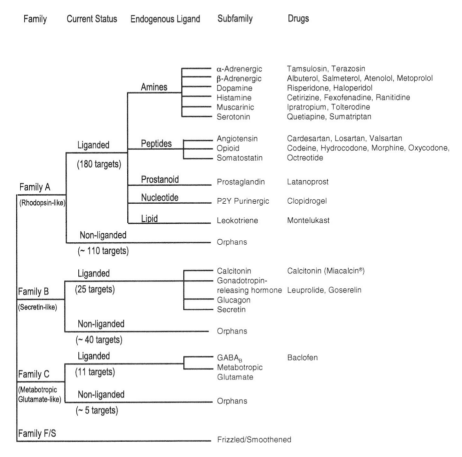

| Family | Current Status | Endogenous Ligand | Subfamily | Drugs |

FIGURE 3.1 GPCR family tree showing representative drugs targeting receptors. (Adapted from Chalmers, D.T. and Behan, D.P. Nat. Rev. Drug Disc. 2002; 1:599–608. With permission.)

GPCRs can be classified into four families based upon sequence similarity (Figure 3.1).[2–5] Family A is the largest of the three represented by the rhodopsin receptor, with many of its members sharing similar sequence motifs such as the third transmembrane DRY and seventh transmembrane NPXXY domains. Family A consists of approximately 180 liganded GPCRs, 110 orphan GPCRs (Figure 3.1), and approximately 350 olfactory GPCRs,[4,5] thus including nearly 70% of non-olfactory GPCRs and nearly 90% of total GPCRs. Family A is represented by the biogenic amine (adrenergic, muscarinic, dopaminergic, histaminergic, etc.), peptide (opioid, angiotensin II, etc.), nucleotide, and prostanoid receptors, among others (see Figure 3.1).

Family B is classified according to structural similarity to the secretin receptor. The structural hallmark of this family is a large extracellular N-terminal domain that contains consensus sites for protein–protein interactions. This family is composed of approximately 25 liganded and 40 orphan GPCRs (Figure 3.1). Calcitonin- and gonadotropin-releasing hormone receptors are representative members of this family

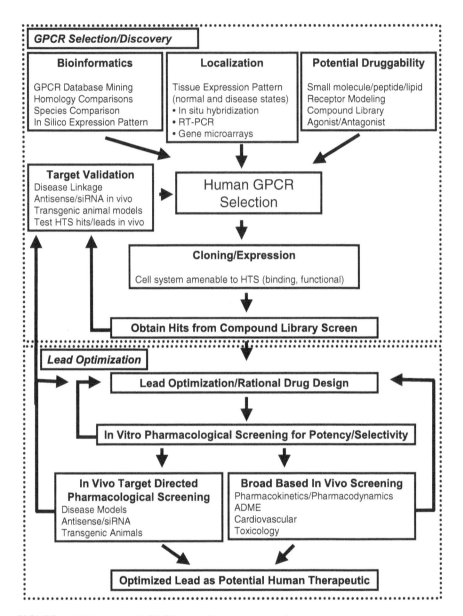

FIGURE 3.2 Flowchart of GPCR drug discovery research.

of GPCRs. Family C shares structural similarity to the metabotropic glutamate receptor and is characterized by an extracellular ligand-binding domain. It is comprised of 11 liganded and about 5 orphan GPCRs and is represented by the metabotropic glutamate and GABA$_B$ receptors. The most recently described and less characterized Family F/S is composed of frizzled and smoothened receptors that play a role in cellular development.[6] A comprehensive description of GPCR classification is provided in Chapter 10.

3.3 GPCR TRACTABILITY: CURRENT THERAPEUTICS

These cell surface receptors have historically been attractive drug targets for pharmaceutical companies. More than 30% of all prescription drugs are directed at fewer than 50 known receptors.[5,7,8] One feature of this receptor family that makes it a tractable drug target is the host of physiological functions including homeostatic mechanisms (cardiovascular, gastrointestinal, and metabolic processes), central nervous system function, and growth that are regulated through the activation of GPCRs by their endogenous ligands, providing a large number of potentially selective targets. Dysregulation of these pathways often leads to pathophysiological consequences that can be ameliorated by selective antagonists to reverse those disease states resulting from excessive stimulation of the GPCR involved. Conversely, selective agonists stimulate GPCR pathways that are hypofunctioning to reverse a disease process physiologically.

The role of GPCRs in a variety of disease states has also been shown through the discovery of naturally occurring mutant GPCRs in humans that contribute to pathologic conditions and via the use of transgenic animals in which genes for both liganded and orphan GPCRs are knocked in or knocked out to allow a better understanding of the *in vivo* phenotype of the GPCR.

Another key aspect of the tractability of GPCRs as drug targets is the ability to design small molecules with high degrees of potency and selectivity. This potency arises from the specificity in binding to domains in the receptor site. The selectivity arises from the diversity in the primary structures of binding sites in other GPCRs across families and isoforms and also in the selective localization of the targeted GPCR in tissues responsible for regulating the targeted physiological function. These features will be discussed in greater depth below, beginning with a brief overview of the current popularity of GPCRs as targets for drugs used in humans today.

Based simply on the sales successes of drugs targeting these receptors, GPCRs are attractive targets. In 2003, total sales for the top 200 prescription drugs were over \$218 billion worldwide. Forty-seven of these drugs target GPCRs and comprise almost 23% of the sales at \$47 billion.[9] Table 3.1 depicts the 20 best-selling GPCR-targeted drugs of 2003 including salmeterol, an antiasthmatic β_2-adrenergic agonist; olanzapine, an antipsychotic 5-HT_2/dopamine receptor antagonist; and clopidrogel, an antithrombotic $P2Y_{12}$-purinergic receptor antagonist, all with annual sales near \$4 billion. Several GPCRs have multiple top-selling drugs targeted at them including H_1-antihistamines (fexofenadine, cetirizine, and desloratadine) and antihypertensive angiotensin II receptor antagonists (losartan, valsartan, cardesartan, and irbesartan). This demonstrates the attractiveness of such targets to pharmaceutical companies.

Another way of determining the impact of GPCRs on therapeutics is to examine the total number of prescriptions issued for drugs targeting GPCRs since many of these drugs are now available as generic versions and, while highly prescribed, do not generate the same revenues as newly introduced drugs normally included in the top 200 in sales. In 2003, total U.S. prescriptions for the top 200 prescribed drugs numbered over 2.2 billion. Fifty-eight of the drugs target GPCRs and included 527 million prescriptions — almost 24% of the total number of prescriptions.[10] Table 3.2 lists the top 15 GPCR targets for which drugs were prescribed in 2003. The

TABLE 3.1
Twenty Top-Selling Drugs Targeting GPCRs in 2003[a]

Trade Name	Brand Name	GPCR	Indication	Sales ($ millions)
Salmeterol	Advair/Seretide/Serevent	β_2-adrenergic	Asthma	4,351
Olanzapine	Zyprexa	$5HT_2$ and Dopamine	Schizophrenia	4,277
Clopidrogel	Plavix	$P2Y_{12}$-purinergic	Thrombosis	3,966
Losartan	Cozaar/Hyzaar/Nu-Lotan	AT_1-angiotensin II	Hypertension	2,939
Risperidone	Risperdal	Dopamine	Schizophrenia	2,512
Valsartan	Diovan/Co-Diovan	AT_1-angiotensin II	Hypertension	2,425
Montelukast	Singulair	$CysLT_1$-leukotriene	Asthma	2,009
Fexofenadine	Allegra/Telfast	H_1-histamine	Rhinitis	1,964
Tamsulosin	Flomax/Harnal	β_1-adrenergic	BPH	1,916
Leuprolide	Leuplin/Lupron	GnRH	Prostate cancer	1,664
Fentanyl	Duragesic	Opioid	Pain	1,631
Cadesartan	Blopress/Atacand	AT_1-angiotensin II	Hypertension	1,629
Irbesartan	Aprovel/Avapro/Avalide	AT_1-angiotensin II	Hypertension	1,530
Quetiapine	Seroquel	$5HT_2$ and dopamine	Schizophrenia	1,487
Cetirizine	Zyrtec	H_1-histamine	Rhinitis	1,353
Metoprolol	Toprol-XL/Seloken	β_1-adrenergic	Hypertension	1,280
Sumatriptan	Imitrex/Imigran	$5HT_1$	Migraine	1,246
Goserelin	Zoladex	GnRH	Prostate cancer	869
Octreotide	Sandostatin	Somatostatin	Acromegaly	695
Desloratadine	Clarinex	H_1-histamine	Rhinitis	694

[a] Based on sales.

opioid receptor was the target most prescribed for (including those for non-GPCRs), with nearly 147 million prescriptions written for a variety of opioid analgesic agonists including hydrocodone, oxycodone, morphine, and codeine. Other highly prescribed drugs include antagonists for the β_1-adrenergic receptor, agonists for the β_2-adrenergic receptor, and H_1 antihistamines.

The vast majority of drugs that target GPCRs are directed at the rhodopsin-like Family A GPCRs, as shown in Figure 3.1, Table 3.1, and Table 3.2. Even in Family A, most of the drugs are further directed at GPCRs that are activated by small-molecule biogenic amine neurotransmitters such as acetylcholine (muscarinic receptors), epinephrine and norepinephrine (adrenergic receptors), dopamine (dopaminergic receptors), histamine (histaminergic receptors), and serotonin (serotonergic receptors). Many of these receptors have long histories of research dating back decades before their actual cloning and were well characterized *in vitro* and *in vivo* as playing roles in a variety of pathophysiological disease states. This basic pharmacological understanding of the receptor and its role in disease and the ability to more easily design small molecules to bind to these biogenic amine GPCRs probably contributed to the preponderance of drugs directed at these targets today.

Additional criteria that play a role in determining the success of a drug directed at a particular GPCR will be described below. Although many peptide-liganded

TABLE 3.2
Fifteen Top GPCR Targets in 2003[a]

GPCR	Function	Indication	Trade Names	Prescriptions (millions)
Opioid	Agonist	Pain	Hydrocodone, propoxyphene, codeine, oxycodone	146.7
β_1-Adrenergic	Antagonist	Hypertension, Glaucoma	Metoprolol, atenolol, propranolol	96.7
H_1-Histamine	Antagonist	Rhinitis, Vertigo	Fexofenadine, cetirizine, meclizine	71.6
β_2-Adrenergic	Agonist	Asthma	Albuterol, salmeterol	49.7
AT_1-Angiotensin II	Antagonist	Hypertension	Valsartan, losartan	38.8
$CysLT_1$-Leukotriene	Antagonist	Asthma	Montelukast	18.5
$P2Y_{12}$-Purinergic	Antagonist	Thrombosis	Clopidrogel	15.2
H_2-Histamine	Antagonist	Acid Reflux	Ranitidine	13.7
Muscarinic Acetylcholine	Antagonist	Asthma, Over-active bladder	Ipratropium, tolterodine	12.0
α_1-Adrenergic	Antagonist	BPH	Tamsulosin, terazosin	10.9
Dopamine	Antagonist	Schizophrenia	Risperidone	8.1
α_2-Adrenergic	Agonist	Hypertension	Clonidine	7.2
Prostanoid FP	Agonist	Glaucoma	Latanoprost	6.9
$5HT_2$ and Dopamine	Antagonist	Schizophrenia	Quetiapine	6.1
$5HT_1$	Agonist	Migraine	Sumatriptan	4.5

[a] Based on number of prescriptions.

Family A GPCRs are known, far fewer drugs on the market target them. However, a number of opioid receptor analgesic agonists such as morphine, oxycodone, and hydrocodone, among others, have been developed as effective narcotic analgesics used for the treatment of pain. In addition, small molecule antagonists to the angiotensin II receptor have been successfully designed and are effective for treating hypertension. Thus, good precedents have been established for creating effective drugs directed at peptide-liganded GPCRs. Indeed, the CXCR4 and CCR5 chemokine receptors are attractive GPCR targets for the development of new antiviral drugs to combat AIDS. Both of these receptors are used by HIV strains as coreceptors for entry into host target immune cells. Peptidic antagonists intended to block HIV entry into these cells are currently under development.[11] Two potentially attractive drug candidates include the orally bioavailable low molecular weight SCH-C CCR5 antagonist that exhibits potent antiviral activity and has been shown to decrease viral load in HIV-infected subjects and the AMD3100 CXCR4 antagonist that blocks CXCR4-dependent viral replication.[11] Among Family B GPCRs, synthetic agonist analogs such as leuprolide and goserelin are used therapeutically to activate the gonadotropin-stimulating hormone receptor for the treatment of prostatic cancer.

3.4 GPCR DRUG DISCOVERY

Selection of a potential GPCR as a target for a preclinical drug discovery effort involves a number of factors viewed from both business and scientific perspectives. This section deals with the scientific criteria used for the selection of a target. It should be stated that a target should obviously fit a company's business plan including alignment with the therapeutic expertise of the company and the capabilities of other preclinical facilities (chemical libraries, high throughput screening, pharmacokinetics, etc.) that will assist in the drug discovery effort.

In order to illustrate some of the scientific criteria involved in a GPCR drug discovery effort, the Figure 3.2 flowchart will be used. In this scheme, the first part of the drug discovery effort requires the prioritization, selection, and compound library screening of GPCR targets. The second phase involves lead optimization involving rational drug design.[12] Important considerations related to GPCR prioritization and discovery research include bioinformatic approaches, receptor localization, druggability, and target validation; the lead optimization phase of drug discovery involves the iterative processes of *in vitro* and *in vivo* target-based pharmacological screening and broad-based *in vivo* pharmacological screening. Each of these points will be discussed in greater detail below and illustrated with specific examples of GPCRs.

3.4.1 GPCR TARGET DISCOVERY AND SELECTION

As depicted in the Figure 3.2 flowchart, this stage of the GPCR drug discovery process can be roughly grouped into five subsets including those dealing directly with the selection of a particular GPCR target (bioinformatics, receptor localization, druggability, and target validation) and the initial processes in discovering chemical entities that may interact with the receptor including the cloning, expression, and high throughput screening of a compound library to identify GPCR target hits amenable to further compound optimization. This latter step is covered in greater depth in other chapters.

There are fewer iterative steps at this stage as compared to the lead optimization stage; however, in some instances it is appropriate to utilize hits obtained from the high throughput screening phase in appropriate *in vivo* experimental models expected to reflect the activity of the GPCR in order to validate the target if no transgenic animal or RNA interference data are available. The results from this approach using early high throughput hits should be interpreted with great caution due to the general lack of potency and selectivity of such early hits.

3.4.1.1 Bioinformatics Approaches for GPCR Discovery

Many GPCR sequences were obtained by the use of homology search analyses such as the pairwise alignment tool known as BLAST[13,14] and degenerate primer pairs in reverse transcriptase/polymerase chain reaction experiments utilizing RNA from tissues of interest.[4,15] However, these approaches are best suited for closely related GPCRs and are not likely to identify unique GPCRs with disparate structures and sequences.

With the cloning of the human genome, the genomic sequences for all GPCRs became available but the problem of identifying unique GPCRs from this mass of information remained. Takeda et al.[16] described the identification of 178 GPCRs from the human genome based upon homology sequence search strategies assuming the absence of introns (many Family A GPCRs lack introns) within the GPCR gene. However, this approach will not detect the many intron-containing GPCRs. Others[4,5,17] employed the hidden Markov model[18,19] — a computational method that provides a statistical representation of the genomic sequence data, weighting the most conserved regions and allowing for insertions and/or deletions within these regions. Although not guaranteed to identify all GPCRs, additional power can be added to this approach by searching against amino acid sequences derived from the DNA sequence (using all three reading frames in both orientations).[4,5,17]

Such strategies led to the identification of 26 previously unidentified human GPCRs and 83 mouse GPCRs.[5] The identification, cloning, expression, and counterscreening of species homologs of human GPCRs are especially important because a number of GPCRs have profound species-dependent pharmacological profile differences that could exert major impacts upon *in vivo* testing of compounds in preclinical (primarily rodent) models. For example, the α_{2A}-adrenergic,[20] histamine H_3,[21,22] and H_4[23] receptors exhibit profound pharmacological differences between rat and human homologs of the receptor, necessitating the discovery of compounds with balanced activities in both species in order to more easily interpret results from *in vivo* studies.

3.4.1.2   GPCR Localization

GPCR drugs are often noted for their high degrees of selectivity. At the molecular level, this selectivity is due in large part to the high affinity of the drug for the GPCR target. At the *in vivo* level, where the response of a whole animal or human is measured, the selectivity of the drug can also be attributed to the specific localization of the GPCR and the relative concentration of the drug in that compartment. Thus, one criterion for the selection of a GPCR for a drug discovery effort is its localization. A GPCR with discrete localization to a target of interest (brain frontal cortex for cognition enhancers, dorsal root ganglia for pain, heart for cardiovascular disorders, etc.) with minimal expression in other tissues would be preferable to a GPCR ubiquitously expressed throughout the body where activation or blockade could result in receptor-mediated side-effect liabilities.

In addition, one needs to consider the localization and relative expression level (up- or down-regulated) of the GPCR in the disease state or states in which it is thought to play a role. A variety of approaches have been taken to determine the tissue expression patterns of GPCRs including (in order of precision) (1) *in situ* hybridization for the detection of the binding of antibodies to the GPCR proteins and/or oligonucleotide probes to tissues, (2) reverse transcriptase/polymerase chain reaction (RT/PCR) of RNA from discrete tissues, and (3) DNA microarrays.[24]

In an interesting experiment examining GPCR localization, Vassilatis et al.[5] described the expression of 100 GPCRs in 26 mouse tissues (9 central and 17 peripheral) using the RT/PCR approach. They were able to show that most GPCRs are expressed in multiple tissues with clearly delineated patterns of expression. However, some GPCRs were expressed only in central tissues (histamine H_3 and galanin type 2 receptors), whereas others (cysteinyl leukotriene 2 and CCR3 chemokine receptors) were expressed only in peripheral tissues. Interestingly, over 93% of GPCRs were detected in the brain, with the largest number of genes detected in the hippocampus. Additional *in situ* hybridization experiments using cRNA probes were in good agreement with the RT/PCR results.

Such expression work can be expanded to an even larger scale by using DNA microarrays in which oligonucleotide GPCR chips that contain all human (or other species) GPCR sequences are hybridized with RNA from a number of tissues from control, disease, drug-treated, and other test cases.[24] Cluster analyses can then be used to identify patterns of gene expression that may then provide insight into the potential physiological role, side-effect liability, and specificity of the GPCR.[4]

3.4.1.3 GPCR Target Validation

Approximately 30 of the 210 liganded GPCRs are currently targets for drugs now on the market. Additionally, approximately 160 non-liganded or orphan receptors exist, resulting in over 300 potential targets for which new chemical entities can be created for potential therapeutic interventions.[8] The challenge is to match the receptor with its physiological function and role in pathophysiological disorders. A number of parallel approaches have been utilized to validate GPCR drug targets including bioinformatic methods as described above. Some predictions can be gained from comparing gene sequences of known GPCRs with orphan GPCRs, such as family types, potential endogenous ligands, and accessory protein interaction domains.

Until actual protein structure information is obtained for GPCRs that will allow the accurate identification of ligand-binding domains, predictions of potential ligands based on sequence homologies and modeling are employed. Several examples of ligand types have been predicted for closely related orphan GPCRs after a ligand for one of the related orphan GPCRs has been elucidated. Examples include the histamine H_4 receptor[25] following identification of the H_3 receptor, the trace amine receptors,[26] and lysophospholipid receptors.[27] However, there are other examples, where sequence homology similarities have resulted in improper classification of orphan GPCRs as receptors for ligands different from their cognate ligands including the histamine H_3 receptor as a muscarinic receptor[28] and the BLT_1 leukotriene B_4 receptor as a purinergic receptor,[29] indicating the limitations to this type of approach.

GPCRs have also been validated as therapeutic targets for a number of diseases based on the finding that naturally occurring mutations of the receptor can result in either a gain or loss of function of the receptor leading to an association with a disease state.[30–32] An excellent review by Seifert and Wenzel-Seifert[32] provides an overview of receptors that exhibit increased constitutive activity (gain of function) including (with associated disease in parentheses): rhodopsin (retinitis pigmentosa), cholecystokinin-2 receptor (gastric carcinoid tumor), KSHV-GPCR (Kaposi's sarcoma), chemokine receptor US28 (atherosclerosis), thyroid-stimulating hormone receptor (hyperthyroidism), luteinizing hormone receptor (precocious male puberty), parathyroid hormone receptor (dwarfism, hypercalcemia, hypophosphatemia), and calcium-sensing receptor (hypocalcemia). Naturally occurring mutations of GPCRs can also lead to loss of function of the receptor with associated pathophysiology.[31] Familial nephrogenic diabetes insipidus has been associated with an inactive truncated human vasopressin V2 receptor[33] as well as a single amino acid mutation (R137H) that results in the rapid desensitization of the receptor.[34]

Other human GPCRs in which a naturally occurring single point mutation results in an inactive receptor and association with a disease[31] are the R16G mutant of the β_2-adrenergic receptor (nocturnal asthma),[35] the W64R mutant of the β_3-adrenergic receptor (obesity),[36] and the R60L mutant of thromboxane A_2 (bleeding disorders).[37] A deletion mutation of the human $P2Y_{12}$ purinergic receptor is associated with bleeding disorders,[38] and an autosomal recessive mutation in the canine orexin-2 receptor was found to underlie narcolepsy in dogs,[39] suggesting a role for orexin in human narcolepsy.

These data suggest that these diseases result from mutations of the GPCRs and also that the wild-type GPCR plays a pivotal role in the normal physiological regulation of the signaling pathway and function deranged by the mutated receptor. Linkage of other inherited diseases such as those discussed above with liganded and orphan GPCRs may reveal novel therapeutic opportunities for particular receptors, leading to the discovery of new drugs to treat the associated disease states.

The identification of naturally occurring GPCR mutants and their altered functions are important to our understanding of the roles of GPCRs in disease processes. However, these mutants are relatively few in number and are generally expressed in a limited number of tissues, with an obvious lack of mutant GPCRs expressed in the central nervous system (CNS) where they are most abundant.

The ability to create transgenic animals in which GPCR genes of interest can be activated or inactivated allows researchers to generate a wide variety of mouse models amenable to testing the role of a single knocked-in or knocked-out GPCR in an array of functional and behavioral models. A recent review by Karasinska et al.[40] lists nearly 90 Family A GPCR gene knock-out mouse models and their associated phenotypes, indicating levels of interest in determining the *in vivo* functions of this important class of receptors. A comparison of the GPCR knock-out phenotype with the efficacy of drugs in development and used in the clinic targeting that receptor demonstrates a very good correlation.[41,42] The therapeutic indications correlate very well with the phenotype of the receptor knock-out animal model for some of the best-selling drugs listed in Table 3.3. [41]

These findings support the idea that GPCR knock-out animals serve an important role in developing new therapeutics. Not only does the inactivation of a GPCR gene reveal insight into the function of that particular receptor, it also allows for the determination of how other GPCRs may interact (heterologous desensitization, shared signaling pathways, up- or down-regulation, protein–protein and pharmacological interactions) with the GPCR in question by comparing the effects of pharmacological agents in wild-type and knock-out animals. GPCR knock-out animals also serve as valuable tools in determining the functions of closely related GPCR subtypes for which there are no selective pharmacological tools available and of orphan GPCRs whose endogenous ligands have yet to be identified.[41]

TABLE 3.3
Correlation of Best-Selling Drugs and Phenotypes

Function	Trade Name	Brand Name	Indication	Knock-Out Phenotype
H_2 receptor antagonist	Ranitidine	Zantac	Gastroesophageal reflux disease	Abolition of histamine-sensitive gastric acid release
Leukotriene antagonist	Montelukast	Singulair	Asthma	Decreased extravasation
H_1 receptor antagonist	Fexofenadine	Allegra	Allergies	Decreased T and B cell responsiveness
μ-opioid receptor agonist	Fentanyl	Duragesic	Pain	Increased pain sensitivity
Angiotensin AT_1 receptor antagonist	Losartan	Cozaar	Hypertension	Low blood pressure
Purinergic $P2Y_{12}$ receptor antagonist	Clopidrogel	Plavix	Thrombosis	Decreased platelet aggregation
Muscarinic receptor antagonist	Tolterodine	Detrol	Overactive bladder	Increased urine retention

Animals with inactivated GPCRs can also be used to test for the selectivities of actions of drugs targeting that specific receptor in order to determine whether the *in vivo* effect of the drug is related to the target GPCR. However, one must be aware of certain cautions when interpreting the phenotypes of GPCR knock-out animals, especially in behavioral tests, including the contribution of the background strain of mouse to the responses and possible disruption of function of other receptors that contribute to the response phenotype that are affected by the inactivated GPCR during pre- and postnatal development and in mature animals.[41] Other limitations of transgenic animals include cost, time, species that can be utilized (primarily mice), and the irreversibility of the inactivation (excepting conditional knock-outs).[43]

Another approach that can be used in the GPCR target validation process is the use of sequence-specific antisense oligonucleotides that target the mRNA of the specific GPCR being interrogated. To date, this technique has been used to effectively inhibit the expression of over 100 GPCR proteins both *in vitro* and *in vivo*.[43] Several advantages of the antisense oligonucleotide approach include (1) the ability to selectively reduce or eliminate expression of GPCRs for which there are no selective ligands to determine the function of the receptor, (2) no receptor up-regulation caused by prolonged reduction in GPCR expression by the antisense oligonucleotides, (3) reversibility, and (4) applicability to any animal species. However, it should be noted that complete suppression of GPCR protein expression is never achieved. About 70% of the studies conducted on GPCRs produced 50% or less suppression of the receptor.[43]

In addition, a number of other disadvantages associated with this approach include sequence-independent effects, limited internalization of the oligonucleotides, narrow dose-range windows to elicit effect, and the need for continuous administration.[43]

It should also be noted that other strategies for inhibiting gene expression related to the antisense oligonucleotide approach include antisense RNA, triplex antigenes, ribozymes, and small interfering RNA (siRNA). The use of siRNA is now the preferred method for target validation *in vitro*. Continued improvements in delivery systems *in vivo* now allow the development of animal models suitable for target validation.[44-46] Fewer reports published to date concern the knock-down of GPCR gene expression by siRNAs, but the expression of chemokine CCR5 and CXCR4 receptors was decreased by siRNAs targeting them, resulting in reduced HIV infection.[47] Additionally, siRNA targeting the lysophospholipid G2A receptor decreased the expression of this receptor in T lymphoid cells, resulting in reduced chemotaxis,[48] demonstrating the utility of this approach for GPCR research.

One final means of GPCR target validation is using pharmacological agents to test for desired therapeutic responses in animal models of disease. These tools can include known standard compounds that demonstrate potent and selective binding to the target in question if one is testing for a novel indication for an already well-characterized GPCR. Conversely, when little is known about the physiological role of the GPCR for which no pharmacological tools are available, compounds that exhibit activity in high throughput screening or from lead optimization testing can be evaluated in animal models to determine the physiological role of the GPCR. A variety of caveats associated with using novel test agents to validate GPCR targets must be considered when interpreting the results including receptor selectivity, pharmacokinetic properties, distribution, and metabolism, among others.

3.4.1.4 GPCR Druggability

The potential to create a potent and selective human therapeutic agent for a GPCR can also influence the selection of that receptor as a target. Biological factors to consider prior to advancing any screening effort for a GPCR target include a number of those already discussed above such as the localization and distribution of the GPCR, the nature of the receptor family type, structure of the receptor, endogenous ligands (i.e., small molecule versus large peptide), and type of the pharmacological effect (agonist or antagonist) desired for the compound. Receptor-binding models can be established from this type of information to allow one to conduct "virtual screening"[49,50] — a computational method that selects from an electronic database the compounds that best fit three-dimensional models of a known pharmacophore and/or receptor, generally based on the protein structure of bovine rhodopsin.[51]

Such approaches have proven useful in reducing the number of compounds required for screening, selecting chemical hits for the dopamine D_3,[52] neurokinin NK1,[53] and somatostatin SST_2[54] receptors that were verified as active by biological screening. The chemical nature of the hits obtained from a chemical library screen can also be used to determine whether a GPCR target is worthy of further investigation from a drug discovery standpoint. Favorable attributes of a collection of screening hits for a target including structural diversity, minimal representation by known toxic structures, low micromolar affinity, and structure amenable to chemical modification would favor advancement of a GPCR target.

3.4.2 LEAD OPTIMIZATION

Lead optimization is an iterative process involving the rational design and synthesis of chemical entities followed by the screening of these compounds for *in vitro* binding and functional potencies at the GPCR target (see Figure 3.2). Results from these *in vitro* studies, which are generally higher throughput assays, are then used to build structure–activity relationships for hundreds to thousands of compounds in order to optimize potency and selectivity. Compounds that exhibit acceptable *in vitro* potency and selectivity are then screened for *in vivo* GPCR target-related activity and tested for other *in vivo* properties such as pharmacokinetics, absorption, distribution, metabolism, excretion, and toxicity (ADMET).

There are a number of important criteria to consider when establishing *in vitro* screens for a GPCR target. Binding assays are useful in quickly establishing potencies of compounds for the target receptor and can also provide information about the nature of the binding (competitive, noncompetitive, number of binding sites the compound recognizes). It is important to screen against both the human form of the GPCR and also the species (often rodent) used in *in vivo* studies to account for any potential species differences. In addition, multiple splice isoforms exist for a number of GPCRs that could influence the pharmacological profile of the receptor.

Ancillary binding screens are often established for closely related receptor subtypes or other targets found to have affinity for early lead compounds in order to further direct rational drug design. A wide host of functional assays measuring a variety of signaling activation pathways are available (see Chapter 9) for determining pharmacological activities (agonist, partial agonist, antagonist, neutral antagonist, inverse agonist) of new compounds at the recombinant or native GPCR.

It is also advantageous to establish *in vitro* functional assays measuring the activity of the natively expressed target GPCR in order to determine whether the results correlate with those from the recombinant GPCR systems. Once *in vitro* activity, potency, and selectivity have been demonstrated for new compounds, they are advanced to *in vivo* animal studies designed to assess their efficacy and potency as drug candidates in disease state models. As with the *in vitro* assays, multiple models, preferably across several species, should be employed to assure efficacy for the therapeutic indication. Testing of the compounds in transgenic animal models of the target GPCR can also help assess the validity of the receptor for the proposed therapeutic indication.[55]

A number of considerations from a chemical perspective also contribute to the synthesis of novel compounds targeting GPCRs. Some of these are basic rules applicable to all drug synthesis operations and some are more specific to designing drugs for GPCRs. The term "privileged structures"[56] referring to structural chemical motifs that serve to provide small molecule compounds that recognize a diverse group of receptor targets seems to apply to a number of classes of GPCR ligands as well and can be employed to optimize lead compound development. Some common GPCR privileged structure motifs include 4-arylpiperidines, 4-arylpiperazines, benzodiazepines, biphenyls, 1,1-diphenylmethane, indoles, spiropiperidines, and xanthines.[57–60]

Focused compound libraries targeting GPCRs can be designed and synthesized based upon such privileged structures using parallel synthesis and higher throughput synthetic methods.[57] Both potency and intrinsic activity (positive for agonists and negative for inverse agonists) must be designed into the chemical leads.

Many recombinantly expressed GPCRs are active in the absence of agonists (constitutively active) and over 60 wild-type GPCRs such as the histamine H_3 receptor[61] exhibit this behavior. Many known antagonists of GPCRs block the effects of agonists at the receptors and also reverse constitutive activity. Based on the large number of GPCRs that are constitutively active in native conditions, designing inverse agonist activity into compounds may optimize the therapeutic potential of chemical leads.

Beyond the efficacy, potency, and selectivity of the compound at the GPCR itself, other factors such as ADMET that influence the drug-like properties of the compound must be considered during the design of a molecule. Models such as the "Rule of Five"[62] (which predicts a compound will be poorly absorbed if it does not possess at least two chemical properties such as a molecular weight below 500 Da, fewer than 5 hydrogen bond donors, fewer than 10 hydrogen bond acceptors, or a logP below 5) have aided chemical optimization of lead compounds.

Some ADMET testing can be done in higher throughput *in vitro* assays such as predictions of metabolic liability with cytochrome P450 inhibition and metabolism profiling, absorption predictions examining Caco-2 cell permeability, and geno-toxicity testing utilizing bacterial reverse mutation assays. Broad-based animal screens, generally across several species, are also used to test the ADMET properties of compounds that demonstrate the best *in vitro* profiles and most drug-like properties. These include pharmacokinetic testing, cardiovascular profiling, acute and chronic toxicity, and *in vivo* metabolism. For example, induction of QT prolongation, a potential precursor to cardiac arrhythmia,[63] is now routinely investigated preclinically because it was observed for several drugs targeting GPCRs, such as the serotonin agonist cisapride, and the histamine H_1 receptor antagonist terfenadine that have been removed from the market. These *in vivo* tests are often conducted in parallel. As with many of the other screens involved in the optimization process, the results help synthetic chemists improve upon the drug-like properties of the lead chemical series.

3.5　NOVEL GPCR FEATURES AND IMPACT ON DRUG DISCOVERY APPROACHES

Recent advances in GPCR research have provided additional opportunities for drug discovery beyond the single conventional binding site approach used to date for all currently marketed drugs that target liganded GPCRs. The discoveries of many yet-to-be defined orphan GPCRs, the finding that GPCRs can form homo- and/or hetero-oligomeric complexes with other GPCRs and the ever-growing number of accessory proteins found to interact with and influence the GPCR signaling complex mean that the number of potential sites at which to target drug action continue to grow and perhaps provide new ways to increase the specificity of action. Advances in drug screening technologies that now allow for screening of orphan GPCRs in the absence of endogenous ligands, more facile determination of effects on protein–protein interactions, continued miniaturization of conventional screening assays, and optimization of homogenous assay systems will undoubtedly speed the rate of discovery at these novel GPCR targets.

3.5.1 ORPHAN GPCRS

Orphan GPCRs represent one area of ongoing drug discovery research targeting approximately 160 receptors across three GPCR families (Figure 3.3, top) in the search for novel therapeutic agents. Several strategies have been used to identify ligands for orphan receptors including reverse pharmacology, the first methodology applied that involves the screening of a panel of known transmitters on recombinantly expressed orphan GPCRs and the measuring of signaling responses. Examples of GPCRs discovered by this approach include the H_3 histamine[21] and LTB_4 leukotriene[64] receptors. A second approach, termed the orphan receptor strategy, also involves the expression of recombinant GPCRs, but the cells expressing the orphan GPCR are exposed to a series of tissue extracts and a variety of signaling responses are monitored for activation responses.[65]

Once a particular extract elicits a positive response, it is fractionated and assayed iteratively until the cognate ligand and, consequently, the GPCR is identified. Nociceptin or orphanin FQ, the ligand for ORL1,[66] was the first ligand identified in this manner. Another approach known as constitutively activating receptor technology (CART™) genetically modifies the orphan GPCR to stabilize the receptor in an active conformation so that it can constitutively activate a number of signaling pathways.[17] Compound libraries can then be screened for both agonists and antagonists.

Since 1995, ligands have been identified for approximately 50 orphan GPCRs (Figure 3.3, bottom) using reverse pharmacology and orphan receptor strategies. However, as previously described, approximately 160 orphan GPCRs for which the endogenous ligands and functions are unknown still remain, and the rate of discovery appears to be decreasing.[67] This suggests that a number of orphan GPCRs may not be "classical" liganded GPCRs and may have other biological functions. These include oligomerization partners with liganded GPCRs, accessory trafficking proteins such as the $GABA_{B(2)}$ receptor with the $GABA_{B(1)}$ receptor,[68–70] signaling through heterotrimeric G protein independent mechanisms, discrete temporal or developmental expression, and highly constitutive "ligandless" GPCRs.[67] Determining the biological functions and therapeutic potentials of the remaining orphan GPCRs will be a challenging endeavor. Orphan receptors are described in more detail in Chapter 16.

3.5.2 GPCR OLIGOMERIZATION

Chapter 15 covers GPCR oligomerization in great depth. The potential implications of this phenomenon on GPCR drug discovery efforts will only be briefly discussed here. An ever-increasing body of evidence accumulated in the past several years demonstrates the existence and function of homo- and hetero-oligomeric complexes of GPCRs in recombinant expression systems and in native tissues. The homo- and

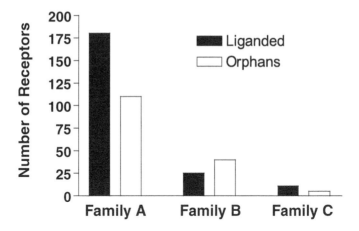

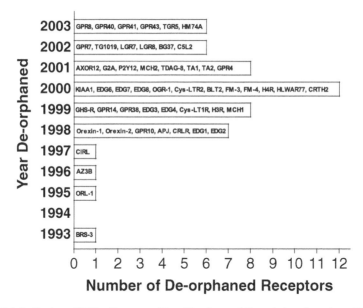

FIGURE 3.3 Orphan GPCR discovery. Top: Numbers of liganded and orphan GPCRs in GPCR families. Bottom: Elucidation timeline of orphan GPCRs since 1995.

hetero-oligomers could account for the complex and subtle changes in activation of different signaling pathways, receptor desensitization and sensitization, and modulation of GPCRs by a variety of agonists.[71,72] This could significantly impact drug discovery efforts by greatly expanding the number of potential drug targets (theoretically over 129,000 possible oligomeric pairs[67]) and produce important implications for the screening and development of new drugs. Additionally, a different mindset for GPCR drug discovery would have to be implemented to take into consideration the possibility of designing drugs that may act at additional sites beyond the conventional monomeric GPCR binding sites.

George et al.[73] proposed several different types of novel drug designs based upon GPCR oligomerization. One design encompasses dimeric ligands, compounds that contain two covalently linked monovalent ligands that may more readily induce or stabilize dimeric conformations of receptors. It may also be possible to design drugs that enhance and/or disrupt oligomerization, potentially by interfering with protein–protein interactions between the two receptors composing the oligomeric GPCR unit. Minimal drug discovery efforts in this area have been reported to date, but the concept appears to have great potential for producing useful therapeutics.

3.6 SUMMARY

In conclusion, GPCRs have long been highly desired drug discovery targets by the pharmaceutical industry, resulting in the development of many of the best-selling drugs on the market today that account for nearly 25% of total drug sales and prescriptions. Drugs targeting these receptors have a diverse variety of therapeutic indications including allergies, cancer, pain management, and diseases affecting cardiovascular, central nervous system, gastrointestinal, and metabolic functions. Developing new GPCR therapeutics begins with the selection of a receptor for a drug discovery effort that involves evaluation of a number of parameters including bioinformatic approaches, receptor localization, potential druggability, and validation as a potential drug target. Advances in GPCR research that offer future opportunities for drug discovery efforts include orphan GPCRs, of which approximately 160 await identification of their cognate ligands and serve as the foci of extensive research to determine their physiological roles. Moreover, GPCR oligomerization offers opportunities for potentially useful therapeutics. Based on their history and future potential, GPCRs should continue to be prime therapeutic targets for drug discovery efforts by the pharmaceutical industry.

REFERENCES

1. Hopkins, A.L. and Groom, C.R. The druggable genome. Nat. Rev. Drug Disc. 2002, 1:727–730.
2. Bockaert, J. and Pin, J.P. Molecular tinkering of G protein-coupled receptors: an evolutionary success. EMBO J. 1999, 18:1723–1729.
3. Foord, S.M. Receptor classification: post genome. Curr. Opin. Pharmacol. 2002, 2:561–566.
4. Chalmers, D.T. and Behan, D.P. The use of constitutively active GPCRs in drug discovery and functional genomics. Nat. Rev. Drug Disc. 2002, 1:599– 608.
5. Vassilatis, D.K., Hohmann, J.G., Zeng, H., Li, F., Ranchalis, J.E., Mortrud, M.T., Brown, A., Rodriguez, S.S., Weller, J.R., Wright, A.C., Bergmann, J.E., Gaitanaris, G.A. The G protein-coupled receptor repertoires of human and mouse. Proc. Natl. Acad. Sci. USA 2003, 100:4903–4908.
6. Malbon, C.C. Frizzleds: new members of the superfamily of G protein-coupled receptors. Frontiers Biosci. 2004, 9:1048–1058.
7. Drews, J. Drug discovery: a historical perspective. Science 2000, 291:1960–1964.

8.   Wise, A., Gearing, K., Rees, S. Target validation of G protein-coupled receptors. Drug Disc. Today 2002, 7:235–246.

9.   Med Ad News Staff. World's best-selling medicines. Med Ad News 2004, 23:60–64.

10. NDC Health. The top 200 prescriptions for 2003 by number of U.S. prescriptions dispensed. RxList, 2003, www.rxlist.com/top200.htm.

11. Schold, D. HIV co-receptors as targets for antiviral therapy. Curr. Top. Med. Chem. 2004, 4:883–893.

12. Kenakin, T. Predicting therapeutic value in the lead optimization phase of drug discovery. Nat. Rev. Drug Disc. 2003, 2:429–438.

13. Altschul, S.F., Gish, W., Miller, W., Myers, E.W., Lipman, D.J. Basic local alignment search tool. J. Mol. Biol. 1990, 215:403–410.

14. Altschul, S.F., Madden, T.L., Schaffer, A.A., Zhang, J., Zhang, Z., Miller, W., Lipman, D.J. Gapped BLAST and PSI-BLAST: a new generation of protein database search programs. Nucleic Acids Res. 1997, 25:3389–3402.

15. Gaulton, A. and Attwood, T.K. Bioinformatics approaches for the classification of G protein-coupled receptors. Curr. Opin. Pharmacol. 2003, 3:114–120.

16. Takeda, S. Kadowaki, S., Haga, T., Takaesu, H., Mitaku, S. Identification of G protein-coupled receptor genes from the human genome sequence. FEBS Lett. 2002, 520:97–101.

17. Menzaghi, F., Behan, D.P., Chalmers, D.T. Constitutively activated G protein-coupled receptors: a novel approach to CNS drug discovery. Curr. Drug Targets CNS Neurol. Dis. 2002, 1:105–121.

18. Karchin, R., Karplus, K., Haussler, D. Classifying G protein-coupled receptors with support vector machines. Bioinformatics 2002, 18:147–159.

19. Krogh, A. Brown, M., Mian, I.S., Sjolander, K., Haussler, D. Hidden Markov models in computational biology: applications to protein modeling. J. Mol. Biol. 1994, 235:1501–1531.

20. O'Rourke, M.F., Iversen, L.J., Lomasney, J.W., Bylund, D.B. Species orthologs of the alpha-2A adrenergic receptor: the pharmacological properties of the bovine and rat receptors differ from the human and porcine receptors. J. Pharmacol. Exp. Ther. 1994, 271:735–740.

21. Lovenberg, T.W., Pyati, J., Chang, H., Wilson, S.J., Erlander, M.G. Cloning of rat histamine H(3) receptor reveals distinct species pharmacological profiles. J. Pharmacol. Exp. Ther. 2000, 293:771–778.

22. Yao, B.B., Hutchins, C.W., Carr, T.L., Cassar, S., Masters, J.N., Bennani, Y.L., Esbenshade, T.A., Hancock, A.A. Molecular modeling and pharmacological analysis of species-related histamine H(3) receptor heterogeneity. Neuropharmacology 2003, 44:773–786.

23. Liu, C., Wilson, S.J., Kuei, C., Lovenberg, T.W.. Comparison of human, mouse, rat, and guinea pig histamine H_4 receptors reveals substantial pharmacological species variation. J. Pharmacol. Exp. Ther. 2001, 299:121–130.

24. Schena, M., Shalon, D., Davis, R.W., Brown, P.O. Quantitative monitoring of gene expression patterns with a complementary DNA microarray. Science 1995, 270:467–470.

25. Liu, C., Ma, X., Jiang, X., Wilson, S.J., Hofstra, C.L., Blevitt, J., Pyati, J., Li, X., Chai, W., Carruthers, N., Lovenberg, T.W. Cloning and pharmacological character-ization of a fourth histamine receptor (H_4) expressed in bone marrow. Mol. Pharmacol. 2001, 59:420–426.

26. Borowsky, B., Adham, N., Jones, K.A., Raddatz, R., Artymyshyn, R., Ogozalek, K.L., Durkin, M.M., Lakhlani, P.P., Bonini, J.A., Pathirana, S., Boyle, N., Pu, X., Kouranova, E., Lichtblau, H., Ochoa, F.Y., Branchek, T.A., Gerald, C. Trace amines: identification of a family of mammalian G protein-coupled receptors. Proc. Natl. Acad. Sci. USA 2001, 98:8966–8971.

27. Xu, Y., Zhu, K., Hong, G., Wu, W., Baudhuin, L.M., Xiao, Y., Damron, D.S. Sphingosylphosphorylcholine is a ligand for ovarian cancer G-protein-coupled receptor 1. Nat. Cell Biol. 2000, 5:261–267.

28. Hancock, A.A., Esbenshade, T.A., Krueger, K.M., Yao, B.B. Genetic and pharmacological aspects of histamine H_3 receptor heterogeneity. Life Sci. 2003, 73:3043–3072.

29. Yokomizo, T., Izumi, T., Chang, K., Takuwa, Y., Shimizu, T. A G-protein-coupled receptor for leukotriene B4 that mediates chemotaxis. Nature 1997, 387:620–624.

30. Johnson, J.A. and Lima, J.J. Drug receptor/effector polymorphisms and pharmacogenetics: current status and challenges. Pharmacogenetics 2003, 13:525–534.

31. Sadee, W. Hoeg, E., Lucas, J., Wang, D. Genetic variations in human G protein-coupled receptors: implications for drug therapy. AAPS PharmSci. 2001, 3:1–26.

32. Seifert, R. and Wenzel-Seifert, K. Constitutive activity of G protein-coupled receptors: cause of disease and common property of wild-type receptors. Naunyn-Schmiedeberg's Arch. Pharmacol. 2002, 366:381–416.

33. Rosenthal, W., Seibold, A., Antaramian, A., Lonergan, M., Arthus, M.F., Hendy, G.N., Birnbaumer, M., Bichet, D.G. Molecular identification of the gene responsible for congenital nephrogenic diabetes insipidus. Nature 1992, 359:233–235.

34. Barak, L.S., Oakley, R.H., Laporte, S.A., Caron, M.G. Constitutive arrestin-mediated desensitization of a human vasopression receptor mutant associated with nephrogenic diabetes insipidus. Proc. Natl. Acad. Sci. USA 2001, 98:93–98.

35. Turki, J., Pak, J., Green, S.A., Martin, R.J., Liggett, S.B. Genetic polymorphisms of the beta 2-adrenergic receptor in nocturnal and nonnocturnal asthma: evidence that Gly16 correlates with the nocturnal phenotype. J. Clin. Invest. 1995, 95:1635–1641.

36. Mitchell, B.D., Blangero, J., Comuzzie, A.G., Almasy, L.A., Shuldiner, A.R., Silver, K., Stern, M.P., MacCluer, J.W., Hixson, J.E. A paired sibling analysis of the β_3-adrenergic receptor and obesity in Mexican Americans. J. Clin. Invest. 1998, 101:584–587.

37. Hirata, T., Kakizuka, A., Ushikubi, F., Fuse, I., Okuma, M., Narumiya, S. Arg 60 to Leu mutation of the human thromboxane A2 receptor in a dominantly inherited bleeding disorder. J. Clin. Invest. 1994, 94:1662–1667.

38. Hollopeter, G., Jantzen, H.M., Vincent, D., Li, G., England, L., Ramakrishnan, V., Yang, R.B., Nurden, P., Nurden, A., Julius, D., Conley, P.B. Identification of the platelet ADP receptor targeted by antithrombotic drugs. Nature 2001, 409:202–207.

39. Lin, L., Faraco, J., Li, R., Kadotani, H., Rogers, W., Lin, X., Qiu, X., de Jong, P.J., Nishino, S., Mignot, E. The sleep disorder canine narcolepsy is caused by a mutation in the hypocretin (orexin) receptor 2 gene. Cell 1999, 98:365–376.

40. Karasinska, J.M., George, S.R., O'Dowd, B.F. Family 1 G protein-coupled receptor function in the CNS: insights from gene knockout mice. Brain Res. Rev. 2003, 41:125–152.

41. Zambrowicz, B.P. and Sands, A.T. Knockouts model the 100 best-selling drugs: will they model the next 100? Nat. Rev. Drug Disc. 2003, 2:38–51.

42. Zambrowicz, B.P., Turner, C.A., Sands, A.T. Predicting drug efficacy: knockouts model pipeline drugs of the pharmaceutical industry. Curr. Opin. Pharmacol. 2003, 3:563–570.

43. Van Oekelen, D., Luyten, W.H.M.L., Leysen, J.E. Ten years of antisense inhibition of brain G protein-coupled receptor function. Brain Res. Rev. 2003, 42:123–142.
44. Dorsett, Y. and Tuschl, T. siRNAs: applications in functional genomics and potential as therapeutics. Nat. Rev. Drug Disc. 2004, 3:318–328.
45. Jain, K.K. RNAi and siRNA in target validation. Drug Disc. Today 2004, 9:307–309.
46. Sioud, M. Therapeutic siRNAs. Trends Pharmacol. Sci. 2004, 25:22–28.
47. Martinez, M.A., Gutierrez, A., Armand-Ugon, M., Blanco, J., Parera, M., Gomez, J., Clotet, B., Este, J.A. Suppression of chemokine receptor expression by RNA interference allows for inhibition of HIV-1 replication. AIDS 2002, 16:2385–2390.
48. Radu, C.G., Yang, L.V., Riedinger, M., Au, M., Witte, O.N. T cell chemotaxis to lysophosphatidylcholine through the G2A receptor. Proc. Natl. Acad. Sci. USA 2004, 101:245–250.
49. Bissantz, C., Bernard, P., Hibert, M., Rognan, D. Protein-based virtual screening of chemical databases. II. Are homology models of G protein-coupled receptors suitable targets? Proteins 2003, 50:5–25.
50. Bleicher, K.H., Green, L.G., Martin, R.E., Rogers-Evans, M. Ligand identification for G-protein-coupled receptors: a lead generation perspective. Curr. Opin. Chem. Biol. 2004, 8:287–296.
51. Palczewski, K., Kumasaka, T., Hori, T., Behnke, C.A., Motoshima, H., Fox, B.A., Le Trong, I., Teller, D.C., Okada, T., Stenkamp, R.E., Yamamoto, M., Miyano, M. Crystal structure of rhodopsin: a G protein-coupled receptor. Science 2000, 289:739–745.
52. Varady, J., Wu, X., Fang, X., Min, J., Hu, Z., Levant, B., Wang, S. Molecular modeling of the three-dimensional structure of dopamine 3 (D3) subtype receptor: discovery of novel and potent D3 ligands through a hybrid pharmacophore- and structure-based database searching approach. J. Med. Chem. 2003, 46:4377–4392.
53. Evers, A. and Klebe, G. Ligand-supported homology modeling of G-protein-coupled receptor sites: models sufficient for successful virtual screening. Angew Chem. Int. Ed. Engl. 2004, 43:248–251.
54. Rohrer, S.P., Birzin, E.T., Mosley, R.T., Berk, S.C., Hutchins, S.M., Shen, D.M., Xiong, Y., Hayes, E.C., Parmar, R.M., Foor, F., Mitra, S.W., Degrado, S.J., Shu, M., Klopp, J.M., Cai, S.J., Blake, A., Chan, W.W., Pasternak, A., Yang, L., Patchett, A.A., Smith, R.G., Chapman, K.T., Schaeffer, J.M. Rapid identification of subtype-selective agonists of the somatostatin receptor through combinatorial chemistry. Science 1998, 282:737–740.
55. Hardy, L.W. and Peet, N.P. The multiple orthogonal tools approach to define molecular causation in the validation of druggable targets. Drug Disc. Today 2004, 9:117–126.
56. Evans, B.E., Rittle, K.E., Bock, M.G., DiPardo, R.M., Freidinger, R.M., Whitter, W.L., Lundell, G.F., Veber, D.F., Anderson, P.S., Chang, R.S, Lotti, V.J., Cerino, D.J., Chen, T.B., Kling, P.J., Kunkel, K.A., Springer, J.P., Hirshfield, J. Methods for drug discovery: development of potent, selective, orally effective cholecystokinin antagonists. J. Med. Chem. 1988, 31:2235–2246.
57. Klabunde, T. and Hessler, G. Drug design strategies for targeting G protein-coupled receptors. Chem. Bio. Chem. 2002, 3:928–944.
58. Bleicher, K.H., Green, L.G., Martin, R.E., Rogers-Evans, M. Ligand identification for G-protein-coupled receptors: a lead generation perspective. Curr. Opin. Chem. Biol. 2004, 8:287–296.
59. Guo, T. and Hobbs, D.W. Privileged structure-based combinatorial libraries targeting G protein-coupled receptors. Assay Drug Dev. Tech. 2003, 1:579–592.

60. Bondensgaard, K., Ankersen, M., Thogersen, H., Hansen, B.S., Wulff, B.S., Bywater, R.P. Recognition of privileged structures by G protein-coupled receptors. J. Med. Chem. 2004, 47: 888–899.

61. Morisset, S., Rouleau, A., Ligneau, X., Gbahou, F., Tardivel-Lacombe, J., Stark, H., Schunack, W., Ganellin, C.R., Schwartz, J.C., Arrang, J.M. High constitutive activity of native H3 receptors regulates histamine neurons in brain. Nature 2000, 408:860–864.

62. Lipinski, C.A., Lombardo, F., Dominy, B.W., Feeney, P.J. Experimental and computational approaches to estimate solubility and permeability in drug discovery and development settings. Adv. Drug Delivery Rev. 2001, 46:3–26.

63. Fernandez, D., Ghanta, A., Kauffman, G.W., Sanguinetti, M.C. Physicochemical features of the HERG channel drug binding site. J. Biol. Chem. 2004, 279:10120–10127.

64. Kamohara, M., Takasaki, J., Matsumoto, M., Saito, T., Ohishi, T., Ishii, H., Furuichi, K. Molecular cloning and characterization of another leukotriene B4 receptor. J. Biol. Chem. 2000, 275:27000–27004.

65. Civelli, O., Nothacker, H.P., Saito, Y., Wang, Z. Lin, S.H., Reinscheid, R.K. Novel neurotransmitters as natural ligands of orphan G protein-coupled receptors. Trends Neurosci. 2001, 24:230–237.

66. Meunier, J.C., Mollereau, C., Toll, L., Suaudeau, C., Moisand, C., Alvinerie, P., Butour, J., Guillemot, J.C., Ferrara, P., Monsarrat, B., Mazarguil, H., Vassart, G., Parmentier, M., Costentin, J. Isolation and structure of the endogenous agonist of opioid receptor like ORL1 receptor. Nature 1995, 377:532–535.

67. Wise, A., Jupe, S.C., Rees, S. The identification of ligands at orphan G protein-coupled receptors Annu. Rev. Pharmacol. Toxicol. 2004, 44:43–66.

68. Jones, K.A., Borowsky, B. Tamm, J.A., Craig, D.A., Durkin, M.M., Dai, M., Yao, W.J., Johnson, M., Gunwaldsen, C., Huang, L.Y., Tang, C., Shen, Q., Salon, J.A., Morse, K., Laz, T., Smith, K.E., Nagarathnam, D., Noble, S.A., Branchek, T.A., Gerald, C. GABAB receptors function as a heteromeric assembly of the subunits of GABABR1 and GABABR2. Nature 1998, 396:674–679.

69. White, J.H., Wise, A., Main, M.J., Green, A., Fraser, N.J., Disney, G.H., Barnes, A.A., Emson, P., Foord, S.M., Marshall, F.H. Heterodimerization is required for the formation of a functional GABAB receptor. Nature 1998, 396:679–682.

70. Kaupmann, K., Malitschek, B., Schuler, V., Heid, J., Froestl, W., Beck, P., Mosbacher, J., Bischoff, S., Kulik, A., Shigemoto, R., Karschin, A., Bettler, B. GABAB-receptor subtypes assemble into functional heteromeric complexes. Nature 1998, 396:683–687.

71. Tallman, J. Dimerization of G-protein-coupled receptors: implications for drug design and signaling. Neuropsychopharmacology 2000, 23:S1–S2.

72. Milligan, G. G protein-coupled receptor dimerization: function and ligand pharmacology. Mol. Pharmacol. 2004, 66:1–7

73. George, S.R., O'Dowd, B.F., Lee, S.P. G protein-coupled receptor oligomerization and its potential for drug discovery. Nat. Rev. Drug Disc. 2002, 1:808–819.

4 G Protein-Coupled Receptors as Cardiovascular Drug Targets

Mika Scheinin and Amir Snapir

CONTENTS

ABSTRACT

At least 55 types of GPCRs are known to directly mediate neuronal and endocrine regulation of cardiac and vascular responses, and many more influence cardiovascular functions indirectly. In spite of this apparent abundance of potential GPCR drug targets, currently only β-adrenoceptor antagonists (β-blockers) and antagonists of angiotensin II type 1 receptors (AT_1 antagonists) are in widespread use in clinical cardiovascular therapeutics. In addition, some other types of receptors for monoamine neurotransmitters and vasoactive peptides are targets for certain classical cardiovascular drugs such as atropine and for newer drug classes with precisely targeted actions such as sumatriptan and related agents that have changed clinical practices in the treatment of migraine. Approximately ten additional types of GPCRs are currently targeted by new drug candidates that are in various phases of clinical development. The genome project and concomitant advances in molecular medicine have, however, recently uncovered new potential drug targets: receptor subtypes with possibilities for drug actions with improved benefit:harm ratios; regulatory mechanisms influencing receptor gene expression and receptor functions; and genetic receptor variants offering opportunities for patient selection, risk profiling, and individualized therapy. More than 100 orphan GPCRs await characterization, and some of them may turn out to be valid cardiovascular drug targets.

4.1 CARDIOVASCULAR PHYSIOLOGY, PHARMACOLOGY AND THERAPEUTICS

GPCRs are involved in cardiovascular regulation in numerous ways. At least 55 types of GPCRs are known to directly mediate neuronal and endocrine regulation of cardiac and vascular responses, and many more influence cardiovascular functions indirectly, *via* primary effects on the neuronal, endocrine, and metabolic regulatory mechanisms that control various aspects of the functioning of cardiac and vascular cells and tissues. These receptors are widely expressed in different cell types of the heart and blood vessels, where they mediate the actions of a variety of hormones growth factors, neurotransmitters, biologically active peptides, and local mediators, and are consequently involved in the regulation of numerous cellular processes and physiological functions. This review is focused mainly on direct receptor-mediated control of myocardial functions and constriction and relaxation responses of the vascular walls.

In addition to such short-term regulation, cardiac and vascular GPCRs also participate in long-term regulation of cardiovascular functions through effects on the development and plasticity of the heart and blood vessels. Figure 4.1 presents a summary of GPCR expression patterns in blood vessel walls, the heart, and the kidneys — tissues that represent the most important direct effectors in cardiovascular physiology, pathology, and therapeutics. Subtype-specific functions have been identified for many genetically defined receptor subtypes, but several pharmacologically defined receptor responses still await clarification of the roles played by different receptor subtypes.

Splice variants also exist for some GPCRs, such as α_{1A}-adrenoceptors[1] and type 1 angiotensin II (AT_1) receptors[2], and their existence further complicates this issue. Closely related receptor subtypes may have very different anatomical distribution patterns and opposing functional effects in the cardiovascular system. In addition to the receptor types shown in Figure 4.1, GPCRs in the central nervous system (such as α_2-adrenoceptors that regulate the sympathetic nervous system), GPCRs located in endocrine organs (through regulation, for example, of renin release) and receptors involved in metabolic regulation exert profound effects on cardiovascular health and disease and may constitute therapeutic targets. The important roles that some types of GPCRs have in cardiovascular diseases through effects mediated by receptors located on blood cells (e.g., many types of prostanoid receptors) or acting via enzymatic mechanisms (e.g., thrombin receptors) will not be discussed here.

As of today, GPCRs constitute important drug targets in cardiovascular therapeutics. Table 4.1 lists examples of clinically important drugs used to treat cardiovascular diseases [derived from the Anatomical Therapeutic Chemical (ATC) Classification System of the World Health Organization]. The examples were chosen to represent different therapeutic indications and different target GPCRs and mechanisms of action. Most current therapeutically employed drugs are directly targeted

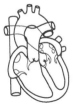

Smooth muscle
Adrenoceptors ($\alpha_{1A/1B/1D}$)
Angiotensin (AT$_1$)
Endothelin (ET$_A$)
Dopamine (D$_{1/3}$)
Serotonin (5-HT$_{1B/1D/2A/2B/4/7}$)
Vasopressin (V$_1$)
Adenosine (A$_{2B}$)
Adrenoceptors ($\alpha_{2A/2B/2C}$, β_2)
Angiotensin (AT$_2$)
CGRP$_1$
Bradykinin (B$_1$/B$_2$)
Chemokine (CCR$_{5/8}$, CXCR$_4$)
Histamine (H$_{1/2/3}$)
Muscarinic (M$_{2/3}$)
Neuropeptide Y (Y$_{1/2/5}$)
Prostanoid
Purinergic (P2Y$_{1/6}$)
Somatostatin

Endothelium
Adenosine (A$_{2A}$)
Adrenoceptors (α_2, $\beta_{2/3}$)
Angiotensin (AT$_2$)
CGRP$_1$
Chemokine (CCR$_{5/8}$)
Dopamine (D$_2$-like)
Endothelin (ET$_B$)
Histamine (H$_{1/2/3}$)
Muscarinic (M$_2$)
Prostanoid
Purinergic (P2Y$_{1/2}$)
Serotonin (5-HT$_{2A/2B}$)
Somatostatin

Nerve endings
Adrenergic ($\alpha_{2A/2C}$)
Serotonin (5-HT$_{1B/1D/4/7}$)
Adenosine (A$_1$)
Dopamine (D$_2$-like)
Muscarinic (M$_2$)
Neuropeptide Y (Y$_{2/3}$)
Purinergic (P2Y$_1$)

Myocard
Adrenoceptors (β_1)
Adenosine (A$_1$)
Angiotensin (AT$_1$)
Muscarinic (M$_2$)
Adrenoceptors ($\alpha_{1A/1B}$, $\beta_{2/3}$)
Angiotensin (AT$_2$)
Dopamine (D$_1$)
Endothelin (ET$_{A/B}$)
Histamine (H$_{1/2}$)
Prostanoid
Serotonin (5-HT$_4$)

Autonomic nervous system
Angiotensin (AT$_1$)
Adrenoceptors ($\alpha_{2A/2C}$)
Muscarinic (M$_2$)
Neuropeptide Y (Y$_3$)

Filtration
Dopamine
Adenosine (A$_{1/2A}$)
Adrenoceptors ($\alpha_{2A/2B/2C}$)
Angiotensin (AT$_2$)
Neuropeptide Y (Y$_2$)
Prostanoid
Vasopressin (V$_1$)

Reabsorption
Angiotensin (AT$_1$)
Adrenoceptors ($\alpha_{2A/2B/2C}$)
Dopamine
Vasopressin (V$_2$)

FIGURE 4.1 Locations of therapeutically relevant types of GPCRs in the blood vessel walls, heart, and kidneys. Targets of clinically approved drugs are shown in **bold**. Note that receptor subtype distributions have not been indicated for all receptor classes (e.g., prostanoid receptors and somatostatin receptors) with complex and incompletely known expression patterns.

TABLE 4.1

Examples of Clinically Approved Drugs Targeted to GPCRs Grouped by ATC Codes of the World Health Organization

Receptor	Drug	Action	Mechanism	Indications and Notes
A03BA — **Belladonna Alkaloids**				
Muscarinic acetylcholine	Atropine	Antagonist	Elimination of parasympathetic tone	Bradycardia; cardiac arrest
C01CA — **Adrenergic and Dopaminergic Agents**				
Adrenoceptors	Norepinephrine	Agonist	Vasoconstriction (α_1, α_2) and cardiac stimulation (β_1)	Rarely used in clinical settings
Adrenoceptors	Epinephrine	Agonist	Vasoconstriction (α_1, α_2), vasodilatation (β_2), and cardiac stimulation (β_1)	Used in cardiac arrest and anaphylactic shock; also used as vasoconstrictor combined with local anesthetic
α_1-Adrenoceptors	Phenylephrine	Agonist	Vasoconstriction	Hypotension
β-Adrenoceptors	Isoprenaline	Agonist	Positive inotropic and chronotropic effects on the myocardium	Not subtype selective
β_1-Adrenoceptors	Dobutamine	Agonist	Positive inotropic and chronotropic effects on the myocardium	Cardiogenic shock
D_1 dopamine and adrenoceptors	Dopamine	Agonist	Vasodilatation (D_1); improved renal function (D_1), cardiac stimulation (β_1)	In high concentrations, vasoconstriction (α_1)
D_1 dopamine	Fenoldopam	Agonist	Vasodilatation (D_1)	Hypertensive crisis
C01E – **Other Cardiac Preparations**				
Adenosine	Adenosine	Agonist	Reduced chronotropy and dromotropy (A_1), and vasodilatation (A_{2A}, A_{2B}, A_3)	Supraventricular tachyarrhythmia; also used as a diagnostic agent to induce coronary vasodilatation
C02A – **Centrally Acting Antiadrenergic Agents**				
α_2-Adrenoceptors	Methyldopa	Agonist	Sympatholysis	Hypertension; metabolized to α-methylnorepinephrine
α_2-Adrenoceptors	Clonidine; guanfacine; moxonidine; rilmenidine	Agonist	Sympatholysis	Hypertension

Receptor		Action	Indication	Drugs
C02C — Peripherally Acting Antiadrenergic Agents				
α_1-Adrenoceptors	Antagonist	Vasodilatation	Hypertension	Prazosin; doxazosin
C02K — Other Antihypertensives				
5-HT_{2A}	Antagonist	Vasodilatation	Hypertension	Ketanserin
ET_A, ET_B endothelin	Antagonist	Vasodilatation	Pulmonary hypertension	Bosentan
C07A — Beta Blocking Agents				
β-Adrenoceptors	Antagonist	Reduction of cardiac output and workload	Hypertension; arrhythmia; ischemic heart disease; chronic heart failure	Propranol; timolol; pindolol
β_1-Adrenoceptors	Antagonist	Reduction of cardiac output and workload	Hypertension; arrhythmia; ischemic heart disease; chronic heart failure	Acebutolol; metoprolol; atenolol; celiprolol
α- and β-Adrenoceptors	Antagonist	Reduction of cardiac output and workload; vasodilatation	Hypertension; chronic heart failure	Labetalol; carvedilol
C09C — Angiotensin II Antagonists				
AT_1 angiotensin	Antagonist	Vasodilatation	Hypertension; chronic heart failure; portal hypertension	Losartan; valsartan
H01BA — Vasopressin and Analogs				
V_1 vasopressin	Agonist	Vasoconstriction	Upper gastrointestinal bleeding; vasoconstrictor combined with local anesthetic	Vasopressin; felypressin
N02C — Antimigraine Preparations				
5-HT, α_1-adrenoceptors	Partial agonists or antagonists	Vasoconstriction and vasodilatation	Acute migraine attacks	Ergot alkaloids
5-HT_{1B} serotonin	Agonist	Vasoconstriction	Migraine	Sumatriptan; naratriptan; zolmitriptan

Source: Data extracted from World Health Organization's Anatomical Therapeutic Chemical Classification System provided online by Wikipedia.org (en.wikipedia.org).

to their corresponding receptors, activating them as agonists or blocking their activity as antagonists. The therapeutic response is often a direct consequence of the drug effect, e.g., relaxation or contraction of vascular smooth muscle cells or alteration of the excitability of cardiac myocytes. More indirect ways of influencing receptor activities or cellular functions may soon emerge as alternative approaches to exert pharmacological control of cardiovascular functions. Regulation of receptor responsiveness through regulation of GPCR kinase (GRK) activity is one such recently proposed mechanism.[3]

Modulation of gene expression through GPCR-linked second messenger pathways represents another promising new approach. Some drugs already in wide clinical use, such as β-adrenoceptor and AT_1 receptor antagonists, produce very important therapeutic effects that are mediated by GPCR-dependent signaling cascades involved in cell proliferation, apoptosis, and tissue remodeling. One drug may actually target several cellular mechanisms through the widespread involvement of its receptors in functions critical for cardiovascular pathophysiology and therapeutics. Antagonists of AT_1 receptors exemplify this phenomenon (see Smit et al.).[4] The most important functions mediated by AT_1 receptors include vasoconstriction, induction of the production and release of aldosterone, renal reabsorption of sodium, cardiac cellular growth, proliferation of vascular smooth muscle, increased activity of the sympathetic nervous system, stimulation of vasopressin release, and inhibition of renin release from the kidney. AT_1 receptor antagonists inhibit the interaction of angiotensin II with its AT_1 receptor, and have proven therapeutic efficacy based on synergism between these mechanisms in many forms of arterial hypertension and chronic heart failure.

4.2   DRUGS IN DEVELOPMENT AND NOVEL DRUG TARGETS

Cardiovascular diseases constitute a major focus area for drug discovery and development. They are very prevalent in all populations, and many therapeutic needs are still unmet despite the many successes of the past 50 years. It is evident from Table 4.1 that relatively few GPCRs are currently exploited as cardiovascular drug targets, even if β-adrenoceptor and AT_1 receptor antagonists represent two of the most widely used drug classes in modern medicine.

Table 4.2 summarizes information on novel drugs targeted to GPCRs that are currently in various phases of clinical development (see www.centerwatch.com). In addition, many types of GPCRs are currently undergoing validation as potential cardiovascular drug targets and numerous drug candidates are in various stages of preclinical development. Even if promising preclinical lead compounds have been identified in many cases, proof of concept can be obtained only after efficacy and safety have been at least preliminarily evaluated in clinical trials. Table 4.2 also indicates that relatively few receptor types are targets for novel drugs that have already entered various phases of their clinical testing programs. Of the drugs and drug targets listed in Table 4.2, some represent novel therapeutic approaches.

Antagonism of cardiac serotonin $5\text{-}HT_4$ receptors represents a novel approach to treat atrial fibrillation by blocking $5\text{-}HT_4$ receptor-mediated enhancement of atrial

TABLE 4.2
Examples of Novel Drugs in Clinical Development, Targeted to Cardiovascular GPCRs

Target Receptor and Action	Drug	Indication	Phase	Company
5-HT$_{1A}$ agonist	Repinotan	Cerebral ischemia	II	Bayer
		Stroke	II	
5-HT$_{1B/2A}$ antagonist	SL 65.0472	Cardiovascular events	I	Sanofi-Synthelabo Pharmaceuticals
5-HT$_4$ antagonist	Piboserod	Atrial fibrillation	II	GlaxoSmithKline
$\alpha_{2B/C}$ adrenoceptor antagonist	OPC-28326	Raynaud's disease	II	Otsuka America Pharmaceutical
		Peripheral vascular disease	II	
A$_1$ antagonist	CVT-124	Congestive heart failure	II	Biogen Idec
A$_1$ agonist	Tecadenoson	Arrhythmia	III	CV Therapeutics
A$_{2A}$ agonist	Binodenoson	Coronary artery disease, diagnosis	III	King Pharmaceuticals
A$_{2A}$ agonist	CVT-3146	Coronary artery disease, diagnosis	II	CV Therapeutics
CGRP antagonist	Olcegepant	Migraine and cluster headaches	I/II	Boehringer Ingelheim
ET$_A$ antagonist	TBC-3711	Congestive heart failure	I	Encysive Pharmaceuticals
		Hypertension	I	
ET$_A$ antagonist	Darusentan	Congestive heart failure	III	Myogen
		Hypertension	II	
ET$_A$ antagonist	Ambrisentan	Congestive heart failure	II	Myogen
		Hypertension	II	
ET$_{A/B}$ antagonist	Tezosentan	Congestive heart failure	III	Actelion Pharmaceuticals
V$_{1/2}$ antagonist	Conivaptan	Congestive heart failure	II	Yamanouchi Pharmaceutical
V$_2$ antagonist	OPC-41061	Congestive heart failure	II	Otsuka America Pharmaceutical
V$_2$ antagonist	SR 121463	Congestive heart failure	II	Sanofi-Synthelabo Pharmaceuticals

Source: Data extracted from CenterWatch online directory of drugs in clinical trials (http://www.centerwatch.com).

contractility and relaxation. α_2-Adrenoceptors are known to mediate constriction of blood vessels, and they have been identified as potential targets in the treatment of Raynaud's disease and other forms of peripheral vascular disease, e.g., intermittent claudication. The unwanted consequences of central α_{2A}-adrenoceptor antagonism have so far invalidated this approach; targeting of peripheral α_{2B}- and α_{2C}-adreno-ceptors may overcome this problem.

Adenosine is currently in limited clinical use to terminate supraventricular arrhythmias and to dilate coronary arteries in diagnostic imaging procedures — two effects mediated by two different receptor subtypes (A_1 and A_{2A}). Both effects are now pursued separately with subtype-selective agonists. In addition, an antagonist of A_1 adenosine receptors is in clinical development for therapy of congestive heart failure, based on its capacity to promote renal excretion of sodium and water.

Receptors for calcitonin gene-related peptide (CGRP) are abundant in vascular smooth muscle and endothelium, and activation of these receptors is known to cause relaxation of blood vessels. Olcegepant is an antagonist of at least some types of CGRP receptors and is in clinical development for migraine and cluster headaches. Bosentan is currently in limited clinical use as a vasodilator in pulmonary hypertension, but other small-molecule endothelin receptor antagonists are now under clinical development for broader therapeutic indications, including hypertension and congestive heart failure. Vasopressin V_2 receptor antagonists are also in clinical development for congestive heart failure, based on their capacity to oppose vasopressin-induced vasoconstriction and renal fluid retention.

In addition to the identification of various GPCR subtypes involved in cardiovascular regulation recently expedited by the efforts and results of the Human Genome Project, exploitation of these receptors as cardiovascular drug targets requires their thorough physiological and pharmacological characterization. Their distributions in tissues and cell types must be mapped; their coupling and signaling mechanisms elucidated; the mechanisms involved in their functional regulation determined; and their roles in cardiovascular development and plasticity identified. Sufficiently selective pharmacological ligands have not in all cases been available to reliably differentiate the functions mediated by different but closely related GPCR subtypes. For instance, until recently, the study of the physiological roles and therapeutic potentials of the three different genetic α_2-adrenoceptor subtypes was difficult.[5] The recent employment of homologous recombination methodologies has produced mouse lines with selective deletions, mutations, or overexpression of individual GPCR subtypes, and has permitted the elucidation of the specific contributions of the different α_2-adrenoceptor subtypes in cardiovascular regulation.[6,7]

It has been demonstrated for most GPCRs that receptor activation initiates a process responsible for receptor desensitization (i.e., waning of stimulated responses) and internalization (i.e., removal from the plasma membrane, thus making the receptors unavailable to extracellular ligands).[8] The process of agonist-dependent desensitization includes uncoupling of the receptors from G proteins as a result of phosphorylation of residues in their intracellular domains (most commonly demonstrated for the third intracellular loop) by specific phosphorylating enzymes, the GPCR kinases (in particular GRK2 and GRK3).[9]

Receptor phosphorylation promotes the binding of β-arrestins that sterically hinder receptor coupling to G proteins and also interact with a variety of proteins involved in endocytosis, acting as scaffolds to transduce and compartmentalize alternative signals. GRK2 activity is regulated by complex protein phosphorylation cascades, including phosphorylation by mitogen-activated protein kinases. Understanding of the mechanisms regulating receptor responsiveness may suggest novel drug targets for cardiovascular therapy, perhaps allowing more physiological and more finely tuned approaches than direct GPCR activation or antagonism.

4.3   RECEPTOR SUBTYPES AS NOVEL TARGETS

Adrenoceptors constitute an established family of therapeutic drug targets and demonstrate how accumulating knowledge of receptor subtypes has gradually led to more refined therapeutic approaches in cardiovascular diseases. Mammals have nine subtypes of adrenoceptors. Norepinephrine, the sympathetic neurotransmitter, and epinephrine, an adrenomedullary hormone, regulate cardiovascular functions through most, or perhaps even all of these subtypes.

The first division of adrenoceptors into two classes was based on the potency of a series of catecholamine derivatives to elicit functional responses in vascular and other smooth muscles and in the heart.[10] α-Adrenoceptor activation elicited contraction of vascular smooth muscle and had little effect on the heart, while β-adrenoceptor activation had positive inotropic and chronotropic effects on the heart and relaxed smooth muscle tissues. Further pharmacological adrenoceptor classifications allowed a distinction between β_1-adrenoceptors predominantly responsible for cardiac stimulation and β_2-adrenoceptors that mediate relaxation of vascular and bronchial smooth muscles.[11] This distinction allowed the development of selective β_1-adrenoceptor antagonists for hypertension and other cardiovascular indications and selective β_2-adrenoceptor agonists for the symptomatic treatment of asthma.

The first subdivision of the α-adrenoceptors into α_1- and α_2-adrenoceptor subtypes was based largely on different functions and anatomical locations.[12,13] Presynaptic α-adrenoceptors regulating neurotransmitter release were termed α_2-adrenoceptors, whereas postjunctional α-receptors were referred to as α_1-adrenoceptors. Further pharmacological experiments revealed that α_2-adrenoceptors are located both pre- and postsynaptically and that they mediate a complex spectrum of pharmacological actions.[14,15] A pharmacological α-adrenoceptor classification was based on actions of selective agonists and antagonists. α_1-Adrenoceptors can be activated by phenylephrine and blocked by low concentrations of prazosin, whereas α_2-adrenoceptors can be activated by clonidine and blocked by low concentrations of yohimbine.[16,17] The current adrenoceptor classification is based on molecular cloning of the nine distinct mammalian genes encoding three α_1-, three α_2- and three β-adrenoceptor subtypes.

α_1-Adrenoceptors regulate many physiological processes, including smooth muscle tone (vascular and other), myocardial inotropy, and hepatic glucose metabolism. The α_{1A}-adrenoceptor subtype is the predominant α_1-adrenoceptor in the heart and in arterial smooth muscle; α_{1B}-adrenoceptors mediate vasoconstriction and elevation of blood pressure.[18] Subtype-nonselective α_1-adrenoceptor agonists (such as phenylephrine) and antagonists (such as prazosin) have been used in cardiovascular therapeutics, but have been largely replaced by new drugs with other mechanisms of action and better tolerability.

The α_2-adrenoceptors mediate a number of physiological and pharmacological responses such as hypotension, sedation, inhibition of insulin release, inhibition of lipolysis, and platelet aggregation.[5] In contrast to the G_s-coupled β-adrenoceptors and the G_q-coupled α_1-adrenoceptors, the α_2-adrenoceptors preferentially couple to the G_i/G_o family of heterotrimeric guanine nucleotide binding proteins and thereby regulate a variety of effector systems including adenylyl cyclases[19] and K^+ and Ca^{++} channels.[20]

The α_{2A}-adrenoceptor is the principal presynaptic inhibitory autoreceptor regulating synaptic release of norepinephrine from central and peripheral sympathetic nerves. This receptor subtype is highly expressed in the brain stem and is abundantly distributed throughout the central nervous system and several peripheral tissues.[21–23] The α_{2A}-adrenoceptor is directly involved in the control of sympathetic outflow, its activation resulting in reduction of blood pressure and heart rate. Actually, most of the classical pharmacological actions in response to α_2-adrenoceptor agonists are now ascribed to the α_{2A}-adrenoceptor subtype. These effects include the sedative, antinociceptive, anesthetic, hypothermic, hypotensive, and bradycardic effects of clonidine and related drugs.[5,24] Compared to wild-type control mice, mice genetically engineered to lack α_{2A}-adrenoceptors exhibit higher resting systemic blood pressure and heart rates that correlate with increased norepinephrine release from cardiac sympathetic nerves.

Expression of the α_{2B}-adrenoceptor subtype in brain is restricted to the thalamus and nucleus of the solitary tract in the brain stem.[21,23,25] Abundant expression of this subtype is seen in rat vascular smooth muscle cells[26] and in human and rat kidney.[21,27] This subtype appears to exert a critical role in the peripheral vasoconstrictor action of α_2-adrenoceptor agonists. Contrary to what was observed in normal control mice, intravenous administration of an α_2-adrenoceptor agonist to α_{2B}-adrenoceptor knockout mice failed to elicit the initial transient increase in blood pressure.[6]

This receptor subtype has also been found to play a dominant role in salt-induced experimental hypertension. Indeed, mice with deletion of only one copy of the α_{2B}-adrenoceptor gene were unable to develop hypertension in response to dietary salt loading, indicating that a full complement of α_{2B}-adrenoceptor genes is necessary to raise blood pressure in response to dietary salt loading.[28] It is not yet known whether this role of α_{2B}-adrenoceptors is dependent on vascular responsiveness, on renal ability to retain sodium and water, on presynaptic regulation of norepinephrine release, or on central neuronal mechanisms.

The physiological roles of α_{2C}-adrenoceptors are less known. It was recently shown that α_{2C}-adrenoceptors mediate presynaptic inhibition of norepinephrine release. The α_{2C}-adrenoceptor subtype inhibits neurotransmitter release at low stimulation frequencies, whereas the α_{2A}-adrenoceptor subtype regulates release at high stimulation frequencies, and regulation in both frequency ranges is physiologically important.[29,30] A very interesting recent finding in this context was that mice with targeted inactivation of the α_{2C}-adrenoceptor gene were more prone to develop congestive heart failure upon aortic banding than control mice. Also, α_{2A}-adrenoceptor gene knock-out sensitized the mice to this form of heart failure induced by increased workload and sympathetic overactivity, and the combined deletion of both the α_{2A}- and the α_{2C}-adrenoceptor subtypes resulted in rapid deterioration in cardiac function and high mortality after aortic banding.[31]

Identification of the physiological functions of the individual α_2-adrenoceptor subtypes and the subsequent availability of genetically modified mouse lines have provided further opportunities for the development and evaluation of new therapeutically useful subtype-selective agonists and antagonists.

4.4   RECEPTOR GENE POLYMORPHISMS

Most human GPCR genes appear to be polymorphic, and the significance of receptor mutations for cardiovascular disease risk and as determinants of therapeutic responses constitutes another rapidly expanding area of research. Genetic polymorphism is defined as the occurrence within a population of two or more allelic variants of a given gene sequence, in such proportions that the rarest cannot be maintained merely by recurrent mutations. Genetic variation may influence one or more aspects of the function of a given GPCR, which may provide the basis for individual variability in clinical phenotypes and pharmacological responses.[32] Several databases have been established to catalogue human gene polymorphisms and make this information available to researchers. One example in cardiovascular medicine is the Gene Canvass (cardiovascular candidate gene polymorphisms) web site: ecgene.net/genecanvas/. The site was set up to facilitate association analysis and research on how single nucleotide polymorphisms (SNPs) and other sequence variations may influence common cardiovascular disorders. It is currently acknowledged[33] that individual SNPs may have poor predictive capacity for disease risk and that more emphasis must be put on interactions of SNPs within haplotypes in determining clinical phenotypes of disease and responses to drug treatment.

Most human GPCR polymorphisms appear as functionally neutral; however, several affect either gene expression (and thus protein abundance) or the structural and functional properties of the encoded protein. Altered abundance or function may be associated with a clinical phenotype. Cardiovascular diseases typically have multifactorial etiologies and result from the synergism of several genes interacting in a complex way with environmental and life style factors. Epidemiological studies have investigated several candidate genes encoding proteins contributing to the control of blood pressure, lipid and energy metabolism, and cardiovascular function, and thus possibly associated with human atherosclerosis, hyperlipidemia, hypertension, and myocardial infarction.

The clinical significance of variant alleles in the LDL receptor gene has already been established. Polymorphisms in the adrenoceptor genes are among the most extensively investigated in the GPCR family. Some of them have already been linked with clinical phenotypes, either involving altered disease risk or outcome or altered drug responses.[34] In principle, any gene known or suspected to be involved in normal or pathological physiological regulation of the cardiovascular system may be considered to represent a candidate gene for cardiovascular diseases.

In drug discovery and development of therapeutics, one outcome of pharmacogenetic and pharmacogenomic research efforts might be a more efficacious use of available drugs. For example, although β-blockers are well recognized as effective therapeutics for patients with congestive heart failure, they are difficult to use and carry relatively high risks for adverse effects. The information that their benefit is largely confined to patients with the angiotensin-converting enzyme DD genotype was obtained retrospectively in a relatively small number of patients.[35] Validation of this finding implies that genotyping could be used to identify patients who are especially likely to benefit from therapy with β-blockers.

A second outcome might be validation or rejection of new drug targets at an early stage of development. A drug target with a functionally significant polymorphism might be rejected in favor of one that does not have such genetic variability. Similarly, as the molecular bases for rarely occurring adverse drug effects such as drug-induced arrhythmias or drug-associated hepatotoxicity become better defined in pharmaco-genetic studies, new screening algorithms could be developed to eliminate drugs likely to be associated with these adverse effects at an early stage of development.[36]

4.5   CONCLUDING REMARKS

The cardiovascular therapeutic applications of most GPCR ligands to date are based on two fundamental premises. First, the drugs target the extracellular domains and specifically the ligand binding sites of the receptors, either mimicking the actions of the natural agonist ligands or blocking their access to the binding sites. Second, the receptor proteins have been considered molecular entities invariable in the human population. Both assumptions have recently been challenged. The clarification of the signal transduction pathways linked to GPCRs now allows the targeting of intracellular effector molecules and mechanisms responsible for receptor regulation. Such interventions are still mostly experimental but may provide more effective and selective, and thus safer, targeting of GPCR-mediated signaling pathways.[37] In addition, it is becoming more and more evident that the interindividual variations in the human population with regard to physiological responses, including responses to GPCR agonists and antagonists are associated with the occurrence of naturally existing genetic variations in GPCRs and their effector molecules.[32]

Genetic tests for more accurate and earlier diagnoses will help classify hetero-genous populations of patients with cardiovascular diseases into genetically distinct groups. Genotyping before prescription of certain drugs may enable clinicians to predict how a patient will respond to a given treatment. This will allow the selection of appropriate therapies with the best benefit to the patient and also prevent the prescription of potentially toxic drugs that are likely to yield dangerous side effects in a genetically susceptible individual.[38] The new molecular knowledge is also expected to facilitate the development of new drugs so that drug interventions may become more individualized.[39]

In spite of the apparent abundance of potential cardiovascular GPCR drug targets, only β-adrenoceptor antagonists and AT_1 angiotensin antagonists are currently in widespread use in clinical cardiovascular therapeutics. Other types of receptors for monoamine neurotransmitters and vasoactive peptides are targets for a few classical cardiovascular drugs, such as atropine, and for newer drug classes with very specific actions, such as sumatriptan and related agents for treatment of migraine. The human genome project and concomitant advances in molecular medicine have, however, recently uncovered new potential drug targets: receptor subtypes with possibilities for drug actions with improved benefit:harm ratios; regulatory mechanisms influencing receptor gene expression and receptor functions; and genetic receptor variants offering opportunities for patient selection, risk profiling, and individualized therapy. More than 100 orphan GPCRs still await characterization, and some of them may turn out to be valid cardiovascular drug targets. It is a major challenge for biomedical research to elucidate their physiological functions and therapeutic potentials.

REFERENCES

1. Chang, D.J., Chang, T.K., Yamanishi, S.S., Salazar, F.H., Kosaka, A.H., Khare, R., Bhakta, S., Jasper, J.R., Shieh, I.S., Lesnick, J.D., Ford, A.P., Daniels, D.V., Eglen, R.M., Clarke, D.E., Bach, C., Chan, H.W. Molecular cloning, genomic characterization and expression of novel human alpha1A-adrenoceptor isoforms. FEBS Lett. 1998; 422:279–283.
2. Warnecke, C., Surder, D., Curth, R., Fleck, E., Regitz-Zagrosek, V. Analysis and functional characterization of alternatively spliced angiotensin II type 1 and 2 receptor transcripts in the human heart. J. Mol. Med. 1999; 77:718–727.
3. Pitcher, J.A., Freedman, N.J., Lefkowitz, R.J. G protein-coupled receptor kinases. Annu. Rev. Biochem. 1998; 67:653–692.
4. Contreras, F., de la Parte, M.A., Cabrera, J., Ospino, N., Israili, Z.H., Velasco, M. Role of angiotensin II AT1 receptor blockers in the treatment of arterial hypertension. Am. J. Ther. 2003; 10:401–408.
5. MacDonald, E., Kobilka, B.K., Scheinin, M. Gene targeting: homing in on alpha 2-adrenoceptor-subtype function. Trends Pharmacol. Sci. 1997; 18:211–219.
6. Link, R.E., Desai, K., Hein, L., Stevens, M.E., Chruscinski, A., Bernstein, D., Barsh, G.S., Kobilka, B.K. Cardiovascular regulation in mice lacking alpha2-adrenergic receptor subtypes b and c. Science 1996; 273:803–805.
7. MacMillan, L.B., Hein, L., Smith, M.S., Piascik, M.T., Limbird, L.E. Central hypotensive effects of the alpha2a-adrenergic receptor subtype. Science 1996; 273:801–803.
8. Ferguson, S.S. Evolving concepts in G protein-coupled receptor endocytosis: the role in receptor desensitization and signaling. Pharmacol. Rev. 2001; 53:1–24.
9. Ferguson, S.S. and Caron, M.G. G protein-coupled receptor adaptation mechanisms. Semin. Cell Dev. Biol. 1998; 9:119–127.
10. Ahlquist, R.P. A study of the adrenotropic receptors. Am. J. Physiol. 1948; 153:586–600.
11. Lands, A.M., Arnold, A., McAuliff, J.P., Luduena, F.P., Brown, T.G., Jr. Differentiation of receptor systems activated by sympathomimetic amines. Nature 1967; 214:597–598.
12. Langer, S.Z. Presynaptic regulation of catecholamine release. Biochem. Pharmacol. 1974; 23:1793–1800.
13. Starke, K., Endo, T., Taube, H.D. Pre- and postsynaptic components in effect of drugs with alpha adrenoceptor affinity. Nature 1975; 254:440–441.
14. Starke, K. Presynaptic alpha-autoreceptors. Rev. Physiol. Biochem. Pharmacol. 1987; 107:73–146.
15. Docherty, J.R. Subtypes of functional alpha1- and alpha2-adrenoceptors. Eur. J. Pharmacol. 1998; 361:1–15.
16. Ruffolo, R.R., Jr., Nichols, A.J., Stadel, J.M., Hieble, J.P. Structure and function of alpha-adrenoceptors. Pharmacol. Rev. 1991; 43:475–505.
17. Piascik, M.T., Soltis, E.E., Piascik, M.M., MacMillan, L.B. Alpha-adrenoceptors and vascular regulation: molecular, pharmacologic and clinical correlates. Pharmacol. Ther. 1996; 72:215–241.
18. Tanoue, A., Koshimizu, T.A., Tsujimoto, G. Transgenic studies of alpha(1)-adrenergic receptor subtype function. Life Sci. 2002; 71:2207–2215.

19. Cotecchia, S., Kobilka, B.K., Daniel, K.W., Nolan, R.D., Lapetina, E.Y., Caron, M.G., Lefkowitz, R.J., Regan, J.W. Multiple second messenger pathways of alpha-adrenergic receptor subtypes expressed in eukaryotic cells. J. Biol. Chem. 1990; 265:63–69.

20. Surprenant, A., Horstman, D.A., Akbarali, H., Limbird, L.E. A point mutation of the alpha 2-adrenoceptor that blocks coupling to potassium but not calcium currents. Science 1992; 257:977–980.

21. Handy, D.E., Flordellis, C.S., Bogdanova, N.N., Bresnahan, M.R., Gavras, H. Diverse tissue expression of rat alpha 2-adrenergic receptor genes. Hypertension 1993; 21:861–865.

22. Rosin, D.L., Zeng, D., Stornetta, R.L., Norton, F.R., Riley, T., Okusa, M.D., Guyenet, P.G., Lynch, K.R. Immunohistochemical localization of alpha 2A-adrenergic receptors in catecholaminergic and other brainstem neurons in the rat. Neuroscience 1993; 56:139–155.

23. Scheinin, M., Lomasney, J.W., Hayden-Hixson, D.M., Schambra, U.B., Caron, M.G., Lefkowitz, R.J., Fremeau, R.T. Distribution of alpha 2-adrenergic receptor subtype gene expression in rat brain. Brain Res. Mol. Brain Res. 1994; 21:133–149.

24. Philipp, M., Brede, M., Hein, L. Physiological significance of alpha(2)-adrenergic receptor subtype diversity: one receptor is not enough. Am. J. Physiol. Regul. Integr. Comp. Physiol. 2002; 283:R287–R295.

25. Tavares, A., Handy, D.E., Bogdanova, N.N., Rosene, D.L., Gavras, H. Localization of alpha 2A- and alpha 2B-adrenergic receptor subtypes in brain. Hypertension 1996; 27:449–455.

26. Richman, J.G. and Regan, J.W. Alpha 2-adrenergic receptors increase cell migration and decrease F-actin labeling in rat aortic smooth muscle cells. Am. J. Physiol. 1998; 274:C654–C662.

27. Eason, M.G. and Liggett, S.B. Human alpha 2-adrenergic receptor subtype distribution: widespread and subtype-selective expression of alpha 2C10, alpha 2C4, and alpha 2C2 mRNA in multiple tissues. Mol. Pharmacol. 1993; 44:70–75.

28. Makaritsis, K.P., Handy, D.E., Johns, C., Kobilka, B.K., Gavras, I., Gavras, H. Role of the alpha 2B-adrenergic receptor in the development of salt- induced hypertension. Hypertension 1999; 33:14–17.

29. Hein, L., Altman, J.D., Kobilka, B.K. Two functionally distinct alpha 2-adrenergic receptors regulate sympathetic neurotransmission. Nature 1999; 402:181–184.

30. Trendelenburg, A.U., Philipp, M., Meyer, A., Klebroff, W., Hein, L., Starke, K. All three alpha(2)-adrenoceptor types serve as autoreceptors in postganglionic sympathetic neurons. Naunyn Schmiedebergs Arch. Pharmacol. 2003; 368:504–512.

31. Brede, M., Wiesmann, F., Jahns, R., Hadamek, K., Arnolt, C., Neubauer, S., Lohse, M.J., Hein, L. Feedback inhibition of catecholamine release by two different alpha 2-adrenoceptor subtypes prevents progression of heart failure. Circulation 2002; 106:2491–2496.

32. Evans, W.E. and Relling, M.V. Pharmacogenomics: translating functional genomics into rational therapeutics. Science 1999; 286:487–491.

33. Drysdale, C.M., McGraw, D.W., Stack, C.B., Stephens, J.C., Judson, R.S., Nandabalan, K., Arnold, K., Ruano, G., Liggett, S.B. Complex promoter and coding region beta 2-adrenergic receptor haplotypes alter receptor expression and predict *in vivo* responsiveness. Proc. Natl. Acad. Sci. USA 2000; 97:10483–10488.

34. Kirstein, S.L. and Insel, P.A. Autonomic nervous system pharmacogenomics: a progress report. Pharmacol. Rev. 2004; 56:31–52.

35. McNamara, D.M., Holubkov, R., Janosko, K., Palmer, A., Wang, J.J., MacGowan, G.A., Murali, S., Rosenblum, W.D., London, B., Feldman, A.M. Pharmacogenetic interactions between beta-blocker therapy and the angiotensin-converting enzyme deletion polymorphism in patients with congestive heart failure. Circulation 2001; 103:1644–1648.

36. Roden, D.M. and George, A.L., Jr. The genetic basis of variability in drug responses. Nat. Rev. Drug Discov. 2002; 1:37–44.

37. Iaccarino, G., Smithwick, L.A., Lefkowitz, R.J., Koch, W.J. Targeting G-beta gamma signaling in arterial vascular smooth muscle proliferation: a novel strategy to limit restenosis. Proc. Natl. Acad. Sci. USA 1999; 96:3945–3950.

38. Housman, D. and Ledley, F.D. Why pharmacogenomics? Why now? Nat. Biotechnol. 1998; 16:492–493.

39. Crooke, S.T. Optimizing the impact of genomics on drug discovery and development. Nat. Biotechnol. 1998; 16 Suppl.: 29–30.

5 G Protein-Coupled Receptors and Cancer

Martine J. Smit and Remko A. Bakker

CONTENTS

ABSTRACT

GPCRs are known to play an essential role in the coordination of cellular communication, and they can activate signaling pathways that control gene expression and cell proliferation. Aberrant GPCR signaling, either through activating inherited or acquired mutations within the GPCRs or due to the overexpression of GPCRs or their respective ligands, may lead to transformation, thus contributing to the onset or progression of oncogenesis. Because the role of GPCRs in the development of cancer is becoming apparent, GPCRs are emerging targets for therapeutic interventions to treat cancer.

5.1   INTRODUCTION

A complex cellular network within and between cells coordinates the balance of growth promoting- and growth-inhibiting mechanisms. In cancer cells, however, these control mechanisms are altered and the cells acquire the ability to grow in an unrestricted manner, migrate from their original sites, invade nearby tissues and form metastases at distant organ sites. When a primary tumor and its metastases invade and disrupt tissues with vital functions, they can become lethal.

Cancer is believed to be a genetic disease because alterations are detected within specific genes — either in (pro)oncogenes or tumor suppressor genes.[1] Increased expression of or activating mutations within (pro)oncogenes or the loss of expression or function of tumor suppressing genes due, for example, to gene silencing through DNA methylation or inactivating mutations may be crucial determinants for the onset of oncogenesis.

In general, more than one activating or inactivating event must occur in an oncogene or tumor suppressor gene for the initiation and progression of cancer. Key molecular mechanisms of cancer involve (trans)activation of receptor tyrosine kinase signaling pathways, small GTPases, including Ras activating the MAP kinase cascades and Rho family GTPases (Rho, Rac, cdc42), focal adhesion kinase, the Wnt signaling cascade, TGF-β, regulation of cell cycle control (Kip family p21, p27, cyclins, cyclin-dependent kinases), transcriptional activation (e.g. CREB, STATs, NF-κB), and the inactivation of tumor suppressor genes (Rb, p53, APC, PTEN).[1] See Figure 5.1. After malignant transformation occurs, microenvironmental factors such as cell adherence to the extracellular matrix, host–tumor interactions, degradation of matrix components, migration, and invasion become essential for tumor progression to the metastatic phenotype.[2,3]

GPCRs are known to play an essential role in the coordination of cellular communication and more particularly they can activate signaling pathways that control expression and proliferation of genes[4] including those mentioned above. Therefore, aberrant GPCR signaling through activating mutations within the GPCRs or due to overexpression of GPCRs or their ligands may lead to transformations that may be able to contribute to the onset or progression of oncogenesis.[5–7] Increasing evidence suggests that some GPCRs are implicated in the pathogenesis of cancer and are therefore to be viewed as potential targets for the development of antitumor agents.

This chapter discusses the different GPCR families with respect to their potential contributions to the onset or progression of cancer. Although numerous *in vitro* GPCRs, in particular those linked to the G_q/phospholipase C (PLC) signaling pathway, induce proliferation, we focus only on those receptors shown to be able to induce tumorigenesis *in vivo* or for which altered GPCR expression has been detected within tumors. Although no detailed studies exclusively address the contributions of GPCRs to cancers, we try to cite examples of different GPCR families that are potentially linked to cancer.

GPCRs that negatively regulate tumor formation or progression are also described since they may also be considered potential drug targets for the treatment of certain cancers. It is beyond the scope of this chapter to address the mechanisms by which GPCRs induce cancer in full detail. For a more detailed description of

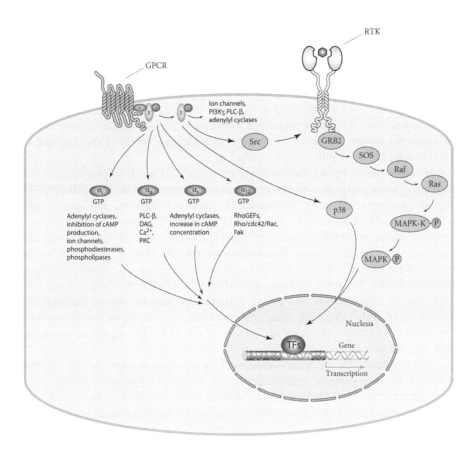

FIGURE 5.1 GPCR-mediated signaling pathways in cancer. Aberrant GPCR signaling through activating mutations within the GPCRs or due to overexpression of GPCRs or their ligands may lead to transformations that may contribute to the onset or progression of oncogenesis. Key molecular mechanisms of cancer that may be activated by GPCRs involve (trans)activation of receptor tyrosine kinase signaling pathways, activation of small GTPases, including Ras activating the MAP kinase cascades and Rho family GTPases (Rho, Rac, cdc42), focal adhesion kinase, and transcriptional activation (e.g., CREB, STATs, NF-κB). PI3K = phosphoinositide-3 kinase. PLC-β = phospholipase C–β. DAG = diacylglycerol. F ak = focal adhesion kinase; MAPK, mitogen-activated protein kinase; TF, transcription factors.

signaling pathways responsible for initiation and progression of cancer, readers are referred to a recent review by Gutkind and colleagues.[1]

5.2 FAMILY A GPCRs

5.2.1 Monoaminergic Receptors

The serotonin 5HT1C and 5HT2B, α_{1b}-adrenergic and muscarinic m1, m3, and m5 acetylcholine receptors appear to harbor oncogenic potential. These receptors are able to activate signaling pathways associated with proliferation. They can transform NIH 3T3 cells in an agonist-dependent manner.[8–11] Injection of cells derived from transformed foci expressing these receptors into nude mice results in the generation of tumors, indicating that these receptors may function as protooncogenes when expressed in NIH 3T3 fibroblasts. Their oncogenic potential can be enhanced when mutational alterations are introduced in sites of the receptor, e.g., the C-terminal region of the third intracellular loop, that render the receptor constitutively active.[10] Activating mutations within the α_{1b}-adrenergic receptors resulted in enhanced ability for tumor generation in nude mice even in the absence of concomitant catecholamine supplementation.

It is less clear whether histaminergic receptors play a role in proliferation and ultimately in oncogenesis. Histamine, a key regulator in a variety of pathophysiological processes, including inflammation, neurotransmission, and gastric acid secretion, has been associated with tumor progression.[4,12] However, both activation of proliferative and antiproliferative signaling by histamine H_1 and H_2 receptors has been described.[13,14] These observations are probably highly dependent on the nature of the cancer cell type. Activation of the H_1 receptor was shown recently to transactivate the β-catenin pathway, providing a molecular explanation for a possible link of inflammation and cancer.[15] Histamine may be able to facilitate the sensitivity of cells to become more prone to cancerous insults by interfering with the β-catenin pathway, elevating the activity of the TCF/LEF-dependent transcription (see Section 5.2.3).

The dopamine D_2 receptor is expressed by pituitary lactotrophs and somatotrophs. Dopamine receptor agonists are potent inhibitors of prolactin (PRL) secretion and are effective in treating prolactinomas, tumors of pituitary lactotrophs that produce prolactin.[16] First generation dopamine receptor agonists such as bromocriptine have been available since the mid-1970s and are used widely to treat pituitary tumors. Cabergoline, a new long-acting safe, high-affinity selective D_2 agonist has been approved to suppress PRL secretion and reduce tumor size.[17] Chimeric ligands that incorporate both intrinsic dopaminergic and somatostatinergic properties (see Section 5.2.1.3.1) by selectively binding to D_2 and SSTR2 receptors have shown increased potency in suppressing PRL and growth hormone (GH) and are currently in the experimental phases of investigation for treatment of pituitary tumors.

5.2.1.1 Lipid Messenger Receptors

5.2.1.1.1 LPA Receptors

Lysophosphatidic acid (LPA) is an extracellular lipid mediator that evokes growth factor-like responses, inducing cell proliferation, migration and survival, indicating a role in the initiation or progression of malignant disease.[18] Increased levels of LPA have been detected in malignant effusions. Its receptors, particularly LPA2 and LPA3, are aberrantly expressed in several human cancers, particularly in ovarian and prostate cancer cells.[19]

In addition, autotaxin, an ectoenzyme known to be involved in tumor invasion, neovascularization, and metastasis, appears to act by producing LPA, further demonstrating a key role for LPA in the metastatic cascade.[20]

LPA signals not only via the classic second messengers but also activates RAS and Rho family GTPases — important switches that control cell proliferation, migration, and morphogenesis. In addition, LPA induces transactivation of the epidermal growth factor receptor (EGFR) through activation of metalloproteinases, thereby modulating the migratory and invasive behaviors of kidney and bladder cancer cells.[21]

Other potent mitogenic GPCR agonists such as thrombin, bombesin, bradykinin, and angiotensin II induce signals that converge on the EGFR to promote migration and invasion, suggesting a common signaling mechanism important in tumorigenesis.

5.2.1.1.2 Platelet-Activating Factor Receptor

PAF (1-O-alkyl-2-acetyl-*sn*-glycero-3-phosphocholine) is an important phospholipid second messenger known to activate platelets, neutrophils, monocytes, and lymphocytes.[22] PAF, through activation of its receptor, induces, among other functions, endothelial cell migration and promotes *in vivo* angiogenesis, thus acting as a mediator of vascularization for tumor growth and metastasis in breast cancer.[23,24] Activation of the $G\alpha_q$-linked PAF receptor in endothelial cells results in stimulation of Src and focal adhesion kinase (Fak), both critical for cell migration.[25]

5.2.1.1.3 Cannabinoid Receptors

Activation of cannabinoid CB1 receptors has been shown to inhibit growth of xenograph tumors of thyroid origin via attenuation of p21[ras] activity.[26] Stimulation of this receptor with the metabolically stable endocannabinoid analog 2-methyl-arachidonyl-2′-fluoro-ethylamide (Met-F-AEA) resulted in inhibition of tumor angiogenesis and metastasis via suppression of VEGF and its receptor (Flt-1) and enhancement of p27[kip1] levels in rat thyroid metastatic cells and in abrogation of formation of metastatic nodules *in vivo*.[26] Effects could be counteracted with the selective CB1 receptor antagonist. These data suggest that the cannabinoid system may be targeted pharmacologically to block molecular and cellular processes to intervening cancer growth and spreading.

5.2.1.2 Peptidergic Receptors

5.2.1.2.1 Somatostatin Receptors

The five somatostatin receptors (SSTR1 through SSTR5) belonging to the family of neuroendocrine receptors are widely expressed in pituitary gland, spleen, and the gastrointestinal tract.[27] Somatostatin exists as two biologically active forms (SS-14 and SS-28) and inhibits secretion of growth hormone (GH) from the pituitary gland. The pituitary gland is crucial for the maintenance of several homeostatic functions including growth, metabolism, and reproduction. Since pituitary tumors are associated with unrestrained secretion and action of trophic hormones, treatment is aimed at the suppression of hormone hypersecretion, inhibition, and prevention of pituitary tumor growth.[16,28] Pituitary tumors account for 15% of intracranial tumors and are usually benign and nonmetastatic.

The high expression levels of SSTR2 and SSTR5 in pituitary adenomas[29] make them attractive targets for drug development. Octreotide and lanreotide, synthetic somatostatin analogs targeting these receptors, have longer half-lives than native somatostatin and are used currently in the treatment of pituitary tumors. Novel somatostatin analogs with affinity for SSTR1 and SSTR3 are also under consideration; SSTR1 has been linked to inhibition of cell cycle progression and SSTR3 signaling can induce apoptosis.[30] Cytotoxic analogs of somatostatin, consisting of radicals conjugated to peptide carriers, can also be used to target these receptors in tumors.[31] However, somatostatin and receptor subtypes are expressed in various normal and neoplastic tissues[32] and require selective targeting.

5.2.1.2.2   Bombesin/Gastrin-Releasing Peptide, Luteinizing-Releasing Hormone Receptor Subtypes, and Luteinizing Hormone Receptor

Bombesin/gastrin-releasing peptide (GRP) and luteinizing-releasing hormone (LHRH) receptor subtypes have been detected in a variety of human malignancies.[31] The presence of these peptidergic receptors permits the localization of certain primary tumors and their metastasis using radiolabeled and bombesin analogs. Cytotoxic analogs of bombesin and LH, a hybrid consisting of a radical conjugated to a peptide carrier, target these tumors and appear promising in the treatment of a variety of cancers including small cell lung carcinomas, prostate, ovarian, and gastrointestinal cancers, and brain tumors expressing the respective receptors.[31] The preclinical work with these cytotoxic conjugates is continuing and clinical trials are pending.

A large number of naturally occurring mutations within the LH receptor, all displaying constitutive activity, have been associated with onset of familial or sporadic precocious puberty and Leydig cell hyperplasia. One mutation that is somatic in nature (D578H in transmembrane helix 6) is found in Leydig cell tumors of boys with precocious puberty.[33-35]

5.2.1.2.3   Thyroid-Stimulating Hormone Receptor

The growth and function of the thyroid are controlled by thyrotropin through activation of the thyroid-stimulating hormone (TSH) receptor.[36] Mutations in TSH receptors are found in approximately 30% of all human thyroid adenomas.[37] In recent years, various constitutively active mutations within the TSH receptor were identified and shown to be associated with continuous stimulation of the cAMP pathway, causing hyperthyroidism and thyroid hyperplasia. Similarly, in Graves' disease, autoantibodies mimic the action of thyrotropin,[38] leading to activation of this receptor. Thyroid carcinogenesis, however, is associated with constitutively activating receptor mutants and an activating mutation within the Ras gene, suggesting that both the Ras and TSH mutations act synergistically to cause the tumor phenotype in some differentiated thyroid carcinomas.[39]

5.2.1.2.4   Cholecytostokinin-2 Receptor

The cholecytostokinin-2 receptor (CCK2R) shows growth-promoting effects on normal and neoplastic gastrointestinal cells when stimulated by gastrin. It is highly expressed in many cancers of different origins including medullary thyroid carcinomas, small cell lung cancers, astrocytomas, and ovarian cancers but is not present in the corresponding tissues.[40] A constitutively active CCK2R mutant in which the conserved glutamate (E) within the (E)–aspartate (D)–arginine (R)–tyrosine (Y) motif (E/DRY motif) was replaced by alanine (A) appeared to have high tumorigenic potential. Injection of NIH 3T3 cells expressing this CAM receptor resulted in the development of rapidly growing tumors.[41] Interestingly, a splice variant of CCK2R isolated from human colorectal cancer contained 69 additional amino acids in the third intracellular loop. This mutant receptor was shown to constitutively activate proliferative signaling pathways including Src tyrosine kinases, suggesting a link between the constitutive activity of this mutant receptor and the development of colorectal cancer.[42]

Expression by transgenic mice of the human CCK2R specifically in pancreatic exocrine cells and stimulation of the receptor induces growth of the pancreas.[43] Moreover, expression of the CCK2 receptor in pancreatic cells over time results in the development of tumors, suggesting a role of CCK2 receptors in the initiation of pancreatic cancer.[43,44]

5.2.1.2.5 Endothelin Receptors

Endothelins and their receptors have been implicated in the development and progression of prostatic, ovarian, and cervical cancers and melanomas[45] through autocrine and paracrine signaling pathways. The three isopeptides that constitute the endothelin family serve as potent mitogens of these tumors. In ovarian carcinoma xenographs, in which the endothelin A receptor ($Et_A R$) is highly expressed, the selective bioavailable antagonist ABT-627 (atrasentan) was shown to significantly inhibit tumor growth associated with a reduction in microvessel density and expression of VEGF and matrix metalloproteinase-2.[46] Combined treatment with paclitaxel led to an additive antitumor effect, suggesting that inhibition of the $Et_A R$ signaling is a promising therapeutic strategy for ovarian carcinoma.

Endothelin receptors are required for the proliferation of normal melanocytes and for the proliferation, adhesion, migration, and invasion of melanoma cells.[47] Endothelin acting on the $ET_A R$,[48] a predominantly expressed receptor on tumor cells, induces vascular endothelial growth factor (VEGF) production by increasing levels of hypoxia-inducible factor 1α.[45] Specific endothelin receptor antagonists affect melanoma tumor growth *in vitro* and *in vivo*.[49,50]

The $ET_B R$ has been associated with cutaneous melanoma and serves as a tumor progression marker. Its activation leads to loss of cell–cell adhesion through decreased expression of cell adhesion molecule E-cadherin and associated cadherin proteins, inhibition of intercellular communication by phosphorylation of gap junctional protein connexin 43, and increased expression of integrins and metalloproteinases.[50] Downstream activation of focal adhesion kinase and p44/p42 MAPK signaling pathways by this receptor is responsible for enhanced cell proliferation, adhesion, migration, and metalloproteinase-dependent invasion. Interestingly, the small molecule A-192621, an orally available nonpeptide $ET_B R$ antagonist, inhibits melanoma growth in nude mice, indicating that interruption of the $ET_B R$ signaling pathway may represent a novel therapeutic approach to target melanoma.

5.2.1.2.6 Melanocortin 1 Receptor

The melanocortin 1 receptor (MC1R) is a major determinant of hair and skin pigmentation. Extensive case-control studies and studies on familial melanoma showed that genetic variations in MC1R play crucial roles in the development of both familial and sporadic melanoma.[51,52] More importantly, certain variants of the MC1R were associated with development of melanoma, independent of pigment synthesis. Therefore, apart from pigment synthesis, the MC1R is involved in hitherto unknown tumorigenic signal transduction pathways. Whether the genetic variants of MC1R harbor constitutively active properties, thus providing a causative link to aberrant proliferation, is yet to be determined.

5.2.1.2.7 Angiotensin Receptors

Angiotensin receptors couple to signal transduction pathways linked to cell proliferation. Increased expression of AT(1) has been reported in human prostate cancer tissue. The angiotensin II (A-II) receptor blocker (ARB), an antihypertensive agent, was shown to inhibit proliferation of prostate cancer cells through suppression of MAPK and STAT3 phosphorylation.[53] Oral administration of ARB inhibited the growth of prostate cancer xerographs in both androgen-dependent and androgen-independent cells. Another report demonstrated a significant role for the A-II–AT1 receptor pathway in tumor angiogenesis and growth *in vivo*.[54]

5.2.1.2.8 Orexin Receptors

Orexins, initially characterized as neuropeptides, appear to be expressed in a few peripheral tissues, including the gastrointestinal tract.[55] Orexins that act via activation of the orexin receptor (OX_1R) were recently shown to suppress the growth of colon cancer cells by promoting apoptosis through cytochrome *c* release from mitochondria and caspase activation.[56]

5.2.1.2.9 Ghrelin Receptors

Ghrelin is a gastrointestinal peptide that stimulates food intake and growth hormone secretion.[57] In certain tumors, specifically pituitary adenomas, gastrointestinal carcinoids, and endocrine tumors of the pancreas, production of ghrelin has been detected.[58,59] The actions of ghrelin are mediated by specific receptors (GHS-R types 1a and b) that regulate GH release-related functions and have also been shown to regulate growth control activities in tumor cells.[61] Recent reports indicate that the GHS-R displays high constitutive activity that may play a causative role in onset of aberrant proliferation.[61]

5.2.1.2.10 Protease-Activated Receptors

A link between coagulation factors and tumor progression and metastasis is recognized, but the molecular basis remains poorly understood.[62] Thrombin, the main effector protease of the coagulation cascade, elicits its response through activation of G protein-coupled protease activated receptors (PARs) designated PAR-1 through PAR-4.[63] Activation of PAR-1 induces proliferate signaling and migratory responses in a variety of cells. Increased PAR-1 mRNA and protein expression has been associated with breast carcinoma cell invasion, whereas it was minimal or absent in noninvasive cell lines and normal breast epithelial tissue specimens.[64,65] PAR-1 is irreversibly proteolytically activated, internalized, and directly sorted to lysosomes critical for signal termination. Aberrant trafficking of PAR-1 associated with constitutive signaling has been suggested to contribute to and to enhance cellular invasion.[66] PAR-1 receptors are also involved in the metastatic potential and angiogenesis of melanoma.[67] Inhibitors of thrombin show encouraging results in clinical trials.

PAR-2, cleaved and activated by trypsin, appears to be highly expressed in human pancreatic and colon cancer.[68,69] The expression of PAR-2 was found to be greater in tissues with infiltrative growth patterns and in those accompanied by severe fibrosis, suggesting that the activation of PAR-2 is involved in cancer invasion and the induction of fibrosis in human pancreatic cancer.[68] Tumor-derived trypsin is believed to contribute to the growth and invasion of colon cancer cells through activation of PAR-2 via transactivation of the EGFR and activation of p44/p42 MAPKs.[69]

5.2.1.3 Chemokine Receptors

The family of chemokines is involved in the control and regulation of the immune system by directing the migration and activation of leukocytes during homeostasis through interactions with their respective GPCRs.[3,70] However, chemokines appear also to be involved in the leukocyte infiltration of tumors, a characteristic of cancer, and to influence metastatic potential and site-specific spread of tumor cells.[3,71]

In addition to their chemotactic properties, chemokines and their receptors play a part in other signaling pathways relevant to oncogenesis, including tumor cell growth, survival, protease induction, and angiogenesis. Local production of chemokines directs the infiltration of macrophages and lymphocytes, expressing respective chemokine receptors in human carcinomas of breast, cervix, and pancreas, as well as sarcomas and gliomas.[72] In ovarian cancer, several chemokines are found at pico- to nanomolar levels in ascites (accumulations of fluid in the peritoneum resulting from cancer) including CCL2, CCL3, CCL4, CCL5, CCL8, CCL22 and CXCL2, CXCL12.[73]

Increased expression of chemokines often correlates with tumor-promoting macrophages and lymphocytes, whereas in some cases chemokines are associated with tumor-inhibiting macrophages and lymphocytes. The role of chemokines in malignant tumors appears complex as some chemokines, through activation of their receptors, show antitumor activity by stimulating immune cells or by inhibiting neovascularization. For example, CXCL10 is suggested to have tumor inhibitory functions since it was shown to limit tumor growth and the establishment of metastasis in different tumor cell systems.[74]

Cancer cells also express chemokine receptors.[75,76] CXCR4 appears to be the most widely expressed chemokine receptor in many tumors, including breast, ovarian, and lung cancer.[71,77] Its corresponding ligand (CXCL12) is strongly expressed in lung, liver, bone marrow, and lymph nodes, suggesting that the CXCL12–CXCR4 axis is responsible for the primary metastatic destination of breast cancer cells.

In normal breast, ovarian, and prostate tissue, epithelial expression of CXCR4 is either limited or absent. Expression of CXCR4 appears to be regulated by autocrine action of vascular endothelial growth factor (VEGF), upregulated by NF-κB, or regulated by hypoxia-inducible factors and indirectly by genes regulating these factors. It is clear that cancer cells with increased expression of CXCR4 are more likely to form metastases.[78]

CCR7 is highly expressed in breast cancer, and its CCL21 ligand strongly expressed in lymph nodes and may be responsible for lymph node metastasis.

Expression of CXCR4 or CCR7 into nonmetastasizing cells allowed these cells to metastasize to lung and lymph nodes, respectively, showing the importance of these receptors in directing organ-specific metastasis *in vivo*.[79,80]

Melanomas and the cells derived from them have been found to express a number of chemokines including CXCL8, CXCL1-3, CCL5, and CCL2 that have been implicated in tumor growth and progression.[81,82] Recent studies have demonstrated organ-specific patterns of melanoma metastasis that correlate with expression of specific chemokine receptors including CXCR3, CXCR4, CCR7, and CCR10.

A role of CXCR3 in melanoma cell metastasis to lymph nodes has been demonstrated recently.[83] Mouse B16F10 cells were shown to constitutively express CXCR3. Induced expression of CXCL9 and CXCL10 facilitated metastasis, while specific antibodies against CXCL9 and CXCL10 and suppression of CXCR3 expression by antisense RNA attenuated metastasis. Through activation of RhoA and Rac1, inducing a reorganization of the actin cytoskeleton, triggering cell chemotaxis and activation of p44/42 and p38 MAPKs in melanoma cells, CXCR3 may contribute to cell motility during invasion and to regulation of cell proliferation and survival.[82]

CXCL1/GROα has been shown to transform melanocytes that have the ability to form tumors in nude mice, most likely via CXCR2 leading to activation of NF-κB, ultimately controlling the transcription of genes that control cell proliferation.[84,85]

5.2.1.4 Purinergic Receptors

The adenosine A_3 receptor was found to be expressed in different tumor cell lines, including Jurkat T, pineal gland, astrocytoma, melanoma, and colon and prostate carcinoma cells,[86] suggesting it may serve as a target for tumor growth inhibition. Activation of the A_3 receptor, however, was shown to attenuate growth of melanocytes both *in vitro* and *in vivo* as well as colon carcinoma growth in mice through modulation of the Wnt signaling pathway linked to the TCF/LEF/β–catenin pathway.[87] Activation of the A_3 receptor may be beneficial for therapeutic intervention of certain tumors.

5.2.2 Family C GPCRs

5.2.2.1 Metabotropic Glutamate Receptors

Studies in a new melanoma mouse model recently implicated the ectopic expression of the $G_{q/11}$-coupled metabotropic glutamate receptor 1 (mGluR1, Grm1), a member of the C family of GPCRs, in melanomagenesis and metastasis.[88,89] Alterations in the expression of Grm1 are sufficient to promote mouse melanocyte transformation *in vivo*.[88] In addition to the coupling to $G_{q/11}$ proteins, Grm1 may act to activate G_s proteins in transfected cells.[90]

5.2.2.2 Calcium Sensing Receptors

The pituitary tumor transforming gene (PTTG) is an oncogene that is normally expressed in few tissues although it is abundantly expressed in the testis. PTTG is widely expressed in cancer,[91] and its expression has been directly correlated with both proliferation and angiogenesis. Recently, extracellular calcium (Ca^{2+}) acting on the calcium-sensing receptor (CaR) has been shown to induce a rapid and prolonged upregulation of PTTG mRNA in a nonmetastasizing model for humoral hypercalcemia of malignancy. In addition, Ca^{2+} also induced the upregulation of VEGF in the same model, indicating that CaR may also induce angiogenesis via VEGF.[92]

5.2.3 FRIZZLED/SMOOTHENED FAMILY OF GPCRs

Signaling through the G protein-coupled frizzled/smoothened receptors plays a crucial role in a diverse array of developmental processes (Figure 5.2). The ten frizzled receptors (FzRs) known in humans may be activated by a plethora of Wingless (Wnt) ligands identified in mammals. These ligands belong to a family of protooncogenes. The "Wnt" denotes the relationship of this family to the Drosophila segment polarity wingless gene and to its vertebrate ortholog, murine mammary tumor virus integration site 1, a mouse protooncogene. Smoothened (Smo) is the sole member of the family of smoothened receptors. Although signaling of Smo is initiated by members of the Hedgehog (Hh) family of ligands, the activity of Smo is unleashed upon binding of Hh to the 12-membrane spanning protein Patched (Ptc) that otherwise keeps Smo in its inactive state. The Hh signaling pathway is intimately linked to cell growth and differentiation, with physiological roles in embryonic pattern formation and adult tissue homeostasis and pathological roles in tumor initiation and growth.

5.2.3.1 Frizzled Receptors

Wnt constitutes a family of genes encoding ligands essential for signaling in early development.[93] These secreted glycoproteins, which resemble other secreted glycoprotein ligands such as gonadotropins, act via members of the Frizzled gene family. The Wnt signaling transduction pathway is essentially a network of a number of separate but interacting pathways. Aberrant activation and upregulation of the Wnt pathway is a key feature of many cancers. Despite extensive investigation of the role of Wnt genes in human cancer,[94] no documentation has been made to date of any mutations or amplifications of genes encoding Wnt ligands or receptors. In contrast, several components of the Wnt pathway have been implicated in carcinogenesis, especially APC and β-catenin.

Mutations that lead to the stabilization of β-catenin have been documented in melanoma cell lines and may be implicated in tumor progression.[95] Aberrant Wnt pathway signaling occurs through mutations mainly of APC and, although less often, of the genes encoding β-catenin or axin-2 (also known as conductin).[96] The nuclear/cytoplasmic localization of β-catenin, rather than at the plasma membrane, is an indicator of possible involvement of the activation of Wnt/Frizzled pathways in melanoma tumors.[97] For example, aberrant Wnt pathway signaling is thought to be an early progression event in 90% of colorectal cancers.[96]

5.2.3.1.1 Initiation of Wnt/Frizzled Signaling
Wnt/Frizzled signaling has been extensively studied in embryogenesis and development of vertebrates and invertebrates.[94] Wnt and its Frizzled receptor govern the

FIGURE 5.2 Simplified scheme of the Hedgehog and Frizzled signal transduction pathways. Patched (Ptc) is an inhibitor of the Smoothened (Smo) protein. In the absence of Hedgehog (Hh) protein-binding to Ptc, the transcription factor Ci is tethered to microtubules by Cos2 and Fused (Fu) proteins, allowing Ci to be cleaved into a transcriptional repressor to block transcription of particular genes. Hh binding to Ptc results in conformational changes and release of the inhibitory action on Smo. Most likely the constitutive activity of Smo then leads to phosphorylation of Cos2 and Fu and release of Ci from microtubules. Ci acts as a transcriptional activator of Hh-response genes, including genes encoding Wingless (Wnt) proteins, which are secreted and diffuse to surrounding cells to interact with Frizzled receptor–low density lipoprotein receptor-related protein (FzR–LPR) complexes. Binding of Wnt to FzR–LRP complexes antagonizes the actions of APC–Axin to stabilise b-catenin by recruitment of Disheveled (Dvl) proteins to FzRs. Accumulated b-catenin may enter the nucleus to bind to the TCF/LEF family of transcription factors to induce transcription of particular genes, including the genes encoding Hh proteins. Hh proteins diffuse from these cells to interact with Ptc receptors to induce Smo signaling. (See also text).

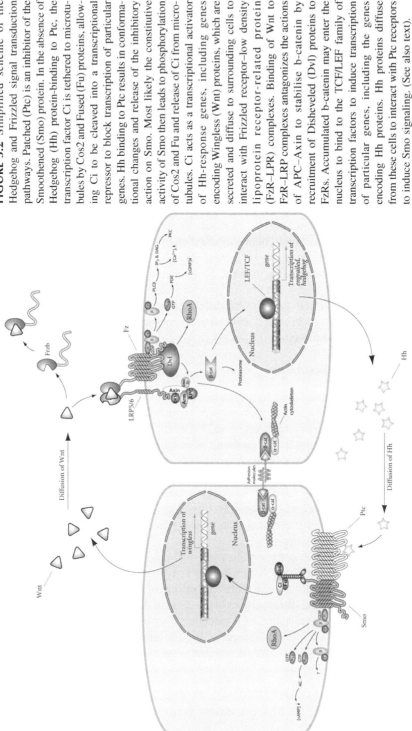

morphogenetic processes of gastrulation. Movement of cells during this process requires the cells to undergo transient transitions that allow them to dissociate and migrate. To this end, FzRs activate various intracellular signaling cascades that affect cellular processes including differentiation, proliferation, motility, and polarity. Cell dissociation and migration are also essential for tumor cell invasion and metastasis and accumulating evidence emphasizes the importance of Frizzled in tumor growth and progression.[98]

In the absence of a Wnt signal, the signal transduction pathway is off, resulting in the destruction of β-catenin by the proteosome. A large multiprotein complex that includes proteins of the APC (the tumor suppressor protein encoded by the adenomalous polyposis coli (APC) gene) and Axin families normally facilitate the phosphorylation of β-catenin by glycogen synthase kinase-3β (GSK3β). Phosphorylated β-catenin is subsequently ubiquitinated and degraded by the proteosome. Binding of secreted Wnt proteins to dedicated FzR–low-density lipoprotein receptor-related protein complexes triggers the intracellular Wnt signaling cascade by antagonizing the actions APC–Axin via an unknown mechanism that requires the Dvl (Dishevelled) protein. In addition to the modulation of the TCF/LEF transcription factors, the recruitment of Dvl to FzRs may also serve to activate the small guanosine triphosphatase (GTPase) RhoA.

5.2.3.1.2 Secreted Frizzled-Related Proteins

The secreted Frizzled-related proteins (sFRPs) comprise a family of five secreted glycoproteins that may antagonize Wnt signaling through the binding and sequestering of Wnt ligands, and thus may have important implications in tumorigenesis. Both the downregulation and upregulation of sFRPs have been correlated with cancers.[99,100] Moreover, in addition to genetic alterations, changes in the status of DNA methylation of the sFRP2 gene, known as epigenetic alterations, can be used as markers for colorectal cancer.[101]

Hypermethylation of the sFRP1 gene on chromosome 8p12 is one of the earliest molecular alterations in colorectal carcinogenesis,[102,103] potentially disrupting the Wnt signaling cascade of cellular growth control.[102] sFRP1 expression is also frequently downregulated in bladder cancer.[104] Restoration of sFRP function in colorectal cancer cells that have epigenic losses of sFRP function has been shown to attenuate Wnt signaling.[96]

5.2.3.1.3 Low Density Lipoprotein Receptor-Related Proteins as Co-Receptors for FzRs

The low density lipoprotein receptor-related proteins 5 and 6 (LPR5 and LPR6) serve as coreceptors with FzRs in the β-catenin pathway. LPR5 and LPR6 may also bind to secreted Dickkopf (Dkk) proteins that act as antagonists on the FzR–LRP complex by the formation of LPR5 and LPR6 complexes with another recently identified transmembrane receptor class (Kremen) which are internalized, making LPR5 and LPR6 unavailable for complexing with FzRs.[105] As a consequence of the binding of Wnt to FzR–LRP complexes, β-catenin is stabilized, i.e., diverted from the proteosomes, and accumulates and enters nuclei where it binds to members of the TCF–LEF family of transcription factors, resulting in the activation of gene transcription.[106] The plasma membrane recruitment of Dvl proteins by FzRs activated by Wnt family members is therefore a key intermediate step leading to the stabilization of β-catenin and activation of TCF/LEF transcription factors.[93]

The identification of both FzRs and LPRs 5 and 6 as components of cell surface receptors for Wnt proteins led to further investigations of their individual functions. Subsequently, it was found that LPR5–LPR6–Arrow proteins constitute distal signal-initiating components and that they are candidate oncogenes because their mutants may activate the Wnt/β-catenin pathway in a Frizzled and ligand-independent manner.[107]

5.2.3.1.4 Dvl Proteins

Dvl proteins serve an important role in the regulation and initiation of Wnt signaling. The upregulation of Dvl has been correlated with cancers.[108,109] The Dvl proteins may not only interact with FzRs to initiate activation of the TCF–LEF transcription factors, they may also activate RhoA. In addition, Dvl proteins may also participate in the regulation of FzR-mediated signaling. Dvl proteins may be phosphorylated by protein kinase C (PKC), allowing the recruitment of β-arrestin-2 to the plasma membrane.[110] β-arrestin is an adaptor protein that, when bound to the receptor, targets receptors to endocytosis that may result in the routing of receptors to lysosomes for receptor degradation.

In addition to β-arrestin-2, β-arrestin-1 has been shown to bind to both Dvl-1 and Dvl-2. Phosphorylation of Dvl enhances this binding, and the coexpression of β-arrestin-1 with Dvl-1 or Dvl-2 synergistically enhances the activity of the LEF transcription factor,[111] suggesting that β-arrestin-1 may also serve as an adapter for FzRs. In addition to serving as adapter proteins, arrestins may serve as scaffolds linking GPCRs to other signaling proteins such as the src family of kinases and members of the mitogen-activated protein kinase cascade.

5.2.4.1.5 FzRs as GPCRs

Although the FzRs exhibit several hallmarks of GPCRs, including presumed heptahelical structures, and exhibit Pertussis toxin- and specific G protein Gα or Gβγ subunit-dependent signaling, their G protein activation capabilities have been investigated only minimally, and the FzRs are often still regarded as new members of the GPCR superfamily.[112] Some FzRs have been shown to activate heterotrimeric guanine nucleotide-binding (G) proteins.[112–122] The receptors appear to couple to members of the $G_{i/o}$ family of G proteins to activate PLCβ (via heterodimeric Gβγ subunits),[115] resulting in increased concentrations of intracellular calcium (Ca^{2+}). They also appear to activate phosphodiesterases (PDEs)[114] through $Gα_{transducin\ 2}$ (G_{t2}) to lower intracellular concentrations of cyclic guanosine monophosphate (cGMP).

Numerous reports discuss the involvement of altered expression levels of both Wnts and FzRs in cancers. For instance, an upregulation of Frizzled-7 (FZD7) has been reported in human gastric cancer,[123] and upregulation in the expression of Wnt5a in carcinomas of the lung, breast, prostate, and melanomas has been documented.[124] Moreover, melanoma cells with both low Wnt5a expression and low *in vitro* invasion show increased invasiveness *in vitro* upon transfection with exogenous

Wnt5a, while treatment of the cells with an antibody to Frizzled-5 inhibited the binding of Wnt5a to the receptor and resulted in decreased invasiveness.[125]

5.2.3.2 Smoothened Receptors

The signaling by lipid-modified secreted glycoproteins of the Hedgehog (Hh) family plays fundamental roles during pattern formation in animal development. Mature Hh proteins are dually modified by cholesteryl and palmitoyl adducts.[126,127] Dysfunction of Hh pathway components is frequently associated with a variety of congenital abnormalities and cancers. Transcriptional regulation of Hh target genes is mediated by the Gli zinc finger transcription factors. Activation of Gli leads, for example, to tissue-specific cell proliferation during embryogenesis, but Gli activity is restricted in adult tissues.

The smoothened GPCR (Smo) is kept in an inactive state by the 12-transmembrane Patched (Ptc) protein, the receptor for Hh. Binding of Hh to Ptc activates Ptc to release the catalytic suppression[128] of the activity of the Smo receptor to induce downstream signaling in a constitutively active manner, i.e., in the absence of ligands bound to Smo (although there may still be a hypothetical endogenous ligand whose access to Smo is controlled by Ptc in a Hh-dependent manner). Downregulation of Ptc, as observed in many cancers, thus results in constitutive activation of the Hh pathway.

In addition to the interaction with Ptc, in Drosophila, Smo has also been shown to interact with a protein complex consisting of the zinc finger transcription factor Cubitus interrruptus (Ci), the serine–threonine kinase Fused (Fu), and the kinesin-like Costal2 (Cos2)[129] recently suggested to serve as a catalytic subunit of protein kinase A (PKA).[130] The protein complex is associated with Smo in membrane fractions both *in vitro* and *in vivo*, and Cos2 binding on Smo is necessary for the Hh-dependent dissociation of Ci from this complex.[129] Activated Ci then translocates to the nucleus to modulate gene expression.

Consistent with the notions that phosphorylation of PKA renders Ci in Drosophila susceptible to proteolytic cleavage and that Hh signaling reduces PKA-dependent cleavage of Ci,[131] $G\alpha_i$ proteins are suggested to participate in Smo-mediated signaling upon stimulation of Ptc with Sonic Hedgehog (SHh).[132] However, intracellular cAMP levels are not altered in mammalian cells stimulated with SHh,[133] suggesting additional signaling pathways are involved in mediating SHh actions.

Members of the ubiquitously expressed G_{12} family of heterotrimeric G proteins that include $G\alpha_{12}$ and $G\alpha_{13}$ have been shown to mediate the actions of SHh to activate RhoA and its downstream effector Rho-associated kinase (Rho-kinase) to activate the Gli transcription factors and inhibit neurite outgrowth of neuroblastoma cells.[134] The signaling pathways activated by Smo/Ptc are depicted in Figure 5.2.[129]

In analogy to the Wnt/Frizzled signaling, the Hh signaling controls many aspects of development and also regulates cell growth and differentiation. Hh and Wnt signaling often act together to control cell growth and tissue morphogenesis. Hh/Smoothened signaling is active in a variety of tumors and is also involved in the maintenance of normal stem cells *in vivo*. Several SHh signal transducers including Ptc, Smo, and the Gli transcription factors are upregulated in basal cell carcinomas

(BCCs).[135] The mRNAs, for example, of Hh, Ptc, and Smo are upregulated in pancreatic cancer tissues.[136]

Despite the prominent role of Smo in activating Hh signaling, experiments using transgenic mice indicated that gain-of-function mutations in Smo are insufficient for the development and/or maintenance of basal cell carcinomas.[137] In contrast, the magnitude of Hh pathway activation appears to be the determinant of skin tumor phenotype.[137]

Altered SHh signaling is crucial for the development of BCCs, the most common human cancers. Mutations in SHh, Ptc, and Smo have been identified — in particular loss-of-function mutations of Ptc, and gain-of-function mutations of Smo.[138] Mutations in SHh are rare,[139] and the identified mutations do not result in SHh proteins with altered activities compared to the wild-type proteins. Two gain-of-function mutations in human Smo that result in alterations in the seventh transmembrane domain and in the intracellular cytoplasmic C-terminal domain of the receptor that do not impair their interactions with Ptc have been identified in BCC sporadic tumors.[138] The two individual mutations in Smo (but not wild-type Smo) can cooperate with adenovirus E1A to transform rat embryonic fibroblast cells in culture. Transgenic mice overexpressing the mutant Smo develop skin abnormalities similar to BCCs, supporting the role of Smo as a protooncogene.[138] These findings suggest that pharmacological inhibition of Smo or its downstream effectors could provide effective treatment for BCCs and possibly for other cancers.

Cyclopamine, a teratogen found in the *Veratrum californicum* plant that blocks the synthesis of cholesterol, is an inhibitor of Hh/Smo signaling by acting on Smo and has recently been shown to lead to a rapid regression of skin tumors, while no adverse effects were noted on normal skin.[140] The effects of oncogenic mutations of Smo and Ptc can be reversed by cyclopamine treatment[141] and the modulation of the activities of Hh pathways has been suggested to offer improvements in the treatment of cancer.[142] Cyclopamine and other small molecule inhibitors of Smo have been shown to be effective in cancer treatment.[143–145] Cyclopamine treatment has been shown to be effective in the inhibition of brain tumorigenesis[146] and medulloblastoma growth[147]; the treatment of skin[140] and colorectal tumors through the induction of apoptosis[145]; the prevention of ultraviolet B-induced BCC formation[148]; the inhibition of the invasiveness of cancer cells[149]; the growth of Hh pathway-activated breast carcinoma cells[150]; and the proliferation of metastatic prostate cancer cell lines.[151] Cyclopamine also inhibits growth of small cell lung cancers and a wide range of foregut-derived malignancies both *in vitro* and *in vivo*.[152] Modulation of the activity of the Hh pathway thus appears a promising avenue for the treatment of cancer. In addition, cyclopamide has recently been shown to be effective in a proof-of-principle study related to rapid clearing of skin lesions in psoriasis patients.[153]

Small non-peptidylic agonists and antagonists for Smo have recently been identified.[154] The Smo agonist designated Ag-1.4 mediates the activation of p44/p42 MAP kinase phosphorylation, which is abrogated in interferon (IFN)-α SHh-responsive cells,[155] possibly explaining the effectiveness of IFN-based therapy in BCC treatment.

5.2.4 VIRALLY ENCODED GPCRS

GPCRs encoded by viruses of the herpes virus family have drawn considerable attention recently because some may provide causative links with respect to the onset or progression of cancer. Various herpes viruses and pox viruses contain DNA sequences encoding proteins with homologies to cellular chemokine receptors.[156] Since GPCRs and chemokine receptors play key roles in cellular communications and appear to play roles in cancer, the virally encoded GPCRs are believed to be of great importance in subverting cellular signaling.

The Kaposi's sarcoma (KS)-associated herpes virus (KSHV or human herpes virus 8) implicated in the pathogenesis of KS, a highly vascularized tumor, encodes a GPCR designated ORF74 — one of the most studied viral chemokine receptors thought to be implicated in the pathology of KS.

ORF74 shows the highest homology to human CXCR2.[157,158] However, in contrast to human chemokine receptors, ORF74 binds several CXC and some CC chemokines[159–161] and is constitutively active.[159] Through promiscuous G protein coupling, ORF74[162–164] is able to constitutively activate a variety of signal transduction cascades in several cell types: activation of phosholipase C, kinase pathways including various MAPKs, Lyn kinase, related adhesion focal tyrosine kinase (RAFTK/Pyk2), phosphatidylinositol 3-kinase and Akt/protein kinase B, RhoA and transcription factors, including hypoxia-inducible factor 1α (HIF-1α), NF-κB, and AP-1 (Figure 5.3).[156]

NIH-3T3 cells transfected with ORF74 cause tumors when injected into nude mice.[165] Moreover, transgenic mice expressing ORF74 within hematopoietic cells develop angioproliferative lesions that morphologically resemble KS lesions[158,166] via a paracrine effect through the expression of inflammatory cytokines and growth factors such as VEGF. Both the constitutive activity of ORF74 and its modulation by endogenous chemokines appear responsible for the induction of a KS-like disease in transgenic mice.[167] It was recently shown that constitutive activation of Akt by ORF74 plays a crucial role in ORF74-mediated sarcomagenesis and pharmacological inhibition of Akt inhibited its oncogenic potential *in vivo*.[168]

These transgenic model systems clearly provide evidence that ORF74 possesses oncogenic potential and represents an important drug target against KS. KSHV has been pirating this gene from its host genome and has modified it both to bind a wide range of chemokines and also to constitutively activate a variety of signaling pathways for its own benefit.

We recently characterized a GPCR known as BILF1, encoded by the Epstein–Barr virus (EBV) associated with many lymphoproliferative diseases such as infectious mononucleosis, Burkitt's lymphoma (BL), and nasopharyngeal carcinoma (NPC).[169] Like ORF74, BILF1 constitutively activates signaling to NF-κB and CRE, both implicated in proliferative signaling. BILF1-specific mRNA was detected in various EBV-positive lymphoblastoid B cell lines as well as in BL and NPC cell lines. Increased activation of signaling pathways was also observed in these cell lines. In addition, reduced levels of phosphorylated RNA-dependent PKR, which is known to function in the cellular antiviral response, were observed in BILF1-expressing cells. Neither of these effects required a ligand to interact with the BILF1

FIGURE 5.3 The KSHV-encoded GPCR ORF74 as a viral oncogene. Through promiscuous G protein coupling, ORF74 constitutively activates a variety of signal transduction cascades in several cell-types, among which activation of phospholipase C, kinase pathways, including various MAPK, Lyn kinase, related adhesion focal tyrosine kinase (RAFTK/Pyk2), phosphatidyl-inositol 3-kinase and Akt/protein kinase B, RhoA and transcription factors, including hypoxia-inducible factor 1α (HIF-1α), nuclear factor kappa-B (NF-κB) and AP-1.[156] ORF74 can be further stimulated or inhibited by CXCL1 and CXCL10, respectively. Activation of these pathways can lead to activation of transcription factors and subsequent expression of VEGF, cytokines, chemokines, and adhesion molecules, resulting in angiogenesis and proliferation of adjacent cells. Activation of these pathways can be highly cell-type dependent and not all events might take place in the same cell as suggested in this picture. See text for details and references.

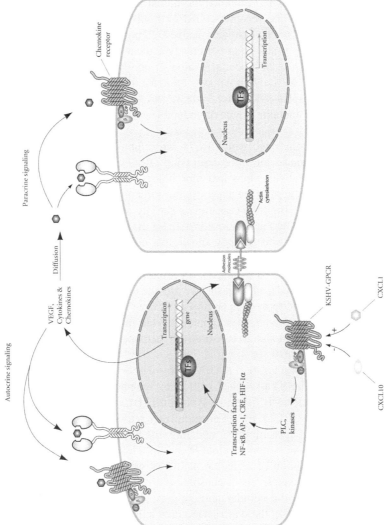

gene product. Taken together, these results indicate that BILF1 encodes a constitutively active viral GPCR, and that this EBV vGPCR plays a role in evasion from the PKR-mediated cellular antiviral response and may be implicated in lymphoproliferative diseases caused by EBV.

Since other virally encoded GPCRs, for example, human cytomegalovirus (HCMV) -encoded receptors US28 and UL33,[170–172] display constitutive activity and show promiscuous G protein coupling, other herpes viruses may contribute to the onset of transformation through expression of virally encoded GPCRs in their infected hosts.

5.2.5 ORPHAN GPCRs

RDC-1, defined as a GPCR, appears to be highly expressed in tumor endothelial cells of both brain and peripheral vasculature and is therefore considered a tumor endothelial marker.[173] Other GPCRs including GPCR mas, GPR35, TDAG8, GPR4, G2A, and D-GPCR are linked to signaling pathways that control proliferation or are reported to be upregulated in certain cancers.[174–177] GPR54, also known as hOT7T175 and AXOR12, is the receptor for the peptide derived from the KISS1 gene, which is known as metastin or kisspentin-54. KISS1 is a suppressor gene that inhibits metastasis in both *in vivo* melanoma and breast cancer carcinoma models.[178]

5.3 CONCLUSIONS

It is apparent that GPCRs are able to modulate signaling pathways that control cell proliferation. Activating mutations within the GPCRs or overexpression of GPCRs or their ligands may lead to transformation and may contribute to the onset or progression of oncogenesis.[5]

Mutational alterations of key amino acids, for example, in the C-terminal region of the third intracellular loop or conserved DRY motif, may render the GPCR constitutively active[179] and confer oncogenic properties to the receptor. Spontaneous mutations may acquire an agonist-independent transformation capability, subvert the normal function of these receptors, and result in disease states associated with uncontrolled cell growth including neoplasia.[10,180] As discussed in this chapter, several constitutively active mutations in GPCRs have been linked to carcinomas including activating mutations in the TSH (thyroid insular carcinoma),[39] MCR1 (melanoma), LH (testicular carcinoma),[35] and smoothened receptors (basal cell carcinoma).[138] Constitutive activity is without doubt also a hallmark of many virally encoded receptors such as ORF74 encoded by KSHV. Transgenic expression of ORF74 induces an angioproliferative disease resembling KS, indicating that ORF74 acts as a viral oncogene.

Besides the inherited or acquired (spontaneously activated) GPCR mutants, GPCRs and their respective ligands may be upregulated in certain disease states, thus enhancing their oncogenic potential. Clear examples include the upregulation of chemokine receptor CXCR4 and its corresponding ligand CXCL12 (breast cancer), LPA and its cognate receptors (ovarian and prostate cancer), CCK2R (colon cancer), PAR-1 (breast cancer, melanoma), endothelin receptors (melanoma and ovarian cancer), and metabotropic mGluR1 (melanoma). Activation of some GPCRs, including adenosine A_3, dopamine D_2, cannabinoid CB1, peptidergic somatostatin SSTR, bombesin and LH receptors were shown to inhibit tumor progression, suggesting that these receptors may represent targets to counteract aberrant proliferation. The use of cytotoxic conjugates targeting the peptidergic receptors appears promising.

In order to effectively target GPCRs in certain cancers, one should determine the molecular basis by which GPCRs play a role in the onset or progression of the disease. A thorough investigation of the ways mutant or overexpressed GPCRs or their ligands affect tumorigenic pathways is essential and will provide novel targets for future therapeutic intervention. Important for the aforementioned constitutively active receptors is the use of inverse agonists, whereas in case of GPCR overexpression or upregulation of endogenous ligands, neutral antagonists are recommended as therapeutic agents targeting these receptors.

The completion of the sequencing of the human genome and genome studies has already improved our understanding of how cancer initiates or progresses and will continue to do so. The use of microarrays helps us dissect genetic events in specific cancer cells. Expression profiles of DNA allow us to predict hereditary cancer, assist in identifying specific disease markers that may serve as new pharmacological targets for therapeutic intervention, and enable us to type the various stages and progression of the disease. Certain orphan GPCRs have been shown to play important roles in tumor biology and thus appear potential targets for drug development. Collections of large datasets, including tumor banks and family, twin, and case-controlled studies, are expected to provide more information on new drug targets with respect to cancer treatment. Because GPCRs have proven to be successful drug targets based on their functionality and membrane localization, and because their role in cancer is becoming apparent, GPCRs are emerging targets for therapeutic intervention in cancer patients.

ACKNOWLEDGMENTS

M.J. Smit is supported by the Royal Netherlands Academy of Arts and Sciences. The authors thank Dr. Rob Leurs of the Vrije Universiteit Amsterdam for critically reviewing this chapter.

REFERENCES

1. Sodhi, A., S. Montaner, and J.S. Gutkind, Molecular mechanisms of cancer, in Signal Transduction and Human Disease, T. Finkel and J.S. Gutkind, Editors. 2003, John Wiley & Sons. p. 71–142.
2. Bogenrieder, T. and M. Herlyn, Axis of evil: molecular mechanisms of cancer metastasis. Oncogene, 2003. 22:6524–6536.
3. Zlotnik, A., Chemokines in neoplastic progression. Semin Cancer Biol, 2004. 14:181–185.
4. Smit, M.J., R.A. Bakker, and E.S. Burstein, G protein-coupled receptors and proliferative signaling. Methods Enzymol, 2002. 343:430–447.

5. Marinissen, M.J. and J.S. Gutkind, G-protein-coupled receptors and signaling networks: emerging paradigms. Trends Pharmacol Sci, 2001. 22:368–376.

6. Heasley, L.E., Autocrine and paracrine signaling through neuropeptide receptors in human cancer. Oncogene, 2001. 20:1563–1569.

7. Dhanasekaran, N., L.E. Heasley, and G.L. Johnson, G protein-coupled receptor systems involved in cell growth and oncogenesis. Endocr Rev, 1995. 16:259–270.

8. Julius, D., T.J. Livelli, T.M. Jessell, and R. Axel, Ectopic expression of the serotonin 1c receptor and the triggering of malignant transformation. Science, 1989. 244:1057–1062.

9. Launay, J.M., G. Birraux, D. Bondoux, J. Callebert, D.S. Choi, S. Loric, and L. Maroteaux, Ras involvement in signal transduction by the serotonin 5-HT2B receptor. J Biol Chem, 1996. 271:3141–3147.

10. Allen, L.F., R.J. Lefkowitz, M.G. Caron, and S. Cotecchia, G-protein-coupled receptor genes as protooncogenes: constitutively activating mutation of the alpha 1B-adrenergic receptor enhances mitogenesis and tumorigenicity. Proc Natl Acad Sci USA, 1991. 88:11354–11358.

11. Gutkind, J.S., E.A. Novotny, M.R. Brann, and K.C. Robbins, Muscarinic acetylcholine receptor subtypes as agonist-dependent oncogenes. Proc Natl Acad Sci USA, 1991. 88:4703–4707.

12. Bartholeyns, J. and M. Bouclier, Involvement of histamine in growth of mouse and rat tumors: antitumoral properties of monofluoromethylhistidine, an enzyme-activated irreversible inhibitor of histidine decarboxylase. Cancer Res, 1984. 44:639–645.

13. Takahashi, K., S. Tanaka, K. Furuta, and A. Ichikawa, Histamine H(2) receptor-mediated modulation of local cytokine expression in a mouse experimental tumor model. Biochem Biophys Res Commun, 2002. 297:1205–1210.

14. Kobayashi, K., S. Matsumoto, T. Morishima, T. Kawabe, and T. Okamoto, Cimetidine inhibits cancer cell adhesion to endothelial cells and prevents metastasis by blocking E-selectin expression. Cancer Res, 2000. 60:3978–3984.

15. Diks, S.H., J.C. Hardwick, R.M. Diab, M.M. van Santen, H.H. Versteeg, S.J. van Deventer, D.J. Richel, and M.P. Peppelenbosch, Activation of the canonical beta-catenin pathway by histamine. J Biol Chem, 2003. 278:52491–52496.

16. Heaney, A.P. and S. Melmed, Molecular targets in pituitary tumours. Nat Rev Cancer, 2004. 4:285–295.

17. Bevan, J.S. and J.R. Davis, Cabergoline: an advance in dopaminergic therapy. Clin Endocrinol (Oxf), 1994. 41:709–712.

18. Chun, J., E.J. Goetzl, T. Hla, Y. Igarashi, K.R. Lynch, W. Moolenaar, S. Pyne, and G. Tigyi, International Union of Pharmacology. XXXIV. Lysophospholipid receptor nomenclature. Pharmacol Rev, 2002. 54:265–269.

19. Mills, G.B. and W.H. Moolenaar, The emerging role of lysophosphatidic acid in cancer. Nat Rev Cancer, 2003. 3:582–591.

20. Umezu-Goto, M., Y. Kishi, A. Taira, K. Hama, N. Dohmae, K. Takio, T. Yamori, G.B. Mills, K. Inoue, J. Aoki, and H. Arai, Autotaxin has lysophospholipase D activity leading to tumor cell growth and motility by lysophosphatidic acid production. J Cell Biol, 2002. 158:227–233.

21. Schafer, B., A. Gschwind, and A. Ullrich, Multiple G-protein-coupled receptor signals converge on the epidermal growth factor receptor to promote migration and invasion. Oncogene, 2004. 23:991–999.

22. Prescott, S.M., G.A. Zimmerman, D.M. Stafforini, and T.M. McIntyre, Platelet-activating factor and related lipid mediators. Annu Rev Biochem, 2000. 69:419–445.

23. Bussolati, B., L. Biancone, P. Cassoni, S. Russo, M. Rola-Pleszczynski, G. Montrucchio, and G. Camussi, PAF produced by human breast cancer cells promotes migration and proliferation of tumor cells and neo-angiogenesis. Am J Pathol, 2000. 157:1713–1725.

24. Camussi, G., G. Montrucchio, E. Lupia, A. De Martino, L. Perona, M. Arese, A. Vercellone, A. Toniolo, and F. Bussolino, Platelet-activating factor directly stimulates *in vitro* migration of endothelial cells and promotes *in vivo* angiogenesis by a heparin-dependent mechanism. J Immunol, 1995. 154:6492–6501.

25. Deo, D.D., N.G. Bazan, and J.D. Hunt, Activation of platelet-activating factor receptor-coupled G alpha q leads to stimulation of Src and focal adhesion kinase via two separate pathways in human umbilical vein endothelial cells. J Biol Chem, 2004. 279:3497–3508.

26. Portella, G., C. Laezza, P. Laccetti, L. De Petrocellis, V. Di Marzo, and M. Bifulco, Inhibitory effects of cannabinoid CB1 receptor stimulation on tumor growth and metastatic spreading: actions on signals involved in angiogenesis and metastasis. FASEB J, 2003. 17:1771–1773.

27. Olias, G., C. Viollet, H. Kusserow, J. Epelbaum, and W. Meyerhof, Regulation and function of somatostatin receptors. J Neurochem, 2004. 89:1057–1091.

28. Asa, S.L. and S. Ezzat, The pathogenesis of pituitary tumours. Nat Rev Cancer, 2002. 2:836–849.

29. Shimon, I., X. Yan, J.E. Taylor, M.H. Weiss, M.D. Culler, and S. Melmed, Somatostatin receptor (SSTR) subtype-selective analogues differentially suppress *in vitro* growth hormone and prolactin in human pituitary adenomas: novel potential therapy for functional pituitary tumors. J Clin Invest, 1997. 100:2386–2392.

30. Bruns, C., I. Lewis, U. Briner, G. Meno-Tetang, and G. Weckbecker, SOM230: a novel somatostatin peptidomimetic with broad somatotropin release inhibiting factor (SRIF) receptor binding and a unique antisecretory profile. Eur J Endocrinol, 2002. 146:707–716.

31. Schally, A.V. and A. Nagy, Chemotherapy targeted to cancers through tumoral hormone receptors. Trends Endocrinol Metab, 2004. 15:300–310.

32. Reubi, J.C., Peptide receptors as molecular targets for cancer diagnosis and therapy. Endocr Rev, 2003. 24:389–427.

33. Themmen, A.P. and H.G. Brunner, Luteinizing hormone receptor mutations and sex differentiation. Eur J Endocrinol, 1996. 134:533–540.

34. Themmen, A.P., J.W. Martens, and H.G. Brunner, Activating and inactivating mutations in LH receptors. Mol Cell Endocrinol, 1998. 145:137–142.

35. Liu, G., L. Duranteau, J.C. Carel, J. Monroe, D.A. Doyle, and A. Shenker, Leydig cell tumors caused by an activating mutation of the gene encoding the luteinizing hormone receptor. New Engl J Med, 1999. 341:1731–1736.

36. Paschke, R. and M. Ludgate, The thyrotropin receptor in thyroid diseases. New Engl J Med, 1997. 337:1675–1681.

37. Gutkind, J.S., Cell growth control by G protein-coupled receptors: from signal transduction to signal integration. Oncogene, 1998. 17:1331–1342.

38. McKenzie, J.M. and M. Zakarija, Clinical review 3: clinical use of thyrotropin receptor antibody measurements. J Clin Endocrinol Metab, 1989. 69:1093–1096.

39. Russo, D., F. Arturi, M. Schlumberger, B. Caillou, R. Monier, S. Filetti, and H.G. Suarez, Activating mutations of the TSH receptor in differentiated thyroid carcinomas. Oncogene, 1995. 11:1907–1911.

40. Reubi, J.C., J.C. Schaer, and B. Waser, Cholecystokinin(CCK)-A and CCK-B/gastrin receptors in human tumors. Cancer Res, 1997. 57:1377–1386.

41. Gales, C., M. Poirot, J. Taillefer, B. Maigret, J. Martinez, L. Moroder, C. Escrieut, L. Pradayrol, D. Fourmy, and S. Silvente-Poirot, Identification of tyrosine 189 and asparagine 358 of the cholecystokinin 2 receptor in direct interaction with the crucial C-terminal amide of cholecystokinin by molecular modeling, site-directed mutagenesis, and structure/affinity studies. Mol Pharmacol, 2003. 63:973–982.

42. Olszewska-Pazdrak, B., K.L. Ives, J. Park, C.M. Townsend, Jr., and M.R. Hellmich, Epidermal growth factor potentiates cholecystokinin/gastrin receptor-mediated Ca^{2+} release by activation of mitogen-activated protein kinases. J Biol Chem, 2004. 279:1853–1860.

43. Clerc, P., C. Saillan-Barreau, C. Desbois, L. Pradayrol, D. Fourmy, and M. Dufresne, Transgenic mice expressing cholecystokinin 2 receptors in the pancreas. Pharmacol Toxicol, 2002. 91:321–326.

44. Clerc, P., S. Leung-Theung-Long, T.C. Wang, G.J. Dockray, M. Bouisson, M.B. Delisle, N. Vaysse, L. Pradayrol, D. Fourmy, and M. Dufresne, Expression of CCK2 receptors in the murine pancreas: proliferation, transdifferentiation of acinar cells, and neoplasia. Gastroenterology, 2002. 122:428–437.

45. Bagnato, A. and F. Spinella, Emerging role of endothelin-1 in tumor angiogenesis. Trends Endocrinol Metab, 2003. 14:44–50.

46. Rosano, L., F. Spinella, D. Salani, V. Di Castro, A. Venuti, M.R. Nicotra, P.G. Natali, and A. Bagnato, Therapeutic targeting of the endothelin a receptor in human ovarian carcinoma. Cancer Res, 2003. 63:2447–2453.

47. Lahav, R., C. Ziller, E. Dupin, and N.M. Le Douarin, Endothelin 3 promotes neural crest cell proliferation and mediates a vast increase in melanocyte number in culture. Proc Natl Acad Sci USA, 1996. 93:3892–3897.

48. Grant, K., M. Loizidou, and I. Taylor, Endothelin-1: a multifunctional molecule in cancer. Br J Cancer, 2003. 88:163–166.

49. Lahav, R., G. Heffner, and P.H. Patterson, An endothelin receptor B antagonist inhibits growth and induces cell death in human melanoma cells *in vitro* and *in vivo*. Proc Natl Acad Sci USA, 1999. 96:11496–11500.

50. Bagnato, A., L. Rosano, F. Spinella, V. Di Castro, R. Tecce, and P.G. Natali, Endothelin B receptor blockade inhibits dynamics of cell interactions and communications in melanoma cell progression. Cancer Res, 2004. 64:1436–1443.

51. Kennedy, C., J. ter Huurne, M. Berkhout, N. Gruis, M. Bastiaens, W. Bergman, R. Willemze, and J.N. Bavinck, Melanocortin 1 receptor (MC1R) gene variants are associated with an increased risk for cutaneous melanoma which is largely independent of skin type and hair color. J Invest Dermatol, 2001. 117:294–300.

52. Valverde, P., E. Healy, S. Sikkink, F. Haldane, A.J. Thody, A. Carothers, I.J. Jackson, and J.L. Rees, The Asp84Glu variant of the melanocortin 1 receptor (MC1R) is associated with melanoma. Hum Mol Genet, 1996. 5:1663–1666.

53. Uemura, H., H. Ishiguro, N. Nakaigawa, Y. Nagashima, Y. Miyoshi, Y. Fujinami, A. Sakaguchi, and Y. Kubota, Angiotensin II receptor blocker shows antiproliferative activity in prostate cancer cells: a possibility of tyrosine kinase inhibitor of growth factor. Mol Cancer Ther, 2003. 2:1139–1147.

54. Egami, K., T. Murohara, T. Shimada, K. Sasaki, S. Shintani, T. Sugaya, M. Ishii, T. Akagi, H. Ikeda, T. Matsuishi, and T. Imaizumi, Role of host angiotensin II type 1 receptor in tumor angiogenesis and growth. J Clin Invest, 2003. 112:67–75.

55. Kirchgessner, A.L. and M. Liu, Orexin synthesis and response in the gut. Neuron, 1999. 24:941–951.

56. Rouet-Benzineb, P., C. Rouyer-Fessard, A. Jarry, V. Avondo, C. Pouzet, M. Yanag-isawa, C. Laboisse, M. Laburthe, and T. Voisin, Orexins acting at native OX1 receptor in colon cancer and neuroblastoma cells or at recombinant OX1 receptor suppress cell growth by inducing apoptosis. J Biol Chem, 2004. 279:45875–45886.

57. Kojima, M., H. Hosoda, Y. Date, M. Nakazato, H. Matsuo, and K. Kangawa, Ghrelin is a growth-hormone-releasing acylated peptide from stomach. Nature, 1999. 402:656–660.

58. Korbonits, M., S.A. Bustin, M. Kojima, S. Jordan, E.F. Adams, D.G. Lowe, K. Kangawa, and A.B. Grossman, The expression of the growth hormone secretagogue receptor ligand ghrelin in normal and abnormal human pituitary and other neuroen-docrine tumors. J Clin Endocrinol Metab, 2001. 86:881–887.

59. Papotti, M., P. Cassoni, M. Volante, R. Deghenghi, G. Muccioli, and E. Ghigo, Ghrelin-producing endocrine tumors of the stomach and intestine. J Clin Endocrinol Metab, 2001. 86:5052–5059.

60. Jeffery, P.L., A.C. Herington, and L.K. Chopin, Expression and action of the growth hormone releasing peptide ghrelin and its receptor in prostate cancer cell lines. J Endocrinol, 2002. 172:R7–R11.

61. Holst, B., N.D. Holliday, A. Bach, C.E. Elling, H.M. Cox, and T.W. Schwartz, Common structural basis for constitutive activity of the ghrelin receptor family. J Biol Chem, 2004. 279:53806–53817.

62. Dvorak, H.F., J.A. Nagy, B. Berse, L.F. Brown, K.T. Yeo, T.K. Yeo, A.M. Dvorak, L. van de Water, T.M. Sioussat, and D.R. Senger, Vascular permeability factor, fibrin, and the pathogenesis of tumor stroma formation. Ann NY Acad Sci, 1992. 667:101–111.

63. Coughlin, S.R., Thrombin signalling and protease-activated receptors. Nature, 2000. 407:258–264.

64. Even-Ram, S., B. Uziely, P. Cohen, S. Grisaru-Granovsky, M. Maoz, Y. Ginzburg, R. Reich, I. Vlodavsky, and R. Bar-Shavit, Thrombin receptor overexpression in malignant and physiological invasion processes. Nat Med, 1998. 4:909–914.

65. Henrikson, K.P., S.L. Salazar, J.W. Fenton, 2nd, and B.T. Pentecost, Role of thrombin receptor in breast cancer invasiveness. Br J Cancer, 1999. 79:401–406.

66. Booden, M.A., L.B. Eckert, C.J. Der, and J. Trejo, Persistent signaling by dysregulated thrombin receptor trafficking promotes breast carcinoma cell invasion. Mol Cell Biol, 2004. 24:1990–1999.

67. Tellez, C. and M. Bar-Eli, Role and regulation of the thrombin receptor (PAR-1) in human melanoma. Oncogene, 2003. 22:3130–3137.

68. Ikeda, O., H. Egami, T. Ishiko, S. Ishikawa, H. Kamohara, H. Hidaka, S. Mita, and M. Ogawa, Expression of proteinase-activated receptor-2 in human pancreatic cancer: a possible relation to cancer invasion and induction of fibrosis. Int J Oncol, 2003. 22:295–300.

69. Darmoul, D., V. Gratio, H. Devaud, and M. Laburthe, Protease-activated receptor 2 in colon cancer: trypsin-induced MAPK phosphorylation and cell proliferation are mediated by epidermal growth factor receptor transactivation. J Biol Chem, 2004. 279:20927–20934.

70. Murphy, P.M., International Union of Pharmacology. XXX. Update on chemokine receptor nomenclature. Pharmacol Rev, 2002. 54:227–229.

71. Balkwill, F., Cancer and the chemokine network. Nat Rev Cancer, 2004. 4:540–550.

72. Balkwill, F. and A. Mantovani, Inflammation and cancer: back to Virchow? Lancet, 2001. 357:539–545.

73. Milliken, D., C. Scotton, S. Raju, F. Balkwill, and J. Wilson, Analysis of chemokines and chemokine receptor expression in ovarian cancer ascites. Clin Cancer Res, 2002. 8:1108–1114.

74. Yang, J. and A. Richmond, The angiostatic activity of interferon-inducible protein-10/CXCL10 in human melanoma depends on binding to CXCR3 but not to glycosaminoglycan. Mol Ther, 2004. 9:846–855.

75. Muller, A., B. Homey, H. Soto, N. Ge, D. Catron, M.E. Buchanan, T. McClanahan, E. Murphy, W. Yuan, S.N. Wagner, J.L. Barrera, A. Mohar, E. Verastegui, and A. Zlotnik, Involvement of chemokine receptors in breast cancer metastasis. Nature, 2001. 410:50–56.

76. Murphy, P.M., Chemokines and the molecular basis of cancer metastasis. New Engl J Med, 2001. 345:833–835.

77. Balkwill, F., The significance of cancer cell expression of the chemokine receptor CXCR4. Semin Cancer Biol, 2004. 14:171–179.

78. Epstein, R.J., Opinion: the CXCL12-CXCR4 chemotactic pathway as a target of adjuvant breast cancer therapies. Nat Rev Cancer, 2004. 4:901–909.

79. Wiley, H.E., E.B. Gonzalez, W. Maki, M.T. Wu, and S.T. Hwang, Expression of CC chemokine receptor-7 and regional lymph node metastasis of B16 murine melanoma. J Natl Cancer Inst, 2001. 93:1638–1643.

80. Murakami, T., W. Maki, A.R. Cardones, H. Fang, A. Tun Kyi, F.O. Nestle, and S.T. Hwang, Expression of CXC chemokine receptor-4 enhances the pulmonary metastatic potential of murine B16 melanoma cells. Cancer Res, 2002. 62:7328–7334.

81. Payne, A.S. and L.A. Cornelius, The role of chemokines in melanoma tumor growth and metastasis. J Invest Dermatol, 2002. 118:915–922.

82. Robledo, M.M., R.A. Bartolome, N. Longo, J.M. Rodriguez-Frade, M. Mellado, I. Longo, G.N. van Muijen, P. Sanchez-Mateos, and J. Teixido, Expression of functional chemokine receptors CXCR3 and CXCR4 on human melanoma cells. J Biol Chem, 2001. 276:45098–45105.

83. Kawada, K., M. Sonoshita, H. Sakashita, A. Takabayashi, Y. Yamaoka, T. Manabe, K. Inaba, N. Minato, M. Oshima, and M.M. Taketo, Pivotal role of CXCR3 in melanoma cell metastasis to lymph nodes. Cancer Res, 2004. 64:4010–4017.

84. Balentien, E., B.E. Mufson, R.L. Shattuck, R. Derynck, and A. Richmond, Effects of MGSA/GRO alpha on melanocyte transformation. Oncogene, 1991. 6:1115–1124.

85. Wang, D. and A. Richmond, Nuclear factor-kappa B activation by the CXC chemokine melanoma growth-stimulatory activity/growth-regulated protein involves the MEKK1/p38 mitogen-activated protein kinase pathway. J Biol Chem, 2001. 276:3650–3659.

86. Fishman, P., S. Bar-Yehuda, G. Ohana, F. Barer, A. Ochaion, A. Erlanger, and L. Madi, An agonist to the A3 adenosine receptor inhibits colon carcinoma growth in mice via modulation of GSK-3 beta and NF-kappa B. Oncogene, 2004. 23:2465–2471.

87. Madi, L., S. Bar-Yehuda, F. Barer, E. Ardon, A. Ochaion, and P. Fishman, A3 adenosine receptor activation in melanoma cells: association between receptor fate and tumor growth inhibition. J Biol Chem, 2003. 278:42121–42130.

88. Pollock, P.M., K. Cohen-Solal, R. Sood, J. Namkoong, J.J. Martino, A. Koganti, H. Zhu, C. Robbins, I. Makalowska, S.S. Shin, Y. Marin, K.G. Roberts, L.M. Yudt, A. Chen, J. Cheng, A. Incao, H.W. Pinkett, C.L. Graham, K. Dunn, S.M. Crespo-Carbone, K.R. Mackason, K.B. Ryan, D. Sinsimer, J. Goydos, K.R. Reuhl, M. Eckhaus, P.S. Meltzer, W.J. Pavan, J.M. Trent, and S. Chen, Melanoma mouse model implicates metabotropic glutamate signaling in melanocytic neoplasia. Nat Genet, 2003. 34:108–112.

89. Marín, Y.E. and S. Chen, Involvement of metabotropic glutamate receptor 1, a G protein coupled receptor, in melanoma development. J Mol Med, 2004. 82:735–749.

90. Aramori, I. and S. Nakanishi, Signal transduction and pharmacological characteristics of a metabotropic glutamate receptor, mGluR1, in transfected CHO cells. Neuron, 1992. 8:757–765.

91. Hamid, T. and S.S. Kakar, PTTG and cancer. Histol Histopathol, 2003. 18:245–251.

92. Tfelt-Hansen, J., P. Schwarz, E.F. Terwilliger, E.M. Brown, and N. Chattopadhyay, Calcium-sensing receptor induces messenger ribonucleic acid of human securin, pituitary tumor transforming gene, in rat testicular cancer. Endocrinology, 2003. 144:5188–5193.

93. Wodarz, A. and R. Nusse, Mechanisms of Wnt signaling in development. Annu Rev Cell Dev Biol, 1998. 14:59–88.

94. Smalley, M.J. and T.C. Dale, Wnt signaling in mammalian development and cancer. Cancer Metastasis Rev, 1999. 18:215–230.

95. Rubinfeld, B., P. Robbins, M. El-Gamil, I. Albert, E. Porfiri, and P. Polakis, Stabilization of beta-catenin by genetic defects in melanoma cell lines. Science, 1997. 275:1790–1792.

96. Suzuki, H., D.N. Watkins, K.W. Jair, K.E. Schuebel, S.D. Markowitz, W. Dong Chen, T.P. Pretlow, B. Yang, Y. Akiyama, M. Van Engeland, M. Toyota, T. Tokino, Y. Hinoda, K. Imai, J.G. Herman, and S.B. Baylin, Epigenetic inactivation of SFRP genes allows constitutive WNT signaling in colorectal cancer. Nat Genet, 2004. 36:417–422.

97. Rimm, D.L., K. Caca, G. Hu, F.B. Harrison, and E.R. Fearon, Frequent nuclear/cytoplasmic localization of β-catenin without exon 3 mutations in malignant melanoma. Am J Pathol, 1999. 154:325–329.

98. Vincan, E., Frizzled/WNT signaling: the insidious promoter of tumour growth and progression. Front Biosci, 2004. 9:1023–1034.

99. Lee, J.L., C.J. Chang, S.Y. Wu, D.R. Sargan, and C.T. Lin, Secreted frizzled-related protein 2 (SFRP2) is highly expressed in canine mammary gland tumors but not in normal mammary glands. Breast Cancer Res Treat, 2004. 84:139–149.

100. Lee, A.Y., B. He, L. You, S. Dadfarmay, Z. Xu, J. Mazieres, I. Mikami, F. McCormick, and D.M. Jablons, Expression of the secreted frizzled-related protein gene family is downregulated in human mesothelioma. Oncogene, 2004. 23:6672–6676.

101. Müller, H.M., M. Oberwalder, H. Fiegl, M. Morandell, G. Goebel, M. Zitt, M. Mühlthaler, D. Öfner, R. Margreiter, and M. Widschwendter, Methylation changes in faecal DNA: a marker for colorectal cancer screening? Lancet, 2004. 363:1283–1285.

102. Oberschmid, B.I., W. Dietmaier, A. Hartmann, E. Dahl, E. Klopocki, B.G. Beatty, N.H. Hyman, and H. Blaszyk, Distinct secreted Frizzled receptor protein 1 staining pattern in patients with hyperplastic polyposis coli syndrome. Arch Pathol Lab Med, 2004. 128:967–973.

103. Caldwell, G.M., C. Jones, K. Gensberg, S. Jan, R.G. Hardy, P. Byrd, S. Chughtai, Y. Wallis, G.M. Matthews, and D.G. Morton, The Wnt antagonist sFRP1 in colorectal tumorigenesis. Cancer Res, 2004. 64:883–888.

104. Stoehr, R., C. Wissmann, H. Suzuki, R. Knuechel, R.C. Krieg, E. Klopocki, E. Dahl, P. Wild, H. Blaszyk, G. Sauter, R. Simon, R. Schmitt, D. Zaak, F. Hofstaedter, A. Rosenthal, S.B. Baylin, C. Pilarsky, and A. Hartmann, Deletions of chromosome 8p and loss of sFRP1 expression are progression markers of papillary bladder cancer. Lab Invest, 2004. 84:465–478.

105. Seto, E.S. and H.J. Bellen, The ins and outs of Wingless signaling. Trends Cell Biol, 2004. 14:45–53.

106. Miller, J.R., A.M. Hocking, J.D. Brown, and R.T. Moon, Mechanism and function of signal transduction by the Wnt/beta-catenin and Wnt/Ca^{2+} pathways. Oncogene, 1999. 18:7860–7872.

107. Brennan, K., J.M. Gonzalez-Sancho, L.A. Castelo-Soccio, L.R. Howe, and A.M. Brown, Truncated mutants of the putative Wnt receptor LRP6/Arrow can stabilize beta-catenin independently of Frizzled proteins. Oncogene, 2004. 23:4873–4784.

108. Uematsu, K., B. He, L. You, Z. Xu, F. McCormick, and D.M. Jablons, Activation of the Wnt pathway in non small cell lung cancer: evidence of dishevelled overexpression. Oncogene, 2003. 22:7218–7221.

109. Uematsu, K., S. Kanazawa, L. You, B. He, Z. Xu, K. Li, B.M. Peterlin, F. McCormick, and D.M. Jablons, Wnt pathway activation in mesothelioma: evidence of Dishevelled overexpression and transcriptional activity of beta-catenin. Cancer Res, 2003. 63:4547–4551.

110. Chen, W., D. ten Berge, J. Brown, S. Ahn, L.A. Hu, W.E. Miller, M.G. Caron, L.S. Barak, R. Nusse, and R.J. Lefkowitz, Dishevelled 2 recruits beta-arrestin 2 to mediate Wnt5A-stimulated endocytosis of Frizzled 4. Science, 2003. 301:1391–1394.

111. Chen, W., L.A. Hu, M.V. Semenov, S. Yanagawa, A. Kikuchi, R.J. Lefkowitz, and W.E. Miller, beta-Arrestin1 modulates lymphoid enhancer factor transcriptional activity through interaction with phosphorylated dishevelled proteins. Proc Natl Acad Sci USA, 2001. 98:14889–14894.

112. Malbon, C.C., Frizzleds: new members of the superfamily of G-protein-coupled receptors. Front Biosci, 2004. 9:1048–1058.

113. Liu, T., A.J. DeCostanzo, X. Liu, H. Wang, S. Hallagan, R.T. Moon, and C.C. Malbon, G protein signaling from activated rat frizzled-1 to the β-catenin-Lef-Tcf pathway. Science, 2001. 292:1718–1722.

114. Ahumada, A., D.C. Slusarski, X. Liu, R.T. Moon, C.C. Malbon, and H.Y. Wang, Signaling of rat Frizzled-2 through phosphodiesterase and cyclic GMP. Science, 2002. 298:2006–2010.

115. Slusarski, D.C., V.G. Corces, and R.T. Moon, Interaction of Wnt and a Frizzled homologue triggers G-protein-linked phosphatidylinositol signaling. Nature, 1997. 390:410–413.

116. Liu, T., X. Liu, H. Wang, R.T. Moon, and C.C. Malbon, Activation of rat frizzled-1 promotes Wnt signaling and differentiation of mouse F9 teratocarcinoma cells via pathways that require Gα$_q$ and Gα$_o$ function. J Biol Chem, 1999. 274:33539–33544.

117. Liu, X., T. Liu, D.C. Slusarski, J. Yang-Snyder, C.C. Malbon, R.T. Moon, and H. Wang, Activation of a frizzled-2/beta-adrenergic receptor chimera promotes Wnt signaling and differentiation of mouse F9 teratocarcinoma cells via Gα$_o$ and Gα$_t$. Proc Natl Acad Sci USA, 1999. 96:14383–14388.

118. Malbon, C.C., H. Wang, and R.T. Moon, Wnt signaling and heterotrimeric G-proteins: strange bedfellows or a classic romance? Biochem Biophys Res Commun, 2001. 287:589–593.

119. Wang, H.Y., WNT-frizzled signaling via cyclic GMP. Front Biosci, 2004. 9:1043–1047.

120. Li, H., C.C. Malbon, and H.Y. Wang, Gene profiling of Frizzled-1 and Frizzled-2 signaling: expression of G-protein-coupled receptor chimeras in mouse F9 teratocarcinoma embryonal cells. Mol Pharmacol, 2004. 65:45–55.

121. Wang, H.Y. and C.C. Malbon, Wnt-frizzled signaling to G-protein-coupled effectors. Cell Mol Life Sci, 2004. 61:69–75.

122. Wang, H.Y. and C.C. Malbon, Wnt signaling, Ca2+, and cyclic GMP: visualizing Frizzled functions. Science, 2003. 300:1529–1530.

123. Kirikoshi, H., H. Sekihara, and M. Katoh, Up-regulation of Frizzled-7 (FZD7) in human gastric cancer. Int J Oncol, 2001. 19:111–115.

124. Iozzo, R.V., I. Eichstetter, and K.G. Danielson, Aberrant expression of the growth factor Wnt-5A in human malignancy. Cancer Res, 1995. 55:3495–3499.

125. Weeraratna, A.T., Y. Jiang, G. Hostetter, K. Rosenblatt, P. Duray, M. Bittner, and J.M. Trent, Wnt5a signaling directly affects cell motility and invasion of metastatic melanoma. Cancer Cell, 2002. 1:279–288.

126. Mann, R.K. and P.A. Beachy, Novel lipid modifications of secreted protein signals. Annu Rev Biochem, 2004. 73:891–923.

127. Gallet, A., R. Rodriguez, L. Ruel, and P.P. Therond, Cholesterol modification of hedgehog is required for trafficking and movement, revealing an asymmetric cellular response to hedgehog. Dev Cell, 2003. 4:191–204.

128. Taipale, J., M.K. Cooper, T. Maiti, and P.A. Beachy, Patched acts catalytically to suppress the activity of Smoothened. Nature, 2002. 418:892–897.

129. Ruel, L., R. Rodriguez, A. Gallet, L. Lavenant-Staccini, and P.P. Therond, Stability and association of Smoothened, Costal2 and Fused with cubitus interruptus are regulated by Hedgehog. Nat Cell Biol, 2003. 5:907–913.

130. Collier, L.S., K. Suyama, J.H. Anderson, and M.P. Scott, Drosophila costal 1 mutations are alleles of protein kinase A that modulate Hedgehog signaling. Genetics, 2004. 167:783–796.

131. Ingham, P.W. and A.P. McMahon, Hedgehog signaling in animal development: paradigms and principles. Genes Dev, 2001. 15:3059–3087.

132. DeCamp, D.L., T.M. Thompson, F.J. de Sauvage, and M.R. Lerner, Smoothened activates Gαi-mediated signaling in frog melanophores. J Biol Chem, 2000. 275:26322–26327.

133. Murone, M., A. Rosenthal, and F.J. de Sauvage, Sonic hedgehog signaling by the patched-smoothened receptor complex. Curr Biol, 1999. 9:76–84.

134. Kasai, K., M. Takahashi, N. Osumi, S. Sinnarajah, T. Takeo, H. Ikeda, J.H. Kehrl, G. Itoh, and H. Arnheiter, The G12 family of heterotrimeric G proteins and Rho GTPase mediate Sonic hedgehog signalling. Genes Cells, 2004. 9:49–58.

135. Tojo, M., H. Kiyosawa, K. Iwatsuki, K. Nakamura, and F. Kaneko, Expression of the GLI2 oncogene and its isoforms in human basal cell carcinoma. Br J Dermatol, 2003. 148:892–897.

136. Kayed, H., J. Kleeff, S. Keleg, J. Guo, K. Ketterer, P.O. Berberat, N. Giese, I. Esposito, T. Giese, M.W. Buchler, and H. Friess, Indian hedgehog signaling pathway: expression and regulation in pancreatic cancer. Int J Cancer, 2004. 110:668–676.

137. Grachtchouk, V., M. Grachtchouk, L. Lowe, T. Johnson, L. Wei, A. Wang, F. de Sauvage, and A.A. Dlugosz, The magnitude of hedgehog signaling activity defines skin tumor phenotype. EMBO J, 2003. 22:2741–2751.

138. Xie, J., M. Murone, S.M. Luoh, A. Ryan, Q. Gu, C. Zhang, J.M. Bonifas, C.W. Lam, M. Hynes, A. Goddard, A. Rosenthal, E.H. Epstein, Jr., and F.J. de Sauvage, Activating Smoothened mutations in sporadic basal-cell carcinoma. Nature, 1998. 391:90–92.

139. Couve-Privat, S., M. Le Bret, E. Traiffort, S. Queille, J. Coulombe, B. Bouadjar, M.F. Avril, M. Ruat, A. Sarasin, and L. Daya-Grosjean, Functional analysis of novel sonic hedgehog gene mutations identified in basal cell carcinomas from xeroderma pigmentosum patients. Cancer Res, 2004. 64:3559–3565.

140. Tabs, S. and O. Avci, Induction of the differentiation and apoptosis of tumor cells in vivo with efficiency and selectivity. Eur J Dermatol, 2004. 14:96–102.

141. Taipale, J., J.K. Chen, M.K. Cooper, B. Wang, R.K. Mann, L. Milenkovic, M.P. Scott, and P.A. Beachy, Effects of oncogenic mutations in Smoothened and Patched can be reversed by cyclopamine. Nature, 2000. 406:1005–1009.

142. Karhadkar, S.S., G. Steven Bova, N. Abdallah, S. Dhara, D. Gardner, A. Maitra, J.T. Isaacs, D.M. Berman, and P.A. Beachy, Hedgehog signalling in prostate regeneration, neoplasia and metastasis. Nature, 2004. 431:707–712.

143. Romer, J.T., H. Kimura, S. Magdaleno, K. Sasai, C. Fuller, H. Baines, M. Connelly, C.F. Stewart, S. Gould, L.L. Rubin, and T. Curran, Suppression of the Shh pathway using a small molecule inhibitor eliminates medulloblastoma in Ptc1$^{+/-}$ p53$^{-/-}$ mice. Cancer Cell, 2004. 6:229–240.

144. Berman, D.M., S.S. Karhadkar, A. Maitra, R. Montes De Oca, M.R. Gerstenblith, K. Briggs, A.R. Parker, Y. Shimada, J.R. Eshleman, D.N. Watkins, and P.A. Beachy, Widespread requirement for Hedgehog ligand stimulation in growth of digestive tract tumours. Nature, 2003. 425:846–851.

145. Qualtrough, D., A. Buda, W. Gaffield, A.C. Williams, and C. Paraskeva, Hedgehog signalling in colorectal tumour cells: induction of apoptosis with cyclopamine treatment. Int J Cancer, 2004. 110:831–837.

146. Dahmane, N., P. Sanchez, Y. Gitton, V. Palma, T. Sun, M. Beyna, H. Weiner, and A. Ruiz i Altaba, The Sonic Hedgehog–Gli pathway regulates dorsal brain growth and tumorigenesis. Development, 2001. 128:5201–5212.

147. Berman, D.M., S.S. Karhadkar, A.R. Hallahan, J.I. Pritchard, C.G. Eberhart, D.N. Watkins, J.K. Chen, M.K. Cooper, J. Taipale, J.M. Olson, and P.A. Beachy, Medulloblastoma growth inhibition by hedgehog pathway blockade. Science, 2002. 297:1559–1561.

148. Athar, M., C. Li, X. Tang, S. Chi, X. Zhang, A.L. Kim, S.K. Tyring, L. Kopelovich, J. Hebert, E.H. Epstein, Jr., D.R. Bickers, and J. Xie, Inhibition of smoothened signaling prevents ultraviolet B-induced basal cell carcinomas through regulation of Fas expression and apoptosis. Cancer Res, 2004. 64:7545–7552.

149. Sheng, T., C. Li, X. Zhang, S. Chi, N. He, K. Chen, F. McCormick, Z. Gatalica, and J. Xie, Activation of the hedgehog pathway in advanced prostate cancer. Mol Cancer, 2004. 3:29.

150. Kubo, M., M. Nakamura, A. Tasaki, N. Yamanaka, H. Nakashima, M. Nomura, S. Kuroki, and M. Katano, Hedgehog signaling pathway is a new therapeutic target for patients with breast cancer. Cancer Res, 2004. 64:6071–6074.

151. Sanchez, P., A.M. Hernandez, B. Stecca, A.J. Kahler, A.M. DeGueme, A. Barrett, M. Beyna, M.W. Datta, S. Datta, and A. Ruiz Altaba, Inhibition of prostate cancer proliferation by interference with Sonic Hedgehog–Gli-1 signaling. Proc Natl Acad Sci USA, 2004. 101:12561–12566.

152. Watkins, D.N. and C.D. Peacock, Hedgehog signalling in foregut malignancy. Biochem Pharmacol, 2004. 68:1055–1060.

153. Tas, S. and O. Avci, Rapid clearance of psoriatic skin lesions induced by topical cyclopamine: a preliminary proof of concept study. Dermatology, 2004. 209:126–131.

154. Frank-Kamenetsky, M., X.M. Zhang, S. Bottega, O. Guicherit, H. Wichterle, H. Dudek, D. Bumcrot, F.Y. Wang, S. Jones, J. Shulok, L.L. Rubin, and J.A. Porter, Small-molecule modulators of Hedgehog signaling: identification and characterization of Smoothened agonists and antagonists. J Biol, 2002. 1:10.

155. Li, C., S. Chi, N. He, X. Zhang, O. Guicherit, R. Wagner, S. Tyring, and J. Xie, IFN alpha induces Fas expression and apoptosis in hedgehog pathway-activated BCC cells through inhibiting Ras–Erk signaling. Oncogene, 2004. 23:1608–1617.

156. Smit, M.J., C. Vink, D. Verzijl, P. Casarosa, C.A. Bruggeman, and R. Leurs, Virally encoded G protein-coupled receptors: targets for potentially innovative anti-viral drug development. Curr Drug Targets, 2003. 4:431–441.

157. Cesarman, E., R.G. Nador, F. Bai, R.A. Bohenzky, J.J. Russo, P.S. Moore, Y. Chang, and D.M. Knowles, Kaposi's sarcoma-associated herpes virus contains G protein-coupled receptor and cyclin D homologs which are expressed in Kaposi's sarcoma and malignant lymphoma. J Virol, 1996. 70:8218–8223.

158. Guo, H.G., P. Browning, J. Nicholas, G.S. Hayward, E. Tschachler, Y.W. Jiang, M. Sadowska, M. Raffeld, S. Colombini, R.C. Gallo, and M.S. Reitz, Jr., Characterization of a chemokine receptor-related gene in human herpes virus 8 and its expression in Kaposi's sarcoma. Virology, 1997. 228:371–378.

159. Arvanitakis, L., E. Geras-Raaka, A. Varma, M.C. Gershengorn, and E. Cesarman, Human herpesvirus KSHV encodes a constitutively active G-protein-coupled receptor linked to cell proliferation. Nature, 1997. 385:347–350.

160. Rosenkilde, M.M., T.N. Kledal, H. Brauner-Osborne, and T.W. Schwartz, Agonists and inverse agonists for the herpes virus 8-encoded constitutively active seven-transmembrane oncogene product, ORF-74. J Biol Chem, 1999. 274:956–961.

161. Smit, M.J., D. Verzijl, P. Casarosa, M. Navis, H. Timmerman, and R. Leurs, Kaposi's sarcoma-associated herpesvirus-encoded G protein-coupled receptor ORF74 constitutively activates p44/p42 MAPK and Akt via G(i) and phospholipase C-dependent signaling pathways. J Virol, 2002. 76:1744–1752.

162. Burger, M., J.A. Burger, R.C. Hoch, Z. Oades, H. Takamori, and I.U. Schraufstatter, Point mutation causing constitutive signaling of CXCR2 leads to transforming activity similar to Kaposi's sarcoma herpes virus-G protein-coupled receptor. J Immunol, 1999. 163:2017–2022.

163. Couty, J.P., E. Geras-Raaka, B.B. Weksler, and M.C. Gershengorn, Kaposi's sarcoma-associated herpes virus G protein-coupled receptor signals through multiple pathways in endothelial cells. J Biol Chem, 2001. 276:33805–33811.

164. Montaner, S., A. Sodhi, S. Pece, E.A. Mesri, and J.S. Gutkind, The Kaposi's sarcoma-associated herpes virus G protein-coupled receptor promotes endothelial cell survival through the activation of Akt/protein kinase B. Cancer Res, 2001. 61:2641–2648.

165. Bais, C., B. Santomasso, O. Coso, L. Arvanitakis, E.G. Raaka, J.S. Gutkind, A.S. Asch, E. Cesarman, M.C. Gershengorn, E.A. Mesri, and M.C. Gerhengorn, G-protein-coupled receptor of Kaposi's sarcoma-associated herpes virus is a viral oncogene and angiogenesis activator. Nature, 1998. 391:86–89.

166. Yang, T.Y., S.C. Chen, M.W. Leach, D. Manfra, B. Homey, M. Wiekowski, L. Sullivan, C.H. Jenh, S.K. Narula, S.W. Chensue, and S.A. Lira, Transgenic expression of the chemokine receptor encoded by human herpes virus 8 induces an angioproliferative disease resembling Kaposi's sarcoma. J Exp Med, 2000. 191:445–454.

167. Holst, P.J., M.M. Rosenkilde, D. Manfra, S.C. Chen, M.T. Wiekowski, B. Holst, F. Cifire, M. Lipp, T.W. Schwartz, and S.A. Lira, Tumorigenesis induced by the HHV8-encoded chemokine receptor requires ligand modulation of high constitutive activity. J Clin Invest, 2001. 108:1789–1796.

168. Sodhi, A., S. Montaner, V. Patel, J.J. Gomez-Roman, Y. Li, E.A. Sausville, E.T. Sawai, and J.S. Gutkind, Akt plays a central role in sarcomagenesis induced by Kaposi's sarcoma herpes virus-encoded G protein-coupled receptor. Proc Natl Acad Sci USA, 2004. 101:4821–4826.

169. Beisser, P.S., D. Verzijl, Y.K. Gruijthuijsen, E. Beuken, P. Koevoets, P. Prickaerts, M.J. Smit, R. Leurs, C.A. Bruggeman, and C. Vink, The Epstein–Barr virus BILF1 gene encodes a G protein-coupled receptor that inhibits phosphorylation of PKR. J Virol, 2005. 79:441–449.

170. Casarosa, P., R.A. Bakker, D. Verzijl, M. Navis, H. Timmerman, R. Leurs, and M.J. Smit, Constitutive signaling of the human cytomegalovirus-encoded chemokine receptor US28. J Biol Chem, 2001. 276:1133–1137.

171. Casarosa, P., Y.K. Gruijthuijsen, D. Michel, P.S. Beisser, J. Holl, C.P. Fitzsimons, D. Verzijl, C.A. Bruggeman, T. Mertens, R. Leurs, C. Vink, and M.J. Smit, Constitutive signaling of the human cytomegalovirus-encoded receptor UL33 differs from that of its rat cytomegalovirus homolog R33 by promiscuous activation of G proteins of the Gq, Gi, and Gs classes. J Biol Chem, 2003. 278:50010–50023.

172. Waldhoer, M., P. Casarosa, M.M. Rosenkilde, M.J. Smit, R. Leurs, J.L. Whistler, and T.W. Schwartz, The carboxyl terminus of human cytomegalovirus-encoded 7 trans-membrane receptor US28 camouflages agonism by mediating constitutive endocytosis. J Biol Chem, 2003. 278:19473–19482.

173. Madden, S.L., B.P. Cook, M. Nacht, W.D. Weber, M.R. Callahan, Y. Jiang, M.R. Dufault, X. Zhang, W. Zhang, J. Walter-Yohrling, C. Rouleau, V.R. Akmaev, C.J. Wang, X. Cao, T.B. St Martin, B.L. Roberts, B.A. Teicher, K.W. Klinger, R.V. Stan, B. Lucey, E.B. Carson-Walter, J. Laterra, and K.A. Walter, Vascular gene expression in nonneoplastic and malignant brain. Am J Pathol, 2004. 165:601–608.

174. Young, D., G. Waitches, C. Birchmeier, O. Fasano, and M. Wigler, Isolation and characterization of a new cellular oncogene encoding a protein with multiple potential transmembrane domains. Cell, 1986. 45:711–719.

175. Okumura, S., H. Baba, T. Kumada, K. Nanmoku, H. Nakajima, Y. Nakane, K. Hioki, and K. Ikenaka, Cloning of a G-protein-coupled receptor that shows an activity to transform NIH3T3 cells and is expressed in gastric cancer cells. Cancer Sci, 2004. 95:131–135.

176. Weigle, B., S. Fuessel, R. Ebner, A. Temme, M. Schmitz, S. Schwind, A. Kiessling, M.A. Rieger, A. Meye, M. Bachmann, M.P. Wirth, and E.P. Rieber, D-GPCR: a novel putative G protein-coupled receptor overexpressed in prostate cancer and prostate. Biochem Biophys Res Commun, 2004. 322:239–249.

177. Sin, W.C., Y. Zhang, W. Zhong, S. Adhikarakunnathu, S. Powers, T. Hoey, S. An, and J. Yang, G protein-coupled receptors GPR4 and TDAG8 are oncogenic and overex-pressed in human cancers. Oncogene, 2004. 23:6299–6303.

178. Harms, J.F., D.R. Welch, and M.E. Miele, KISS1 metastasis suppression and emergent pathways. Clin Exp Metastasis, 2003. 20:11–18.

179. Leurs, R., M.J. Smit, A.E. Alewijnse, and H. Timmerman, Agonist-independent reg-ulation of constitutively active G-protein-coupled receptors. Trends Biochem Sci, 1998. 23:418–422.

180. Schöneberg, T., A. Schulz, H. Biebermann, T. Hermsdorf, H. Römpler, and K. Sangkuhl, Mutant G-protein-coupled receptors as a cause of human diseases. Phar-macology & Therapeutics, 2004. 104:173–206.

6 G Protein-Coupled Receptors in Metabolic Disease

Regina M. Reilly and Christine A. Collins

CONTENTS

ABSTRACT

The identification of primary molecular targets for metabolic diseases has been confounded because such diseases typically develop over protracted time lines and ultimately are manifest in multiple organ systems. This review focuses on current understanding of a subset of G protein-coupled receptors (GPCRs) whose roles in nutrient and energy sensing, satiety, and associated feeding behaviors suggest that they may be modulated pharmacologically to affect glucose and energy homeostasis and ameliorate conditions such as type 2 diabetes and obesity. Receptor function will be examined at two levels: first, energy homeostasis by integration of neuro-endocrine signals, primarily in the central nervous system (CNS), and second, fuel sensing and glucose homeostasis in the periphery.

6.1   INTRODUCTION

GPCRs play essential roles in establishing mechanisms that monitor energy expenditure, nutrient partitioning and metabolism, and ultimately determine body weight and composition. Energy homeostasis is, in large part, the outcome of a complex network of neural and hormonal signals exchanged between the gastrointestinal and nervous systems.

Although efferent signals can originate in a number of peripheral organs such as liver, stomach, intestine, skeletal muscle, and adipose tissue, their sites of action in the brain are more restricted. The hypothalamus serves as a center for integrating signaling events; responses are coordinated through GPCRs with discrete anatomical distribution, particularly in the arcuate nucleus (ARC) and the paraventricular nucleus (PVN). Such neuroendocrine signaling regulates metabolic events associated with both short-term (e.g., meal-to-meal) and long-term (e.g., body weight maintenance) ingestive behaviors. Ultimately, the imbalance between energy intake and expenditure determines excessive weight gain or loss.

6.2   CENTRAL MEDIATION OF FEEDING AND ENERGY HOMEOSTASIS

6.2.1   MELANOCORTIN SYSTEM

The hypothalamic melanocortin system is perhaps the most extensively documented family of GPCRs (melanocortin receptors or MCRs) and ligands implicated in the central regulation of feeding behavior and energy homeostasis.[1,2] Mounting pharmacologic and genetic data indicate that MC3R and MC4R are the key modulators of this system.[3,4] The melanocortin system is distinctive in that both natural agonists and antagonists have been identified.

Bioactive peptides (melanocyte-stimulating hormones or MSHs) derived from the pro-opiomelanocortin prohormone are synthesized and processed in hypothalamic neurons[5,6] and act as endogenous ligands for these receptors,[7] activating the signaling cascade and eliciting anorexigenic (weight loss) responses.

Agouti-related peptide (AGRP), an endogenous antagonist of MC3R and MC4R, produces orexigenic effects through the PVN. The melanocortin system can be viewed as an integration point for several pathways that determine energy expenditure, including responses to leptin, the key indicator of body fat stores. This polypeptide hormone, produced in adipose tissue, acts to stimulate expression of α-MSH and suppresses expression of AGRP through its non-GPCR receptor.[8] Furthermore, correlations between the competence of MC4R signaling and expression of the anorexigenic brain-derived neurotrophic factor (BDNF) suggest that MC4R may, in fact, control energy balance through BDNF and its receptor.[9]

In rodents, pharmacologic blockade of the MC4R or haploinsufficiency of the gene increases feeding and leads to obesity. The phenotypes of MC3R- and MC4R-knockout mice indicate that each receptor plays a similar role in energy balance.[10–16] Although targeted deletion of either or both receptors results in increased adiposity, their functions are not wholly redundant; MC4R-null mice are hyperphagic and hyperinsulinemic, whereas MC3R-null mice are hypophagic and only mildly hyperinsulinemic. Chen et al.[13] and Cummings and Schwartz[17] have proposed that MC4R primarily regulates food intake and possibly energy expenditure, whereas MC3R influences feed efficiency and the deposition of fat. The role of the melanocortin system in the regulation of body weight has been validated in humans. A mutation in MC4R has been found to be the most common genetic cause of severe obesity in children.[18]

6.2.2 Cannabinoid System

The cannabinoids affect both feeding responses and energy balance through mechanisms that are distinctive from those of the melanocortins. The cannabinoid system[19] includes two GPCRs, cannabinoid receptors-1 (CB1) and -2 (CB2); transporters and hydrolyzing enzymes responsible for metabolizing lipophilic natural ligands; and the endocannabinoids, anandamide and 2-arachidonoyl glycerol.[20,21] These endogenous ligands exert potent orexigenic effects and impact reward centers in the brain, thereby eliciting responses that enhance feeding stimuli and palatability. CB1 is expressed centrally in areas responsible for hedonic responses (e.g., nucleus accumbens and hippocampus) and energy homeostasis (hypothalamus and nucleus of the solitary tract) and in epidydimal fat depots where it stimulates lipogenesis.[22]

The endocannabinoid system is an integral part of the neural circuitry regulated by leptin.[23] Whereas administration of leptin reduces the levels of anandamide and 2-arachidonoyl glycerol in the hypothalamus of normal rats, endocannabinoid levels are elevated in leptin-deficient mice and in diet-induced obese mice in which leptin levels are also elevated. CB1-knockout mice are lean, compared to their wild-type littermates, but mechanisms that underlie this leanness appear to differ with age. Young CB1-null mice are hypophagic, while altered metabolism appears to be responsible for the lean phenotype of their adult counterparts.[24,25]

6.2.3 Ghrelin Receptor

Ghrelin is an octanoylated 28-amino acid, gut-secreted neuroendocrine peptide that integrates energy homeostasis and body growth.[26–28] Ghrelin signals the brain by binding to a GPCR, the growth hormone secretagogue receptor (GHSR), to increase metabolic efficiency by stimulating the orexigenic neuropeptide Y (NPY)–AGRP pathway. GHSR is expressed in the anterior pituitary and various hypothalamic and thalamic nuclei.[29] As the first circulating hormone known to stimulate food intake in humans, ghrelin has generated significant interest. Its infusion acutely enhances appetite and increases food consumption,[30] and elevated levels of circulating ghrelin are characteristic of Prader–Willi Syndrome,[31] in which patients exhibit insatiable appetites.[32] In healthy individuals, endogenous plasma ghrelin levels rise before meals or upon fasting and decline postprandially,[33,34] thereby suggesting a complementary relationship with leptin.[35,36] Infusion of ghrelin into rodents increases feeding and weight gain and decreases energy expenditure.[37] Weight gain resulting from chronic administration of ghrelin has been shown to be the consequence of reduced fat utilization and the fact that ghrelin is adipogenic.[38]

Intercerebroventricular (ICV) administration of anti-ghrelin IgG into rats suppresses feeding in both acute and chronic models.[37] Further evidence that ghrelin antagonism leads to weight reduction derives from rodents lacking GHSR.[39] Ghrelin receptor-knockout mice weigh 10% less than their wild-type littermates, although ghrelin-knockout mice exhibit no overt phenotype.[40] In addition, transgenic rats expressing GHSR antisense mRNA specifically in the ARC weigh 10% less and have 80% less fat than their wild-type counterparts by 3 months of age.[41]

6.2.4 MELANIN-CONCENTRATING HORMONE RECEPTORS

Melanin-concentrating hormone (MCH) is a 19-amino acid cyclic, orexigenic neuropeptide, expressed predominantly in the lateral hypothalamus where it binds a cognate GPCR, MCHR1, and signals through complex neural networks to promote feeding and reduce energy expenditure.[42–46] Infusion of MCH into rodents stimulates feeding and, on a chronic level, causes moderate weight gain.[47–52] Increased expression of MCH in genetic models of obesity and the finding that MCH neurons express the leptin receptor suggest that leptin regulates MCH expression. Results of confirmatory *in vitro* and *in vivo* studies continue to emerge.[53–56]

Within the brain, high levels of MCHR1 mRNA have been found in areas that regulate food intake, olfaction, and motivated behavior, and expression levels increase upon fasting. In aggregate, these observations suggest a role for MCH, acting through MCHR1, in integration of taste, olfaction, and positive reward aspects of feeding behavior. MCH-transgenic mice are glucose intolerant, insulin resistant, and prone to weight gain on a standard chow diet.[57] MCH-null mice are mildly hypophagic and lean and exhibit increased oxygen consumption and metabolic rate, reduced fat stores, and resistance to diet-induced obesity.[58] Data for MCHR1-null mice[59,60] are largely consistent with the phenotypic analysis of mice lacking the MCH ligand, although MCHR1-knockout mice are mildly hyperphagic on normal chow, suggesting they maintain lean body mass via increased basal metabolic rate.

In summary, phenotypic data for MCH and MCHR1-transgenic and -null mice indicate that both these gene products interact to modulate feeding behavior and energy balance. If the increase in resting energy expenditure in MCHR1-null mice can be replicated upon pharmacological blockade, this could be a key component in explaining mechanism of action and affording a therapeutic approach to obesity.

A second receptor, MCHR2, with similarly high affinity for MCH has been identified in the human brain.[61–64] MCHR2 is distributed throughout the brain, especially in cortical areas; expression levels in the feeding centers of the hypothalamus are low relative to other brain regions. Due to differences in expression pattern from MCHR1, it has been suggested that MCHR2 may be involved in physiological effects of MCH other than feeding behavior and neuroendocrine modulation. Because rodent genomes do not encode MCHR2, it will be difficult to determine a role for this receptor subtype in the human brain.

6.2.5 SEROTONERGIC SYSTEM

The human serotonergic system includes at least twelve GPCRs, four ion channel subtypes (5-HT3A, 5-HT3B, 5-HT3C, and 5-HT3E), one transporter, and three enzymes involved in the biosynthesis of serotonin (5-hydroxytryptophan or 5-HT), a monoamine neurotransmitter implicated in such diverse conditions as sleep, anxiety, and obesity. Although serotonin is synthesized in the CNS, it is largely distributed throughout the periphery in neurons of the intestinal myenteric plexus, enterochromaffin cells, and platelets. The small proportion retained centrally resides in specific neural nuclei. This serotonergic circuit projects throughout the CNS and constitutes the most extensive neurochemical system in the brain.

Pleiotropic responses of serotonin are defined and specificity is determined at the level of distinct GPCRs. The hypothalamic 5HT1B and 5HT2C receptors are principal determinants of feeding behavior and weight control.[65] Each receptor appears to serve a discrete role in establishing or sensing satiety through interactions with serotonin, an inhibitor of appetite and feeding. Identification of specific receptor functions has relied on use of agonists[66–68] and phenotypic analysis of receptor-null mice.[69–72] The 5HT1B-knockout mice eat and grow more than their wild-type counterparts. Absence of the 5HT1B receptor increases weights of select organs but does not promote obesity in the null mice. The 5HT2C-knockout mice are chronically hyperphagic, but they remain relatively lean until developing late-onset obesity. By 10 months, 5HT2C-null mice weighed ~30% more, increased adiposity 40%, and had elevated plasma insulin and leptin levels, 7- and 2.5-fold, respectively, relative to their wild-type counterparts.

It appears that reduced oxygen consumption and energy expenditure contribute to the phenotypic outcomes of the 5HT2C-null mice. Because 5HT1B-knockout mice exhibit stress-related anxiety and 5HT2C-knockout mice are excitable and have disrupted sleep cycles, additional studies are required to elucidate mechanisms underlying the relationship of the serotonergic system, feeding, and energy homeostasis.

6.2.6 OREXIN SYSTEM

Orexins modulate food intake, energy balance, and arousal and sleep patterns as a consequence of activation of two GPCRs (OX1R and OX2R) by two ligands (OXA and OXB).[73,74] Components of the orexin system are similarly distributed anatomically but remain functionally distinct. The ultimate physiological outcomes are determined by the specific receptor-ligand pairings; OX1R appears to bind OXA exclusively, whereas both orexin ligands have affinity for OX2R. Extensive studies have been conducted that detail expression patterns for the receptors.[75–79] They are expressed in the lateral and posterior regions of the hypothalamus, known to modulate feeding behavior, where OXR neurons are proximal to, but distinct from, MCH neurons.[77–80]

As expected for receptors that coordinate complex feeding behavior, an extensive network of projections extends throughout the CNS. Transcriptional profiling and immunohistochemical staining[79] indicate that OXA is expressed in the periphery, including ganglia and endocrine cells in the kidney, adrenals, stomach, and intestine. In the pancreas, the majority of cells that are insulin positive also express OXA. OX1R is selective for OXA and appears to explain the preponderant data implicating the orexins in feeding behavior. Accumulating data from animal models[81,82] confirm that the OXA ligand stimulates appetite, but experimental outcomes have been found to vary, depending upon site and time of delivery. Xu and co-workers[83] suggest that apparent inconsistencies could derive from responsiveness of the orexin system to circadian rhythm and its involvement in determining wakefulness and, thereby, activity levels. Whereas the hypophagia observed in orexin ligand-knockout mice[84] supports a role for the orexins in appetitive behavior, the absence of a related phenotype in OX1R-null mice[85] suggests that the system requires further analysis to define its potential as a therapeutic target for the treatment of obesity.

6.2.7 POTENTIAL EMERGING TARGETS FOR DRUG DISCOVERY IN METABOLIC DISEASE

Novel therapies targeting GPCRs will continue to emerge from studies that identify ligands for orphan receptors and those that discover novel disease-relevant activities for "known'" receptors. Among the latter group of receptors, several have been implicated in obesity and are listed in Table 6.1. Specifically, these include the dopamine D2 receptor,[86] the bombesin-like receptor H3,[87] the histamine 3 receptor,[88] and the galanin receptor 1.[89]

6.3 PERIPHERAL SIGNALS AFFECTING NUTRIENT SENSING AND UTILIZATION

6.3.1 SATIETY AND GLUCOSE HOMEOSTASIS

Satiety signals are produced in the gastrointestinal (GI) tract in response to the presence of nutrients. Several satiety signals (cholecystokinin, pancreatic polypeptide, peptide YY, and amylin) are peptides that interact with their cognate GPCRs to induce a sense of fullness — resulting in decreased food intake and, in some cases, modulation of nutrient metabolism. The receptors for satiety-inducing peptides may reside locally in the GI tract, on peripheral nerves running from the GI tract to the hindbrain (particularly the vagus nerve), and in some cases in the CNS itself.[90] As is true for centrally produced peptide hormones, the ARC in the hypothalamus is responsible for integrating local and distal signals to regulate food intake, nutrient metabolism, and energy expenditure.

6.3.1.1 Cholecystokinin System

Cholecystokinin (CCK) is the best-studied satiety hormone. It is produced in intestinal duodenal cells in response to the presence of nutrients in the lumen of the GI

TABLE 6.1
Select Human GPCRs Relevant to Metabolic Disease

Abbreviation[a]	Description	Entrez Gene ID[a]	Ligand	Physiological Effect
5HT1B	Serotonin 1B receptor	3351	Serotonin	Anorexigenic
5HT2C	Serotonin 2C receptor	3358	Serotonin	Anorexigenic
ADRB3	β3-adrenergic receptor	155	Noradrenaline	Lipolysis
BRS3	Bombesin-like receptor 3	680	Bombesin	Anorexigenic
CALCR	Calcitonin	799	IAPP	Anorexigenic
CALCRL	Calcitonin gene-related peptide receptor	10203	IAPP	Anorexigenic
CB1	Cannabinoid receptor 1	1268	Endocannabinoid	Orexigenic
CB2	Cannabinoid receptor 2	1269	Endocannabinoid	Orexigenic
CCKAR	Cholecystokinin A receptor	886	Cholecystokinin	Anorexigenic
CCKBR	Cholecystokinin B receptor	887	Cholecystokinin	Anorexigenic
DRD2	Dopamine receptor D2	1813	Dopamine	Anorexigenic
GALR1	Galanin receptor 1	2587	Galanin	Orexigenic
GCGR	Glucagon receptor	2642	Glucagon	Anti-incretin
GHSR	Ghrelin receptor	2693	Ghrelin	Orexigenic
GIPR	Gastric inhibitory polypeptide receptor	2696	GIP	Incretin
GLP1R	Glucagon-like peptide-1 receptor	2740	GLP1; OXM	Anorexigenic; incretin
GLP2R	Glucagon-like peptide-2 receptor	9340	GLP2	Anorexigenic
GPR40	G protein-coupled receptor 40	2864	LCFA	TBD[c]
GPR41	G protein-coupled receptor 41	2865	Short-chain carboxylic acids	TBD[c]
GPR43	G protein-coupled receptor 43	2867	Short-chain carboxylic acids	TBD[c]
HRH3	Histamine receptor 3	11255	Histamine	Orexigenic
MC3R	Melanocortin 3 receptor	4159	α-, β-MSH AGRP[b]	Anorexigenic Orexigenic
MC4R	Melanocortin 4 receptor	4160	α-, β-MSH AGRP[b]	Anorexigenic Orexigenic
MCHR1	Melanin-concentrating hormone receptor 1	2847	MCH	Orexigenic
MCHR2	Melanin-concentrating hormone receptor 2	84539	MCH	Orexigenic
NPY1R	Neuropeptide Y receptor Y1	4886	PP; PYY; NPY	Anorexigenic
NPY2R	Neuropeptide Y receptor Y2	4887	PP; PYY; NPY	Anorexigenic

TABLE 6.1 (Continued)
Select Human GPCRs Relevant to Metabolic Disease

Abbreviation[a]	Description	Entrez Gene ID[a]	Ligand	Physiological Effect
NPY4R	Neuropeptide Y receptor Y4; PPYR1	5540	PP; PYY; NPY	Anorexigenic
NPY5R	Neuropeptide Y receptor Y5	4889	PP; PYY; NPY	Anorexigenic
OX1R	Orexin 1 receptor	3061	orexin A	Orexigenic
OX2R	Orexin 2 receptor	3062	orexin A and B	Orexigenic

[a] Abbreviations, as primary citations or aliases, and EntrezGene designations adopted from the National Center for Biotechnology Information (NCBI) site, [http://www.ncbi.nlm.nih.gov/entrez/query.fcgi? db=gene].

[b] Ligands are agonists at stated receptors, except as noted for AGRP, the endogenous antagonist of MC3R and MC4R.

[c] Physiologic consequences to be determined.

tract. CCK interacts locally with receptors on the sensory fibers of the vagus, where the signal is passed on to the hindbrain and then to integrative feeding centers in the brain. This peptide also acts at the level of the exocrine pancreas and gall bladder to stimulate secretion of digestive enzymes. Administration of CCK to humans leads to a dose-dependent decrease in the meal size, whereas administration of a CCK receptor antagonist causes an increase in meal size and a reduction in the sensation of fullness.[91,92] The two known CCK receptors are CCKAR and CCKBR (also called CCK-1 and CCK-2 receptors, respectively). Specific antagonists are known for each receptor subtype, and rodents lacking CCKAR function have been studied, providing good evidence for distinct physiological contributions of each receptor type. CCKAR is expressed in pancreas, gall bladder, vagus nerve, and CNS, whereas CCKBR is expressed in the stomach, pancreas, vagus nerve, and CNS.[93,94] Rodents demonstrate 60% intraspecies homology between the CCKAR and CCKBR receptor subtypes. Specific CCKAR antagonists block meal-induced gall bladder contraction, gastric emptying, and lead to increased food intake, whereas CCKAR agonists potently inhibit food intake.[95,96]

Otsuka Long Evans Tokushima Fatty (OLETF) rats were identified as possible animal models for diabetes and obesity.[97–99] They carry a spontaneous mutation of the CCKAR gene, with deletions in the promoter and first and second exons. OLETF rats do not respond to exogenously supplied CCK, tend to eat very large meals, and show increases in body weight over their lifetimes. In outbred Long Evans rats, administration of CCK over short periods was effective in reducing meal size but had little impact on total daily food intake and body weight.[100] Likewise, mice with targeted mutations of the CCKAR gene were not obese.[101] When CCK was administered chronically to mice, tolerance to the food intake effects of the peptide developed.[102] Nevertheless, several pharmaceutical companies are investigating the use of CCKAR-specific agonists for the treatment of human obesity. The phenotype of CCKBR-deficient mice does not support a role for this receptor in regulation of food intake. Mice null for CCKBR exhibited defects in CNS regulation of memory, pain sensation, and anxiety.[103]

6.3.1.2 PP-Fold Peptide System

Pancreatic polypeptide (PP), peptide YY (PYY), and the potent orexigen, neuropeptide Y (NPY), all belong to the PP-fold peptide family. PP levels in the circulation are increased by food intake, adrenergic stimulation, and activities of other pancreatic and GI hormones, and also exhibit circadian changes.[104] PP is secreted by cells within the endocrine and exocrine pancreas and inhibits secretion of digestive enzymes and bicarbonate from the pancreas.[105] PP may play a role in regulation of food intake, as genetically obese rodents have decreased PP levels, and administration of PP improves hyperphagia and hyperinsulinemia in certain of these models.[106] Children with Prader–Willi syndrome have reduced secretion of PP,[107] and circulating levels of this peptide are increased in individuals with anorexia nervosa.[108] PP overexpression in transgenic mice leads to a lean phenotype and decreased food intake.[109] Peripherally administered PP has been found to decrease food intake in rodents and humans.[110,111] PP binds to the NPY1-5R family of GPCRs, with highest affinity for NPY4R and NPY5R.[112] As PP appears unable to cross the blood–brain barrier, its effects seem to be mediated through areas in the CNS that have incomplete barriers, for example, the area postrema where NPY4R is known to be highly expressed.[104]

PYY is secreted from endocrine cells in the small intestine, colon, and pancreas in response to food intake. PYY exerts multiple effects on the GI tract, including inhibition of gastric acid secretion, gastric emptying, pancreatic enzyme secretion, and gastric motility.[113] PYY is present in the circulation in two forms, PYY(1-36) and a proteolytically truncated form designated PYY(3-36).[114] Whereas PYY(1-36) binds with similarly high affinity to all Y receptors, PYY(3-36) binds most tightly to NPY2R and NPY5R.[112] Unlike PP, PYY is capable of crossing the blood–brain barrier and the actions of PYY(3-36) on decreasing food intake appear to be mediated by the receptors in the ARC. However, this peptide is not effective in NPY2R-null mice, pointing out a specific role for this receptor in modulating appetite.[115] PYY(3-36) has been reported to inhibit food intake when administered peripherally in rodents and humans.[115,116] In contrast, administration of the peptide directly to the CNS potently stimulates food intake in mice, an effect that is reduced in animals deficient in both NPY1R and NPY5R.[117]

NPY is expressed throughout the central and peripheral nervous systems. It potently increases food intake and decreases energy expenditure in rodents when delivered ICV.[118] Decreasing NPY activity by antisense oligonucleotide methodology[119] or immunoneutralization[120] results in decreased feeding. However, the NPY-null mouse does not exhibit an altered body weight phenotype,[121] suggesting that redundant pathways in regulation of food intake may bypass the NPY circuitry, possibly through the melanocortin system. Expression of NPY is downstream from leptin,[122] as this fat cell hormone decreases expression of NPY and AGRP co-expressed in ARC neurons. Studies with selective pharmacological agents indicate that the relevant receptors mediating the orexigenic effects of NPY are NPY1R and NPY5R.[123] Mice that lack these receptors demonstrate mild obesity phenotypes, and only mice lacking NPY5R showed reduced feeding responses to NPY.[124,125]

6.3.1.3 Calcitonin Receptors

Amylin (also known as islet amyloid polypeptide or IAPP) is a pancreatic peptide hormone secreted along with insulin in response to a meal. This peptide is known to reduce food intake, decrease adiposity, and inhibit gastric acid secretion and gastric emptying.[126] Individuals with type 1 diabetes are characterized by amylin deficiency, and animal studies have identified a role for this peptide in glucose homeostasis through inhibition of postprandial glucagon secretion.[127] Pramlintide, a stable analog of amylin, has been evaluated in diabetic subjects and may have a therapeutic benefit.[127] High affinity binding sites have been identified for amylin in brain, kidney, and skeletal muscle.

Amylin is structurally related to calcitonin, calcitonin gene-related peptide, and adrenomedullin. Two GPCRs, the calcitonin receptor and the calcitonin gene-related peptide receptor, form the basis of all the receptors for this peptide family.[128,129] Differentiating receptor specificities are determined through association of the GPCR with one of three receptor activity-modifying proteins (RAMPs). RAMPs are thought to be involved in transport of the GPCRs to the plasma membrane, converting them to active forms of the receptor.[130] Upon association with RAMP1 or RAMP3, the calcitonin receptor becomes a high affinity binding site for amylin. Antagonism of the RAMP1- or RAMP3-calcitonin receptor complex increases food intake.[131]

6.3.1.4 Glucagon Receptor Family

Several peptide hormones produced from the preproglucagon gene exert effects on food intake and glucose metabolism. These include glucagon-like peptide-1 (GLP1), GLP2, oxyntomodulin (OXM), and glucagon itself. Preproglucagon is cleaved through the action of prohormone convertases 1 and 2, generating different products in specific tissues. In the pancreas, glucagon is produced, whereas the segment containing GLP1 and GLP2 is secreted as an inactive peptide. In the gut and brain, the glucagon sequence remains in a larger peptide (glicentin) thought to be inactive, while the two GLP peptides are cleaved and secreted separately. OXM is produced from glicentin. OXM and GLP1 are secreted from L cells of the small intestine following a meal. These peptides induce satiety and are involved in inhibition of gastric motility, both acting on the same receptor, GLP1R.[132] GLP1 is one of the primary incretin factors responsible for stimulating secretion of insulin following a meal. It is also known to stimulate insulin synthesis and inhibit glucagon secretion.[133]

Peptide homologs are undergoing clinical trials as anti-diabetic agents.[134–136] GLP1R is expressed in pancreatic beta cells, lung, stomach, kidney, and heart.[137] GLP1R is expressed in several brain regions as well, and ICV administration of GLP1 decreased food intake in animal models.[138,139]

GLP1R-knockout mice are glucose intolerant and exhibit deficient insulin secretion in response to orally delivered glucose.[140] Null mice also exhibited deficits in learning, consistent with possible involvement of this receptor in memory.[141]

GLP2 is produced in the gut by neuroendocrine cells in response to nutrients. It regulates gastric acid secretion and gastric motility, stimulates enterocyte glucose uptake, and functions as a trophic factor for the intestinal epithelium.[142] GLP2R is expressed in the GI tract and in the brain.[143] GLP1R and GLP2R are closely related GPCRs with ~50% amino acid identity and are related to other members of the glucagon–secretin GPCR superfamily. Peripherally administered GLP2 showed no effect on food intake,[144] although the peptide administered directly to the CNS in rats was able to reduce food intake.[145]

Glucagon is a peptide hormone produced in the alpha cells of the pancreatic islets.[146] The major biological actions of glucagon involve regulation of glucose homeostasis through actions in the liver to increase glycogenolysis and gluconeogenesis. It is a counter-regulatory hormone to insulin and serves to maintain blood levels of glucose during stress and between meals. In diabetic patients, high levels of glucagon relative to insulin are thought to contribute to hyperglycemia and other metabolic perturbations.[147] Development of glucagon receptor antagonists represents an approach to decrease hepatic glucose output in diabetes.[148]

Although the physiological effects of glucagon acting on its hepatocyte receptor have been studied for decades,[149] the receptor was cloned and characterized relatively recently.[150] It shares ~42% identity with the GLP1R and is expressed in liver, kidney, brain, and adipose tissue.

Expression in pancreatic islet beta cells is thought to contribute to glucose sensitivity of insulin release.[151] Targeted disruption of the glucagon receptor in mice leads to viable animals with mild hypoglycemia and improved glucose tolerance relative to wild-type animals.[152,153] In addition, glucagon receptor-null animals showed greatly increased circulating levels of glucagon and GLP1 and exhibited increases in total pancreas weight with alpha cell hyperplasia, reflecting the increased glucagon production.

6.3.1.5   Glucose-Dependent Insulinotropic Polypeptide Receptor

Glucose-dependent insulinotropic polypeptide (GIP; also known as gastric inhibitory polypeptide) is related in sequence to the glucagon and secretin family of hormones. It is synthesized and released from K cells in the duodenum and jejunum in response to a meal, particularly in response to dietary fat.[154] GIP and GLP1 are considered major incretin factors, leading to meal-induced insulin secretion. The GIP receptor (GIPR) shows 41% sequence identity to GLP1R and has a wide tissue distribution, including expression in pancreas, gut, adipose tissue, heart, adrenal cortex, and brain. GIP has been reported to stimulate fatty acid synthesis, increase fatty acid incorporation into triglycerides, and increase sensitivity of insulin-stimulated glucose transport. A decrease in the number of pancreatic islet GIPRs in diabetes may contribute to poor glucose control in these patients.[155,156] As diminished GIP responsiveness occurs in non-diabetic relatives of type 2 diabetic patients, a defect in this signaling pathway may contribute to diminished beta cell function and development of glucose intolerance.[157]

In support of this hypothesis, mice with targeted disruptions of the GIPR gene had higher blood glucose levels and impaired insulin secretion after oral glucose challenge.[158,159] GIPR-null mice were also protected from obesity and insulin resistance when fed a high-fat diet.[160] Because wild-type mice fed a high fat-diet exhibited increased levels of GIP, it was concluded that GIPR links overeating to increased adiposity and may therefore serve as a potential target for anti-obesity drugs.

6.3.2 PERIPHERAL ENERGY METABOLISM

6.3.2.1 β3-Adrenergic Receptor

Many of the peptide hormones described exert significant effects on energy metabolism via modulation of nutrient utilization or partitioning of fuel stores, and through effects on the hypothalamic feeding and energy circuits. Peripheral regulation of energy metabolism also occurs; the most significant example is the effect of sympathetic activation on β-adrenergic receptors. A number of well-characterized adrenergic receptor subtypes exist, and the signal transduction pathways activated following receptor activation have been well studied.

Sympathetic activation leads to two functional responses in fat cells mediated through the β3-adrenergic receptor.[161] Increased lipolysis (breakdown of triglyceride stores to release fatty acids) occurs following receptor activation, adenylyl cyclase stimulation, and cAMP-dependent phosphorylation of hormone-sensitive lipase. In brown fat in particular, sympathetic nervous activation in response to cold or overnutrition leads to increased oxygen consumption, food intake, and heat production via non-shivering thermogenesis.[162] This effect is also mediated via the β3-adrenergic receptor, through production of transcription factors that coordinate synthesis of mitochondrial uncoupling proteins, electron transport components, and eventually, mitochondrial biogenesis.[163]

To demonstrate the requirement for β-adrenergic receptors in diet-induced thermogenesis, mice lacking the β3-adrenergic receptor were generated and shown to have modestly increased fat stores.[164] Mice lacking β-adrenergic receptors 1, 2, and 3 were also produced.[165] They were mildly obese, showed lowered metabolic rates, and were cold intolerant. On a high-fat diet, they became massively obese compared with wild-type controls.

Pharmaceutical companies have long been interested in the potential therapeutic use of β3-adrenergic agonists for the treatment of obesity and insulin sensitivity. Although such agents have produced beneficial effects in animal studies, the differences in receptor selectivity, tissue distribution, and subtype content in humans compared with rodents has been a stumbling block to rapid development in this area.[166]

6.3.3 NUTRIENT SENSING RECEPTORS

A small family of GPCRs with low homologies to other receptors has recently been identified.[167] GPRs 40, 41, and 43 share 30 to 40% sequence identities to one another. Ligands for these receptors have been identified as long-chain fatty acids

(GPR40) and short-chain carboxylic acids (GPR41 and GPR43).[168–171] Long-chain fatty acids serve as energy sources and also have the potential to be converted to signaling molecules within cells. Excess long-chain fatty acids are known to induce insulin resistance and contribute to metabolic derangements in diabetes and obesity.[172] The identification of fatty acid receptors presents the intriguing possibility that some of the recognized metabolic effects of circulating free fatty acids may occur via these GPCRs.

GPR40 is expressed most highly in brain and in pancreatic beta cells. It has been shown that long-chain free fatty acids enhance glucose-dependent insulin secretion from beta cells through activation of GPR40.[173] Some selectivity for specific fatty acids was found, with saturated free fatty acids of chain length C12–16 and unsaturated free fatty acids of chain length C18–20 preferred. Some eicosanoids also showed receptor stimulatory activities comparable to those of the long-chain fatty acids. GPR41 is expressed in a number of tissues including white fat. Activation of this receptor in fat cells and in intact animals by short-chain fatty acids (C2–C6) led to stimulation of leptin production.[171] Because circulating leptin levels are reflective of both adipose mass and nutritional status, an involvement of GPR41 in leptin secretion identifies a signaling role for short-chain fatty acids from the diet in this nutrient-sensing pathway. GPR43 has a limited distribution, and its selective expression in leukocytes suggests a role for this receptor in recruitment of these cells toward sites of infection rather than in energy homeostasis.[169]

6.4 CONCLUSIONS

Critical analysis of GPCRs implicated in metabolic disease reveals that, despite apparent selectivity at the level of ligand binding, few receptors prove exquisitely specific when subjected to exhaustive pharmacological analysis in cellular and animal models. This review has highlighted several targets with demonstrated potential for therapeutic intervention, by virtue of medical need and tractability. Ultimate success, as measured by amelioration of human maladies, will depend upon the ability of the scientific community to elucidate mechanisms of action and further augment understanding of the molecular basis and promise of targeting GPCRs for the treatment of disease.

REFERENCES

1. Zimanyi, I.A.; Pelleymounter, M.A. The role of melanocortin peptides and receptors in regulation of energy balance. Curr. Pharm. Des. 2003, 9, 627–641.
2. Yang, Y.K.; Harmon, C.M. Recent developments in our understanding of the melanocortin system in the regulation of food intake. Obes. Rev. 2003, 4, 239–248.
3. Adage, T.; Scheurink, A.J.; deBoer, S.F.; de Vries, K.; Konsman, J.P.; Kuipers, F.; Adan, R.A.; Baskin, D.G.; Schwartz, M.W.; van Dijk, G. Hypothalamic, metabolic, and behavioral responses to pharmacological inhibition of CNS melanocortin signaling in rats. J. Neurosci. 2001, 21, 3639–3645.

4. Foster, A.C.; Joppa, M.; Markison, S.; Gogas, K.R.; Fleck, B.A.; Murphy, B.J.; Wolff, M.; Cismowski, M.J.; Ling, N.; Goodfellow, V.S.; Chen, C.; Saunders, J.; Conlon, P.J. Body weight regulation by selective MC4 receptor agonists and antagonists. Ann. N.Y. Acad. Sci. 2003, 994, 103–110.

5. Pritchard, L.E.; Turnbull, A.V.; White, A. Pro-opiomelanocortin processing in the hypothalamus: impact on melanocortin signalling and obesity. J. Endocrinol. 2002, 172, 411–421.

6. Kishi, T.; Aschkenasi, C.J.; Lee, C.E.; Mountjoy, K.G.; Saper, C.B.; Elmquist, J.K. Expression of melanocortin 4 receptor mRNA in the central nervous system of the rat. J. Comp. Neurol. 2003, 457, 213–235.

7. Harrold, J.A.; Widdowson, P.S.; Williams, G. beta-MSH: a functional ligand that regulated energy homeostasis via hypothalamic MC4-R? Peptides. 2003, 24, 397–405.

8. Bjorbaek, C.; Hollenberg, A.N. Leptin and melanocortin signaling in the hypothalamus. Vitam. Horm. 2002, 65, 281–311.

9. Xu, B.; Goulding, E.H.; Zang, K.; Cepoi, D.; Cone, R.D.; Jones, K.R.; Tecott, L.H.; Reichardt, L.F. Brain-derived neurotrophic factor regulates energy balance downstream of melanocortin-4 receptor. Nat. Neurosci. 2003, 6, 736–742.

10. Huszar, D.; Lynch, C.A.; Fairchild-Huntress, V.; Dunmore, J.H.; Fang, Q.; Berkemeier, L.R.; Gu, W.; Kesterson, R.A.; Boston, B.A.; Cone, R.D.; Smith, F.J.; Campfield, L.A.; Burn, P.; Lee, F. Targeted disruption of the melanocortin-4 receptor results in obesity in mice. Cell. 1997, 88, 131–141.

11. Marsh, D.J.; Hollopeter, G.; Huszar, D.; Laufer, R.; Yagaloff, K.A.; Fisher, S.L.; Burns, P.; Palmiter, R.D. Response of melanocortin-4 receptor-deficient mice to anorectic and orexigenic peptides. Nat. Genet. 1999, 21, 119–122.

12. Butler, A.A.; Cone, R.D. The melanocortin receptors: lessons from knockout models. Neuropeptides. 2002, 36, 77–84.

13. Chen, A.S.; Marsh, D.J.; Trumbauer, M.E.; Frazier, E.G.; Guan, X.M.; Yu, H.; Rosenblum, C.I.; Vongs, A.; Feng, Y.; Cao, L.; Metzger, J.M.; Strack, A.M.; Camacho, R.E.; Mellin, T.N.; Nunes, C.N.; Min, W.; Fisher, J.; Gopal-Truter, S.; MacIntyre, D.E.; Chen, H.Y.; Van der Ploeg, L.H. Inactivation of the mouse melanocortin-3 receptor results in increased fat mass and reduced lean body mass. Nat. Genet. 2000, 26, 97–102.

14. Ste. Marie, L.; Miura, G.I.; Marsh, D.J.; Yagaloff, K.; Palmiter, R.D. A metabolic defect promotes obesity in mice lacking melanocortin-4 receptors. Proc. Natl. Acad. Sci. U.S.A. 2000, 97, 12339–12344.

15. Butler, A.A.; Kesterson, R.A.; Khong, K.; Cullen, M.J.; Pelleymounter, M.A.; Dekoning, J.; Baetscher, M.; Cone, R.D. A unique metabolic syndrome causes obesity in the melanocortin-3 receptor-deficient mouse. Endocrinology. 2000, 141, 3518–3521.

16. Weide, K.; Christ, N.; Moar, K.M.; Arens, J.; Hinney, A.; Mercer, J.G.; Eiden, S.; Schmidt, I. Hyperphagia, not hypometabolism, causes early onset obesity in melanocortin-4 receptor knockout mice. Physiol. Genomics. 2003, 13, 47–56.

17. Cummings, D.E.; Schwartz, M.W. Melanocortins and body weight: a tale of two receptors. Nat. Genet. 2000, 26, 8–9.

18. Farroqi, I.S.; Keogh, J.M.; Yeo, G.S.; Lank, E.J.; Cheetham, T.; O'Rahilly, S. Clinical spectrum of obesity and mutations in the melanocortin 4 receptor gene. New Engl. J. Med. 2003, 348, 1085–1095.

19. Di Marzo, V.; Bifulco, M.; De Petrocellis, L. The endocannabinoid system and its therapeutic exploitation. Nat. Rev. Drug Discov. 2004, 3, 771–784.

20. Mechoulam, R.; Ben-Shabat, S.; Hanus, L.; Ligumsky, M.; Kaminski, N.E.; Schatz, A.R.; Gopher, A.; Almog, S.; Martin, B.R.; Compton, D.R.; Pertwee, R.G.; Griffin, G.; Bayewitch, M.; Barg, J.; Vogel, Z. Identification of an endogenous 2-monoglyceride, present in canine gut, that binds to cannabinoid receptors. Biochem. Pharmacol. 1995, 50, 83–90.

21. Stella, N.; Schweitzer, P.; Piomelli, D. A second endogenous cannabinoid that modulates long-term potentiation. Nature. 1997, 388, 773–778.

22. Harrold, J.A.; Elliott, J.C.; King, P.J.; Widdowson, P.S.; Williams, G. Down-regulation of cannabinoid-1 (CB-1) receptors in specific extrahypothalamic regions of rats with dietary obesity: a role for endogenous cannabinoids in driving appetite for palatable food? Brain Res. 2002, 952, 232–238.

23. Di Marzo, V.; Goparaju, S.K.; Wang, L.; Liu, J.; Batkai, S.; Jarai, Z.; Fezza, F.; Miura, G.I.; Palmiter, R.D.; Sugiura, T.; Kunos, G. Leptin-regulated endocannabinoids are involved in maintaining food intake. Nature. 2001, 410, 822–825.

24. Cota, D.; Marsicano, G.; Tschop, M.; Grubler, Y.; Flachskamm, C.; Schubert, M.; Auer, D.; Yassouridis, A.; Thone-Reineke, C.; Ortmann, S.; Tomassoni, F.; Cervino, C.; Nisoli, E.; Linthorst, A.C.; Pasquali, R.; Lutz, B.; Stalla, G.K.; Pagotto, U. The endogenous cannabinoid system affects energy balance via central orexigenic drive and peripheral lipogenesis. J. Clin. Invest. 2003, 112, 423–431.

25. Ravinet-Trillou, C.; Delgorge, C.; Menet, C.; Arnone, M.; Soubrie, P. CB1 cannabinoid receptor knockout in mice leads to leanness, resistance to diet-induced obesity and enhanced leptin sensitivity. Int. J. Obes. Relat. Metab. Disord. 2004, 28, 640–648.

26. Horvath, T.L.; Castaneda, T.; Tang-Christensen, M.; Pagotto, U.; Tschop, M.H. Ghrelin as a potential anti-obesity target. Curr. Pharm. Des. 2003, 9, 1383–1395.

27. Inui, A.; Asakawa, A.; Bowers, C.Y.; Mantovani, G.; Laviano, A.; Meguid, M.M.; Fujimiya, M. Ghrelin, appetite, and gastric motility: the emerging role of the stomach as an endocrine organ. FASEB J. 2004, 18, 439–456.

28. Smith, R.G.; Sun, Y.; Betancourt, L.; Asnicar, M. Growth hormone secretagogues: prospects and potential pitfalls. Best Pract. Res. Clin. Endocrinol. Metab. 2004, 18, 333–347.

29. Lu, S.; Guan, J.L.; Wang, Q.P.; Uehara, K.; Yamada, S.; Goto, N.; Date, Y.; Nakazato, M.; Kojima, M.; Kangawa, K.; Shioda, S. Immunocytochemical observation of ghrelin-containing neurons in the rat arcuate nucleus. Neurosci. Lett. 2002, 321, 157–160.

30. Wren, A.M.; Seal, L.J.; Cohen, M.A.; Brynes, A.E.; Frost, G.S.; Murphy, K.G.; Dhillo, W.S.; Ghatei, M.A.; Bloom, S.R. Ghrelin enhances appetite and increases food intake in humans. J. Clin. Endocrinol. Metab. 2001, 86, 5992–5995.

31. Goldstone, A.P. Prader–Willi syndrome: advances in genetics, pathophysiology and treatment. Trends Endocrinol. Metab. 2004, 15, 12–20.

32. Cummings, D.E.; Clement, K.; Purnell, J.Q.; Vaisse, C.; Foster, K.E.; Frayo, R.S.; Schwartz, M.W.; Basdevant, A.; Weigle, D.S. Elevated plasma ghrelin levels in Prader–Willi syndrome. Nat. Med. 2002, 8, 643–644.

33. Cummings, D.E.; Purnell, J.Q.; Frayo, R.S.; Schmidova, K.; Wisse, B.E.; Weigle, D.S. A preprandial rise in plasma ghrelin levels suggests a role in meal initiation in humans. Diabetes. 2001, 50, 1714–1719.

34. Cummings, D.E.; Frayo, R.S.; Marmonier, C.; Aubert, R.; Chapelot, D. Plasma ghrelin levels and hunger scores in humans initiating meals voluntarily without time- and food-related cues. Am. J. Physiol. Endocrinol. Metab. 2004, 287, E297–E304.

35. Toshinai, K.; Mondal, M.S., Nakazato, M.; Date, Y.; Murakami, N.; Kojima, M.; Kangawa, K.; Matsukura, S. Upregulation of ghrelin expression in the stomach upon fasting, insulin-induced hypoglycemia, and leptin administration. Biochem. Biophys. Res. Commun. 2001, 281, 1220–1225.

36. Nogueiras, R.; Tovar, S.; Mitchell, S.E.; Rayner, D.V.; Archer, Z.A.; Dieguez, C.; Williams, L.M. Regulation of growth hormone secretagogue receptor gene expression in the arcuate nuclei of the rat by leptin and ghrelin. Diabetes. 2004, 53, 2552–2558.

37. Nakazato, M.; Murakami, N.; Date, Y.; Kojima, M.; Matsuo, H.; Kangawa, K.; Matsukura, S. A role for ghrelin in the central regulation of feeding. Nature. 2001, 409, 194–198.

38. Tschop, M.; Smiley, D.L.; Heiman, M.L. Ghrelin induces adiposity in rodents. Nature. 2000, 407, 908–913.

39. Sun, Y.; Wang, P.; Zheng, H.; Smith, R.G. Ghrelin stimulation of growth hormone release and appetite is mediated through the growth hormone secretagogue receptor. Proc. Natl. Acad. Sci. U.S.A. 2004, 101, 4679–4684.

40. Sun, Y.; Ahmed, S.; Smith, R.G. Deletion of ghrelin impairs neither growth nor appetite. Mol. Cell. Biol. 2003, 23, 7973–7981.

41. Shuto, Y.; Shibasaki, T.; Otagiri, A.; Kuriyama, H.; Ohata, H.; Tamura, H.; Kamegai, J.; Sugihara, H.; Oikawa, S.; Wakabayashi, I. Hypothalamic growth hormone secretagogue receptor regulates growth hormone secretion, feeding, and adiposity. J. Clin. Invest. 2002, 109, 1429–1436.

42. Bittencourt, J.C.; Presse, F.; Arias, C.; Peto, C.; Vaughan, J.; Nahon, J.L.; Vale, W.; Sawchenko, P.E. The melanin-concentrating hormone system of the rat brain: an immuno- and hybridization histochemical characterization. J. Comp. Neurol. 1992, 319, 218–245.

43. Bittencourt, J.C.; Frigo, L.; Rissman, R.A.; Casatti, C.A.; Nahon, J.L.; Bauer, J.A. The distribution of melanin-concentrating hormone in the monkey brain (*Cebus apella*). Brain Res. 1998, 804, 140–143.

44. Casatti, C.A.; Elias, C.F.; Sita, L.V.; Frigo, L.; Furlani, V.C.; Bauer, J.A.; Bittencourt, J.C. Distribution of melanin-concentrating hormone neurons projecting to the medial mammillary nucleus. Neuroscience. 2002, 115, 899–915.

45. Qu, D.; Ludwig, D.S.; Gammeltoft, S.; Piper, M.; Pelleymounter, M.A.; Cullen, M.J.; Mathes, W.F.; Przypek, R.; Kanarek, R.; Maratos-Flier, E. A role for melanin-concentrating hormone in the central regulation of feeding behaviour. Nature. 1996, 380, 243–247.

46. Chambers, J.; Ames, R.S.; Bergsma, D.; Muir, A.; Fitzgerald, L.R.; Hervieu, G.; Dytko, G.M.; Foley, J.J.; Martin, J.; Liu, W.S.; Park, J.; Ellis, C.; Ganguly, S.; Konchar, S.; Cluderay, J.; Leslie, R.; Wilson, S.; Sarau, H.M. Melanin-concentrating hormone is the cognate ligand for the orphan G-protein-coupled receptor SLC-1. Nature. 1999, 400, 261–265.

47. Saito, Y.; Nothacker, H.P.; Wang, Z.; Lin, S.H.; Leslie, F.; Civelli, O. Molecular characterization of the melanin-concentrating-hormone receptor. Nature. 1999, 400, 265–269.

48. Lembo, P.M.; Grazzini, E.; Cao, J.; Hubatsch, D.A.; Pelletier, M.; Hoffert, C.; St-Onge, S.; Pou, C.; Labrecque, J.; Groblewski, T.; O'Donnell, D.; Payza, K.; Ahmad, S.; Walker, P. The receptor for the orexigenic peptide melanin-concentrating hormone is a G-protein-coupled receptor. Nat. Cell. Biol. 1999, 1, 267–271.

49. Della-Zuana, O.; Presse, F.; Ortola, C.; Duhault, J.; Nahon, J.L.; Levens, N. Acute and chronic administration of melanin-concentrating hormone enhances food intake and body weight in Wistar and Sprague–Dawley rats. Int. J. Obes. Relat. Metab. Disord. 2002, 26, 1289–1295.

50. Ito, M.; Gomori, A.; Ishihara, A.; Oda, Z.; Mashiko, S.; Matsushita, H.; Yumoto, M.; Ito, M.; Sano, H.; Tokita, S.; Moriya, M.; Iwaasa, H.; Kanatani, A. Characterization of MCH-mediated obesity in mice. Am. J. Physiol. Endocrinol. Metab. 2003, 284, E940–E945.

51. Hakansson, M.L.; Brown, H.; Ghilardi, N.; Skoda, R.C.; Meister, B. Leptin receptor immunoreactivity in chemically defined target neurons of the hypothalamus. J. Neurosci. 1998, 18, 559–572.

52. Sahu, A. Leptin decreases food intake induced by melanin-concentrating hormone (MCH), galanin (GAL) and neuropeptide Y (NPY) in the rat. Endocrinology. 1998, 139, 4739–4742.

53. Huang, Q.; Viale, A.; Picard, F.; Nahon, J.; Richard, D. Effects of leptin on melanin-concentrating hormone expression in the brain of lean and obese Lep(ob)/Lep(ob) mice. Neuroendocrinology. 1999, 69, 145–153.

54. Kokkotou, E.G.; Tritos, N.A.; Mastaitis, J.W.; Slieker, L.; Maratos-Flier. E. Melanin-concentrating hormone receptor is a target of leptin action in the mouse brain. Endocrinology. 2001, 142, 680–686.

55. Gomori, A.; Ishihara, A.; Ito, M.; Mashiko, S.; Matsushita, H.; Yumoto, M.; Ito, M.; Tanaka, T.; Tokita, S.; Moriya, M.; Iwaasa, H.; Kanatani, A. Chronic intracerebroventricular infusion of MCH causes obesity in mice: melanin-concentrating hormone. Am. J. Physiol. Endocrinol. Metab. 2003, 284, E583–E588.

56. Segal-Lieberman, G.; Bradley, R.L.; Kokkotou, E.; Carlson, M.; Trombly, D.J.; Wang, X.; Bates, S.; Myers, M.G. Jr.; Flier, J.S.; Maratos-Flier, E. Melanin-concentrating hormone is a critical mediator of the leptin-deficient phenotype. Proc. Natl. Acad. Sci. U.S.A. 2003, 100, 10085–10090.

57. Ludwig, D.S.; Tritos, N.A.; Mastaitis, J.W.; Kulkarni, R.; Kokkotou, E.; Elmquist, J.; Lowell, B.; Flier, J.S.; Maratos-Flier, E. Melanin-concentrating hormone overexpression in transgenic mice leads to obesity and insulin resistance. J. Clin. Invest. 2001, 107, 379–386.

58. Shimada, M,; Tritos, N.A.; Lowell, B.B.; Flier, J.S.; Maratos-Flier, E. Mice lacking melanin-concentrating hormone are hypophagic and lean. Nature. 1998, 396, 670–674.

59. Chen, Y.; Hu, C.; Hsu, C.K.; Zhang, Q.; Bi, C.; Asnicar, M.; Hsiung, H.M.; Fox, N.; Slieker, L.J.; Yang, D.D.; Heiman, M.L.; Shi, Y. Targeted disruption of the melanin-concentrating hormone receptor-1 results in hyperphagia and resistance to diet-induced obesity. Endocrinology. 2002, 143, 2469–2477.

60. Marsh, D.J.; Weingarth, D.T.; Novi, D.E.; Chen, H.Y.; Trumbauer, M.E.; Chen, A.S.; Guan, X.-M.; Jiang, M.M.; Feng, Y.; Camacho, R.E.; Shen, Z.; Frazier, E.G.; Yu, H.; Metzger, J.M.; Kuca, S.J.; Shearman, L.P.; Gopal-Truter, S.; MacNeil, D.J.; Strack, A.M.; MacIntyre, D.E.; Van der Ploeg, L.H.T.; Qian, S. Melanin-concentrating hormone 1 receptor-deficient mice are lean, hyperactive, and hyperphagic and have altered metabolism. Proc. Natl. Acad. Sci. U.S.A. 2002, 99, 3240–3245.

61. An, S.; Cutler, G.; Zhao, J.J.; Huang, S.G.; Tian, H.; Li, W.; Liang, L.; Rich, M.; Bakleh, A.; Du, J.; Chen, J.L.; Dai, K. Identification and characterization of a melanin-concentrating hormone receptor. Proc. Natl. Acad. Sci. U.S.A. 2001, 98, 7576–7581.

62. Sailer, A.W.; Sano, H.; Zeng, Z.; McDonald, T.P.; Pan, J.; Pong, S.S.; Feighner, S.D.; Tan, C.P.; Fukami, T.; Iwaasa, H.; Hreniuk, D.L.; Morin, N.R.; Sadowski, S.J.; Ito, M.; Ito, M.; Bansal, A.; Ky, B.; Figueroa, D.J.; Jiang, Q.; Austin, C.P.; MacNeil, D.J.; Ishihara, A.; Ihara, M.; Kanatani, A.; Van der Ploeg, L.H.; Howard, A.D.; Liu, Q. Identification and characterization of a second melanin-concentrating hormone receptor, MCH-2R. Proc. Natl. Acad. Sci. U.S.A. 2001, 98, 7564–7569.

63. Tan, C.P.; Sano, H.; Iwaasa, H.; Pan, J.; Sailer, A.W.; Hreniuk, D.L.; Feighner, S.D.; Palyha, O.C.; Pong, S.S.; Figueroa, D.J.; Austin, C.P.; Jiang, M.M.; Yu, H.; Ito, J.; Ito, M.; Ito, M.; Guan, X.M.; MacNeil, D.J.; Kanatani, A.; Van der Ploeg, L.H.; Howard, A.D. Melanin-concentrating hormone receptor subtypes 1 and 2: species-specific gene expression. Genomics. 2002, 79, 785–792.

64. Wang, S.; Behan, J.; O'Neill, K.; Weig, B.; Fried, S.; Laz, T.; Bayne, M.; Gustafson, E.; Hawes, B.E. Identification and pharmacological characterization of a novel human melanin-concentrating hormone receptor, mch-r2. J. Biol. Chem. 2001, 276, 34664–34670.

65. Vickers, S.P.; Dourish, C.T. Serotonin receptor ligands and the treatment of obesity. Curr. Opin. Invest. Drugs. 2004, 5, 377–388.

66. Hayashi, A.; Sonoda, R.; Kimura, Y.; Takasu, T.; Suzuki, M.; Sasamata, M.; Miyata, K. Antiobesity effect of YM348, a novel 5-HT2C receptor agonist, in Zucker rats. Brain Res. 2004, 1011, 221–227.

67. De Vry, J.; Schreiber, R. Effects of selected serotonin 5-HT(1) and 5-HT(2) receptor agonists on feeding behavior: possible mechanisms of action. Neurosci. Biobehav. Rev. 2000, 24, 341–353.

68. Bickerdike, M.J. 5-HT2C receptor agonists as potential drugs for the treatment of obesity. Curr. Top. Med. Chem. 2003, 3, 885–897.

69. Bouwknecht, J.A.; van der Gugten, J.; Hijzen, T.H.; Maes, R.A.; Hen, R.; Olivier, B. Male and female 5-HT(1B) receptor knockout mice have higher body weights than wild types. Physiol. Behav. 2001, 74, 507–516.

70. Heisler, L.K.; Tecott, L.H. Knockout corner: neurobehavioural consequences of a serotonin 5-HT(2C) receptor gene mutation. Int. J. Neuropsychopharmacol. 1999, 2, 67–69.

71. Lopez-Gimenez, J.F.; Tecott, L.H.; Palacios, J.M.; Mengod, G.; Vilaro, M.T. Serotonin 5-HT (2C) receptor knockout mice: autoradiographic analysis of multiple serotonin receptors. J. Neurosci. Res. 2002, 67, 69–85.

72. Nonogaki, K.; Abdallah, L.; Goulding, E.H.; Bonasera, S.J.; Tecott, L.H. Hyperactivity and reduced energy cost of physical activity in serotonin 5-HT(2C) receptor mutant mice. Diabetes. 2003, 52, 315–320.

73. Sakurai, T.; Amemiya, A.; Ishii, M.; Matsuzaki, I.; Chemelli, R.M.; Tanaka, H.; Williams, S.C.; Richardson, J.A.; Kozlowski, G.P.; Wilson, S.; Arch, J.R.; Buckingham, R.E.; Haynes, A.C.; Carr, S.A.; Annan, R.S.; McNulty, D.E.; Liu, W.S.; Terrett, J.A.; Elshourbagy, N.A.; Bergsma, D.J.; Yanagisawa, M. Orexins and orexin receptors: a family of hypothalamic neuropeptides and G protein-coupled receptors that regulate feeding behavior. Cell. 1998, 92, 573–585.

74. de Lecea, L.; Kilduff, T.S.; Peyron, C.; Gao, X.; Foye, P.E.; Danielson, P.E.; Fukuhara, C.; Battenberg, E.L.; Gautvik, V.T.; Bartlett, F.S. 2nd; Frankel, W.N.; van den Pol, A.N.; Bloom, F.E.; Gautvik, K.M.; Sutcliffe, J.G. The hypocretins: hypothalamus-specific peptides with neuroexcitatory activity. Proc. Natl. Acad. Sci. U.S.A. 1998, 95, 322–327.

75. Blanco, M.; Lopez, M.; Garcia-Caballero, T.; Gallego, R.; Vazquez-Boquete, A.; Morel, G.; Senaris, R.; Casanueva, F.; Dieguez, C.; Beiras, A. Cellular localization of orexin receptors in human pituitary. J. Clin. Endocrinol. Metab. 2001, 86, 1616–1619.

76. Blanco, M.; Gallego, R.; Garcia-Caballero, T.; Dieguez, C.; Beiras, A. Cellular localization of orexins in human anterior pituitary. Histochem. Cell Biol. 2003, 120, 259–264.

77. Volgin, D.V.; Swan, J.; Kubin, L. Single-cell RT-PCR gene expression profiling of acutely dissociated and immunocytochemically identified central neurons. J. Neurosci. Methods. 2004, 136, 229–236.

78. Peyron, C.; Tighe, D.K.; van den Pol, A.N.; de Lecea, L.; Heller, H.C.; Sutcliffe, J.G.; Kilduff, T.S. Neurons containing hypocretin (orexin) project to multiple neuronal systems. J. Neurosci. 1998, 18, 9996–10015.

79. Nakabayashi, M.; Suzuki, T.; Takahashi, K.; Totsune, K.; Muramatsu, Y.; Kaneko, C.; Date, F.; Takeyama, J.; Darnel, AD.; Moriya, T.; Sasano, H. Orexin-A expression in human peripheral tissues. Mol. Cell. Endocrinol. 2003, 205, 43–50.

80. Drazen, D.L.; Coolen, L.M.; Strader, A.D.; Wortman, M.D.; Woods, S.C.; Seeley, R.J. Differential effects of adrenalectomy on melanin-concentrating hormone and orexin A. Endocrinology. 2004, 145, 3404–3412.

81. Sweet, D.C.; Levine, A.S.; Billington, C.J.; Kotz, C.M. Feeding response to central orexins. Brain Res. 1999, 821, 535–538.

82. Dube, M.G.; Kalra, S.P.; Kalra, P.S. Food intake elicited by central administration of orexins/hypocretins: identification of hypothalamic sites of action. Brain Res. 1999, 842, 473–477.

83. Xu, Y.L.; Jackson, V.R.; Civelli, O. Orphan G protein-coupled receptors and obesity. Eur. J. Pharmacol. 2004, 500, 243–253.

84. Willie, J.T.; Chemelli, R.M.; Sinton, C.M.; Yanagisawa, M. To eat or to sleep? Orexin in the regulation of feeding and wakefulness. Annu. Rev. Neurosci. 2001, 24, 429–458.

85. Willie, J.T.; Chemelli, R.M.; Sinton, C.M.; Tokita, S.; Williams, S.C.; Kisanuki, Y.Y.; Marcus, J.N.; Lee, C.; Elmquist, J.K.; Kohlmeier, K.A.; Leonard, C.S.; Richardson, J.A.; Hammer, R.E.; Yanagisawa, M. Distinct narcolepsy syndromes in orexin receptor-2- and orexin-null mice: molecular genetic dissection of non-REM and REM sleep regulatory processes. Neuron. 2003, 38, 715–730.

86. Pijl, H. Reduced dopaminergic tone in hypothalamic neural circuits: expression of a "thrifty" genotype underlying the metabolic syndrome? Eur. J. Pharmacol. 2003, 480, 125–131.

87. Yamada, K.; Wada, E.; Santo-Yamada, Y.; Wada K. Bombesin and its family of peptides: prospects for the treatment of obesity. Eur. J. Pharmacol. 2002, 440, 281–290; erratum published in Eur. J. Pharmacol. 2002, 448, 269.

88. Hancock, A.A. H3 receptor antagonists/inverse agonists as anti-obesity agents. Curr. Opin. Invest. Drugs. 2003, 4, 1190–1197.

89. Gundlach, A.L. Galanin/GALP and galanin receptors: role in central control of feeding, body weight/obesity and reproduction? Eur. J. Pharmacol. 2002, 440, 255–268.

90. Woods, S.C. Gastrointestinal satiety signals. I. An overview of gastrointestinal signals that influence food intake. Am. J. Physiol. Gastrointest. Liver Physiol. 2004, 286, G7–G13.

91. Beglinger, C.; Degen, L.; Matzinger, D.; D'Amato, M.; Drewe, J. Loxiglumide, a CCK-A receptor antagonist, stimulates calorie intake and hunger feeling in humans. Am. J. Physiol. Regul. Integr. Comp. Physiol. 2001, 280, R1149–R1154.

92. Moran, T.H.; Kinzig, K.P. Gastrointestinal satiety signals. II. Cholecystokinin. Am. J. Physiol. Gastrointest. Liver Physiol. 2004, 286, G183–G188.

93. Kopin, A.S.; Lee, Y.M.; McBride, E.W.; Miller, L.J.; Lu, M.; Lin, H.Y.; Kolakowski, L.F. Jr.; Beinborn, M. Expression cloning and characterization of the canine parietal cell gastrin receptor. Proc. Natl. Sci. U.S.A. 1992, 89, 3605–3609.

94. Pisegna, J.R.; DeWeerth, A.; Huppi, K.; Wank, S.A. Molecular cloning of the human brain and gastric cholecystokinin receptor: structure, functional expression and chromosomal localization. Biochem. Biophys. Res. Commun. 1992, 189, 296–303.

95. Asin, K.E.; Bednarz, L.; Nikkel A.L.; Gore, P.A. Jr.; Nadzan A.M. A-71623, a selective CCK-A receptor agonist, suppresses food intake in the mouse, dog, and monkey. Pharmacol. Biochem. Behav. 1992, 42, 699–704.

96. Moran, T.H.; Ameglio, P.J.; Schwartz, G.J.; McHugh, P.R. Blockade of type A, not type B, CCK receptors attenuates satiety actions of exogenous and endogenous CCK. Am. J. Physiol. Regul. Integr. Comp. Physiol. 1992, 262, R46–R50.

97. Funakoshi, A.; Miyasaka, K.; Shinozaki, H.; Masuda, M.; Kawanami, T.; Takata, Y.; Kono, A. An animal model of congenital defect of gene expression of cholecystokinin (CCK)-A receptor. Biochem. Biophys. Res. Commun. 1995, 210, 787–796.

98. Moran, T.H.; Katz, L.F.; Plata-Salaman, C.R.; Schwartz, G.J. Disordered food intake and obesity in rats lacking cholecystokinin A receptors. Am. J. Physiol. Regul. Integr. Comp. Physiol. 1998, 274, R618–R625.

99. Bi, S.; Moran, T.H. Actions of CCK in the controls of food intake and body weight: lessons from the CCK-A receptor deficient OLETF rat. Neuropeptides. 2002, 36, 171–181.

100. West, D.B.; Fey, D.; Woods, S.C. Cholecystokinin persistently suppresses meal size but not food intake in free-feeding rats. Am. J. Physiol. Regul. Integr. Comp. Physiol. 1984, 246, R776–R787.

101. Kopin, A.S.; Mathes, W.F.; McBride, E.W.; Nguyen, M.; Al-Haider, W.; Schmitz, F.; Bonner-Weir, S.; Kanarek, R.; Beinborn, M. The cholecystokinin-A receptor mediates inhibition of food intake yet is not essential for the maintenance of body weight. J. Clin. Invest. 1999, 103, 383–391.

102. Crawley, J.N.; Beinfeld, M.C. Rapid development of tolerance to the behavioural actions of cholecystokinin. Nature. 1983, 302, 703–706.

103. Noble, F.; Roques, B.P. Phenotypes of mice with invalidation of cholecystokinin (CCK_1 or CCK_2) receptors. Neuropeptides. 2002, 36, 157–170.

104. Wynne, K.; Stanley, S.; Bloom, S. The gut and regulation of body weight. J. Clin. Endocrinol. Metab. 2004, 89, 2576–2582.

105. Schwartz, T.W. Pancreatic polypeptide: a hormone under vagal control. Gastroenterology. 1983, 85, 1411–1425.

106. Gates, R.J.; Lazarus, N.R. The ability of pancreatic polypeptides (APP and BPP) to return to normal the hyperglycaemia, hyperinsulinaemia and weight gain of New Zealand obese mice. Horm. Res. 1977, 8, 189–202.

107. Zipf, W.B.; O'Dorisio, T.M., Cataland, S.; Sotos, J. Blunted pancreatic polypeptide responses in children with obesity of Prader–Willi syndrome. J. Clin. Endocrinol. Metab. 1981, 52, 1264–1266.

108. Uhe, A.M.; Szmukler, G.I.; Collier, G.R.; Hansky, J.; O'Dea, K.; Young, G.P. Potential regulators of feeding behavior in anorexia nervosa. Am. J. Clin. Nutr. 1992, 55, 28–32.

109. Ueno N.; Inui, A.; Iwamoto, M.; Kaga, T.; Asakawa, A.; Okita, M.; Fujimiya, M.; Nakajima, Y.; Ohmoto, Y.; Ohnaka, M.; Nakaya, Y.; Miyazaki, J.I.; Kasuga, M. Decreased food intake and body weight in pancreatic polypeptide-overexpressing mice. Gastroenterology. 1999, 117, 1427–1432.

110. Asakawa, A.; Inui, A.; Yuzuriha, H.; Ueno, N.; Katsuura, G.; Fujimiya, M.; Fujino, M.A., Niijima, A.; Meguid, M.M.; Kasuga, M. Characterization of the effects of pancreatic polypeptide in the regulation of energy balance. Gastroenterology. 2003, 124, 1325–1336.

111. Batterham, R.L.; Le Roux, C.W.; Cohen, M.A.; Park, A.J.; Ellis, S.M.; Patterson, M.; Frost, G.S.; Ghatei, M.A.; Bloom, S.R. Pancreatic polypeptide reduces appetite and food intake in humans. J. Clin. Endocrinol. Metab. 2003, 88, 3989–3992.

112. Larhammar D. Structural diversity of receptors for neuropeptide Y, peptide YY and pancreatic polypeptide. Regul. Pept. 1996, 65, 165–174.

113. Leiter, A.B.; Toder, A.; Wolfe, H.J.; Taylor, I.L.; Cooperman, S.; Mandel, G.; Goodman, R.H. Peptide YY: structure of the precursor and expression in exocrine pancreas. J. Biol. Chem. 1987, 262, 12984–12988.

114. Eberlein, G.A.; Eysselein, V.E.; Schaeffer, M.; Layer, P.; Grandt, D.; Goebell, H.; Niebel, W.; Davis, M.; Lee, T.D.; Shively, J.E.; Reeve, J.R. A new molecular form of PYY: structural characterization of human PYY(3-36) and PYY(1-36). Peptides. 1989, 10, 797–803.

115. Batterham, R.L.; Cowley, M.A.; Small, C.J.; Herzog, H.; Cohen, M.A.; Dakin, C.L.; Wren, A.M.; Brynes, A.E.; Low M.J.; Ghatei, M.A.; Cone, R.D.; Bloom, S.R. Gut hormone $PYY_{(3-36)}$ physiologically inhibits food intake. Nature. 2002, 418, 650–654.

116. Batterham, R.L.; Cohen, M.A.; Ellis, S.M.; Le Roux, C.W.; Withers, D.J.; Frost, G.S.; Ghatei, M.A.; Bloom, S.R. Inhibition of food intake in obese subjects by peptide YY_{3-36}. N. Engl. J. Med. 2003, 349, 941–948.

117. Kanatani, A.; Mashiko, S.; Murai, N.; Sugimoto, N.; Ito, J.; Fukuroda, T.; Fukami, T.; Morin, N.; MacNeil, D.J.; Van der Ploeg, L.H.; Saga, Y.; Nishimura, S.; Ihara, M. Role of the Y1 receptor in the regulation of neuropeptide Y-mediated feeding: comparison of wild-type, Y1 receptor-deficient, and Y5 receptor-deficient mice. Endocrinology. 2000, 141, 1011–1016.

118. Stanley, B.G.; Leibowitz, S.F. Neuropeptide Y: stimulation of feeding and drinking by injection into the paraventricular nucleus. Life Sci. 1984, 35, 2635–2642.

119. Akabayashi, A.; Wahlestedt, C.; Alexander, J.T.; Leibowitz, S.F. Specific inhibition of endogenous neuropeptide Y synthesis in arcuate nucleus by antisense oligonucleotides suppresses feeding behavior and insulin secretion. Brain Res. Mol. Brain Res. 1994, 21, 55–61.

120. Shibasaki, T.; Oda, T.; Imaki, T.; Ling, N.; Demura, H. Injection of anti-neuropeptide Y gamma-globulin into the hypothalamic paraventricular nucleus decreases food intake in rats. Brain Res. 1993, 601, 313–316.

121. Erickson, J.C.; Clegg, K.E.; Palmiter, R.D. Sensitivity to leptin and susceptibility to seizures of mice lacking neuropeptide Y. Nature. 1996, 381, 415–421.

122. Stephens, T.W.; Basinski, M.; Bristow, P.K.; Bue-Valleskey, J.M.; Burgett, S.G.; Craft, L.; Hale, J.; Hoffmann, J.; Hsiung, H.M.; Kriauciunas, A.; MacKellar, W.; Rosteck, P.R.; Schoner, B.; Smith, D.; Tinsley, F.C.; Zhang, X-Y.; Heiman, M. The role of the neuropeptide Y in the antiobesity action of the obese gene product. Nature. 2002, 377, 530–532.

123. Gerald, C.; Walker, M.W.; Criscione, L.; Gustafson, E.L.; Batzl-Hartmann, C.; Smith, K.E.; Vaysse, P.; Durkin, M.M.; Laz, T.M.; Linemeyer, D.L.; Schaffhauser, A.O.; Whitebread, S.; Hofbauer, K.G.; Taber, R.I.; Branchek, T.A.; Weinshank, R.L. A receptor subtype involved in neuropeptide-Y-induced food intake. Nature. 1996, 382, 168–171.

124. Kushi, A.; Sasai, H.; Koizumi, H.; Takeda, N.; Yokoyama, M.; Nakamura, M. Obesity and mild hyperinsulinemia found in neuropeptide Y-Y1 receptor-deficient mice. Proc. Natl. Acad. Sci. U.S.A. 1998, 95, 15659–15664.

125. Marsh, D.J.; Hollopeter, G.; Kafer, K.E.; Palmiter, R.D. Role of the Y5 neuropeptide Y receptor in feeding and obesity. Nat. Med. 1998, 4, 718–721.

126. Höppener, J.W.M.; Ahren, B.; Lips, C.J.M. Islet amyloid and type 2 diabetes mellitus. N. Engl. J. Med. 2000, 343, 411–419.

127. Nyholm, B.; Brock, B.; Ørskov, L.; Schmitz, O. Amylin receptor agonists: a novel pharmacological approach in the management of insulin-treated diabetes mellitus. Expert Opin. Invest. Drugs. 2001, 10, 1641–1652.

128. Muff, R.; Born, W.; Fischer, J.A. Calcitonin, calcitonin gene-related peptide, adrenomedullin and amylin: homologous peptides, separate receptors and overlapping biological actions. Eur. J. Endocrinol. 1995, 133, 17–20.

129. Poyner, D.R.; Sexton, P.M.; Marshall, I.; Smith, D.M.; Quirion, R.; Born, W.; Muff, R.; Fischer, J.A.; Foord, S.M. International Union of Pharmacology. XXXII. The mammalian calcitonin gene-related peptides, adrenomedullin, amylin, and calcitonin receptors. Pharmacol. Rev. 2002, 54, 233–246.

130. Born, W.; Fischer, J.A.; Muff, R. Receptors for calcitonin gene-related peptide, adrenomedullin, and amylin: the contributions of novel receptor-activity-modifying proteins. Receptors Channels. 2002, 8, 201–209.

131. Reidelberger, R.D.; Haver, A.C.; Arnelo, U.; Smith, D.D.; Schaffert, C.S.; Permert, J. Amylin receptor blockade stimulates food intake in rats. Am. J. Physiol. Regul. Integr. Comp. Physiol. 2004, 287, R568–R574.

132. Stanley, S.; Wynne, K.; Bloom, S. Gastrointestinal satiety signals. III. Glucagon-like peptide 1, oxyntomodulin, peptide YY, and pancreatic polypeptide. Am. J. Physiol. Gastrointest. Liver Physiol. 2004, 286, G693–G697.

133. Wang, Z.; Wang, R.M.; Owji, A.A.; Smith, D.M.; Ghatei, M.A.; Bloom, S.R. Glucagon-like peptide-1 is a physiological incretin in rat. J. Clin. Invest. 1995, 95, 417–421.

134. Meier, J.J.; Nauck, M.A. The potential role of glucagon-like peptide 1 in diabetes. Curr. Opin. Invest. Drugs. 2004, 5, 402–410.

135. Holst, J.J. Treatment of type 2 diabetes mellitus with agonists of the GLP1 receptor or DPP-IV inhibitors. Expert. Opin. Emerg. Drugs. 2004, 9, 155–166.

136. Zander, M.; Madsbad, S.; Madsen, J.L.; Holst, J.J. Effect of 6-week course of glucagon-like peptide 1 on glycaemic control, insulin sensitivity, and β-cell function in type 2 diabetes: a parallel-group study. Lancet. 2002, 359, 824-830.

137. Dillon, J.S.; Tanizawa, Y.; Wheeler, M.B.; Leng, X.-H.; Ligon, B.B.; Rabin, D.U.; Yoo-Warren, H.; Permutt, M.A.; Boyd, A.E., III. Cloning and functional expression of the human glucagon-like peptide-1 (GLP1) receptor. Endocrinology. 1993, 133, 1907–1910.

138. Turton, M.D.; O'Shea, D.; Gunn, I.; Beak, S.A.; Edwards, C.M.B.; Meeran, K.; Choi, S.J.; Taylor, G.M.; Heath, M.M.; Lambert, P.D.; Wilding, J.P.H.; Smith, D.M.; Ghatei, M.A.; Herbert, J.; Bloom, S.R. A role for glucagon-like peptide-1 in the central regulation of feeding. Nature. 1996, 379, 69–72.

139. Kinzig, K.P.; D'Alessio, D.A.; Seeley, R.J. The diverse roles of specific GLP1 receptors in the control of food intake and the response to visceral illness. J. Neurosci. 2002, 22, 10470–10476.

140. Scrocchi, L.A.; Brown, T.J.; MacLusky, N.; Brubaker, P.L.; Auerbach, A.B.; Joyner, A.L.; Drucker, D.J. Glucose intolerance but normal satiety in mice with a null mutation in the glucagon-like peptide 1 receptor gene. Nat. Med. 1996, 2, 1254–1258.

141. During, M.J.; Cao, L.; Zuzga, D.S.; Francis, J.S.; Fitzsimons, H.L.; Jiao, X.; Bland, R.J.; Klugmann, M.; Banks, W.A.; Drucker, D.J.; Haile, C.N. Glucagon-like peptide-1 is involved in learning and neuroprotection. Nat. Med. 2003, 9, 1173–1179.

142. Drucker, D.J. Glucagon-like peptide 2. J. Clin. Endocrinol. Metab. 2001, 86, 1759–1764.

143. Yusta, B.; Huang, L.; Munroe, D.; Wolff, G.; Fantaske, R.; Sharma, S.; Demchyshyn, L.; Asa, S.L.; Drucker, D.J. Enteroendocrine localization of GLP2 receptor expression in humans and rodents. Gastroenterology. 2000, 119, 744–755.

144. Sørensen, L.B.; Flint, A.; Raben, A.; Hartmann, B.; Holst, J.J.; Astrup, A. No effect of physiological concentrations of glucagon-like peptide-2 on appetite and energy intake in normal weight subjects. Int. J. Obes. Relat. Metab. Disord. 2003, 27, 450–456.

145. Tang-Christensen, M.; Larsen, P.J.; Thulesen, J.; Rømer, J.; Vrang, N. The progluca-gon-derived peptide, glucagon-like peptide-2, is a neurotransmitter involved in the regulation of food intake. Nat. Med. 2000, 6, 802–807.

146. Unger, R.H.; Orci, L. Glucagon and the A cell: physiology and pathophysiology. N. Engl. J. Med. 1981, 304, 1518–1524; 1575–1580.

147. Shah, P.; Vella, A.; Basu, A.; Basu, R.; Schwenk, W.F.; Rizza, R.A. Lack of suppres-sion of glucagon contributes to postprandial hyperglycemia in subjects with type 2 diabetes mellitus. J. Clin. Endocrinol. Metab. 2000, 85, 4053–4059.

148. Djuric, S.W.; Grihalde, N.; Lin, C.W. Glucagon receptor antagonists for the treatment of type II diabetes: current prospects. Curr. Opin. Invest. Drugs. 2002, 3, 1617–1623.

149. Burcelin, R.; Katz, E.B.; Charron, M.J. Molecular and cellular aspects of the glucagon receptor: role in diabetes and metabolism. Diabetes Metab. 1996, 22, 373–396.

150. Jelinek, L.J.; Lok, S.; Rosenberg, G.B.; Smith, R.A.; Grant, F.J.; Biggs, S.; Bensch, P.A.; Kuijper, J.L.; Sheppard, P.O.; Sprecher, C.A. et al. Expression cloning and signaling properties of the rat glucagon receptor. Science. 1993, 259, 1614–1616.

151. Huypens, P.; Ling, Z.; Pipeleers, D.; Schuit, F. Glucagon receptors on human islet cells contribute to glucose competence of insulin release. Diabetologia. 2000, 43, 1012–1019.

152. Parker, J.C.; Andrews, K.M.; Allen, M.R.; Stock, J.L.; McNeish, J.D. Glycemic control in mice with targeted disruption of the glucagon receptor gene. Biochem. Biophys. Res. Commun. 2002, 290, 839–843.

153. Gelling, R.W.; Du, X.Q.; Dichmann, D.S.; Rømer, J.; Huang, H.; Cui, L.; Obici, S.; Tang, B.; Holst, J.J.; Fledelius, C.; Johansen, P.B.; Rossetti, L.; Jelicks, L.A.; Serup, P.; Nishimura, E.; Charron, M.J. Lower blood glucose, hyperglucagonemia, and pancreatic α cell hyperplasia in glucagon receptor knockout mice. Proc. Natl. Acad. Sci. U.S.A. 2003, 100, 1438–1443.

154. Yip, R.G.C.; Wolfe, M.M. GIP biology and fat metabolism. Life Sci. 2000, 66, 91–103.

155. Holst, J.J.; Gromada, J.; Nauck, M.A. The pathogenesis of NIDDM involves a defective expression of the GIP receptor. Diabetologia. 1997, 40, 984–986.

156. Vilsbøll, T.; Krarup, T.; Madsbad, S.; Holst, J. Defective amplification of the late phase insulin response to glucose by GIP in obese type II diabetic patients. Diabe-tologia 2002, 45, 1111–1119.

157. Meier, J.J.; Hücking, K.; Holst, J.J.; Deacon, C.F.; Schmiegel, W.H.; Nauck, M.A. Reduced insulinotropic effect of gastric inhibitory polypeptide in first-degree relatives of patients with type 2 diabetes. Diabetes. 2001, 50, 2497–2504.

158. Miyawaki, K.; Yamada, Y.; Yano, H.; Niwa, H.; Ban, N.; Ihara, Y.; Kubota, A.; Fujimoto, S.; Kajikawa, M.; Kuroe, A.; Tsuda, K.; Hashimoto, H.; Yamashita, T.; Jomori, T.; Tashiro, F.; Miyazaki, J.; Seino, Y. Glucose intolerance caused by a defect in the entero-insular axis: a study in gastric inhibitory polypeptide receptor knockout mice. Proc. Natl. Acad. Sci. U.S.A. 1999, 96, 14843–14847.

159. Hansotia, T.; Baggio, L.L.; Delmeire, D.; Hinke, S.A.; Yamada, Y.; Tsukiyama, K.; Seino, Y.; Holst, J.J.; Schuit, F.; Drucker, D.J. Double incretin knockout (DIRKO) mice reveal an essential role for the enteroinsular axis in transducing the glucoregu-latory actions of DPP-IV inhibitors. Diabetes. 2004, 53, 1326–1335.

160. Miyawaki, K.; Yamada, Y.; Ban, N.; Ihara, Y.; Tsukiyama, K.; Zhou, H.; Fujimoto, S.; Oku, A.; Tsuda, K.; Toyokuni, S.; Hiai, H.; Mizunoya, W.; Fushiki, T.; Holst, J.J.; Makino, M.; Tashita, A.; Kobara, Y.; Tsubamoto, Y.; Jinnouchi, T.; Jomori, T.; Seino, Y. Inhibition of gastric inhibitory polypeptide signaling prevents obesity. Nat. Med. 2002, 8, 738–742.

161. Robidoux, J.; Martin, T.L.; Collins, S. β-adrenergic receptors and regulation of energy expenditure: a family affair. Annu. Rev. Pharmacol. Toxicol. 2004, 44, 297–323.

162. Atgie, C.; D'Allaire, F.; Bukowiecki, L.J. Role of β_1- and β_3-adrenoreceptors in the regulation of lipolysis and thermogenesis in rat brown adipocytes. Am. J. Physiol. 1997, 273, C1136–C1142.

163. Lowell, B.B.; Bachman, E.S. β-adrenergic receptors, diet-induced thermogenesis, and obesity. J. Biol. Chem. 2003, 278, 29385–29388.

164. Susulic, V.S.; Frederich, R.C.; Lawitts, J.; Tozzo, E.; Kahn, B.B.; Harper, M.E.; Himms-Hagen, J.; Flier, J.S.; Lowell, B.B. Targeted disruption of the β_3-adrenergic receptor gene. J. Biol. Chem. 1995, 270, 29483–29492.

165. Bachman, E.S.; Dhillon, H.; Zhang, C.-Y.; Cinti, S.; Bianco, A.C.; Kobilka, B.K.; Lowell, B.B. β_{AR} signaling required for diet-induced thermogenesis and obesity resistance. Science. 2002, 297, 843–845.

166. Crowley, V.E.F.; Yeo, G.S.H.; O'Rahilly, S. Obesity therapy: altering the energy intake-and-expenditure balance sheet. Nat. Rev. Drug Discov. 2002, 1, 276–286.

167. Sawzdargo, M.; George, S.R.; Nguyen, T.; Xu, S.; Kolakowski, L.F.; O'Dowd, B.F. A cluster of four novel human G protein-coupled receptor genes occurring in close proximity to CD22 gene on chromosome 19q13.1. Biochem. Biophys. Res. Commun. 1997, 239, 543–547.

168. Briscoe, C.P.; Tadayyon, M.; Andrews, J.L.; Benson, W.G.; Chambers, J.K.; Eilert, M.M.; Ellis, C.; Elshourbagy, N.A.; Goetz, A.S.; Minnick, D.T.; Murdock, P.R.; Sauls, H.R. Jr.; Shabon, U.; Spinage, L.D.; Strum, J.C.; Szekeres, P.G.; Tan, K.B.; Way, J.M.; Ignar, D.M.; Wilson, S.; Muir, A.I. The orphan G protein-coupled receptor GPR40 is activated by medium and long chain fatty acids. J. Biol. Chem. 2003, 278, 11303–11311.

169. Le Poul, E.; Loison, C.; Struyf, S.; Springael, J-Y.; Lannoy, V.; Decobecq, M-E.; Brezillon, S.; Dupriez, V.; Vassart, G.; Van Damme, J.; Parmentier, M.; Detheux, M. Functional characterization of human receptors for short chain fatty acids and their role in polymorphonuclear cell activation. J. Biol. Chem. 2003, 278, 25481–25489.

170. Brown, A.J.; Goldsworthy, S.M.; Barnes, A.A.; Eilert, M.M.; Tcheang, L.; Daniels, D.; Muir, A.I.; Wigglesworth, M.J.; Kinghorn, I.; Fraser, N.J.; Pike, N.B.; Strum, J.C.; Steplewski, K.M.; Murdock, P.R.; Holder, J.C.; Marshall, F.H.; Szekeres, P.G.; Wilson, S.; Ignar, D.M.; Foord, S.M.; Wise, A.; Dowell, S.J. The orphan G protein-coupled receptors GPR41 and GPR43 are activated by propionate and other short chain carboxylic acids. J. Biol. Chem. 2003, 278, 11312–11319.

171. Xiong, Y.; Miyamoto, N.; Shibata, K.; Valasek, M.A.; Motoike, T.; Kedzierski, R.M.; Yanagisawa, M. Short-chain fatty acids stimulate leptin production in adipocytes through the G protein-coupled receptor GPR41. Proc. Natl. Acad. Sci. U.S.A. 2004, 101, 1045–1050.

172. Boden, G.; Shulman, G.I. Free fatty acids and type 2 diabetes: defining their role in the development of insulin resistance and beta-cell dysfunction. Eur. J. Clin. Invest. 2002, 32, 14–23.

173. Itoh, Y.; Kawamata, Y.; Harada, M.; Kobayashi, M.; Fujii, R.; Fukusumi, S.; Ogi, K.; Hosoya, M.; Tanaka, Y.; Uejima, H.; Tanaka, H.; Maruyama, M.; Satoh, R.; Okubo, S.; Kizawa, H.; Komatsu, H.; Matsumura, F.; Noguchi, Y.; Shinohara, T.; Hinuma, S.; Fujisawa, Y.; Fujino, M. Free fatty acids regulate insulin secretion from pancreatic cells through GPR40. Nature. 2003, 422, 173–176.

7 G Protein-Coupled Receptors in CNS Drug Discovery

Rita Raddatz and Deborah S. Hartman

CONTENTS

ABSTRACT

G protein-coupled receptors (GPCRs) are widely expressed in the central nervous system (CNS), where they mediate and modulate synaptic transmission in the brain and spinal cord. A large percentage of CNS drugs target GPCRs, and these compounds have been exploited scientifically to further explore the molecular natures and physiological functions of these target proteins. Large GPCR groups such as the dopamine, serotonin, and opioid receptor families show indications of both redundancy and exquisite receptor subtype selectivity of biological responses. Structural studies are now adding to our understanding of GPCR activation at the molecular level, although the influences of the neuronal cell environment and synaptic architecture remain key determinants in the functioning of GPCRs in the CNS. This chapter provides an overview of the roles of neuronal GPCRs in normal and disease states in the areas of psychiatry, pain, neurodegeneration, and neuroendocrine function.

7.1 INTRODUCTION

GPCRs play a critical role in developmental processes and synaptic transmission in the central nervous system (CNS), as well as in learning and memory, thought and emotional state, motor and hormonal control, and pain sensation. Analysis of the human genome sequence revealed a repertoire of 367 nonsensory GPCRs for endogenous ligands,[1] over 100 of which remain "orphan receptors" without known naturally occurring ligands. Expression profiling predicts that an unexpectedly large number (over 90%) of nonsensory GPCRs are found in the brain, each with a unique distribution pattern and a high degree of overlap, producing thousands of distinct cell-specific combinations.

Additional sequence variation is generated in the CNS through receptor polymorphisms, alternative splicing, and RNA editing, and receptor function is regulated by a wide array of post-translational modifications including glycosylation and phosphorylation. This chapter profiles several prominent GPCR families in the CNS area with respect to structure and function, with an intentional focus on receptors of relevance to modern drug discovery.

7.1.1 SYNAPTIC TRANSMISSION AND NEUROTRANSMITTER PATHWAYS

The function of a GPCR in the CNS must be seen in the context of its neuronal cell environment, which is specialized to receive, integrate, and transmit electrical and chemical signals, and its neuroanatomical localization. A wide range of neurotransmitter (NT) molecules that signal through GPCRs have been identified and include amino acids, biogenic amines, peptides, and lipids. NT pathways in the CNS have been mapped according to distribution and connectivity of cells signaling through these molecules and by overall functions and behaviors controlled. Understanding of overall CNS function requires knowledge of how a given protein, e.g., a GPCR, functions within these pathways under normal and pathological conditions.

GPCRs in the CNS are expressed in various types of neurons and in astrocytes and glial cells. The genotype of a GPCR, defined as its protein sequence, is influenced by its specific cell environment to produce the observed receptor phenotype, defined as its pharmacological and signaling properties.[2] Cell-specific factors that influence receptor phenotype include differences in receptor reserve, available G protein complement, accessory proteins, lipid environment, and the presence of other GPCRs.

These phenotypic differences may be used to therapeutic advantage to create selective ligands, but also need to be taken into account when characterizing ligands in recombinant systems.

Each individual neuron forms thousands of discrete synapses with other cells, and the resulting networks enable functional effects at individual GPCRs to produce consequences in distant regions of the brain. A typical synapse consists of three basic elements: a presynaptic nerve terminal that releases an NT when the neuron is depolarized, a postsynaptic cell that contains the NT receptor and its effector mechanisms, and a physical distance between the two cells of approximately 200 Angstroms containing specialized extracellular matrix and synaptic machinery proteins, across which the NT diffuses.

GPCRs mediate a relatively slow modulation of neuronal activity, in contrast to the rapid neurotransmissions through ion channels. Postsynaptic GPCR-mediated effector systems include changes in intracellular cAMP levels, increases in phophoinositol turnover, and indirect modulation of ion channel activity by activation of kinases via release of G-protein β/γ subunits. GPCRs also play major roles in presynaptic cells, acting as CNS autoreceptors to decrease release of the same NT or as heteroreceptors, modulating releases of neurotransmitters other than their cognate ligands.

The serotonin 5HT1A receptor is a typical example of an autoreceptor, expressed on cell bodies in the raphe nucleus which, when activated, reduces 5HT release from the terminals of these neurons in the cortex.[3-5] There are also more complex negative feedback loops, for example, the activation of postsynaptic metabotropic glutamate receptors (mGluRs) in various CNS regions results in release of endocannabinoid signaling molecules that activate presynaptic cannabinoid CB1 receptors, reducing glutamate release from those neurons.[6,7] GPCRs modulate the outflow of CNS circuits in a region- and cell-specific fashion, providing a wealth of options for therapeutic intervention.

7.1.2 NEURONAL CELL ENVIRONMENT

Neuronal cell environments are highly differentiated and exert strong influences on GPCR function. The study of ligand–receptor interactions in a system that closely mimics the native environment of the receptor *in vivo* represents an ongoing challenge. Overexpression of GPCRs in recombinant systems can exaggerate agonist efficacy of ligands such that compounds with partial agonist activity *in vivo* appear to be full agonists. Therefore, the true differentiation of full and partial agonism must be determined *in vivo*.

Constitutive activity of GPCRs and subsequently the ability of ligands to demonstrate inverse agonism (i.e., decreased basal receptor activity in the absence of an agonist) is also heavily cell context-specific. Apparent constitutive activity is often seen for heterologously expressed GPCRs and has been demonstrated *in vivo*,[8,9] suggesting that inverse agonists will have therapeutic profiles different from those of antagonists.

Although traditionally thought to function as monomeric units, mounting evidence indicates that GPCRs function as oligomers and that pharmacological phenotypes of hetero-oligomers may differ from those of homo-oligomers (see Chapter 15, this volume).[10]

In fact, GPCRs themselves can function as cell-specific accessory proteins, as seen with the GABA-B receptor.[11] As shown in Table 7.1, most of the known GPCR families in the CNS contain multiple receptor types, with significantly overlapping expression patterns in the six key CNS regions shown, both within and between GPCR families. Thus, a wide range of hetero-oligomers may be possible in the CNS, providing much greater diversity in functional receptor units than is possible through receptor variants alone.[67]

Additional non-GPCR proteins may also alter receptor specificity or signaling directly or indirectly.[68] The nature and physiological relevance of receptor inter-actions are outside the scope of this chapter, but it is recognized that our under-standing of drug effects and activities of endogenous ligands will be expanded in the coming decade to include the functional diversity of protein complexes.

7.2 PSYCHIATRIC DISEASES

Psychiatric disorders include schizophrenia, depression, anxiety, and bipolar disease and represent a major focus of drug discovery efforts in the CNS area. The most effective treatments have highly complex pharmacologies, but emerging therapies now target specific GPCRs in the hope of increasing efficacy while reducing unwanted side effects. It is widely accepted that the major mental illnesses are polygenetic and result from the interactions of both genetic and environmental components. The dopamine and serotonin GPCR families have figured prominently in therapeutic approaches for many of these indications.

Each disease term encompasses a spectrum of disorders, and despite recent pharmacological and genetic advances, current knowledge remains insufficient to explain underlying biochemical mechanisms. The high rates of comorbidity involved with these disorders[69–71] and the abilities of drugs developed for one indication to show efficacy in treating another highlight the importance of an integrated approach in advancing our understanding of disease mechanisms and in discovering novel therapeutics.

7.2.1 SCHIZOPHRENIA

Schizophrenia and schizoaffective disorders are characterized by an imbalance of the dopaminergic system resulting from or consequently affecting glutaminergic, GABA, and serotonergic pathway transmission, with genetic, environmental, and develop-mental components.[72] GPCRs play a major role in the symptomatic treatment of schizophrenia and form the basis for the only available drug treatment, but these drugs unfortunately have no demonstrated abilities to affect disease progression. Most of our knowledge about biochemical processes underlying these disorders comes from compounds found serendipitously to be therapeutically effective in humans.

Treatment of schizophrenia is in part a detective story spanning over 50 years (Figure 7.1), during which researchers found that currently available drugs exhibit a broad range of receptor interactions that also produce a range of undesired side

TABLE 7.1
Distribution of Prominent GPCR Families with Known Endogenous Ligands in the Central Nervous System

Receptor Type	Receptor Family	Subtype	Endogenous Ligands	Amygdala	Cerebral Cortex	Hippocampus	Hypothalamus	Spinal Cord	Basal Ganglia	Other	Ref.
Bioamine	Adrenergic	α1A	Epinephrine, norepinephrine	+		+	+	+	+		12–15
		α1B		+	+	+	+	+	+		
		α1D		+	+	+		+	+		
		α2A		+	+	+	+	+			
		α2B							+		
		α2C		+	+	+	+	+	+		
		b1			+	+		+	+		
		b2			+	+	+	+	+		
		b3								Adipose	
	Muscarinic	M1	Acetylcholine	+	+	+			+		16–20
		M2			+	+	+	+	+		
		M3			+	+	+	+	+		
		M4			+	+			+		
		M5			+	+	+	+	+		
	Dopamine	D1	Dopamine	+	+	+	+	+	+		21–28
		D2		+	+	+	+	+	+		
		D3		+	+	+	+	+	+		
		D4		+	+	+		+		Retina	
		D5				+		+			
	Histamine[b]	H1	Histamine	+	+	+	+	+	+		29–34
		H2		+	+	+	+	+	+		
		H3		+	+	+	+		+		
		H4				+				Immune cells	

TABLE 7.1 (Continued)
Distribution of Prominent GPCR Families with Known Endogenous Ligands in the Central Nervous System

Receptor Type	Receptor Family	Subtype	Endogenous Ligands	Amygdala	Cerebral Cortex	Hippocampus	Hypothalamus	Spinal Cord	Basal Ganglia	Other	Ref.
	Serotonin	5HT1A	Serotonin	+	+	+	+				35–36
		5HT1B		+	+	+		+	+		
		5HT1D				+			+		
		5HT1F			+	+	+	+	+		
		5HT2A			+	+		+	+		
		5HT2B		+			+			Human, absent in rat brain	
		5HT2C		+	+	+	+	+	+		
		5HT4			+	+			+		
		5HT5		+		+			+		
		5HT6			+	+			+		
		5HT7		+	+	+	+			SCN	
Peptide	Chemokine	4 major families	~50 Peptide chemokines							Glia, neurons, astrocytes	37
	Galanin	GALR1	Galanin	+	+	+	+	+			38–40
		GALR2		+	+	+	+	+			
		GALR3		+			+	+			
	Melanocortin	MC1, 2	ACTH and MSH							Peripheral	41–43
		MC3		+	+	+	+				
		MC4		+	+	+	+		+		
		MC5			+				+		

Family	Receptor	Endogenous ligand							Distribution	Reference
Neuromedin U	NMU1	NMU	+	+	+	+		+	Predominantly peripheral	44–45
	NMU2									
Neuropeptide Y	NPY1	NPY	+	+	+	+	+	+		46–47
	NPY2		+	+	+	+	+	+		
	NPY4		+	+	+	+	+	+		
	NPY5		+	+		+	+			
Neurotensin	NTS1	Neurotensin	+	+	+	+	+	+		48–49
	NTS2									
Opioid	DOR	Enkephalins	+	+	+	+	+	+		50–51
	KOR	Dynorphins		+		+		+		
	MOR	Endorphins, enkephalins	+	+	+	+	+	+		
	ORL-1	Nociceptin/ orphanin FQ	+	+	+	+	+	+		
Tachykinin	NK1	Substance P	+	+	+	+	+	+	Peripheral	52
	NK2	NKA								
	NK3	NKB	+	+	+	+	+	+		
Lipid	Cannabinoids	CB1	Anandamide, 2-AG	+	+	+	+	+		Peripheral, retina	53–54
	CB2		+	+	+	+	+	+		
Other	Adenosine	A1	Adenosine	+	+	+		+			55–57
	A2a							+		
	A2b							+		
	A3					+				
Melatonin	MT1	Melatonin	+	+	+	+			SCN, retina	58–59
	MT2		+		+				retina	

TABLE 7.1 (Continued)
Distribution of Prominent GPCR Families with Known Endogenous Ligands in the Central Nervous System

Receptor Type	Receptor Family	Subtype	Endogenous Ligands	Amygdala	Cerebral Cortex	Hippocampus	Hypothalamus	Spinal Cord	Basal Ganglia	Other	Ref.
	mGluRs	mGluR1	Glutamate		+	+		+	+		60–65
		mGluR5			+	+		+	+		
		mGluR2			+	+		+	+		
		mGluR3			+	+		+	+		
		mGluR4		+	+	+		+	+		
		mGluR6								Retina	
		mGluR7		+	+	+	+	+	+		
		mGluR8			+	+					
	GABA	GABA-B	GABA		+			+	+		66

a Based on published rat and human mRNA distribution.
b Includes autoradiography data.

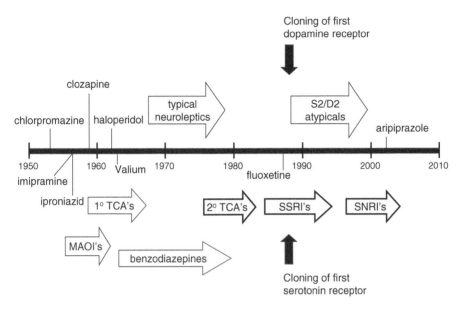

FIGURE 7.1 History of psychiatric drug development. Timeline shows emergence of selected drug mechanisms (solid arrows) and therapeutic agents (plain text) with relevance to schizophrenia (above the line) and depression/anxiety (below the line). S2/D2 = serotonin 5HT2A + dopamine D2 receptor antagonists. TCA = tricyclic antidepressants. MAOI = monoamine oxidase inhibitor. SSRI = selective serotonin reuptake inhibitor. SNRI = selective serotonin and norepinephrine reuptake inhibitor.

effects. Human genetics studies are beginning to define susceptibility genes for schizophrenic disorders,[73] and molecular pharmacology approaches to development of more selective compounds may provide the next wave of breakthroughs in disease understanding and treatment.

7.2.1.1 Characteristics of Antipsychotic Drugs

Chlorpromazine was the first compound identified to have therapeutic benefit for schizophrenia, and surprisingly it emerged from the field of antihistamine research.[74] It has high affinity for adrenergic α-1A, -1B, -2B, and -2C, histamine H1, dopamine D2, D3, and D4, serotonin 5HT6 and 5HT7, and muscarinic M1, M3, and M5 receptors.[75] It has strong sedative properties and was quickly superseded by drugs such as haloperidol and clozapine that maintained broad receptor interaction profiles and lower frequencies of adverse motor and cardiovascular effects. Compound structures are shown in Figure 7.2.

Haloperidol is considered a "typical" antipsychotic, showing efficacy in treating positive symptoms, e.g., delusions, paranoia, and hallucinations, in 70 to 80% of schizophrenic patients.[76] It has high affinity for dopamine D2 and adrenergic α-1B receptors, with somewhat lower affinity for dopamine D3 and D4 and adrenergic α-1A receptors. Typical antipsychotics, including haloperidol, are still associated with

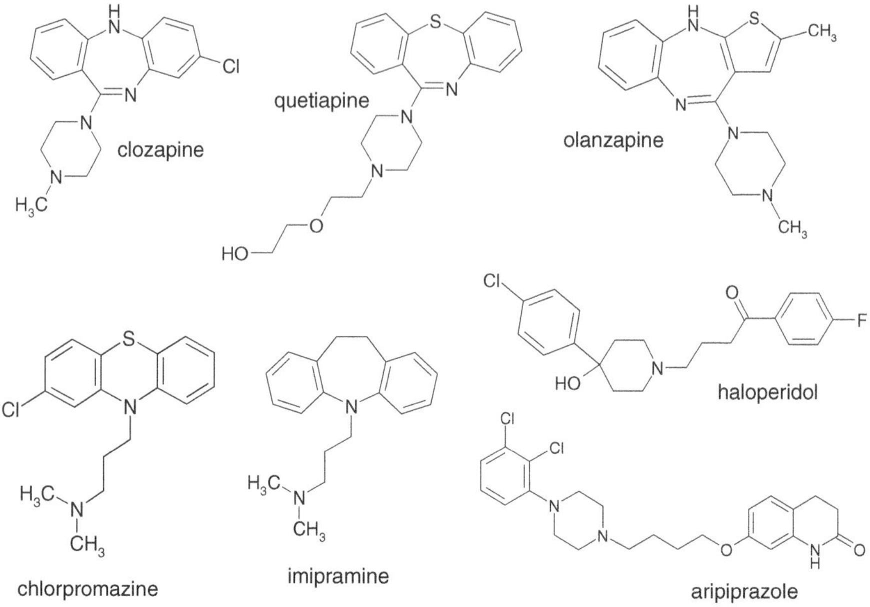

FIGURE 7.2 Selected structures of psychiatric drugs. Atypical antipsychotics include cloz-apine and its close analogs, quetiapine and olanzapine, plus the D2 partial agonist, aripipra-zole. Haloperidol and chlorpromazine are typical antipsychotics, also known as neuroleptics. Imipramine, a derivative of chlorpromazine, is an antidepressant.

high incidences of extrapyramidal symptoms (EPS), weight gain, sedation, hyper-prolactinemia, and impairment of cognitive function.[77]

Clozapine is different in that it is effective for positive disease symptoms as well as negative symptoms such as withdrawal, diminished emotions, and decreased social interaction, and it presents greatly reduced risks of movement disorders. The combination of efficacy in treating positive and negative disease symptoms and low incidences of EPS defines the class of "atypical" antipsychotic drugs, but the phar-macological basis for atypicality is still unclear.[78]

Interestingly, clozapine also has a broad receptor interaction profile, with lower affinity binding compared to haloperidol at most GPCRs. It has highest affinity overall for adrenergic α-1A and 1B, moderate affinity for the muscarinic M1 through M5 and serotonin 5HT2A, 5HT2C, 5HT6 and 5HT7 receptors, and 10-fold higher affinity for dopamine D4 versus dopamine D1, D2, D3, and D5 receptors.[75] Some experts believe that D2 receptor blockade may be sufficient for antipsychotic effi-cacy,[79] whereas others propose that nonselective drugs with custom-designed recep-tor profiles will be required to treat complex psychiatric disorders.[80]

Despite the introduction of new atypical drugs, clozapine remains the "gold standard" of antipsychotic efficacy, as the only compound to show benefit for the population of patients (over 30%) who do not respond to other drug treatment options[81] and for reducing emergent suicidal behavior[82] — a major cause of death in schizophrenic patients. Interestingly, at efficacious doses of clozapine and its close

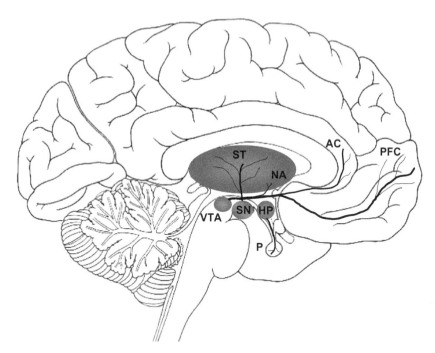

FIGURE 7.3 Dopaminergic pathways in a sagittal section of brain. The nigrostriatal pathway containing about 75% of the DA in the brain originates in the substantia nigra (SN) and projects primarily to the basal ganglia or striatum (ST), the center of movement control. This pathway is implicated in Parkinson's disease and in the extrapyramidal (short term) and tardive dyskinesia (long term) side effects observed during antipsychotic therapy. The mesolimbic pathway originates in the ventral tegmental area (VTA) and projects to the medial components of the limbic system such as the nucleus accumbens (NA) and anterior cingulated (AC) cortex. This pathway is implicated in emotion, memory, and the positive symptoms of schizophrenia. Neurons in the mesocortical pathway also originate in the VTA and project primarily to the PFC. This pathway is implicated in motivation and planning, attention, social behavior, and the negative symptoms of schizophrenia. The tuberoinfundibular pathway originates in the hypothalamus and projects to the pituitary gland where DA release controls prolactin secretion, influencing lactation and fertility.

analog, quetiapine (Figure 7.2), only transient D2 receptor occupancy is observed in brain-imaging studies.[83]

The existence of multiple dopaminergic systems in the brain is key to understanding the pathology and current treatment strategies for schizophrenia (Figure 7.3). The dopamine hypothesis of schizophrenia, first proposed in 1965,[84] now proposes that excess dopamine observed in the striatum leads to positive symptoms and a dopamine deficiency in prefrontal cortex contributes to negative symptoms. Whereas D2 receptor blockade in the mesolimbic pathway is believed to account for efficacy against positive symptoms, D2 blockade in the nigrostriatal pathway results in acute dyskinesia, akinesia, and tardive dyskinesia, motor side effects collectively referred to as EPS. Atypical antipsychotics selectively block dopamine receptors in limbic brain areas, consistent with their lower side-effect profiles.

Recent developments include identification of new modes of action for antipsychotics, including the launch of a dopamine D2 receptor partial agonist, aripiprazole,[85] and the finding that olanzapine and risperidone show inverse agonism at serotonin and/or dopamine receptors.[86] Inverse agonist activity has not been shown to correlate with atypicality,[87] but most atypical drugs have moderate to high affinity 5HT2A receptor antagonism, which is believed to facilitate dopaminergic transmission in the cortex.[88,89] A selective serotonin 5HT2A and 5HT2C receptor antagonist (SR46349B) and the neurokinin 3 (NK3) antagonist (SR142801) have shown antipsychotic efficacy in a first clinical trial.[90] However, sonepiprazole, a D4-selective compound, failed to show efficacy in a separate placebo-controlled trial.[91] 5HT2 receptors are unique in that they are downregulated when chronically exposed to either agonists or antagonists,[92] and 5HT2C transcripts are now known to undergo RNA editing that may be reduced in schizophrenic patients.[93] RNA editing can produce up to 14 different receptor isoforms that vary with respect to receptor coupling to G proteins, agonist trafficking, and level of constitutive activity.[94]

Muscarinic receptors are also implicated in schizophrenia, based on the activity of the M1 and M4 agonist, xanomeline, in reducing psychotic and behavioral aspects of dementia.[95]

Cannabinoid receptors are localized in areas where they may indirectly modulate dopaminergic transmission relevant to both positive and negative symptoms.[96] The CB1 receptor is reported to be upregulated in the prefrontal cortices of schizophrenic patients.[97]

One clinical trial with the SR141716A CB1 antagonist has been reported, but it did not demonstrate significant improvement in positive or negative symptoms of schizophrenia.[90] The adenosine A2a receptor is also of interest based on the activity of CGS21680 in *Cebus paella* monkeys.[98] Specifically, A2a–D2 heterodimers localized in the dendritic spines of the striatopallidal GABAergic neurons have been implicated as potential targets for schizophrenia and other disorders.[99] Research related to finding novel antipsychotic drugs includes efforts aimed at GPCRs, ion channels, neurotrophic factors, and even cell-signaling mechanisms[100] that may achieve better efficacy against positive and negative symptoms and therapeutic benefit for cognitive deficits such as reduced ability to focus attention and deficiencies in executive function. Preclinical models have implicated serotonin 5HT2A,[101] 5HT6,[102] dopamine D1[103] and D2,[104] and adrenergic α-2A[105] receptors as targets to enhance cognitive function.

7.2.1.2 Exploiting GPCR Families: A Closer Look at Dopamine Receptors

GPCRs have evolved as receptor families with protein sequence homologies, similar pharmacologies, and often-overlapping patterns of expression. In addition, multiple receptor isoforms are produced through genetic polymorphism and alternate splicing that can differ in functional properties. This allows a single endogenous agonist to exert a wide range of effects, differentially regulated based on receptor sequence, cell type, and tissue localization. This section provides specific illustrations of heterogeneity among dopamine receptors, one of the best-characterized GPCR families in the CNS, and one that has been successfully exploited in drug development.

Five dopamine receptors have been identified and classified as D1-like (D1 and D5) and D2-like (D2, D3, and D4) based on their similarities to pharmacologically defined sites identified in brain tissue.[106–108] The first family member cloned, D2, was identified using low stringency hybridization with a β-adrenergic receptor sequence.[109] The human D1, D3, D4, and D5 subtypes show 41, 75, 52, and 44% amino acid identities, respectively, with D2. D1 and D2 receptors are highly expressed in the nigrostriatal pathway where they play key roles in motor control.[110] D3 and D4 receptors are expressed in mesolimbic and mesocortical areas of the brain, with lower levels in striatum. The D4 receptor protein is found at high levels in the prefrontal cortex, a region believed to be important for cognition and attention,[111] and it has been found to modulate GABA release in the thalamic reticular nucleus, a region that controls attention, sensory processing, and rhythmic activity during slow wave sleep.[112] Consistent with this localization, D4 has shown activity in models of cognitive behavior in rats.[113]

Multiple isoforms have been identified for all D2-like DA receptors.[110] Alternative splicings of 29 and 21 amino acids in the predicted third cytoplasmic loop were found for D2 and D3, respectively. No splice variants were identified for D4, but the DRD4 gene locus is highly polymorphic.[114] All dopamine receptors activate multiple signaling pathways, e.g., D2 receptors inhibit adenylate cyclase (AC), stimulate phospholipase C, mobilize intracellular calcium, and activate Na^+/H^+ exchange. All dopamine receptors contain consensus protein kinase A phosphorylation sites in putative intracellular domains that may regulate function. D2 receptor phosphorylation by protein kinase C directs preferential coupling away from AC inhibition and toward potentiation of arachidonic acid release.[115]

D3, D4, and D5 receptors bind dopamine with high affinity, whereas D1 and D2 receptors show both high- and low-affinity-binding constants that differ over 100-fold.[116] The orthosteric binding site on a dopamine receptor consists of an aromatic microdomain formed by conserved residues within transmembrane (TM) domains 3, 5, and 7. Despite the high homology across DA receptor subtypes in these TM regions, the substitution of only four to six residues in the TM binding domain of D2 is sufficient to confer a D4-specific binding profile, and vice versa.[117]

The D4 receptor is unique in that it shows high-affinity binding for all three catecholamines [dopamine (DA), norepinephrine (NE), and epinephrine (EP)], suggesting that it may constitute a molecular integration point among the three catecholamine pathways in the brain.[118,119] The discovery that clozapine binds with approximately 10-fold higher affinity to D4 than D2 generated significant novel drug discovery interest in the D4 receptor in the 1990s, which led to development of D4-specific compounds and surprising findings regarding the D4 receptor gene structure.

The D4 receptor gene (DRD4) located near the telomere of chromosome 11p is one of the most polymorphic human genes known. One D4 variant is produced by a 12-base-pair repeat in exon 1 and correlated with a psychotic condition called delusional disorder.[120,121] A more dramatic variation occurs in the third cytoplasmic loop, where over 100 distinct haplotypes[122,123] involving a 48-base-pair tandem repeat sequence (VNTR) with 2 to 11 copies per allele (2R through 11R) have been identified. 4R is the most common variant in humans and the most effective for inhibiting AC activity *in vitro*. VNTRs encode a proline-rich sequence with consensus SH3 binding domains that may mediate receptor interactions with cytoskeletal proteins and serine, threonine, and tyrosine kinases.[124] VNTRs are present in both humans and primates, but not in rodents.

The D4 7R allele, which is highly heterogeneous even within the VNTR region, has shown association with the human behavioral traits of "novelty seeking,"[125,126] attention-deficit hyperactivity disorder (ADHD),[123,127] and Tourette's syndrome.[128] This is consistent with the reduced ability of the D4 7R variant to inhibit AC, and presumably neuronal firing, compared to the 4R variant.[129]

The 7R allele has evolved very recently in evolutionary terms and has been suggested to confer a "response-ready" advantage to human populations in resource-restricted, rapidly changing social environments. Interestingly, the general population contains about 2% heterozygous carriers of the D4 receptor-null mutation. A homozygous individual suffering from somatic ailments including acoustic neuroma, obesity, and autonomic nervous system disturbances, but no major psychiatric illness has been reported.[130] Taken together, these findings exemplify the molecular diversity found within a GPCR family and its functional complexity in the CNS.

7.2.2 DEPRESSION

The clinical description of depression is complex, covering a broad range of symptoms that lack a unifying biological hypothesis. Depression has both genetic and environmental components, with linkage studies suggesting it is a polygenic disorder.[131] Here again, effective therapeutics are contributing to an understanding of its underlying molecular mechanisms.[132] Modern treatment for depression, which focuses exclusively on agents that modulate monoamine neurotransmission, began with a monoamine oxidase inhibitor (MAOI) originally developed to treat tuberculosis. MAOIs increase serotonin and norepinephrine concentrations in the brain by inhibiting the MAO enzyme. They are highly effective in treating depression but are used only rarely due to potentially dangerous drug interaction effects.

A second breakthrough in depression treatment came from derivatives of chlorpromazine, developed as potential antipsychotics. One such derivative, imipramine (Figure 7.2), failed to show benefits in schizophrenic patients but was remarkably effective in patients who had severe depression.[133] Imipramine, still in use today, is a tricyclic antidepressant (TCA) that acts by inhibiting cellular reuptake mechanisms for norepinephrine and serotonin to increase activity within these GPCR families.[134] Imipramine retains some activity at GPCRs, including antagonism at histamine H1, muscarinic, and α-1 adrenergic receptors, but this activity contributes to an unattractive side-effect profile. Recent developments include selective serotonin reuptake inhibitors (SSRIs), serotonin + norepinephrine reuptake inhibitors (SNRIs), and norepinephrine + dopamine reuptake inhibitors (NDRIs), implicating multiple NT pathways in the spectrum of disorders that comprise major depression.[135]

Until now, GPCRs played only minor roles as direct drug targets in the treatment of depression, mediating the effects of some atypical or "second generation" anti-depressants such as mirtazapine, which has antagonist activity at 5HT2 and α-2 receptors and at the 5HT3 ligand-gated ion channel receptor, and nefazodone, which blocks 5HT2 receptors in addition to 5HT reuptake.[135] 5HT1A receptor agonism has been proposed to help alleviate depression, anxiety, negative symptoms, and cognitive dysfunction.[136] 5HT2C receptor ligands may provide antidepressive activity through disinhibition of the mesolimbic dopamine system.[92] GPCRs feature more prominently, however, in a new wave of emerging therapies aiming to address deficits of current therapies including slow onset of action, inability to achieve full remission, difficulty targeting significant populations of nonresponding patients, and minimization of residual side effects including sexual dysfunction.

Several new therapeutics target neuropeptide GPCRs that mediate stress responses.[137] Previous interest was focused on the NK1 neurokinin receptor, which mediates highly diverse effects of substance P.[138,139] A direct association of NK1 receptors with depression has been demonstrated in guinea pig and gerbil models[140] and by behavioral studies in NK1 receptor knockout mice.[141] In recent clinical trials, two highly selective NK1 receptor antagonists, aprepitant[142] and L-759274,[143] have shown good tolerability and efficacy for major depression. Sustained activation of stress systems has been implicated in pathophysiology in some depressed patients.[144] Overactivity of the hypothalamic–pituitary–adrenal (HPA) axis has been observed in depressive disorders,[145] with excess levels of the cortisol stress hormone detected in about 50% of patients with moderate to severe depression. This led to interest in the corticotrophin-releasing factor (CRF) receptor antagonist R-121919 that has shown efficacy in preclinical models and in a small clinical trial for major depression.[146]

Other neuropeptide receptors including neuropeptide Y, melanocortin-concentrating hormone, and vasopressin V1b receptors, in addition to dopamine, cannabinoid, and δ-opioid receptors, have also been implicated in the pathophysiology of depression.[134] New approaches within the monoaminergic systems may be possible.[147] For example, modulation of serotonin autoreceptors may provide a quicker onset of action for reuptake inhibitors. Neuroprotection, neurogenesis, and neuroimmune system modulation are additional research areas for major depression in which GPCRs may emerge as targets.[137]

7.2.3 ANXIETY

The pharmacological treatment of anxiety also began with a serendipitous discovery involving a compound synthesized for a completely different indication. Anti-infective research in the late 1940s produced a set of compounds with unexpected "calming effects" in rats and led quickly to the development of the meprobamate derivative introduced in 1955. Although not very effective, this compound played an important role in demonstrating the feasibility of pharmacological interventions in anxiety disorders.

The discovery of the benzodiazepines, which act on GABA-A receptor ion channels, revolutionized the treatment of anxiety with the launch of Valium in 1963. Today, the SSRIs developed for depression represent first-line therapy for anxiety disorders, supporting shared molecular mechanisms underlying anxiety and depression.[148] Although GPCRs have not historically been targets for anxiolytic drugs, they are well represented in emerging therapies, where there is strong overlap with approaches to depression. For example, CRF antagonists represent a promising new approach to treating both anxiety and depression.[149] Neurokinin receptor antagonists are of interest due to NK1 receptor localization in relevant areas of the brain and anxiolytic activity of NK1 antagonists in preclinical models.[150]

Novel anxiolytic approaches under investigation include allosteric GABA-B receptor modulators, group II and group III mGluR agonists, selective adrenergic and serotonin receptor ligands, and cannabinoid and NPY receptor ligands, among others.[151] Importantly, the discovery that atypical antipsychotic drugs are effective in treating anxiety disorders including obsessive–compulsive disorder, post-traumatic stress disorder, and generalized anxiety disorder provides opportunity for investigation of the mechanisms common to anxiety and schizophrenia in both preclinical and clinical settings.[71]

7.3 PAIN AND ANALGESIA

Many known drugs used to alleviate pain also modulate the activities of enzymes involved in inflammatory processes in peripheral tissues. However, some of the most efficacious analgesics known, including morphine and related compounds, interact with the opioid family of GPCRs within the CNS. The quest for analgesics as efficacious as morphine that lack side effects such as tolerance, addiction, and respiratory depression continues even today, 200 years after the purification of morphine from opium.

Development of better analgesics requires understanding the complexity of the CNS components involved in what we perceive as pain under different conditions. Specialized nerve endings respond to various noxious stimuli in a process known as nociception. Through multisynaptic pathways, these primary afferent nociceptors send information to the dorsal horn of the spinal cord and on to the thalamus and somatosensory cortex. However, not all nociceptive responses are perceived as pain. Perception of pain is modified by emotional states including stress and other inputs. This modulation is thought to occur under various conditions via multisynaptic descending modulatory circuits linking limbic brain regions such as the amygdala to the dorsal horn of the spinal cord.[152]

Different modalities such as acute and chronic pain may respond to modulation at different points in these pathways. Opioid receptors are expressed in the ascending nociceptive pathways including the dorsal horn of the spinal cord, thalamus, and cortex as well as in the descending antinociceptive pathways. Complex interactions of opioid receptor activation at the different levels of these pathways may explain the remaining challenges to perfecting opioid-based therapies.

7.3.1 OPIOID RECEPTORS

GPCRs interacting with morphine and other analgesic opiate ligands were first defined pharmacologically in various tissues. The cloning of the μ, δ, and κ opioid receptors (designated MOR, DOR and KOR by International Union of Pharmacology nomenclature) in the 1990s gave real promise that selective agents with fewer side effects would be developed. Morphine is a MOR-selective ligand with very low potencies at both KOR and DOR. Overwhelming evidence, including lack of mor-phine-induced analgesia in MOR knockout mice[153] identified the recombinant MOR protein as necessary for analgesic actions of opioid compounds.

Some aspects of the involvement of opioid receptors in analgesia, such as the significance of multiple splice variants of the MOR[154] and identification and charac-terization of two new human μ opioid receptor splice variants (hMOR-1O and hMOR-1X) remain to be clarified. Whereas selective DOR ligands have been developed with the promise of analgesia with fewer side effects[155] in acute pain states, DOR antag-onism demonstrates less efficacy than morphine.[156] However, DOR antagonists poten-tiate morphine-induced analgesia when coadministered,[157,158] and this effect is likely due to direct interaction of MOR and DOR proteins in hetero-oligomers.[159] Indeed, MOR and DOR receptors are coexpressed in neurons of DRG and dorsal spinal cord,[160] suggesting this heterodimer may be a relevant target for analgesics.

Another related receptor, opioid receptor-like 1 (ORL-1),[161] was cloned based on homology with the known opioid receptors. It appears to have low affinity for most opioid ligands and high affinity for an endogenous peptide ligand, nociceptin or orphanin F/Q.[162,163] As the name implies, nociceptin was first described as a peptide that induces hyperalgesia when injected into the spinal cord, suggesting this receptor may also be involved in pain modulation and represent a useful therapeutic target. However, it can act in both spinal cord and higher CNS centers to induce analgesia and hyperalgesia, respectively, further adding to the complexity of mod-ulation within the opioid receptor family.[164]

7.3.2 OTHER GPCRS

The NK1 receptor and its endogenous peptide agonist, substance P (SP), have been studied as potential targets for the treatment of pain for many years. Although strong evidence from animal models demonstrated analgesic properties of the available selective NK1 antagonists, they have not proven effective in humans. SP is found in small unmyelinated sensory fibers associated with nociception.[165] NK1 receptors are expressed in the dorsal horn of the spinal cord where SP is released following nociceptive stimuli, and NK1 receptor expression is modulated during specific pain states in animal models.[166]

Although NK1 antagonists did not consistently demonstrate analgesia in acute models of pain, a number of positive results involving neuronal injury and induction of inflammation were reported in animal models. The lack of success of a number of clinical trials in demonstrating analgesic effects in humans (except for some efficacy in dental pain)[167] has been a major disappointment.[168] Although the NK1–SP system in the spinal cord is very similar in humans and rodents, differences in supraspinal systems may attribute to the species differences. Alternatively, achieving sufficient receptor occupancy *in vivo* and selection of appropriate pain states have been suggested as issues to address in future clinical trails of NK1 antagonists.

Cannabinoid receptor subtypes CB1 and CB2 are activated by endogenous lipids such as anandamide, known collectively as endocannabinoids. Similar to the opioid system, the effects of exogenous ligands for these receptors, including the active components of marijuana (cannabis), were studied long before identification of the receptors. Although involvement of CB receptors in pain modulation has been suggested for many years, the lack of selective small molecules and the stigma of association with drugs of abuse slowed progress in this field.

It is now becoming clear that effective analgesia can be separated from psychotropic effects by selective modulation of CB receptors. Data from animal models demonstrated high efficacy analgesia in response to all types of nociception (acute and inflammatory states) tested with a number of agents currently under investigation.[169,170] Clinical trials using synthetic cannabinoids and extracts of whole plant cannabis (including delta-9-tetrahydocannabinol and cannabidiol) have shown mixed results for efficacy in treating pain and muscle spasticity associated with multiple sclerosis and various neuropathic pain conditions.[170–174] The results of future trials using more selective agents with properties that allow oral dosing are eagerly awaited.

More speculative, early stage research on the involvement of other GPCRs in pain perception is based largely on localization of receptors in nociception-related pathways. Further studies with selective pharmacological agents are needed to reveal the involvement of GPCRs such as the sensory neuron-specific receptors (SNSRs), galanin receptors, and neuromedin U (NMU) receptors in pain modulation.[175]

7.4 NEURODEGENERATION

The development and progression of neurodegenerative diseases, including Alzheimer's disease (AD) and Parkinson's disease (PD), are poorly addressed by known drug therapies and represent major unmet medical needs. Although most research efforts in this area are focused on multiple enzymes as drug targets, ligands that interact with a number of GPCRs show promise for novel therapeutic mechanisms.

7.4.1 Parkinson's Disease

Parkinson's disease is a progressive neurodegenerative disease characterized by loss of dopaminergic neurons in the substantia nigra. Although the cause of neuronal loss is not understood, the resulting loss of motor control can be temporarily ameliorated by treatment with the dopamine precursor, L-3,4-dihydroxyphenyl-alanine (L-DOPA). However, motor fluctuations limit this therapeutic approach for most patients. Adjunctive therapy with dopamine receptor agonists with half-lives longer than that of L-DOPA may limit side effects including tremor and limb rigidity. Furthermore, dopamine agonists are now used as monotherapy in early-stage PD patients after demonstrating neuroprotective effects and reduced risk of dyskinesias.[176]

Other approaches that address sparing of the affected neurons would provide more long lasting and safer treatment. One theory suggests that the loss of dopaminergic neurons in PD and other conditions such as AD and amyotrophic lateral sclerosis (ALS) is secondary to diminished levels of neurokinin peptides including substance P,[177] suggesting that NK1 receptor ligands could be useful in preventing neurodegeneration. Adenosine A2a receptor antagonists have shown therapeutic benefit for dyskinesia induced by L-DOPA,[178] and both A2a[179] and adenosine A2a–dopamine D2 heteromers[67,99] have been proposed as relevant targets for this disease. As genetic studies shed new light on pathways involved in neurodegenerative diseases, additional GPCRs may emerge as targets for drugs that can halt or reverse disease progression, facilitate disease prevention, or provide more effective symptomatic relief.[180]

7.4.2 ALZHEIMER'S DISEASE

AD is marked by a progressive loss of cognitive and other CNS functions and is a growing problem in our aging society. AD targets the cholinergic neurons of the basal forebrain that provide the major cholinergic innervation to the hippocampus, amygdala, and cortex that are involved in many aspects of memory, cognition, and behavior.[181] The current mainstays of AD treatment are cholinesterase inhibitors that block the enzyme that degrades acetylcholine (ACh) released from these neurons, thereby helping maintain cholinergic tone. Another approach to enhancing cholinergic tone is to mimic ACh actions, such as with M1 muscarinic receptor agonists.[182]

This mechanism may also modulate production of β-amyloid tangles that represent neuropathological hallmarks of the AD process. Muscarinic agonists such as xanomeline have shown efficacy in improving cognitive and behavioral deficits in AD[183] but produce intolerable cholinergic side effects, perhaps due to lack of selectivity. The development of positive modulators of muscarinic receptors that interact with allosteric sites on the receptors may provide more opportunity for selectivity, and continues to be an active area of research for treating Alzheimer's disease.[184]

Increasing the release of ACh from neuronal terminals may achieve a similar increase in cholinergic tone and is a potential therapeutic mechanism for GPCR ligands currently in development for the treatment of cognitive disorders. For example, 5HT6 antagonists increase the release of neurotransmitters, including ACh, in rat frontal cortex,[185] 5HT6 antagonists enhance spatial learning retention in the rodent water maze model[102,186] and reverse anticholinergic-induced memory deficits in rodent learning models.[185,187,188] This suggests that modulation of 5HT6 may be effective in enhancing cognitive functions, although failure to reproduce these effects has also been reported.[189] The 5HT6 antagonist, SB-271046, also improved cognitive deficits in aged rats, further suggesting a role for 5HT6 antagonists in the treatment of dementia.[187] 5HT1A antagonists also reverse anticholinergic-induced deficits in performance of learning and memory tasks in some, but not all, animal models. This suggests that other members of the 5HT receptor family may provide additional therapeutic approaches to cognition enhancement in AD.[190]

7.4.3 CHEMOKINE RECEPTORS AND NEURODEGENERATION

The blood–brain barrier isolates the healthy CNS from circulating cells of the immune system. Within the CNS, microglia act as immune cells and when activated they are capable of releasing proinflammatory and cytotoxic substances including chemokines. More than 50 human chemokines have been identified and grouped into four subfamilies (α, β, γ, and δ) based on the pattern of two cysteine residues in the N-terminal domain, each with a corresponding family of GPCRs (Table 7.2). Receptors within each subfamily show some promiscuity for chemokine ligands.

TABLE 7.2
Chemokine Receptors and Their Ligands

Ligands	Chemokine subfamily	Alpha	Beta	Gamma	Delta
	Cysteine motif	CXC	CC	C	CX3C
	Number of chemokine ligands	13	17	1	1
Receptors	Number of receptors	6	10	1	1
	Receptor nomenclature	CXCR1 through 6	CCR1 through 10	XCR1	CX3CR1

C = cysteine residue. X = any amino acid residue.

Chemokine receptors are expressed on neurons, microglia, and glia and may modulate various aspects of neuroinflammation and neurodegeneration.[191] For example, neuronal presynaptic chemokine receptors have been shown to inhibit glutamate release and may thereby modulate excitotoxicity.[192] Additional effects of chemokines directly on neurons may also play a role in disease progression. In AD, the chemokines IP-10 and Gro-α are elevated. Upon binding to CXCR3 and CXCR2, their respective receptors in neurons, they activate intracellular signaling pathways that may cause τ-hyperphosphorylation.[193] Chemokine release in the CNS likely also plays a more traditional inflammatory role as lymphocytes are able to penetrate the CNS in response to chemotactic agents released from activated astrocytes and glia, thereby amplifying the inflammatory response. Indeed, neuroinflammation is recognized as a significant component of neurodegenerative disease pathology.

7.5   NEUROENDOCRINE FUNCTION

In the CNS, hormone secretion from the pituitary is regulated by neurosecretory cells in the hypothalamus. GPCRs are abundant in the hypothalamus, many of which interact with peptide neurotransmitters to regulate hunger, thirst, water balance, body temperature, and blood pressure. Table 7.1 shows a large number of receptors expressed in hypothalamus, and additional examples include somatostatin, orexin, and angiotensin receptors. The hypothalamus has been an area of major interest in the study of orphan receptors, and two examples of recently deorphanized GPCRs in the hypothalamic–neurohypophysial system are neuromedin U-2 and galanin-like peptide receptors,[194] Many of the remaining human orphan GPCRs have been detected in the hypothalamus by *in situ* hybridization and await further characterization.[1]

7.6   ORPHAN GPCRs IN THE CENTRAL NERVOUS SYSTEM

The successful history of CNS drugs targeting GPCRs led to interest in quickly identifying orphan GPCRs with promise as new drug targets. There are approximately 100 GPCRs, identified by sequence homology and predicted 7-transmembrane (7TM) domain topology, for which endogenous ligands have not been identified that are known as orphan GPCRs. Although most of these GPCRs are expressed widely in the CNS, the future impact of orphan GPCRs on CNS drug discovery is a subject of debate.[195,196]

Researchers are in agreement that increased knowledge of physiology, including new signaling systems involving GPCRs, will ultimately result in new medications, but it is difficult to predict the timeframes required. Certain recently deorphanized receptors such as the orexin receptors were validated rapidly as drug targets for the treatment of CNS disease. Orexin receptors OX1R and OX2R were deorphanized in 1998 by identifying the components of a rat brain extract that activated the receptors in a screening assay, peptides now known as orexin-A and orexin-B.[197]

These peptides were initially shown to exert activity in appetite control, but orexin-knockout mice soon displayed disrupted sleep patterns similar to narcolepsy.[198] At the same time, positional cloning in a narcoleptic dog colony revealed that the *canarc-1* gene corresponds to the orexin B receptor.[199] The autosomal recessive *canarc-1* mutation in Doberman pinschers and Labrador retrievers produces narcolepsy that is pharmacologically and physiologically validated with the human condition. The easily detectable activity of orexin peptides has substantiated interest in their receptors as potential CNS drug targets. However, other GPCRs with unknown or less well-characterized ligands represent major challenges for drug discovery.[200]

7.6.1 CHALLENGES IN EVALUATING ORPHAN GPCRS IN THE CNS

Receptor localization can provide the first clues to orphan GPCR function in normal and disease states, but it can be misleading in the brain. High throughput expression profiling is widely used to map orphan GPCR mRNA transcripts in the CNS, but receptor expression measured at the protein level often reveals different distributions because many receptors such as CB1 are localized to neuronal axon terminals distant from their origins of synthesis in cell bodies.[201]

Receptor distribution can also differ significantly across species. A number of CNS GPCRs recently deorphanized, including the trace amine receptors[202] and the sensory neuron-specific receptors (SNSRs),[203] differ in the number of alleles for highly homologous receptor subtypes in rodents and humans, making alignment of species homologs very difficult. Knockout mice have been used widely to gain insight into the functions of orphan GPCRs, but this has met with mixed success for CNS receptors because of compensatory mechanisms and confounding behavioral consequences of gene deletions.

Advances in RNAi technology and development of receptor-selective ligands may allow the study of receptor functions in challenging neurochemical and behavioral assays.[200] Overall, the process of evaluating orphan GPCR function is still guided by a combination of convenience, serendipity, and intuition. Orphan receptors are described in more detail in Chapter 16.

7.7 CONCLUSIONS

GPCRs have proven highly successful as drug targets in the CNS. In 2002, nearly half of the 13 new CNS drugs approved by the U.S. Food & Drug Administration targeted these receptors.[204] However, the development of drugs for CNS indications presents several unique and formidable challenges. The etiology, diagnosis, and treatment of mental illnesses include emotional and social aspects that may never be fully understood in terms of receptor pharmacology, making it particularly difficult to develop predictive preclinical disease models.

Drugs targeting GPCRs and other protein classes expressed in the CNS must be designed to cross the blood–brain barrier, which is a highly effective filter protecting the brain from exogenous compounds. In addition, system modulation or "fine tuning" of neurotransmitter circuits in the brain may be more desirable than complete blockade or sustained activation of receptor targets, perhaps through small molecule allosteric modulators of GPCRs.

Other important areas for GPCR-related research in the CNS include mapping of receptor expression at the single-cell and synaptic levels, analysis of cell-specific and synapse-specific receptor signaling pathways, and better understanding of the roles of receptor protein complexes. Receptor interaction sites for peptide ligands and endogenous amino acid ligands such as glutamate may not be suitable for pharmacological agents that can penetrate the CNS. Exploiting the potential of allosteric modulators, inverse agonists, and partial agonists may provide valuable tool compounds for research into the functions of GPCRs along with important new CNS disease therapies.

REFERENCES

1. Vassilatis, D.K.; Hohmann, J.G.; Zeng, H.; Li, F.; Ranchalis, J.E.; Mortrud, M.T.; Brown, A.; Rodriguez, S.S.; Weller, J.R.; Wright, A.C.; Bergmann, J.E.; Gaitanaris, G.A. The G protein-coupled receptor repertoires of human and mouse. *Proc. Natl. Acad. Sci. U.S.A.* 2003, 100, 4903–4908.
2. Kenakin, T. Predicting therapeutic value in the lead optimization phase of drug discovery. *Nat. Rev. Drug Disc.* 2003, 2, 429–438.
3. Sprouse, J.S.; Aghajanian, G.K. (–)-Propranolol blocks the inhibition of serotonergic dorsal raphe cell firing by 5-HT1A selective agonists. *Eur. J. Pharmacol.* 1986, 128 (3), 295–298.
4. Sprouse, J.S.; Reynolds, L.S.; Braselton, J.P.; Rollema, H.; Zorn, S.H. Comparison of the novel antipsychotic ziprasidone with clozapine and olanzapine: inhibition of dorsal raphe cell firing and the role of 5-HT1A receptor activation. *Neuropsychopharmacology* 1999, 21, 622–631.
5. Blier, P.; de Montigny, C. Modification of 5-HT neuron properties by sustained administration of the 5-HT1A agonist gepirone: electrophysiological studies in the rat brain. *Synapse* 1987, 1, 470–480.
6. Doherty, J.; Dingledine, R. Functional interactions between cannabinoid and metabotropic glutamate receptors in the central nervous system. *Curr. Opin. Pharmacol.* 2003, 3, 46–53.
7. Varma, N.; Carlson, G.C.; Ledent, C.; Alger, B.E. Metabotropic glutamate receptors drive the endocannabinoid system in hippocampus. *J. Neurosci.* 2001, 21, RC188.
8. Morisset, S.; Rouleau, A.; Ligneau, X.; Gbahou, F.; Tardivel-Lacombe, J.; Stark, H.; Schunack, W.; Ganellin, C.R.; Schwartz, J.C.; Arrang, J.M. High constitutive activity of native H3 receptors regulates histamine neurons in brain. *Nature* 2000, 408, 860–864.

9. De Deurwaerdere, P.; Navailles, S.; Berg, K.A.; Clarke, W.P.; Spampinato, U. Constitutive activity of the serotonin2C receptor inhibits *in vivo* dopamine release in the rat striatum and nucleus accumbens. *J. Neurosci.* 2004, 24, 3235–3241.

10. Bai, M. Dimerization of G-protein-coupled receptors: roles in signal transduction. *Cell Signal.* 2004, 16, 175–186.

11. Bettler, B.; Kaupmann, K.; Mosbacher, J.; Gassmann, M. Molecular structure and physiological functions of GABA(B) receptors. *Physiol. Rev.* 2004, 84, 835–867.

12. Nicholas, A.P; Hökfely, T.; Pieribone, V.A. The distribution and significance of CNS adrenoceptors examined with *in situ* hybridization. *TiPS.* 1996, 17, 245.

13. Smith, M.S.; Schambra, U.B.; Wilson, K.H.; Page, S.O.; Schwinn, D.A. Alpha 1-adrenergic receptors in human spinal cord: specific localized expression of mRNA encoding alpha1-adrenergic receptor subtypes at four distinct levels. *Brain Res. Mol. Brain Res.* 1999, 63, 254–261.

14. Hirasawa, A.; Horie, K.; Tanaka, T.; Takagaki, K.; Murai, M.; Yano, J.; Tsujimoto, G. Cloning, functional expression and tissue distribution of human cDNA for the alpha -1C adrenergic receptor. *Biochem. Biophys. Res. Commun.* 1993, 195, 902–909.

15. Scheinin, M.; Lomasney, J.W.; Hayden-Hixson, D.M.; Schambra, U.B.; Caron, M.G.; Lefkowitz, R.J.; Fremeau, R.T. Jr. Distribution of alpha 2-adrenergic receptor subtype gene expression in rat brain. *Brain Res. Mol. Brain Res.* 1994, 21, 133–149.

16. Vilaro, M.T.; Palacios, J.M.; Mengod, G. Multiplicity of muscarinic autoreceptor subtypes: comparison of the distribution of cholinergic cells and cells containing mRNA for five subtypes of muscarinic receptors in the brain. *Brain Res. Mol. Brain Res.*, 1994, 21, 30–46.

17. Buckley, N.J.; Bonner, T.I.; Brann, M.R. Localization of a family of muscarinic receptor mRNAs in rat brain. *J. Neurosci.* 1988, 8, 4646–4652.

18. Wei, J.; Walton, E.A.; Milici, A.; Buccafusco, J.J. m1–m5 Muscarinic receptor distribution in rat CNS by RT-PCR and HPLC. *J. Neurochem.* 1994, 63, 815–821.

19. Weiner, D.M.; Levey, A.I.; Brann, M.R. Expression of muscarinic acetylcholine and dopamine receptor mRNAs in rat basal ganglia. *Proc. Natl. Acad. Sci. U.S.A.* 1990, 87, 7050–7054.

20. Ellis, J. Muscarinic receptors, in *Understanding G Protein-Coupled Receptors and Their Role in the CNS*. Pangalos, M.N., Davies, C.H., Eds., Oxford University Press, New York, 2002; pp. 349–371.

21. Valerio, A.; Belloni, M.; Gorno, M.L.; Tinti, C.; Memo, M.; Spano, P. Dopamine D_2, D_3, and D_4 receptor mRNA levels in rat brain and pituitary during aging. *Neurobiol. Aging.* 1994, 15, 713–719.

22. Fremeau, R.T.; Duncan, G.E.; Fornaretto, M.G.; Dearry, A.; Gingrich, J.A.; Brees, G.R.; Caron, M.G. Localization of D_1dopamine receptor mRNA in brain supports a role in cognitive, affective and neuroendocrine aspects of dopaminergic neurotransmission. *Proc. Natl. Acad. Sci. U.S.A.* 1991, 88, 3772–3776.

23. van Dijken, H.; Dijk, J.; Voom, P.; Holstege, J.C. Localization of dopamine D2 receptor in rat spinal cord identified with immunocytochemistry and *in situ* hybridization. *Eur. J. Neurosci.* 1996, 8, 621–628.

24. Mansour, A.; Meador-Woodruff, J.H.; Bunzow, J.R.; Civelli, O.; Akil, H.; Watson, S.J. Localization of dopamine D_2 receptor mRNA and $D_1$and D_2 receptor binding in the rat brain and pituitary: an *in situ* hybridization receptor autoradiographic analysis. *J. Neurosci.* 1990, 10, 2587–2601.

25. Bouthenet, M.-L.; Souil, E.; Martres, M.-P.; Sokoloff, P.; Giros, B.; Schwartz, J.-C. Localization of dopamine D_3 receptor mRNA in the rat brain using *in situ* hybridization histochemistry: comparison with D_2 receptor mRNA. *Brain Res.* 1991, 564, 203–219.

26. Matsumoto, M.; Hidaka, K.; Tada, S.; Tasaki, Y.; Yamaguchi, T. Low levels of mRNA for dopamine D_4 receptor in human cerebral cortex and striatum. *J. Neurochem.* 1996, 66, 915–919.

27. Xie, G.X.; Jones, K.; Peroutka, S.J.; Palmer, P.P. Detection of mRNAs and alternatively spliced transcripts of dopamine receptors in rat peripheral sensory and sympathetic ganglia. *Brain Res.* 1998, 785, 129–135.

28. Meador-Woodruff, J.H.; Mansour, A.; Grandy, D.K.; Damask, S.P.; Civelli, O.; Watson, S.J., Jr. Distribution of D_5 dopamine receptor mRNA in the rat brain. *Neurosci. Lett.* 1992, 145, 209–212.

29. Brown, R.E.; Stevens, D.R.; Haas, H.L. The physiology of brain histamine. *Prog. Neurobiol.* 2001, 63, 637–672.

30. Lovenberg, T.W.; Roland, B.L.; Wilson, S.J.; Jiang, X.; Pyati, J.; Huvar, A.; Jackson, M.R.; Erlander, M.G. Cloning and functional expression of the human histamine H3 receptor. *Mol. Pharmacol.* 1999, 55, 1101–1107.

31. Kashiba, H.; Fukui, H.; Morikawa, Y.; Senba, E. Gene expression of histamine H1 receptor in guinea pig primary sensory neurons: a relationship between H1 receptor mRNA-expressing neurons and peptidergic neurons. *Mol. Brain Res.* 1999, 66, 24–34.

32. Pillot, C.; Heron, A.; Cochois, V.; Tardivel-Lacombe, J.; Ligneau, X.; Schwartz, J.C.; Arrang, J.M. A detailed mapping of the histamine H(3) receptor and its gene transcripts in rat brain. *Neuroscience* 2002, 114, 173–193.

33. Murakami, H.; Sun-Wada, G.H.; Matsumoto, M.; Nishi, T.; Wada, Y.; Futai, M. Human histamine H2 receptor gene: multiple transcription initiation and tissue-specific expression. *FEBS Lett.* 1999, 451, 327–331.

34. Oda, T.; Morikawa, N.; Saito, Y.; Masuho, Y.; Matsumoto, S. Molecular cloning and characterization of a novel type of histamine receptor preferentially expressed in leukocytes. *J. Biol. Chem.* 2000, 275, 36781–36786.

35. Saudou, F.; Hen, R. 5-Hydroxytryptamine receptor subtypes in vertebrates and invertebrates. *Neurochem. Int.* 1994, 25, 503–532.

36. Roberts, C.; Price, G.W.; Middlemiss, D.N. Serotonin receptors, in *Understanding G Protein-Coupled Receptors and Their Role in the CNS.* Pangalos, M.N., Davies, C.H., Eds., Oxford University Press, New York, 2002; pp. pp. 439–468.

37. Ambrosini, E.; Aloisi, F. Chemokines and glial cells: a complex network in the central nervous system. *Neurochem. Res.* 2004, 29, 1017–1038.

38. Gustafson, E.L.; Smith, K.E.; Durkin, M.M.; Gerald, C.; Branchek, T.A. Distribution of a rat galanin receptor mRNA in rat brain. *Neuroreport.* 1996, 7, 953–957.

39. O'Donnell, D.; Ahmad, S.; Wahlestedt, C.; Walker, P. Expression of the novel galanin receptor subtype GALR2 in the adult rat CNS: distinct distribution from GALR1. *J. Comp. Neurol.* 1999, 409, 469–481.

40. Mennicken, F.; Hoffert, C.; Pelletier, M.; Ahmad, S.; O'Donnell, D. Restricted distribution of galanin receptor 3 (GalR3) mRNA in the adult rat central nervous system. *J. Chem. Neuroanat.* 2002, 24, 257–268.

41. Roselli-Rehfuss, L.; Mountjoy, K.G.; Robbins, L.S.; Mortrud, M.T.; Low, M.J.; Tatro, J.B.; Entwistle, M.L.; Simerly, R.B.; Cone, R.D. Identification of a receptor for γ-melanotropin and other propiomelanocortin peptides in the hypothalamus and limbic system. *Proc. Natl. Acad. Sci. U.S.A.* 1993, 90, 8856–8860.

42. Mountjoy, K.G.; Mortrud, M.T.; Low, M.J.; Simerly, R.B.; Cone, R.D. Localization of the melanocortin-4 receptor (MC4-R) in neuroendocrine and autonomic control circuits in the brain. *Mol. Endocrinol.* 1994, 8, 1298–1308.

43. Griffon, N.; Mignon, V.; Facchinetti, P.; Diaz, J.; Schwartz, J. C.; Sokoloff, P. Molecular cloning and characterization of the rat fifth melanocortin receptor. *Biochem. Biophys. Res. Commun.* 1994, 200, 1007–1014.

44. Howard, A.D.; Wang, R.; Pong, S.S.; Mellin, T.N.; Strack, A.; Guan, X.M.; Zeng, Z.; Williams, D.L. Jr.; Feighner, S.D.; Nunes, C.N.; Murphy, B.; Stair, J.N.; Yu, H.; Jiang, Q.; Clements, M.K.; Tan, C.P.; McKee, K.K.; Hreniuk, D.L.; McDonald, T.P.; Lynch, K.R.; Evans, J.F.; Austin, C.P.; Caskey, C.T.; Van der Ploeg, L.H.; Liu, Q. Identification of receptors for neuromedin U and its role in feeding. *Nature* 2000, 406, 70–74.

45. Raddatz, R.; Wilson, A.E.; Artymyshyn, R.; Bonini, J.A.; Borowsky, B.; Boteju, L.W.; Zhou, S.; Kouranova, E.V.; Nagorny, R. Guevarra, M.S.; Dai, M.; Lerman, G.S.; Vaysse, P.J.; Branchek, T.A.; Gerald, C.; Forray, C.; Adham, N. Identification and characterization of two neuromedin U receptors differentially expressed in peripheral tissues and the central nervous system. *J. Biol. Chem.* 2000, 275, 32452–32459.

46. Parker, R.M.C.; Herzog, H. Regional distribution of Y-receptor subtype mRNAs in rat brain. *Eur. J. Neurosci.* 1999, 11, 1431.

47. Landry, M.; Holmberg, K.; Zhang, X.; Hökfelt, T. Effect of axotomy on expression of NPY, galanin, and NPY Y1 and Y2 receptors in dorsal root ganglia and the superior cervical ganglion studied with double-labeling *in situ* hybridization and immunohistochemistry. *Exp. Neurol.* 2000, 162, 361–384.

48. Vincent, J.P.; Mazella, J.; Kitabgi, P. Neurotensin and neurotensin receptors. *Trends Pharmacol. Sci.* 1999, 20, 302–309.

49. Walker, N.; Lepee-Lorgeoux, I.; Fournier, J.; Betancur, C.; Rostene, W.; Ferrara, P.; Caput, D. Tissue distribution and cellular localization of the levocabastine-sensitive neurotensin receptor mRNA in adult rat brain. *Brain Res. Mol. Brain Res.* 1998, 57, 193–200.

50. Mansour, A.; Fox, C.A.; Akil, H.; Watson, S.J. Opioid-receptor mRNA expression in the rat CNS: anatomical and functional implications. *Trends Neurosci.* 1995, 18, 22–29.

51. Neal, C.R.; Mansour, A.; Reinscheid, R.; Nothacker, H. P.; Civelli, O.; Akil, H.; Watson, S.J. Opioid receptor-like (ORL1) receptor distribution in the rat central nervous system: comparison of ORL1 receptor mRNA expression with ^{125}I-[^{14}Tyr]-orphanin FQ binding. *J. Comp. Neurol.* 1999, 412, 563–605.

52. Tsuchida, K.; Shigemoto, R.; Yokota, Y.; Nakanishi, S. Tissue distribution and quantitation of the mRNAs for three rat tachykinin receptors. *Eur. J. Biochem.* 1990, 193, 751–757.

53. Lu, Q.; Straiker, A.; Lu, Q.; Maguire, G. Expression of CB2 cannabinoid receptor mRNA in adult rat retina. *Vis. Neurosci.* 2000, 17, 91–95.

54. Mailleux, P.; Vanderhaeghen, J.-J. Distribution of neuronal cannabinoid receptor in the adult rat brain: a comparative receptor binding radioautography and *in situ* hybridization histochemistry. *Neuroscience.* 1992, 48, 655–668.

55. Latini, S.; Pazzagli, M; Pepeu, G.; Pedata, F. A2 adenosine receptors: their presence and neuromodulatory role in the central nervous system. Gen. Pharmacol. 1996, 27, 925–933.

56. Dixon, A.K.; Gubitz, A.K.; Sirinathsinghji, D.J.; Richardson, P.J.; Freeman, T.C. Tissue distribution of adenosine receptor mRNAs in the rat. *Br. J. Pharmacol.* 1996, 118, 1461–1468.

57. Fredholm, B.B. Adenosine receptors, in *Understanding G Protein-Coupled Receptors and Their Role in the CNS.* Pangalos, M.N., Davies, C.H., Eds., Oxford University Press, New York, 2002; pp. 191–204.

58. Mazzucchelli, C.; Pannacci, M.; Nonno, R.; Lucini, V.; Fraschini, F.; Stankov, B.M. The melatonin receptor in the human brain: cloning experiments and distribution studies. *Brain Res. Mol. Brain Res.* 1996, 39, 117–126.

59. Reppert, S.M.; Godson, C.; Mahle, C.D.; Weaver, D.R.; Slaugenhaupt, S.A.; Gusella, J.F. Molecular characterization of a second melatonin receptor expressed in human retina and brain: the Mel~b melatonin receptor. *Proc. Natl. Acad. Sci. U.S.A.* 1995, 92, 8734–8738.

60. Valerio, A.; Paterlini, M.; Boifava, M.; Memo, M.; Spano, P.F. Metabotropic glutamate receptor mRNA expression in rat spinal cord. *Neuroreport.* 1997, 8, 2695–2699.

61. Ohishi, H.; Akazawa, C.; Shigemoto, R.; Nakanishi, S.; Mizuno, N. Distributions of the mRNAs for L-2-amino-4-phosphonobutyrate-sensitive metabotropic glutamate receptors, mGluR4 and mGluR7, in the rat brain. *J. Comp. Neurol.* 1995, 360, 555–570.

62. Testa, C.M.; Standaert, D.G.; Young, A.B.; Penney, J.B. Jr. Metabotropic glutamate receptor mRNA expression in the basal ganglia of the rat. *J. Neurosci.* 1994, 14, 3005–3018.

63. Fotuhi, M.; Standaert, D.G.; Testa, C.M.; Penney, J.B., Jr.; Young, A.B. Differential expression of metabotropic glutamate receptors in the hippocampus and entorhinal cortex of the rat. *Brain Res. Mol. Brain Res.* 1994, 21, 283–292.

64. Tanabe, Y.; Nomura, A.; Masu, M.; Shigemoto, R.; Mizuno, N.; Nakanishi, S. Signal transduction, pharmacological properties, and expression patterns of two rat metabotropic glutamate receptors, mGluR3 and mGluR4. *J. Neurosci.* 1993, 13, 1372–1378.

65. Pin, J.-P.; Bockaert, J. Metabotropic glutamate receptors, in *Understanding G Protein-Coupled Receptors and Their Role in the CNS*. Pangalos, M.N., Davies, C.H., Eds., Oxford University Press, New York, 2002; pp. 588–616.

66. Benke, D.; Honer, M.; Michel, C.; Bettler, B.; Mohler, H. γ-Aminobutyric acid type B receptor splice variant proteins GBR1a and GBR1b are both associated with GBR2 *in situ* and display differential regional and subcellular distribution *J. Biol. Chem.* 1999, 274, 27323–27330.

67. Agnati, L.F.; Ferre, S.; Lluis, C.; Franco, R.; Fuxe, K. Molecular mechanisms and therapeutical implications of intramembrane receptor/receptor interactions among heptahelical receptors with examples from the striatopallidal GABA neurons *Pharmacol. Rev.* 2003, 55, 509–550.

68. Neubig, R.P.; Siderovski, D.P. Regulators of G-protein signaling as new central nervous system drug targets. *Nat. Rev. Drug Dis.* 2002, 1, 187–197.

69. Cosoff, S.J.; Hafner, R.J. The prevalence of comorbid anxiety in schizophrenia, schizoaffective disorder, and bipolar disorder. *Austral. NZ J. Psychiatr.* 1998, 32, 67–72.

70. Wetherell, J.L.; Palmer, B.W.; Thorp, S.R.; Patterson, T.L. Golshan, S.; Jeste, D.V. Anxiety symptoms and quality of life in middle-aged and older outpatients with schizophrenia and schizoaffective disorder. *J. Clin. Psychiatr.* 2003, 64,1476–1482.

71. McIntyre, R.; Katzman, M. The role of atypical antipsychotics in bipolar depression and anxiety disorders. *Bipolar Disord.* 2003, 5 (Suppl. 2), 20–35.

72. Benes, F.M.; Tamminga, C.A. Neurobiology of schizophrenia, in *Psychiatry as a Neuroscience*, John Wiley & Sons, New York, 2002; pp. 197–236.

73. Owen, M.J.; Williams, N.M.; O'Donovan, M.C. The molecular genetics of schizophrenia: new findings promise new insights. *Mol. Psychiatr.* 2004, 9, 14–27.

74. Capuano, B.; Crosby, I.T.; Lloyd, E.J. Schizophrenia: genesis, receptorology, and current therapeutics. *Curr. Med. Chem.* 2002, 9, 521–548.

75. National Institutes of Mental Health Psychoactive Drug Screening Program http://pdsp.cwru.edu/; PDSP Ki database: http://kidb.cwru.edu/pdsp.php

76. Tamminga, C.A. Similarities and differences among antipsychotics. *J. Clin. Psychiatr.* 2003, 64 (Suppl. 17), 7–10.

77. Arana, G.W. An overview of side effects caused by typical antipsychotics. *J. Clin. Psychiatr.* 2000, 61 (Suppl. 8), 5–11.

78. Roth, B.L.; Sheffler, D.; Potkin, S.G. Atypical antipsychotic drug actions: unitary or multiple mechanisms for 'atypicality'? *Clin. Neurosci. Res.* 2003, 31, 108–117.

79. Kapur, S.; Remmington, G. Dopamine D(2) receptors and their role in atypical antipsychotic action: still necessary and may even be sufficient. *Biol. Psychiatr.* 2001, 50, 873–883.

80. Roth, B.L.; Sheffler, D.J.; Kroeze, W.K. Magic shotguns versus magic bullets: selectively non-selective drugs for mood disorders and schizophrenia. *Nat. Rev. Drug Dis.* 2004, 3, 353–359.

81. Kane, J.; Honigfield, G.; Singer, J.; Meltzer, H.Y. Clozapine for the treatment-resistant schizophrenic: a double-blind comparison with chlorpromazine. *Arch. Gen. Psychiatr.* 1988, 45, 789–796.

82. Meltzer, H.Y.; Alphs, L.; Green, A.I.; Altamura, A.C.; Anand, R.; Bertoldi, A.; Bourgeois, M.; Chouinard, G.; Islam, M.Z.; Kane, J.; Krishnan, R.; Lindenmayer, J.P.; Potkin, S. Clozapine treatment for suicidality in schizophrenia: International Suicide Prevention Trial. *Arch. Gen. Psychiatr.* 2003, 60, 82–91.

83. Kapur, S.; Zipursky, R.; Jones, C. et al. A positron emission tomography study of quetiapine in schizophrenia: a preliminary finding of an antipsychotic effect with only transiently high dopamine D2 receptor occupancy. *Arch. Gen. Psychiatr.* 2000, 57, 553–559.

84. Carlsson, A. Current status of the dopamine hypothesis of schizophrenia. *Neuropsychopharmacology.* 1988, 1, 179–186.

85. Shapiro, D.; Renock, S.; Arrington, E.; Chiodo, L.A.; Liu, L.-X.; Sibley, D.R.; Roth, B.L.; Mailman, R. Aripiprazole: a novel antipsychotic drug with a unique and robust pharmacology. *Neuropsychopharmacology.* 2003, 28, 1400–1411.

86. Milligan, G. Constitutive activity and inverse agonists of G protein coupled receptors: a current perspective. *Mol. Pharmacol.* 2003, 64, 1271–1276.

87. Rauser, L.; Savage, J.E.; Meltzer, H.Y.; Roth, B.L. Inverse agonist actions of typical and atypical antipsychotic drugs at the human 5-hydroxytryptamine(2C) receptor. *J. Pharmacol. Exp. Ther.* 2001, 299, 83–89.

88. Roth, B.L.; Hanizavareh, S.M.; Blum, A.E. Serotonin receptors represent highly favorable molecular targets for cognitive enhancement in schizophrenia and other disorders. *Psychopharmacology.* 2004, 174,17–24.

89. Mortimer, A.M.; Barnes, T.R.E. *Serotonergic Mechanisms in Antipsychotic Treatment.* Marcel Dekker, New York, 1996; pp. 311–330.

90. Meltzer, H.Y.; Arvanitis, L.; Bauer, D.; Rein, W. Placebo-controlled evaluation of four novel compounds for the treatment of schizophrenia and schizoaffective disorder. *Am. J. Psychiatr.* 2004, 161, 975–984.

91. Corrigan, M.H.; Gallen, C.C.; Bonura, M.L.;, Merchant, K.M. Effectiveness of the selective D4 antagonist sonepiprazole in schizophrenia: a placebo-controlled trial. *Biol. Psychiatr.* 2004, 55, 445–451.

92. Serretti, A.; Artioli, P.; De Ronchi, D. The 5HT2C receptor as a target for mood disorders. *Expert Opin. Ther. Targets.* 2004, 8, 15–23.

93. Sodhi, M.S.; Burnet, P.W.; Makoff, A.J.; Kerwin, R.W.; Harrison, P.J. RNA editing of the 5HT2C receptor is reduced in schizophrenia. *Mol. Psychiatr.* 2001, 6, 373–379.

94. Burns, C.M.; Chu, H.; Rueter, S.M.; Hutchinson, L.K.; Canton, H.; Sanders-Bush, E.; Emeson, R.B. Regulation of serotonin 2C receptor G-protein coupling by RNA editing. *Nature.* 1997, 387, 303–308.

95. Mirza, N.R.; Peters, D.; Sparks, R.G. Xanomeline and the antipsychotic potential of muscarinic receptor subtype selective agonists. *CNS Drug Rev.* 2003, 9, 159–186.

96. van der Stelt, M.; Di Marzo, V. The endocannabinoid system in the basal ganglia and in the mesolimbic reward system: implications for neurological and psychiatric disorders. *Eur. J. Pharmacol.* 2003, 480, 133–150.

97. Dean, B.; Sundram, S.; Bradbury, R; Scarr, E.; Copolov, D. Studies on [3H]CP-55940 binding in the human central nervous system: regional specific changes in density of cannabinoid-1 receptors associated with schizophrenia and cannabis use. *Neuroscience.* 2001, 103 (1), 9–15.

98. Andersen, M.B.; Fuxe, K.; Werge, T.; Gerlach, J. The adenosine A2a receptor agonist CGS21680 exhibits antipsychotic-like activity in *Cebus paella* monkeys. *Behav. Pharmacol.* 2002, 13, 639–644.

99. Ferre, S.; Ciruela, F.; Canals, M.; Marcellino, D.; Burgueno, J,; Casado, V.; Hillion, J.; Torvinen, M.; Fanelli, F.; de Benedetti, P.; Goldberg, S.R.; Bouvier, M.; Fuxe, K.; Agnati, L.F.; Lluis, C. Franco, R.; Woods, A. Adenosine A2a-dopamine D2 receptor–receptor heteromers: targets for neuro-psychiatric disorders. *Parkinsonism Rel. Dis.* 2004, 10, 265–271.

100. Mortimer, A.M. Novel antipsychotics in schizophrenia. *Exp. Opin. Invest. Drugs.* 2004, 13, 315–329.

101. Williams, G.V.; Rao, S.G.; Goldman-Rakic, P.S. The physiological role of 5-HT2A receptors in working memory. *J. Neurosci.* 2002, 22, 2843–2854.

102. Woolley, M.L.; Bentley, J.C.; Sleight, A.J.; Marsden, C.A.; Fone, K.C. A role for 5-HT6 receptors in retention of spatial learning in the Morris water maze. *Neuropharmacology.* 2001, 41, 210–219.

103. Goldman-Rakic, P.S.; Castner, S.A.; Svensson, T.H.; Siever, L.J.; Williams, G.V. Targeting the dopamine D1 receptor in schizophrenia: insights for cognitive dysfunction. *Psychopharmacology.* 2004, 174, 3–16.

104. Wang, M.; Vijayraghavan, S.; Goldman-Rakic, P.S. Selective D2 receptor actions on the functional circuitry of working memory. *Science.* 2004, 303, 853–856.

105. Svensson, T.H. Alpha-adrenoceptor modulation hypothesis of antipsychotic atypicality. *Prog. Neuropsychopharmacol. Biol. Psychiatr.* 2003, 27, 1145–1158.

106. Seeman, P.; Lee, T.; Chan-Wong, M.; Wong, K. Antipsychotic drug doses and neuroleptic/dopamine receptors. *Nature.* 1976, 261, 717–719.

107. Creese, I.; Burt, D.R.; Snyder, S.H. Dopamine receptor binding predicts clinical and pharmacological potencies of antischizophrenic drugs. *Science.* 1976, 192, 481–483.

108. Kebabian, J.W.; Calne, D.B. Multiple receptors for dopamine. *Nature.* 1979, 277, 93–96.

109. Bunzow, J.R.; Van Tol, H.H.; Grandy, D.K.; Albert, P.; Salon, J.; Christie, M.; Machida, C.A.; Neve, K.A.; Civelli, O. Cloning and expression of a rat D2 dopamine receptor cDNA. *Nature.* 1998, 336, 783–787.

110. Emilien, G.; Maloteaux, J.M.; Geurts, M.; Hoogenberg, K.; Cragg, S. Dopamine receptors: physiological understanding to therapeutic intervention potential. *Pharmacol. Ther.* 1999, 84, 133–156.

111. Nieoullon, A. Dopamine and the regulation of cognition and attention. *Prog. Neurobiol.* 2002, 67, 53–83.

112. Floran, B.; Floran, L.; Erlij, D.; Aceves, J. Activation of dopamine D4 receptors modulates [3H]GABA release in slices of the rat thalamic reticular nucleus. *Neuropharmacology.* 2004, 46, 497–503.

113. Powell, S.B.; Paulus, M.P.; Hartman, D.S.; Godel, T.; Geyer, M.A. RO-10-5824 is a selective dopamine D4 receptor agonist that increases novel object exploration in C57 mice. *Neuropharmacology*. 2003, 44, 473–481.

114. Oak, J.N.; Oldenhof, J.; Van Tol, H.H. The dopamine D(4) receptor: one decade of research *Eur. J. Pharmacol*. 2000, 405, 303–327.

115. Jackson, D.M.; Westlind-Danielsson, A. Dopamine receptors: molecular biology, biochemistry and behavioural aspects. *Pharmacol. Ther*. 1994, 64, 291–370.

116. Hartman, D.S.; Civelli, O. Molecular attributes of dopamine receptors: new potential for antipsychotic drug development. *Ann. Med*. 1996, 28, 211–219.

117. Simpson, M.M.; Ballesteros, J.A.; Chiappa, V.; Chen, J.; Suehiro, M.; Hartman, D.S.; Godel, T.; Snyder, L.A.; Sakmar, T.P.; Javitch, J.A. Dopamine D4/D2 receptor selectivity is determined by a divergent aromatic microdomain contained within the second, third, and seventh membrane-spanning segments. *Mol. Pharmacol*. 1999, 56, 1116–1126.

118. Newman-Tancredi, A.; Audinot-Bouchez, V.; Gobert, A.; Millan, M.J. Noradrenaline and adrenaline are high affinity agonists at dopamine D_4 receptors. *Eur. J. Pharmacol*. 1997, 319, 379–383.

119. Lanau, F.; Zenner, M.T.; Civelli, O.; Hartman, D.S. Epinephrine and norepinephrine act as potent agonists at the recombinant human dopamine D4 receptor. *J. Neurochem*. 1997, 68, 804–812.

120. Catalano, M.; Nobile, M.; Novelli, E.; Nothen, M.M.; Smeraldi, E. Distribution of a novel mutation in the first exon of the human dopamine D4 receptor gene in psychotic patients. *Biol. Psychiatr*. 1993, 34, 459–464.

121. Zenner, M.T.; Nobile, M.; Henningsen, R.; Smeraldi, E.; Civelli, O.; Hartman, D.S.; Catalano, M. Expression and characterization of a dopamine D4R variant associated with delusional disorder. *FEBS Lett*. 1998, 422, 146–150.

122. Ding, Y.C.; Chi, H.C.; Grady, D.L.; Morishima, A.; Kidd, J.R.; Kidd, K.K.; Flodman, P.; Spence, M.A.; Schuck, S.; Swanson, J.M.; Zhang, Y.P.; Moyzis, R.K. Evidence of positive selection acting at the human dopamine receptor D4 gene locus. *Proc. Natl. Acad. Sci. U.S.A*. 2002, 99, 309–314.

123. Grady, D.L.; Chi, H.C.; Ding, Y.C.; Smith, M.; Wang, E.; Schuck, S.; Flodman, P.; Spence, M.A.; Swanson, J.M.; Moyzis, R.K. High prevalence of rare dopamine receptor D4 alleles in children diagnosed with attention-deficit hyperactivity disorder. *Mol. Psychiatr*. 2003, 8, 536–545.

124. Lichter, J.B.; Barr, C.L.; Kennedy, J.L.; Van Tol, H.H.; Kidd, K.K.; Livak, K.J. A hypervariable segment in the human dopamine receptor D4 (DRD4) gene. *Hum. Mol. Genet*. 1993, 2, 767–773.

125. Benjamin, J.; Li, L.; Patterson, C.; Greenberg, B.D.; Murphy, D.L.; Hamer, D.H. Population and familial association between the D4 dopamine receptor gene and measures of novelty seeking. *Nat. Genet*. 1996, 12, 81–84.

126. Ebstein, R.P.; Nemanov, L.; Klotz, I.; Gritsenko, I.; Belmaker, R.H. Additional evidence for an association between the dopamine D4 receptor (D4DR) exon III repeat polymorphism and the human personality trait of novelty seeking. *Mol. Psychiatr*. 1997, 2, 472–477.

127. Langley, K.; Marshall, L.; van den Bree, M.; Thomas, H.; Owen, M.; O'Donovan, M.; Thapar, A. Association of the dopamine D4 receptor gene 7-repeat allele with neuropsychological test performance of children with ADHD. *Am. J. Psychiatr*. 2004, 61, 133–138.

128. Diaz-Anzaldua, A.; Joober, R.; Riviere, J.B.; Dion, Y.; Lesperance, P.; Richer, F.; Chouinard, S.; Rouleau, G.A. Tourette syndrome and dopaminergic genes: a family-based association study in the French Canadian founder population. *Mol. Psychiatr*. 2004, 9, 272–277.

129. Wang, E.; Ding, Y.C.; Flodman, P.; Kidd, J.R.; Kidd, K.K.; Grady, D.L.; Ryder, O.A.; Spence, M.A.; Swanson, J.M.; Moyzis, R.K. The genetic architecture of selection at the human dopamine receptor D4 (DRD4) gene locus. *Am. J. Hum. Genet.* 2004, 74, 931–944.

130. Nothen, M.M.; Cichon, S.; Hemmer, S.; Hebebrand, J.; Remschmidt, H.; Lehmkuhl, G.; Poustka, F.; Schmidt, M.; Catalano, M.; Fimmers, R. Human dopamine D4 receptor gene: frequent occurrence of a null allele and observation of homozygosity. *Hum. Mol. Genet.* 1994, 3, 2207–2212.

131. Zubenko, G.S.; Maher, B.S.; Hughes, H.B., III; Zubenko, W.N.; Scott-Stiffler, J.; Marazita, M.L. Genome-wide linkage survey for genetic loci that affect the risk of suicide attempts in families with recurrent, early-onset, major depression. *Am. J. Med. Genet.* 2004, 15, 47–54.

132. Wong, M.L.; Licinio, J. From monoamines to genomic targets: a paradigm shift for drug discovery in depression. *Nat. Rev. Drug Discov.* 2004, 3, 136–151.

133. Kuhn, R. Uber die behandlung depressives zustande mit einem iminobenzylderivat. *Schwiez Med.Wochenschr.* 1957, 87, 1135–1140.

134. Pacher P.; Kecskemeti, I. Trends in development of new antidepressants: is there a light at the end of the tunnel? *Curr. Med. Chem.* 2004, 11, 925–943.

135. Kent, J.M. SNaRIs, NaSSAs, and NaRIs: new agents for the treatment of depression. *Lancet* 2000, 355, 911–918.

136. Millan, M.J. Improving the treatment of schizophrenia: focus on serotonin (5-HT)1A receptors. *J. Pharmacol. Exp. Ther.* 2000, 295, 853–861.

137. Hindmarch, I. Beyond the monoamine hypothesis: mechanisms, molecules, and methods. *Eur. Psychiatr.* 2002, 17 (Suppl. 3), 294–299.

138. Duffy, R. Potential therapeutic targets for neurokinin-1 receptor antagonists. *Exp. Opin. Emerg. Drugs.* 2004, 9, 9–21.

139. Stout, S.C.; Owens, M.J.; Nemeroff, C.B. Neurokinin 1 receptor antagonists as potential antidepressants. *Annu. Rev. Pharmacol. Toxicol.* 2001, 41, 877–906.

140. Rupniak, N.M.; Kramer, M.S. Discovery of the anti-depressant and anti-emetic efficacy of substance P receptor (NK1) antagonists. *Trends Pharmacol. Sci.* 1999, 20, 485–490.

141. Rupniak, N.M.; Carlson, E.C.; Harrison, T.; Oates, B.; Seward, E.; Owen, S.; de Felipe, C.; Hunt, S.; Wheeldon, A. Pharmacological blockade or genetic deletion of substance P (NK1) receptors attenuates neonatal vocalization in guinea pigs and mice. *Neuropharmacology.* 2000, 39, 1413–1421.

142. Kramer, M.S.; Cutler, N.; Feighner, J.; Shrivastava, R.; Carman, J.; Sramek, J.J.; Reines, S.A.; Liu, G.; Snavely, D.; Wyatt-Knowles, E.; Hale, J.J.; Mills, S.G.; Mac-Coss, M.; Swain, C.J.; Harrison, T.; Hill, R.G.; Hefti, F.; Scolnick, E.M.; Cascieri, M.A.; Chicchi, G.G.; Sadowski, S.; Williams, A.R.; Hewson, L.; Smith, D.; Carlson, E.J.; Hargreaves, R.J.; Rupniak, N.M.J. Distinct mechanism for antidepressant activity by blockade of central substance P receptors. *Science.* 1998, 281, 1640–1645.

143. Kramer, M.S.; Winokur, A.; Kelsey, J.; Preskorn, S.H.; Rothschild, A.J.; Snavely, D.; Ghosh, K.; Ball, W.A.; Reines, S.A.; Munjack, D.; Apter, J.T.; Cunningham, L.; Kling, M.; Bari, M.; Getson, A.; Lee, Y. Demonstration of the efficacy and safety of a novel substance P (NK1) receptor antagonist in major depression. *Neuropsychopharmacology.* 2004, 29, 385–392.

144. Gold, P.W.; Wong, M.L.; Chrousos, G.; Licinio, J. Stress system abnormalities in melancholic and atypical depression: molecular, pathophysiological, and therapeutic implications. *Mol. Psychol.* 1996, 1, 257–264.

145. Strohle, A.; Holsboer, F. Stress responsive neurohormones in depression and anxiety. *Pharmacopsychiatry* 2003, 36 (Suppl. 3), S207–S214.

146. Zobel, A.W.; Nickel, T.; Kunzel, H.E.; Ackl, N.; Sonntag, A.; Ising, M.; Holsboer, F. Effects of the high-affinity corticotrophin-releasing hormone receptor 1 antagonist R-121919 in major depression: the first 20 patients. *J. Psych. Res.* 2000, 34, 171–181.

147. Bymaster, F.P.; McNamara, R.K.; Tran, P.V. New approaches to developing antidepressants by enhancing monoaminergic neurotransmission. *Exp. Opin. Invest. Drugs.* 2003, 12, 531–543.

148. Gorwood, P. Generalized anxiety disorder and major depressive disorder comorbidity: an example of genetic pleiotropy? *Eur. Psychiatr.* 2004, 19, 27–33.

149. Zorrilla, E.P.; Koob, G.F. The therapeutic potential of CRF1 antagonists for anxiety. *Exp. Opin. Invest. Drugs.* 2004, 13, 799–828.

150. Blier, P.; Gobbi, G.; Haddjeri, N.; Santarelli, L.; Mathew, G.; Hen, R. Impact of substance P receptor antagonism on the serotonin and norepinephrine systems: relevance to the antidepressant–anxiolytic response. *Rev. Psychiatr. Neurosci.* 2004, 29, 208–218.

151. Millan, M.J. The neurobiology and control of anxious states. *Prog. Neurobiol.* 2003, 70, 83–244.

152. Fields, H. State-dependent opioid control of pain. *Nat. Rev. Neurosci.* 2004, 5, 565–575.

153. Gaveriaux-Ruff, C.; Kieffer, B.L. Opioid receptor genes inactivated in mice: the highlights. *Neuropeptides.* 2002, 36, 62–71.

154. Pan, Y.X.; Xu, J.; Mahurter, L.; Xu M.; Gilbert, A.K.; Pasternak, G.W. Identification and characterization of two new human mu opioid receptor splice variants, hMOR-1O and hMOR-1X. *Biochem. Biophys. Res. Commun.* 2003, 301, 1057–1061.

155. Rapaka, R.S.; Porreca, F. Development of delta opioid peptides as nonaddicting analgesics. *Pharm. Res.* 1991, 8, 1–8.

156. Scherrer, G.; Befort, K.; Contet, C.; Becker, J.; Matifas, A.; Kiefferk B. The delta agonists DPDPE and deltorphin II recruit predominantly mu receptors to produce thermal analgesia: a parallel study of mu, delta and combinatorial opioid receptor knockout mice. *Eur. J. Neurosci.* 2004, 19, 2239–2248.

157. Traynor, J.R.; Elliott, J. δ-Opioid receptor subtypes and cross-talk with μ-receptors. *TiPS.* 1993, 14, 84–86.

158. He, L.; Lee, N.M. Delta opioid receptor enhancement of mu opioid receptor-induced antinociception in spinal cord. *J. Pharmacol. Exp. Ther.* 1998, 285, 1181–1186.

159. Gomes, I.; Gupta, A.; Filipovska, J.; Szeto, H.H.; Pintar, J.E.; Devi, L.A. A role for heterodimerization of mu and delta opiate receptors in enhancing morphine analgesia. *Proc. Natl. Acad. Sci. U.S.A.* 2004, 101, 5135–5159.

160. Cheng, P.Y.; Liu-Chen, L.Y.; Pickel, V.M. Dual ultrastructural immunocytochemical labeling of mu and delta opioid receptors in the superficial layers of the rat cervical spinal cord. *Brain Res.* 1997, 778, 367–380.

161. Mollereau, C.; Parmentier, M.; Mailleux, P.; Butour, J.L.; Moisand, C.; Chalon, P.; Caput, D.; Vassart, G.; Meunier, J.C. ORL1, a novel member of the opioid receptor family. Cloning, functional expression and localization. *FEBS Lett.* 1994, 341, 33–38.

162. Meunier, J.-Cl.; Mollereau, C.; Toll, L.; Suaudeau, Ch.; Moisand, Ch.; Alvinerie, P.; Butour, J.-L.; Guillemot, J. Cl.; Ferrara, P.; Monsarrat, B.; Mazarguil, H.; Vassart, G.; Parmentier, M.; Costentin, J. Isolation and structure of the endogenous agonist of opioid receptor-like ORL₁ receptor. *Nature.* 2002, 377, 532–535.

163. Reinscheid, R.K.; Nothacker, H.P.; Bourson, A.; Ardati, A.; Henningsen, R.A.; Bunzow, J.R.; Grandy, D.K.; Langen, H.; Monsma, F.J. Jr; Civelli, O. Orphanin FQ: a neuropeptide that activates an opioid-like G protein-coupled receptor. *Science.* 1995, 270, 792–794.

164. Henderson, G.; McKnight A.T., The orphan opioid receptor and its endogenous ligand: nociceptin/orphanin FQ. *Trends Pharmacol. Sci.* 1997, 18, 293–300.

165. Salt, T.E.; Hill, R.G. Neurotransmitter candidates of somatosensory primary afferent fibres. *Neuroscience.* 1983, 10, 1083–1103.

166. Taylor, B.K.; McCarson, K.E. Neurokinin-1 receptor gene expression in the mouse dorsal horn increases with neuropathic pain. *J. Pain.* 2004, 5, 71–76.

167. Dionne, R.A.; Max, M.B.; Gordon, S.M.; Parada, S.; Sang, C.; Gracely, R.H.; Sethna, N.F.; MacLean, D.B. The substance P receptor antagonist CP-99,994 reduces acute postoperative pain. *Clin. Pharmacol. Ther.* 1998, 64, 562–568.

168. Hill, R. NK1 (substance P) receptor antagonists: why are they not analgesic in humans? *TiPS.* 2000, 21, 244–246.

169. Walker, J.M.; Huang, S.M. Endocannabinoids in pain modulation. *Prostaglandins Leukot. Essent. Fatty Acids.* 2002, 66, 235–242.

170. Baker, D.; Pryce, G.; Giovannoni, G.; Thompson, A.J. The therapeutic potential of cannabis. *Lancet Neurol.* 2003, 2, 291–298.

171. Svendsen, K.B.; Jensen, T.S.; Bach, F.W. Does the cannabinoid dronabinol reduce central pain in multiple sclerosis? Randomised double-blind placebo-controlled cross-over trial. *Br. Med. J.* 2004, 329, 253.

172. Karst, M.; Salim, K.; Burstein, S.; Conrad, I.; Hoy, L.; Schneider, U. Analgesic effect of the synthetic cannabinoid CT-3 on chronic neuropathic pain: a randomized controlled trial. *JAMA.* 2003, 290, 1757–1762.

173. Wade, A.; Crawford, G.M.; Angus, M.; Wilson, R.; Hamilton, L. A randomized, double-blind, 24-week study comparing the efficacy and tolerability of mirtazapine and paroxetine in depressed patients in primary care. *Int. Clin. Psychopharmacol.* 2003, 18, 133–141.

174. Zajicek, J.; Fox, P.; Sanders, H.; Wright, D.; Vickery, J.; Nunn, A.; Thompson, A. Cannabinoids for treatment of spasticity and other symptoms related to multiple sclerosis (CAMS): multicentre randomised placebo-controlled trial. *Lancet* 2003, 362, 1517–1526.

175. Ahmad S.; Dray, A. Novel G protein-coupled receptors as pain targets. *Curr. Opin. Invest. Drugs.* 2004, 5, 67–70.

176. Schwarz, J. Rationale for dopamine agonist use as monotherapy in Parkinson's disease. *Curr. Opin. Neurol.* 2003, 16, S27–S33.

177. Chen, L.-W.; Yung, K.K.L.; Chan, Y.S. Neurokinin peptides and neurokinin receptors as potential therapeutic intervention targets of basal ganglia in the prevention and treatment of Parkinson's disease. *Curr. Drug Ther.* 2004, 5, 197–206.

178. Fox, S.H.; Henry, B.; Hill, M.P.; Peggs, D.; Crossman, A.R.; Brotchie, J.M. Neural mechanisms underlying peak-dose dyskinesia induced by levodopa and apomorphine are distinct: evidence from the effects of the alpha(2) adrenoreceptor antagonist idazoxan. *Movement Dis.* 2001, 16, 642–650.

179. Bezard, E.; Brotchie, J.M.; Gross, C.E. Pathophysiology of levodopa-induced dyskinesia: potential for new therapies. *Nat. Rev. Neurosci.* 2001, 2, 577–588.

180. Huang, Y.; Cheung, L.; Rowe, D.; Halliday, G. Genetic contributions to Parkinson's disease. *Brain Res. Brain Res. Rev.* 2004, 46, 44–70.

181. Mufson, E.J.; Ginsberg, S.D.; Ikonomovic, M.D.; DeKosky, S.T. Human cholinergic basal forebrain: chemoanatomy and neurologic dysfunction. *J. Chem. Neuroanat.* 2003, 26, 233–242.

182. Fisher, A.; Pittel, Z.; Haring, R.; Bar-Ner, N.; Kliger-Spatz, M.; Natan, N.; Egozi, I.; Sonego, H.; Marcovitch, I.; Brandeis, R. M1 muscarinic agonists can modulate some of the hallmarks in Alzheimer's disease: implications in future therapy. *J. Mol. Neurosci.* 2003, 20, 349–356.

183. Bodick, N.C.; Offen, W.W.; Levey, A.I.; Cutler, N.R.; Gauthier, S.G.; Satlin, A.; Shannon, H.E.; Tollefson, G.D.; Rasmussen, K.; Bymaster, F.P.; Hurley, D.J.; Potter, W.Z.; Paul, S.M. Effects of xanomeline, a selective muscarinic receptor agonist, on cognitive function and behavioral symptoms in Alzheimer disease. *Arch. Neurol.* 1997, 54, 465–473.

184. Lazareno, S.; Popham, A.; Birdsall, N.J. Progress toward a high-affinity allosteric enhancer at muscarinic M1 receptors. *J. Mol. Neurosci.* 2003, 20, 363–367.

185. Riemer, C.; Borroni, E.; Levet-Trafit, B.; Martin, J.R.; Poli, S.; Porter, R.H.P.; Bös, M. Influence of the 5-HT6 receptor on acetylcholine release in the cortex: pharmacological characterization of 4-(2-bromo-6-pyrrolidin-1-ylpyridine-4-sulfonyl)phenylamine, a potent and selective 5-HT6 receptor antagonist. *J. Med. Chem.* 2003, 46, 1273–1276.

186. Rogers, D.C.; Hagan, J.J. 5-HT6 receptor antagonists enhance retention of a water maze task in the rat. *Psychopharmacology.* 2001, 158, 114–119.

187. Foley, A.G.; Murphy, K.J.; Hirst, W.D.; Gallagher, H.C.; Hagan, J.J.; Upton, N.; Walsh, F.S.; Regan C.M. The 5-HT6 receptor antagonist SB-271046 reverses scopolamine-disrupted consolidation of a passive avoidance task and ameliorates spatial task deficits in aged rats. *Neuropsychopharmacology.* 2004, 29, 93–100.

188. Woolley, M.L.; Marsden, C.A.; Sleight, A.J.; Fone, K.C.F. Reversal of a cholinergic-induced deficit in a rodent model of recognition memory by the selective 5-HT6 receptor antagonist, Ro 04-6790. *Psychopharmacology.* 2003, 170, 358–367.

189. Lindner, M.D.; Hodges, D.B., Jr.; Hogan, J.B.; Orie, A.F.; Corsa, J.A.; Barten, D.M.; Polson, C.; Robertson, B.J.; Guss, V.L.; Gillman, K.W.; Starrett, J.E., Jr.; Gribkoff, V.K. Assessment of the effects of serotonin 6 (5-HT6) receptor antagonists in rodent models of learning. *J. Pharmacol. Exp. Ther.* 2003, 307, 682–691.

190. Schechter, L.E.; Dawson, L.A.; Harder, J.A. The potential utility of 5-HT1A receptor antagonists in the treatment of cognitive dysfunction associated with Alzheimer's disease. *Curr. Pharm. Design.* 2002, 8, 139–145.

191. Tran, P.B.; Miller, R.J. Chemokine receptors: signposts to brain development and disease. *Nat. Rev. Neurosci.* 2003, 4, 444–455.

192. Ragozzino, D.; Renzi, M.; Giovanneli, A.; Eusebi, F. Stimulation of chemokine CXC receptor 4 induces synaptic depression of evoked parallel fibers inputs onto Purkinje neurons in mouse cerebellum. *Neuroimmunology.* 2002, 127, 30–36.

193. Xia, M.; Hyman, B.T. GROα/KC, a chemokine receptor CXCR2 ligand, can be a potent trigger for neuronal ERK1/2 and PI-3 kinase pathways for tau hyperphosphoryaltion: a role in Alzheimer's disease? *J. Neuroimmunol.* 2002, 122, 55–64.

194. Ueta, Y.; Ozaki, Y.; Saito, J. Novel G-protein coupled receptor ligands and neurohypophysial hormones. *J. Neuroendocrinol.* 2004, 16, 378–382.

195. Ellis, C. The state of GPCR research in 2004. *Nat. Rev. Drug Disc.* 2004, 3, 577–626.

196. Ma, P.; Zemmel, R. Value of novelty? *Nat. Rev. Drug Disc.* 2002, 1, 571–572.

197. Sakurai, T.; Amemiya, A.; Ishii, M.; Matsuzaki, I.; Chemelli, R.M.; Tanaka, H.; Williams, S.C.; Richardson, J.A.; Kozlowski, G.P.; Wilson, S.; Arch, J.R.; Buckingham, R.E.; Haynes, A.C.; Carr, S.A.; Annan, R.S.; McNulty, D.E.; Liu, W.S.; Terrett, J.A.; Elshourbagy, N.A.; Bergsma, D.J.; Yanagisawa, M. Orexins and orexin receptors: a family of hypothalamic neuropeptides and G protein-coupled receptors that regulate feeding behavior. *Cell.* 1998, 92, 573–585.

198. Chemelli, R.M.; Willie, J.T.; Sinton, C.M.; Elmquist, J.K.; Scammell, T.; Lee, C.; Richardson, J.A.; Williams, S.C.; Xiong, Y.; Kisanuki, Y.; Fitch, T.E.; Nakazato, M.; Hammer, R.E.; Saper, C.B.; Yanagisawa, M. Narcolepsy in orexin knockout mice: molecular genetics of sleep regulation. *Cell.* 1999, 98, 437–451.

199. Lin, L.; Faraco, J.; Li, R.; Kadotani, H.; Rogers, W.; Lin, X.; Qiu, X.; de Jong, P.J.; Nishino, S.; Mignot, E. The sleep disorder canine narcolepsy is caused by a mutation in the hypocretin (orexin) receptor 2 gene. *Cell*. 1999, 98, 365–376.

200. Wise, A.; Jupe, S.C.; Rees, S. The identification of ligands at orphan G-protein coupled receptors. *Annu. Rev. Pharmacol. Toxicol*. 2004, 44, 43–66.

201. Pickel, V.M.; Chan, J.; Kash, T.L.; Rodriguez, J.J.; MacKie, K. Compartment-specific localization of cannabinoid 1 (CB1) and mu-opioid receptors in rat nucleus accumbens. *Neuroscience*. 2004, 127, 101–112.

202. Borowsky, B.; Adham, N.; Jones, K.A.; Raddatz, R.; Artymyshyn, R.; Ogozalek, K.L.; Durkin, M.M.; Lakhlani, P.P.; Bonini, J.A.; Pathirana, S.; Boyle, N.; Pu, X.; Kouranova, E.; Lichtblau, H.; Ochoa, F.Y.; Branchek, T.A.; Gerald, C. Trace amines: identification of a family of mammalian G protein-coupled receptors. *Proc. Natl. Acad. Sci. U.S.A*. 2001, 98, 8966–8971.

203. Lembo, P.M.; Grazzini, E.; Groblewski, T.; O'Donnell, D.; Roy, M.O.; Zhang, J.; Hoffert, C.; Cao, J.; Schmidt, R.; Pelletier, M.; Labarre, M.; Gosselin, M.; Fortin, Y.; Banville, D.; Shen, S.H.; Strom, P.; Payza, K.; Dray, A.; Walker, P.; Ahmad, S. Proenkephalin A gene products activate a new family of sensory neuron-specific GPCRs. *Nat. Neurosci*. 2002, 5, 201–209.

204. http://www.centerwatch.com/patient/drugs/drugls02.html (accessed July 2004).

8 Recombinant G Protein-Coupled Receptors for Drug Discovery

Kenneth H. Lundstrom

CONTENTS

ABSTRACT

Recombinant GPCR expression supports various needs in drug discovery. Over-expressed receptors can be applied for binding assays to screen chemical libraries as a means of finding hits and improving affinity and selectivity of synthesized compounds. Another application is to provide overexpressed GPCRs for structural characterization as the basis of structure-based drug design. All potential expression systems have been evaluated for GPCRs with mixed results. These include cell-free *Escherichia coli*-based translation systems, prokaryotic expression in *E. coli* and *Halobacterium salinarum*, and eukaryotic expression in yeast, insect and mammalian cells. Several GPCRs have been expressed at relatively high levels, in the range of 1 to 10 mg/L culture. Material from large-scale receptor production has been subjected to purification attempts and further structural biology characterization.

8.1 INTRODUCTION

Recombinant protein expression has been a central tool in basic and applied research for the past 20 years. Despite fairly good methods developed for both prokaryotic and eukaryotic expression of many soluble cytoplasmic proteins, integral transmembrane proteins have presented major challenges for obtaining high yields in every expression system. The low expression levels are mainly related to the topologies of these proteins. GPCRs, in particular, with their seven transmembrane (7TM) domains, have generated disappointingly low expression levels due to the requirements for folding and transport and the cellular toxicity related to heterogenous expression in membranes. Much effort has been dedicated to improving expression, and the efforts have evolved along the three main paths described below.

8.1.1 Evaluation of Various Expression Systems in Parallel

Expression systems including cell-free translation systems, prokaryotic and eukaryotic vectors have been used for the expression of many different GPCRs. These studies indicated that no universal system is suitable for high level expression of all GPCRs. See Figure 8.1.

8.1.2 Modification of Coding Regions of Target GPCR Genes

One effect of sequence alterations has been to enhance the transcriptional activity that might have a favorable outcome on expression levels. Additionally, deletions of the transmembrane regions have in several cases significantly improved expression levels. In the case of the metabotropic glutamate receptor 1 (mGluR1), the expression of only the extracellular domain generated very high yields and resulted in successful crystallization and a high resolution structure of the domain, which is known to contain the ligand-binding region.[1] However, for many GPCRs, the binding pocket is located within one or several of the transmembrane regions. Alternatively, site-directed and random mutagenesis techniques have been applied in attempts to improve the expressability of recombinant proteins.[2] Although single point mutations generally resulted in reduced levels or complete loss of expression,[3] certain mutations can be advantageous, leading to improved expression levels.[4]

8.1.3 Engineering of Fusion Proteins

Particularly for bacterial expression, fusion to prokaryotic proteins such as the maltose binding protein (MBP),[5] glutathione S-transferase (GST),[6] and bacteriophage PRD1 membrane protein[7] has produced favorable results because the fusion partner has been demonstrated to improve yields and provide more stable proteins.

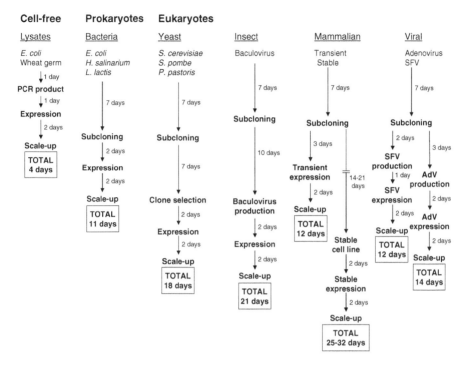

FIGURE 8.1 Procedure and time requirements for recombinant GPCR production in various expression systems. Key steps in different processes are shown.

8.1.4 RECEPTOR PRODUCTION FOR STRUCTURAL BIOLOGY APPLICATIONS

In this chapter, the various expression systems generally applied for overexpression of recombinant proteins are described. Most of these systems have also been used for GPCR expression and their special properties and achievements are presented below. Recombinantly expressed GPCRs provide the material for assays to monitor direct ligand binding activity as well as functional coupling to G proteins; both types of assays are used widely for drug screening programs.

As a general rule, high expression levels are desired for binding assays because high receptor densities in isolated membranes allow screening of a larger number of compounds. In contrast, functional assays seem to work better for lower receptor concentrations since the ratio of GPCRs to G proteins becomes more favorable. Because the drug discovery aspects of GPCRs are described elsewhere, the main focus of this chapter is on receptor production for structural biology applications.

8.2 CELL-FREE TRANSLATION

E. coli lysates containing coupled transcription–translation reaction reagents with continuous supplies of nucleotides, amino acids, and other essential components have been developed for cell-free translation.[8] The system applied for rapid translation of simple soluble proteins can generate yields of tens of milligrams.[9] Technology development has allowed the production of isotopically labeled proteins for nuclear magnetic resonance (NMR) and x-ray crystallography in cell-free translation systems.[10] It is possible to apply the technology directly on polymerase chain reaction (PCR) fragments, which enables the analysis of many constructs in parallel.[11] However, the cell-free translation system still has certain severe limitations. Expression of membrane proteins, and particularly GPCRs, has not been very successful.[12]

Attempts to overexpress the human β2 adrenergic receptor (β2 AR) resulted in extremely low binding activity compared to other recombinant systems (unpublished results, Dr. Mikako Shirouzu, Genomics Sciences Center, RIKEN Institute, Yokohama, Japan). Continuous technology development may result in a breakthrough; in particular, the cell-free system based on a wheat germ lysate may have the potential to be applied successfully to membrane proteins.[13] The addition or co-expression of chaperon proteins may enhance the production of membrane proteins.

8.3   *E. coli* EXPRESSION

Because of the simple and rapid procedure for expression and scale-up of recombinant protein production, GPCRs have been evaluated in *E. coli*. As bacterial cells lack many of the post-translational modification mechanisms in mammalian cells such as glycosylation, the functionality of GPCRs expressed in bacteria is of certain concern. Two alternative approaches involve expressing the recombinant receptors in the plasma membrane or in bacterial inclusion bodies. Both methods have stringent requirements for achieving success.

The insertion of foreign and especially mammalian receptors in bacterial membranes appears toxic to host cells and results in slow growth rates of bacteria, which significantly reduces yields. To overcome these problems, deletions have been introduced in the GPCR sequences and fusion constructs engineered.[5] Expression in bacterial inclusion bodies generally produces relatively high yields, but the requirement for refolding of receptors prior to purification and crystallography has hampered progress. Although some success has been achieved for the refolding as described below, it still represents a major bottleneck in structural biology.

One of the most successful GPCR expression studies in *E. coli* membranes was obtained for the rat neurotensin receptor (NTR).[5] The N-terminus of the NTR was deleted and replaced by the MBP and a C-terminal bio-tag (*in vivo* biotinylated) that interacts with avidin. The MBP–NTR fusion protein was expressed in a functional form in milligram quantities and was subjected to a two-step purification procedure consisting of a monomeric avidin column followed by a neurotensin affinity column. Purification of the rat MBP–NTR from *E. coli* produced milligram amounts of a functional receptor that was reconstituted in lipid vesicles and applied for solid state NMR (described in detail in Chapter 13) bound to a high-affinity neurotensin peptide agonist.[14]

In another study, the human adenosine A2a receptor (hA2aR) was overexpressed as a fusion protein with MBP in the inner membranes at levels of 10 to 20 nmol receptor/L culture.[15] A truncation at Ala316 at the C-terminus made the receptor protease resistant, although still functional, which allowed establishment of a three-step ligand affinity purification procedure of fully functional hA2aR. Additionally, when the human serotonin 5-HT1A receptor was fused to the signal sequence of the maltose binding protein MalE, active receptor expression was obtained in the inner bacterial membrane measured by ligand binding.[16]

Expression in bacterial inclusion bodies has been studied for both truncated and full-length GPCRs. The N-terminal extracellular domain of the human parathyroid hormone receptor 1 (PTHR1) was expressed at high levels in *E. coli* inclusion bodies.[17] A stable, soluble monomeric receptor was obtained after oxidative refolding for which ligand binding was demonstrated by surface plasmon resonance spectroscopy and isothermal titration calorimetry. Circular dichroism indicated that the refolded material contained secondary structures of approximately 25% α-helices and 23% β-sheets.

The extracellular region of the human glucagon-like peptide 1 receptor (GLP-1R) was expressed in bacterial inclusion bodies, for which renaturation was obtained from guanidinium-solubilized material.[18] After ion exchange chromatography and gel filtration, the activity of GLP-1R was measured by cross-linking, surface plasmon resonance, and isothermal titration calorimetry. The extracellular domain of the GLP-1R was purified and refolded and monitored for binding activity. Interestingly, high affinity binding was observed for the peptide antagonist ^{125}I-exendin-4 (9–39), whereas ^{125}I-labelled GLP-1 showed little binding, suggesting that the endogenous agonist GLP-1 might require the presence of extracellular loops and transmembrane regions.[19] In another study, 11 GPCRs were expressed with N-terminal GST- and C-terminal His tags in *E. coli* inclusion bodies at levels varying from 0.1 to 10% of the total cellular protein.[20] Refolding protocol technologies have been established for several GPCRs although the yields of activated receptors are relatively poor.[21]

The expression of the human leukotriene B4 (LTB4) receptor BLT1 in bacterial inclusion bodies was optimized by replacing the codons with the lowest abundance in the *E. coli* genome with the most frequent codons.[22] This procedure led to high expression levels but the receptor was insoluble and misfolded in bacterial inclusion bodies. The C-terminally tagged BLT1 receptor was purified by metal affinity chromatography in urea and refolding was obtained by removing urea from the Ni–NTA-bound material in the presence of the lauryldimethylamino oxide (LDAO) detergent. Circular dichroism analysis suggested a structure with approximately 50% α-helices. Further structural studies indicated that both the LTB4 agonist and the BLT1 receptor undergo conformational changes during their interaction.

8.4   OTHER PROKARYOTIC SYSTEMS

Halobacterium salinarum, a member of Archaebacteria, has a purple-colored plasma membrane that contains a complex of the bacterio-opsin protein (Bop) and its chromophore retinal in a 1:1 ratio.[23] This complex, known as bacteriorhodopsin, forms highly ordered two-dimensional structures for which a high resolution structure was determined.[24] Heterologous expression systems based on bacteriorhodopsin have been established for *H. salinarum*.[25] Introduction of the human muscarinic M1 receptor and the human serotonin 5-HT2C receptor into this expression system disappointingly produced no signals with Western blotting.[26] However, engineering of chimeric receptors between Bop and the yeast Ste2 receptor resulted in detectable expression levels.

Lactococcus lactis is a Gram-positive bacterium frequently applied to overexpression of integral membrane proteins.[27] The most frequently used system is based on the nisin NisA promoter and the two regulatory trans-acting factors known as NisR and NisK.[28] Most of the membrane proteins expressed to date are of prokaryotic origin, typically ABC transporters and major facilitator subfamily (MSF) efflux pumps.[27] Only a few eukaryotic targets have been expressed in *L. lactis*. The human Lys-Asp-Glu-Leu (KDEL) receptor with a 7TM topology has been expressed in a functional form although at moderate levels of less than 0.1% of the total membrane protein. In contrast, the yields for the 6TM yeast mitochondrial carrier proteins were in the range of 5%. No reports have been published for *L. lactis*-based GPCR expression but this approach may be worth investigating.

8.5   YEAST EXPRESSION

Various yeasts representing lower eukaryotic organisms are commonly used for recombinant protein expression. For instance, Baker's yeast, *Saccharomyces cerevisiae*, has been subjected to GPCR expression. The yeast α-factor Ste2p was expressed with C-terminal FLAG and His_6 tags at yields up to 1 mg.[29] Purified Ste2p receptor could be reconstituted into artificial phospholipid vesicles. Nevertheless, restoration of ligand-binding activity required the presence of solubilized yeast membranes. In addition, the human dopamine D1A receptor was successfully expressed and purified with a His_6 tag before reconstitution.[30]

Another yeast strain employed for recombinant protein expression is the fission yeast *Schizosaccharomyces pombe*, which has glycosylation patterns different from those of *S. cerevisiae*. Two types of *S. pombe* vectors have been engineered: for stable expression, a vector for chromosomal gene integration[31] and for transient expression, an episomal vector.[32] *S. pombe* is more similar to mammalian cells than other yeast strains possessing mammalian-like signal transduction systems.[11] The human dopamine D2 receptor overexpressed in fission yeast was localized to the plasma membrane and the yields were three times higher than in *S. cerevisiae*.[33]

The *Pichia pastoris* methylotrophic yeast has become very attractive as a host for recombinant protein expression because of its capacity to perform many post-translational modifications including glycosylation, disulfide bond formation, and proteolytic processing.[34] *P. pastoris* utilizes the methanol-regulated alcohol oxidase promoter I (AOX1) and the vector DNA can integrate as several copies in the host genome. Various GPCRs have been successfully expressed from these vectors.[35] Three-fold enhanced expression of the mouse serotonin 5HT5A receptor (m5HT5AR) and the human β2 AR was obtained after fusion to the prepropeptide sequence of the *S. cerevisiae* α-factor. Addition of the antagonist alprenolol to the culture medium increased the binding activity for the β2 AR. A similar effect was seen for the m5HT5AR in the presence of yohimbine. B_{max} values of 25 and 40 pmol/mg were obtained for β2 AR and m5-HT5AR, respectively.

The GPCRs expressed in *P. pastoris* showed pharmacological profiles similar to those observed in mammalian cells. Expression of 100 GPCRs in the MePNet structural genomics program (see Chapter 14) attained a success rate of >95% in *P. pastoris.* Expression optimization by temperature shifts, additives, and ligand supplements to culture media increased the B_{max} values to >100 pmol/mg for certain GPCRs. Finally, another methylotrophic yeast strain, *Hansenula polymorpha*, was evaluated as a host for chemokine receptor CCR5 and CXCR4 expression, but the yields were much lower than for *P. pastoris* (unpublished results, Dr. Renaud Wagner, University of Strasbourg, France).

8.6 INSECT CELLS

Insect cells obviously resemble mammalian cells in many aspects and are therefore attractive for recombinant protein production. Development of baculovirus vectors for heterologous gene expression has been very successful.[36] Generally, the strong polyhedrin promoter from *Autographa californica* has been applied for expression in several insect cell lines from organisms such as *Spodoptera frugiperda* (Sf9 and Sf21 cells) and *Trichoplusia ni* (Tni and High Five™ cells).[37] Insect cells possess many of the mammalian post-translational modifications, although certain differences between mammalian and insect cell pathways exist.

For example, N-glycosylation in insect cells is simpler and of high mannose type.[38] In addition to insect cells, baculovirus is also capable of infecting mammalian cells. In this context, replacement of the polyhedrin promoter with a CMV promoter allowed expression of recombinant proteins from baculovirus vectors also in mammalian cells.[39] However, one drawback of this approach was the requirement for very high virus concentrations — MOI (multiplicity of infection) levels in the range of 500 — which made large-scale production unfeasible.

Baculovirus vectors have been widely used for GPCR expression in insect cells resulting in high expression levels with B_{max} values in the range of 20 to 80 pmol/mg.[40] They include a large number of different GPCRs such as adrenergic, dopamine, muscarinic, opioid, tachykinin, and serotonin receptors. In attempts to further improve folding and transport to the plasma membrane and thus enhance expression levels, various signal sequences were fused to the N-termini of GPCRs. The mellitin sequence from *Apis melifera* improved the expression level of the human dopamine D2S receptor approximately two-fold.[41] Similar observations were made for the introduction of the influenza virus hemagglutinin signal sequence in front of the human β2 AR.[42]

The signal sequence from the baculovirus gp64 protein enhanced the expression levels of the human μ opioid receptor[43] and of various serotonin receptors.[44] Another approach is the engineering of the host insect cells to facilitate the folding process by co-expression of chaperon proteins such as BiP (immunoglobulin heavy chain binding protein) in the endoplasmic reticulum (ER) and calnexin residing in ER membranes.[45] Appropriate attention has also been paid to cell culture conditions. When Sf9 cells were grown in serum-free medium, yields for the human μ opioid receptor doubled.[43]

The locations of purification tags play crucial roles in relation to expression levels. Typically, engineering of His_6 tags at the N-termini of opioid receptors reduced the expression levels substantially, whereas the effect was less dramatic when introduced at the C-terminus.[46] Deletions in gene constructs were also shown to affect expression levels. A truncated turkey β-adrenergic receptor with deletions in both the N- and C-terminals to remove all N-glycosylation sites and an additional point mutation Cys116Leu improved yields more than five-fold in Tni cells.[47] This led to record high expression levels with B_{max} values of 360 pmol/mg that from large-scale production provided 10 mg of β-AR for crystallization.

As an alternative to using baculovirus for GPCR expression in insect cells, stable inducible expression has been developed under the control of the metallothionein promoter in *Drosophila* Schneider-2 cells. The expression levels of human opioid receptors have been relatively low, in the range of 20,000 to 30,000 receptors/cell.[48] However, fusion of GFP to the C-terminus allowed quantitative fluorescence intensity analysis, suggesting that the total receptor level was eight-fold higher than the diprenorphine binding data indicated. This may be due to problems with folding and transport of recombinant receptors. GFP quantification and monitoring could prove a versatile tool to develop further technology to improve expression levels in insect cells.

8.7 MAMMALIAN EXPRESSION

The optimal hosts for mammalian GPCR expression are obviously mammalian cells due to the capabilities of their more native post-translational modification mechanisms and the presence of mammalian G proteins for functional activity assessments. Transient expression has been achieved for many GPCRs, such as adrenergic, dopamine, muscarinic, serotonin, and somatostatin receptors with B_{max} values in the range of 10 to 20 pmol/mg. Although the expression levels have been acceptable, the transfection procedure has been inefficient and adaptation to large-scale production has been especially difficult.[50] Recent development of a modified calcium–phosphate co-precipitation method has generated yields of hundreds of milligrams of a monoclonal antibody in 100 L suspension cultures of HEK293 EBNA cells.[51] Although this bodes well for future receptor expression, the yields of the topologically more demanding GPCRs may continue to be less encouraging.

BHK, CHO, and HEK293 are the most frequently used cell lines for generation of stable expression of GPCRs. Stable expression in mammalian cells has been hampered by time-consuming processes and also by relatively low expression levels. However, stable overexpression of the human β2 AR in CHO and HeLa cell lines from vectors containing the gene for dihydrofolate reductase allowed a stepwise increase of methotrexate concentration, which substantially enhanced expression levels.[52] Another problem has been the concern of instability of stable constructs as there is, despite antibiotic selection, a pressure to remove integrated foreign gene sequences. To address this question, inducible expression vectors based on tetracycline regulation have been constructed.[53]

Another approach involved a cold-inducible expression system based on the Sindbis virus replicon that generates no expression at 37°C and high levels below 34°C.[54] Inducible systems have allowed expression of various GPCRs in the range of 5 to 20 pmol receptor/mg protein.

Viral vectors have proven useful due to their broad host ranges and strong promoters that often generate extreme expression levels. Adenovirus vectors have been applied for GPCR expression in rabbit ventricular myocytes, where binding activity in the range of 30 to 40 pmol/mg was obtained for the human $\beta2$ AR.[55] Poxviruses, especially the hybrid bacteriophage vaccinia virus vector with the T7 promoter, have been used widely.[56] Vaccinia-based GPCR expression of the neuropeptide Y (NPY) receptor generated 5 to 10 million receptors/cell.[57]

Probably the most frequently used heterologous viral gene expression system for membrane proteins is based on the Semliki Forest virus (SFV), a member of the single-stranded RNA alphaviruses. SFV vectors have been applied for more than 100 GPCRs and the expression levels in most cases were extremely high when measured by metabolic labeling and Western blotting.[58]

The binding activity of GPCRs on both whole cells and isolated membranes has been impressive, with B_{max} values up to 200 pmol/mg protein. The broad host range of SFV has allowed expression studies in parallel in different mammalian host cells. Interestingly, the human neurokinin-1 receptor (hNK1R) was best expressed in CHO-K1 cells,[59] whereas the μ opioid receptor showed the highest binding activity in the rat C6 glioma cell line.[60]

The optimal time of expression varied strongly from one receptor to another. A large number of mutants of hNK1R[61] and human dopamine D3 receptors (hD3R)[62] expressed from SFV vectors were analyzed for binding and functional activities in comparison to wild-type receptors. The results were verified to the molecular models for hNK1R and hD3R, respectively, based on the high resolution structure of bacteriorhodopsin. Interestingly, although the binding results for many mutants were in agreement with the postulated model, others suggested that the models need to be refined. The SFV system has been applied for large-scale receptor production in suspension cultures in fermenters and spinner flasks.[56]

Production of hNK1R as a fusion protein with the autocatalytic SFV capsid protein in CHO suspension cultures resulted in 10 mg/L of recombinant receptor. Purification of the C-terminally His$_6$-tagged hNK1R generated homogenous active receptor. Within the MePNet structural genomics program, 100 GPCRs were expressed from SFV with a success rate of 96% (see Chapter 14).

8.8 COMPARISON OF EXPRESSION SYSTEMS

As indicated in the introduction, no single expression system is ideal for all GPCRs. Table 8.1 summarizes the different expression systems and lists the advantages, disadvantages, and examples of GPCRs expressed. Generally, the cell-free expression system has provided only poor expression of GPCRs; the only examples to date are unpublished studies of human $\beta2$ AR. However, the simplicity and speed of cell-free translation and the flexibility that allows evaluation of expression from PCR products quickly and directly makes this an attractive approach. The system requires further development and the scale-up costs continue to be of concern. *E. coli* expression definitely has advantages in that it is simple and rapid to use and scale-up costs are reasonable. The possibility of isotope labeling for NMR and x-ray crystallography studies has some value. GPCRs have been difficult to express at

TABLE 8.1
Expression Systems for GPCRs

Vector	Advantages	Disadvantages	Target GPCR	Ref.
Cell-Free Systems				
E. coli; wheat germ	Rapid, PCR applicable	Expensive, for soluble proteins	Human β2 AR	[a]
Prokaryotes — Bacteria				
E. coli MB	Rapid, inexpensive, scaleable	Toxic to cells, no post-translational modifications Low yields	Human A2aR Rat NTR Human 5HT1AR	15 5 16
E. coli IB	Rapid, inexpensive, high yields, easy scale-up	Requires refolding	Human ΛPTHR1 Human ΛGLP-1 Human leukotriene BLT1	17 18, 19 20
Halobacterium salinarium	Rapid, scaleable, colorimetric expression	Cloning complicated, no post-translational modifications, fusion proteins	Human muscarinic M1R Human 5HT2CR Yeast Ste2R	26 26 26
Lactococcus lactis	Rapid, scaleable, inexpensive	No post-translation modifications	No GPCRs ABC transporters	27
Eukaryotes — Yeasts				
Saccharomyces cerevisiae	Relatively easy to use, high biomass, scaleable	Hyperglycosylation, clone selection, thick cell wall	Yeast Ste2R Dopamine D1AR	29 30
Schizosaccharomyces pombe	Relatively inexpensive	Thick cell wall	Dopamine D2R	32
Hansenula polymorpha	Inducible	Low yields	Chemokine receptors	[b]
Pichia pastoris	High yields	Thick cell wall	Human β2 AR 5HT5A receptor	34 34
Eukaryotes — Insect Cells				
Baculovirus	Mammalian-like glycosylation	More expensive, virus stock prep slow	Human NK1R Human β2 AR	40 40
Drosophila S2	GFP fusion	Time-consuming	Human opioid receptor	48
Eukaryotes — Mammalian				
Transient	Native processing	Expensive	Muscarinic M1, M4R	49
Stable	Inducible	Time-consuming, low yields	Human β2 AR	52

TABLE 8.1 (Continued)
Expression Systems for GPCRs

Vector	Advantages	Disadvantages	Target GPCR	Ref.
Eukaryotic — Viral, Mammalian				
Adenovirus	Broad host range	Slow virus production	β2 AR	55
Vaccinia virus	High expression	Safety concerns	NPY	57
SFV	Broad host range Extreme expression	Relatively expensive Safety concerns	>100 GPCRs	58

MB, membrane bound; IB, inclusion bodies; SFV, semliki forest virus.

[a] Unpublished results, M. Shirouzu, RIKEN Institute, Yokohama, Japan.
[b] Unpublished results, R. Wagner, University of Strasbourg, France.

satisfactory levels in bacteria due to the toxic effects they cause to host cells and the difficulties in refolding when expressed in inclusion bodies. Despite these problems, some success has been achieved for expression of hA2aR and rNTR in bacterial inner membranes and refolding of LT1R from inclusion bodies.

Nevertheless, these achievements have required many years of hard work, and it seems that the methods developed are relatively receptor specific and not easily applicable to a large number of GPCRs. One shortcoming of bacterial expression of mammalian GPCRs is the lack of appropriate post-translational modification capacity. For instance, many GPCRs are glycosylated, a mechanism not present in *E. coli*. On the other hand, glycosylation may not be required for successful crystallization and may even interfere with appropriate crystal formation.

Expression of mammalian recombinant proteins in various yeast strains has been very successful due to their eukaryotic status with post-translational mechanisms similar to those present in mammalian cells. However, yeast cells are quite different from their mammalian counterparts in respect to glycosylation patterns and hyperglycosylation is a typical phenomenon. GPCR expression has been achieved at relatively high levels in *S. cerevisiae* and *S. pombe*, but clearly *P. pastoris* generated the best expression levels. Several GPCRs have been expressed at high levels and clearly the very high biomasses obtained in fermenter cultures provide good bases for obtaining large quantities of receptors. The value of the purification procedure has to some extent been hampered by the difficulties in breaking the thick walls of yeast cells.

Insect cells have provided robust expression of GPCRs, especially by application of baculovirus vectors. The advantages include the presence of mammalian-like post-translational mechanisms, although certain differences exist between insect and mammalian cells. Because insect cell lines grow in semi-attached cultures, scale-up has been straightforward, albeit more expensive than for bacteria and yeasts. As previously described, record high binding activity could be obtained for the truncated non-glycosylated form of the turkey β-adrenergic receptor. Time will tell whether glycosylation requirements are essential in crystallography. The establishment of stable cell lines in *Drososphila* Schneider-2 cells looks also potentially promising.

Recombinant expression in mammalian cells is the closest approach to native mammalian GPCRs, but it has been hampered by low yields and time-consuming and expensive procedures. Improvements in transfection methods for transient expression and novel inducible vectors for stable expression have produced some progress. Alternatively, viral vectors, especially SFV vectors have provided both rapid high level expression in a broad range of host cells and an established method for large scale production. The drawbacks with SFV are the relatively high costs for virus production and the need to address safety concerns related to application of infectious, albeit replication-deficient, recombinant virus particles.

8.9 CONCLUSIONS

GPCR expression has been evaluated in all potentially applicable expression systems available today. Several examples demonstrate successful expression of several GPCRs to provide materials from bacteria, yeast, insect, and mammalian cells. Supplying drug screening programs with GPCRs for binding assays can be considered a routine process because the requirements are fairly easy to meet. The toughest task is to provide material for structural biology and particularly crystallography. As demonstrated, both the quantity and quality of expressed GPCRs play crucial roles. Other components such as the lipid compositions of membranes may be extremely important and these issues must be addressed in more depth. The need to develop additional technology to achieve success in determining and applying the structural biologies of GPCRs is clear.

REFERENCES

1. Kunishima N, Shimada Y, Tsuji Y, Sati T, Yamamoto M Kumasaka T, Nakakishi S, Jingami H, Morikawa K. Structural basis of glutamate recognition by a dimeric metabotropic glutamate receptor. Nature 2000; 407:971–977.
2. Lundstrom K, Ahti H, Ulmanen I. *In vitro* mutagenesis of rat catechol-o-methyltransferase. In: El-Gewely MR, Ed. Site-Directed Mutagenesis and Protein Engineering. Amsterdam: Elsevier Science Publishers, 1991:119–122.
3. Lundstrom K, Hawcock AB, Vargas A, Ward P, Thomas P, Naylor A. Effect of single point mutations of the human tachykinin NK1 receptor on antagonist affinity. Eur J Pharmacol 1997; 337:73–81.
4. Berthold DA, Stenmark P, Nordlund P. Screening for functional expression and overexpression of a family of diiron-containing interfacial membrane proteins using the univector recombination system. Protein Sci 2003; 12:124–134.
5. Tucker J, Grisshammer R. Purification of a rat neurotensin receptor expressed in *Escherichia coli*. Biochem J 1996; 317:891–899.
6. Sigrell JA, Heijbel A. Purification of GST fusion proteins, on-column cleavage, and sample clean-up. Sci World J 2002; 2:39–40.
7. Hanninen AL, Bamford DH, Grisshammer R. Expression in *Escherichia coli* of rat neurotensin receptor fused to membrane proteins from the membrane-containing bacteriophage PRD1. Biol Chem Hoppe Seyler 1994; 375:833–836.

8. Spirin AS, Baranov VI, Ryabova LA, Ovodov SY, Alakhov YB. A continuous cell-free translation system capable of producing polypeptides in high yield. Science 1988; 242:1162–1164.
9. Kim DM, Swartz JR. Prolonging cell-free protein synthesis by selective reagent additions. Biotechnol Prog 2000; 16:385–390.
10. Kigawa T, Muto Y, Yokoyama S. Cell-free synthesis and amino acid-selective stable isotope labeling of proteins for NMR analysis. J Biomol NMR 1995; 6:129–134.
11, Patzelt H, Goto N, Iwai H, Lundstrom K, Fernholz E. Modern methods for the expression of proteins in isotopically enriched form. In: Zerbe O, Ed. BioNMR in Drug Research. Weinheim: Wiley-VCH, 2003:1–38.
12. Klammt C, Lohr F, Schafer B, Haase W, Dotsch V, Ruterjans H, Glaubitz C, Bernhard F. High level cell-free expression and specific labeling of integral membrane proteins. Eur J Biochem 2004; 271:568–80.
13. Madin K, Sawasaki T, Ogasawara T, Endo Y. A highly efficient and robust cell-free protein synthesis system prepared from wheat embryos: plants apparently contain a suicide system directed at ribosomes. Proc Natl Acad Sci USA 2000; 97:559–564.
14. Luca S, White JF, Sohal AK, Filippov DV, van Boom JH, Grisshammer R, Baldus M. The conformation of neurotensin bound to its G protein-coupled receptor. Proc Natl Acad Sci USA 2003; 100:10706–10711.
15. Weiss HM, Grisshammer R. Purification and characterization of the human adenosine A2a receptor functionally expressed in *Escherichia coli*. Eur J Biochem 2002; 269:82–92.
16. Bertin B, Freissmuth M, Breyer RM, Schutz W, Strosberg AD, Marullo S. Functional expression of the human serotonin 5-HT1A receptor in *Escherichia coli*. Ligand binding properties and interaction with recombinant G protein alpha-subunits. J Biol Chem 1992; 267:8200–8206.
17. Grauschopf U, Lilie H, Honold K, Wozny M, Reusch D, Esswein A, Schafer W, Rucknagel KP, Rudolph R. The N-terminal fragment of human parathyroid hormone receptor 1 constitutes a hormone binding domain and reveals a distinct disulfide pattern. Biochemistry 2000; 39:8878–8887.
18. Bazarsuren A, Grauschopf U, Wozny M, Reusch D, Hoffmann E, Schaefer W, Panzner S, Rudolph R. In vitro folding, functional characterization, and disulfide pattern of the extracellular domain of human GLP-1 receptor. Biophys Chem 2002; 96:305–318.
19. Lopez de Maturana R, Willshaw A, Kuntzsch A, Rudolph R, Donnelly D. The isolated N-terminal domain of the glucagon-like peptide-1 (GLP-1) receptor binds exendin peptides with much higher affinity than GLP-1. J Biol Chem 2003; 278:10195–10200.
20. Kiefer H, Vogel R, Maier K. Bacterial expression of G-protein-coupled receptors: prediction of expression levels from sequence. Receptors Channels 2000; 7:109–119.
21. Kiefer H. *In vitro* folding of alpha-helical membrane proteins. Biochim Biophys Acta 2003; 1610:57–62.
22. Baneres JL, Martin A, Hullot P, Girard JP, Rossi JC, Parello J. Structure-based analysis of GPCR function:conformational adaptation of both agonist and receptor upon leukotriene B4 binding to recombinant BLT1. J Mol Biol 2003; 329:801–814.
23. Murata K, Mitsuoka K, Hirai T, Walz T, Agre P, Heymann JB, Engel A, Fujiyoshi Y. Structural determinants of water permeation through aquaporin-1. Nature 2000; 407:599–605.
24. Luecke H, Schobert B, Cartailler JP, Richter HT, Rosengarth A, Needleman R, Lanyi JK. Coupling photoisomerization of retinal to directional transport in bacteriorhodopsin. J Mol Biol 2000; 300:1237–1255.

25. Turner GJ, Reusch R, Winter-Vann AM, Martinez L, Betlach MC. Heterologous gene expression in a membrane-protein-specific system. Protein Expr Purif 1999; 17:312–23.

26. Winter-Vann AM, Martinez L, Bartus C, Levay A, Turner GJ. G protein-coupled expression in *Halobacterium salinarum*. In: Kuehne SR, de Groot H, Eds. Perspectives on Solid State NMR in Biology. Dordrecht: Kluwer, 2001:141–159.

27. Kunji ER, Slotboom DJ, Poolman B. *Lactococcus lactis* as host for overproduction of functional membrane proteins. Biochim Biophys Acta 2003; 1610:97–108.

28. de Ruyter PG, Kuipers OP, de Vos WM. Controlled gene expression systems for *Lactococcus lactis* with the food-grade inducer nisin. Appl Environ Microbiol 1996; 62:3662–3667.

29. David NE, Gee M, Andersen B, Naider F, Thorner J, Stevens RC. Expression and purification of the *Saccharomyces cerevisiae* alpha-factor receptor (Ste2p), a 7-trans-membrane-segment G protein-coupled receptor. J Biol Chem 1997; 272: 15553–15561.

30. Andersen B, Stevens RC. The human D1A dopamine receptor: heterologous expression in *Saccharomyces cerevisiae* and purification of the functional receptor. Protein Expr Purif 1998; 13:111–119.

31. Grallert B, Nurse P, Patterson TE. A study of integrative transformation in *Schizosaccharomyces pombe*. Mol Gen Genet 1993; 238:26–32.

32. Maundrell K, Hutchison A, Shall S. Sequence analysis of ARS elements in fission yeast. EMBO J. 1988; 7:2203–2209.

33. Sander P, Grunewald S, Bach M, Haase W, Reilander H, Michel H. Heterologous expression of the human D2S dopamine receptor in protease-deficient *Saccharomyces cerevisiae* strains. Eur J Biochem 1994; 226:697–705.

34. Cereghino JL, Cregg JM. Heterologous protein expression in the methylotrophic yeast *Pichia pastoris*. FEMS Microbiol Rev 2000; 24:45–66.

35. Weiss HM, Haase W, Michel H, Reilander H. Comparative biochemical and pharmacological characterization of the mouse 5HT5A 5-hydroxytryptamine receptor and the human β2-adrenergic receptor produced in the methylotrophic yeast *Pichia pastoris*. Biochem J 1998; 330:1137–1147.

36. Luque T, O'Reilly DR. Generation of baculovirus expression vectors. Mol Biotechnol 1999; 13:153–163.

37. Possee RD. Baculoviruses as expression vectors. Curr Opin Biotechnol 1997; 8:569–572.

38. Jarvis DL, Finn EE. Biochemical analysis of the N-glycosylation pathway in baculovirus-infected lepidopteran insect cells. Virology 1995; 212:500–511.

39. Condreay JP, Witherspoon SM, Clay WC, Kost TA. Transient and stable gene expression in mammalian cells transduced with a recombinant baculovirus vector. Proc Natl Acad Sci USA 1999; 96:127–132.

40. Massotte D. G protein-coupled receptor overexpression with the baculovirus–insect cell system: a tool for structural and functional studies. Biochim Biophys Acta 2003; 1610:77–89.

41. Grunewald S, Haase W, Reilander H, Michel H. Glycosylation, palmitoylation, and localization of the human D2S receptor in baculovirus-infected insect cells. Biochemistry 1996; 35:15149–15161.

42. Guan XM, Kobilka TS, Kobilka BK. Enhancement of membrane insertion and function in a type IIIb membrane protein following introduction of a cleavable signal peptide. J Biol Chem 1992; 267:21995–21998.

43. Massotte D, Pereira CA, Pouliquen Y, Pattus F. Parameters influencing human mu opioid receptor overexpression in baculovirus-infected insect cells. J Biotechnol 1999; 69:39–45.

44. Brys R, Josson K, Castelli MP, Jurzak M, Lijnen P, Gommeren W, Leysen JE. Reconstitution of the human 5-HT(1D) receptor-G-protein coupling: evidence for constitutive activity and multiple receptor conformations. Mol Pharmacol 2000; 57:1132–1141.

45. Ailor E, Betenbaugh MJ. Modifying secretion and post-translational processing in insect cells. Curr Opin Biotechnol 1999; 10:142–145.

46. Massotte D, Baroche L, Simonin F, Yu L, Kieffer B, Pattus F. Characterization of delta, kappa, and mu human opioid receptors overexpressed in baculovirus-infected insect cells. J Biol Chem 1997; 272:19987–19992.

47. Warne T, Chirnside J, Schertler GF. Expression and purification of truncated, non-glycosylated turkey beta-adrenergic receptors for crystallization. Biochim Biophys Acta 2003; 1610:133–140.

48. Perret BG, Wagner R, Lecat S, Brillet K, Rabut G, Bucher B, Pattus F. Expression of EGFP-amino-tagged human mu opioid receptor in *Drosophila* Schneider-2 cells: a potential expression system for large-scale production of G-protein coupled receptors. Protein Expr Purif 2003; 31:123–132.

49. Migeon JC, Nathanson NM. Differential regulation of cAMP-mediated gene transcription by m1 and m4 muscarinic acetylcholine receptors: preferential coupling of m4 receptors to Gi alpha-2. J Biol Chem 1994; 269:9767–9773.

50. Gu H, Wall SC, Rudnick G. Stable expression of biogenic amine transporters reveals differences in inhibitor sensitivity, kinetics, and ion dependence. J Biol Chem 1994; 269:7124–7130.

51. Girard P, Derouazi M, Baumgartner G, Bourgeois M, Jordan M, Wurm FM. 100 liter transient transfection. In: Lindner-Olsson E, Chatissavidou E, Lillau E, Eds. Animal Cell Technology: From Target to Market. Dordrecht: Kluwer Academic Publishers, 2001; 37–42.

52. Lohse MJ. Stable overexpression of human beta 2-adrenergic receptors in mammalian cells. Naun Schmiede Arch Pharmacol 1992; 345:444–451.

53. Fux C, Moser S, Schlatter S, Rimann M, Bailey JE, Fussenegger M. Streptogramin- and tetracycline-responsive dual regulated expression of p27(Kip1) sense and anti-sense enables positive and negative growth control of Chinese hamster ovary cells. Nucleic Acids Res 2001; 29:E19.

54. Boorsma M, Nieba L, Koller D, Bachmann MF, Bailey JE, Renner WA. A temperature-regulated replicon-based DNA expression system. Nat Biotechnol 2000; 18:429–432.

55. Drazner MH, Peppel KC, Dyer S, Grant AO, Koch WJ, Lefkowitz RJ. Potentiation of beta-adrenergic signaling by adenoviral-mediated gene transfer in adult rabbit ventricular myocytes. J Clin Invest 1997; 99:288–296.

56. Ward GA, Stover CK, Moss B, Fuerst TR. Stringent chemical and thermal regulation of recombinant gene expression by vaccinia virus vectors in mammalian cells. Proc Natl Acad Sci USA 1995; 92:6773–6777.

57. Walker P, Munoz M, Combe MC, Grouzmann E, Herzog H, Selbie L, Shine J, Brunner HR, Waeber B, Wittek R. High level expression of human neuropeptide Y receptors in mammalian cells infected with a recombinant vaccinia virus. Mol Cell Endocrinol 1993; 91:107–112.

58. Lundstrom K. Semliki Forest virus vectors for rapid and high-level expression of integral membrane proteins. Biochim Biophys Acta 2003; 1610:90–96.

59. Lundstrom K, Schweitzer C, Rotmann D, Hermann D, Schneider EM, Ehrengruber MU. Semliki Forest virus vectors: efficient vehicles for *in vitro* and *in vivo* gene delivery. FEBS Lett 2001; 504:99–103.
60. Lundstrom K, Henningsen R. Semliki Forest virus vectors applied to receptor expression in cell lines and primary neurons. J Neurochem 1998; 71 (Suppl.):S50D.
61. Lundstrom K, Hawcock AB, Vargas A, Ward P, Thomas P, Naylor A. Effect of single point mutations of the human tachykinin NK1 receptor on antagonist affinity. Eur J Pharmacol 1997; 337:73–81.
62. Lundstrom K, Turpin MP, Large C, Robertson G, Thomas P, Lewell XQ. Mapping of dopamine D3 receptor binding site by pharmacological characterization of mutants expressed in CHO cells with the Semliki Forest virus system. J Recept Signal Transduct Res 1998; 18:133–150.

9 High Throughput Screening Assays for G Protein-Coupled Receptors

Usha Warrior, Sujatha Gopalakrishnan, Jurgen Vanhauwe, and David Burns

CONTENTS

ABSTRACT

High throughput screening has become a necessary part of drug discovery for identifying lead molecules for specific disease targets. The pharmaceutical industry invests considerable effort and resources in screening compounds against G protein-coupled receptors (GPCRs) because they are considered sources of attractive therapeutic targets. Formats used for identifying modulators of GPCR function range from simple receptor–ligand-binding assays to high content screening platforms. Each format has its own advantages and disadvantages and the decision to utilize a specific screening methodology depends on the availability and costs of reagents and equipment as well as the expected properties of lead molecules. As the understanding of signal transduction and GPCR regulation increases, new assay formats that provide improved sensitivity, ease of use and increased throughput become available. Most of the screening technologies available to date rely on classic GPCR mechanisms such as ligand binding, GTP binding, cAMP formation, and cytosolic Ca^{2+} influx. The introduction of high content imaging technology makes it now possible to screen compounds and elucidate the detailed molecular mechanisms involved with GPCR trafficking.

9.1 INTRODUCTION

G protein-coupled receptors (GPCRs) are members of the best known family of validated pharmaceutical research drug targets. More than 50% of approved drugs obtain their therapeutic effects by selectively targeting members of this family. Therefore, the pharmaceutical industry invests considerable resources into studying GPCR family members because they are justifiably perceived as sources of attractive therapeutic targets. Depending on targets and disease models, compounds that display agonist, antagonist, or allosteric properties can be sought.

Because the binding and signaling characteristics of GPCRs have been addressed elsewhere, this chapter will focus on current technologies available for investigating compounds that modulate the behavior of GPCRs. Indeed, over the last two decades high throughput screening (HTS) has become a very important tool in the discovery of lead compounds for all target classes including GPCRs. New technology developments have enabled screening to take great strides in recent years, evolving from high throughput to ultra-high throughput modes.

The competitive environment for the discovery of new drugs for disease targets necessitated the development of cost-effective and reliable high throughput screening strategies in order to accelerate the process of identifying lead compounds.[1] Although the industry standard today is the use of 384-well plates, several pharmaceutical companies are developing 1536- or 3456-well-plate screens and screens without wells to increase their daily screening throughput.[2–4] The applicability of GPCR assays to miniaturization will also be discussed throughout this chapter.

9.2 SOURCE MATERIAL AND ASSAY DIVERSITY

Historically, drug discovery focused on native tissues which involved assays such as neurotransmitter or hormone release measurements and radioactive binding assays. Researchers now have a plethora of reagents and assays available to screen for selective lead molecules. As sources of receptors, researchers can choose native tissue, recombinant or nonrecombinant cell lines (see Chapter 8). Issues such as specificities of native tissues due to expression of multiple receptor subtypes, difficulties in obtaining large amounts of native tissues, inabilities to achieve sufficiently high expression levels, and nonhuman pharmacologies have led most screening laboratories to utilize recombinant cell lines.[5]

In most cases, receptor expression levels in nonrecombinant cell lines will be too low for effective use in screening assays (<1 pmol/mg). Because recombinant cell lines usually express receptors of interest at high levels, they are more suitable for HTS. For pharmacological reasons, some researchers prefer utilizing human cell lines instead of nonhuman cell lines [e.g., human embryonic kidney (HEK293) versus Chinese hamster ovary (CHO cells)]. The use of insect of cell lines in a functional assay for GPCRs has also been reported.[6]

If the receptor of interest is unavailable, screening with a native tissue or non-recombinant cell line as a receptor source may be the most practical or only alternative. To overcome the inability to screen with nonrecombinant cell lines due to low endogenous expression levels, certain techniques have been developed to allow for boosting receptor expression without the use of cDNA. Zinc finger proteins (ZFPs) that recognize novel DNA sequences serve as the basis of a technology platform developed by Sangamo Biosciences (Richmond, California) that can be used to upregulate GPCR gene expression at a level that allows the cell line to be used for screening ligands without the use of exogenous cDNA.[7] GeneDriver® technology (Invitrogen, Carlsbad, California) is based on homologous recombination between the endogenous target GPCR gene and a synthesized corresponding DNA fragment that contains a strong cytomegalovirus (CMV) promoter.[8,9]

Whether to use a cell-based or membrane-based assay is a matter of personal choice that depends predominantly on the equipment availability for HTS and on the assay technology. Ligand binding should be measured in membranes, whereas calcium signaling is primarily measured in live cells using kinetic readouts [(e.g., fluorescence imaging plate reader (FLIPR)] or gene reporter assays. For pharmacological reasons, it is important to note that differences in potency can be seen when utilizing membrane- or cell-based assays.[10] These differences can be explained by changes in the environment of the receptor during preparation of membranes from whole cells. Also, compound interferences are different in membranes compared to cell-based assays; especially when compound concentration is reduced due to avid uptake in the cells or binding to serum components. Other compounds can elicit cytotoxic cell responses and/or interfere with downstream signaling when utilizing a detection method distant from receptor binding.

9.3 HIGH THROUGHPUT SCREENING

The development of robust and sensitive assays for finding inhibitors or allosteric modulators of ligand binding and/or receptor activity is critical to screening that will allow the discovery of new leads. Traditional screening methods used membrane preparations and specific, radio-labeled ligands. Compounds that competed with a labeled ligand were identified after the separation of the unbound radioactivity by filtration. Even though receptor–radioligand binding screens are robust, they do not provide information on the intrinsic activities of a compound; all active compounds are identified only as inhibitors of the ligand–receptor interaction in these primary screens.

Biochemical or cell-based functional assays constitute information-rich platforms capable of distinguishing agonists, antagonists, partial agonists, inverse agonists, allosteric modulators, and allosteric enhancers (potentiators). They can identify novel leads with desirable properties for a therapeutic target. Quality control is an important consideration when implementing an HTS campaign. The robustness of the HTS assay is measured by the Z′ factor determined by analyzing multiple wells of total binding and nonspecific binding.[11] The Z′ factor is calculated using the following equation:

$$Z' = 1 - (3\sigma_s + 3\ \sigma_C)/\mu_s - \mu_C$$

where σ_s = total binding mean standard deviation; σ_C = nonspecific binding mean standard deviation; μ_s = mean of total binding; and μ_C = mean of nonspecific binding.

The Z′ factor determines the inherent noise of the assay. If the Z′ factor is good (>0.5), further development will be conducted to ensure a robust screen:

1. The affinity of the known benchmark compounds will be determined and compared with the published values.
2. As the test compounds are solubilized in dimethyl sulfoxide (DMSO), the sensitivity of the assay to DMSO will be ascertained in the presence of different concentrations of DMSO.
3. Drug concentration to be used in the final screening will be determined by analysis of scatter plots produced with varying compound concentrations; this includes determining the Z factor.

The Z factor, similar to the Z′ factor, is a screening window coefficient calculated by analyzing wells containing drugs and wells defined for nonspecific binding. The screening concentration is chosen so that the Z factor produced is in the range of an excellent assay and the final DMSO concentration in the well does not negatively impact the quality of the assay. Usually a Z factor of 0.5 to 1 is considered an excellent assay; 0 to 0.5 is a "doable" assay.

After the primary screen, the inhibitory wells from the primary screens are retested and IC_{50} and EC_{50} values are determined after performing a concentration-dependent titration. Screening libraries are usually maintained as DMSO-solubilized liquid stocks and often stored for convenience at room temperature. Several recent studies indicate that these storage conditions produce precipitation and partial degradation of compounds.[12–14] In light of these studies, many screening laboratories consider stability as well as throughput and are converting their stores into inert low temperature facilities. It is always advisable to reconfirm the activities of the powder sources and confirm the structures of the compounds.

9.4 RECEPTOR BINDING ASSAYS: GENERAL CONSIDERATIONS

Receptor binding assays have become increasingly popular since the introduction of multiple filtration platforms for automation. Recent advances have also enabled researchers to perform receptor binding assays in a homogeneous format so that no additional separation steps are required. The selection of an ideal radioligand for a binding assay depends on ligand stability, specific activity and selectivity; higher values for all of these properties are preferred. The affinity of the ligand for the receptor also dictates the choice of the radio label.

Suppliers such as Amersham Biosciences (Piscataway, New Jersey) and Perkin-Elmer (Boston, Massachusetts), sell iodinated and tritiated ligands for several GPCRs. If the receptor density is high and the ligand has high affinity for the receptor, either ^{125}I- or ^{3}H-labeled ligands can be used. If the receptor density is high and the ligand has low affinity, a ^{3}H-labeled ligand is a better choice, especially for scintillation proximity assay (SPA). If the receptor density is low and a high affinity ligand is available, a ^{125}I- labeled ligand is better for screening. Typical expression levels above 50,000 receptors per cell are required for ^{125}I-labeled ligands and 10-fold higher receptor densities are required for ^{3}H-labeled ligands.

Another important step is deciding whether to use an agonist or antagonist ligand for an HTS campaign. Indeed, it is generally accepted that GPCRs can exist in low- and high-affinity states, depending on whether they interact with G proteins.[15,16] While this has been a topic of discussion for many years, a favorable explanation may be that GPCRs are in a high affinity state when interacting with a guanosine nucleotide-free G protein. Agonists bind with a higher affinity to a high affinity state of a receptor, whereas antagonists bind with equal affinity to either state of a receptor.

There are exceptions to this generalization, for example, ergoline derivatives such as LSD and lisuride are serotonergic or dopaminergic agonists that do not readily distinguish between low- and high-affinity states of these receptors. In whole cell binding assays, agonists activate the receptors and can lead to homologous or heterologous desensitization of receptors. Also, agonists tend to label only portions of the receptors. As a result, the signal may be too low, especially when the receptor is expressed at a low level. Labeled antagonists are generally preferred in HTS receptor binding assays when the option is available unless the biology requires an agonist screen.

9.4.1 HETEROGENEOUS RECEPTOR BINDING ASSAYS

Heterogeneous binding assays can be performed with radio-labeled or lanthanide-labeled ligands. Virtually every heterogeneous binding assay utilizes a filtration step to remove unbound labeled ligand. This technology has the advantage of removing compounds that could interfere with the detection of the bound tracer. The main disadvantage of this nonhomogeneous assay is the difficulty in automating the separation step, which results in lower throughput of the screen.

Fluorescent labeling of ligands can prove quite a challenge especially when small molecules are involved; maintaining functional activity or affinity after labeling requires additional testing. A limited number of europium-labeled ligands including galanin, bombesin, neurokinin-A, neurotensin and substance P are available from PerkinElmer for use with its DELFIA® assay platform. DELFIA is a heterogeneous time-resolved fluorometric assay method in which an enhancement step assures high sensitivity and wide response range.

After binding equilibrium of receptor and ligand is attained, the contents of the wells are filtered into an Acrowell® (Pall Life Sciences, Ann Arbor, Michigan), a patented low fluorescent background membrane. An enhancement solution is added to the wells and the fluorescence is measured in a time-resolved fluorescence reader. This assay format eliminates the need to use radioactive ligands although few receptor ligands tagged with europium are available for HTS, resulting in very limited use of this technology.

9.4.2 HOMOGENEOUS RECEPTOR BINDING ASSAYS

Filtration assays are nonhomogeneous, difficult to automate, and the throughput is considerably reduced compared to homogeneous formats. We will first discuss homogeneous radioactive binding assays because they represent the largest diversity of assays for screening GPCRs.

9.4.2.1 Homogeneous Radioactive Binding Assays

Examples of multiple platforms for homogeneous radioactive binding assays include SPA® beads (Amersham Biosciences), LEADseeker® beads (Amersham Biosciences), FlashBlue® beads (PerkinElmer), and FlashPlate® (PerkinElmer). All these formats are scintillation proximity assays (SPAs). In this type of assay, an isotope-labeled ligand is brought into close proximity to a scintillant embedded in beads or plate plastics by binding to a receptor immobilized onto the beads or plates.[17]

SPA is a simple assay format and is well suited to automation because it requires no separation step. The scintillant in SPA beads is incorporated into fluomicro-spheres. SPA beads are made of either polyvinyl toluene (PVT) or yttrium silicate (Ysi). Beads are available with several surface coatings to immobilize membranes including PVT–WGA (wheat germ agglutinin), Ysi–WGA, PVT–PEI (polyethylen-imine)–WGA, and Ysi–poly-D-lysine. Beads coated with WGA bind cell membranes through surface carbohydrates.

Beads coated with poly-D-lysine, which rely on the negative charge of the membrane, are also used for certain receptors. The choice of bead type depends on the nature of the radioligand. If the ligand is a glycoprotein, WGA-coated SPA beads should be avoided because they interact with the ligand and the background noise in the assay will increase. Similarly, poly-D-lysine-coated beads should not be used with negatively charged ligands.

The capacity of WGA beads for crude membranes is approximately 10 to 30 µg membrane protein/mg bead; 1 mg of poly-D-lysine-coated beads can bind 10 µg membrane protein. Ysi beads are denser than PVT and PS (polystyrene) beads and settle very quickly in aqueous buffers; care must be taken when pipetting to ensure the even distribution of solids. Assays using Ysi beads must be shaken in order to facilitate equilibrium. SPA beads are typically added to the assay pre-coupled with membranes. Alternatively, beads can be added together with receptor and ligand or they can also be added after the reaction has reached equilibrium.

SPA beads emit light in the blue region of the emission spectrum between 400 and 450 nm and the light is measured using a standard photomultiplier tube-based reader such as a PerkinElmer TopCount® or MicroBeta®. An SPA-based receptor binding assay should be allowed to reach equilibrium before counting is performed and this results in a longer incubation period. Because the incubation time is fairly long, care should be taken during assay development to ensure the stability of the assay components.

Unfortunately, colored compounds can cause a reduction in counting efficiency during screening and this produces quenching of the signal. This color quenching is in addition to chemical quenching (interference in the process of energy transfer between solvent and fluorescent molecules) and physical quenching (orientation of bound labeled ligands is far away from bead). Quench correction protocols that use quenching agents have been developed, but these protocols do not fully correct for this problem.

An improvement over the first generation of SPA bead is the LEADseeker® bead of Amersham Biosciences.[18] LEADseeker beads are optimized to read in CCD-based imagers such as Amersham's LEADseeker or the Viewlux® of PerkinElmer.

Yellow and red compounds do not quench the light emitted in the red region. Blue compounds can absorb the red light output, but the number of blue compounds in the compound libraries is relatively rare — and hence less of an issue for screening. The LEADseeker beads consist of two types, one composed of polystyrene (PS) and the other of yttrium oxide (Yox). For receptor binding assays, these beads typically have wheat germ agglutinin coatings. Assay development using LEADseeker beads is similar to the traditional SPA format. Importantly, the final readout for these beads is in a CCD-based imaging platform, so reading takes less time than required for a PMT-based system. Plates containing LEADseeker beads must be dark-adapted before reading to eliminate the phosphorescence innate to the plastic material of the plates. Also, the LEADseeker system is much more amenable to miniaturization and assays can be easily converted into 1536-well plates.

FlashPlates (PerkinElmer) are similar to SPA beads but instead of incorporating the scintillant directly into beads, it is incorporated into the plastic of the plate well. Available plates contain 96 or 384 wells with different coatings, similar to SPA beads. PerkinElmer also provides ScreenReady Target® FlashPlates for certain receptors, where the receptor is already immobilized onto the FlashPlate. ScreenReady Targets are ready-to-use, fully validated microtiter plate assays for homogeneous GPCR receptor binding assays using ^{125}I or ^{3}H radioligands. Also like SPA imaging beads, imaging flash plates (Imageplate®) are available from PerkinElmer. Note that these homogeneous formats are not suitable for high throughput screening if both receptor density and ligand affinity are low.

9.4.2.2 Homogeneous Non-Radioactive Binding Assays

Fluorescent assays offer the advantage of eliminating the use of radioactivity. Peptide and protein ligands are most suitable for fluorescent labeling, whereas maintaining a functional fluorescently labeled small molecule can prove quite a challenge. Nonetheless, certain fluorescent small molecules have been synthesized (e.g., muscarinic receptor ligands are available from Invitrogen). A wide range of fluorophores is available for labeling, for example, BODIPY, Alexa and Cy dyes. BODIPY and Cy dyes are hydrophobic in nature, whereas Alexa dyes tend to be more hydrophilic. The physicochemical properties of these dyes can be important for maintaining functional activities of ligands.

Fluorescent applications for receptor binding involve fluorescence intensity, fluorescence polarization, and time-resolved fluorescence resonance energy transfer (TR-FRET). In some HTS labs, FMAT® (Applied Biosystems, Foster City, California) is also utilized for measuring receptor binding in 96-, 384-, or 1536-well plates.[19] FMAT is a fluorescence macroconfocal, biological binding technology that enables mix-and-read assays with live cells and beads.

The most common nonradioactive format for HTS receptor binding assays is fluorescence polarization (FP) — a homogeneous format in which the ligand is labeled with a fluorescent tag. For each FP data point, two separate intensities (parallel and perpendicular) are measured and the polarization signal is calculated.[20] All polarization values are expressed in millipolarization units (mP). The mP values of the FP assay are calculated based on the following equation:

$$mP = 1000 * [(I_S - I_P)/(I_S + I_P)]$$

where I_S is the parallel emission intensity measurement and I_P is the perpendicular emission intensity measurement. FP assays have a major advantage in that they eliminate the need for radioactivity. Another advantage of FP assays is that they involve no need to immobilize or capture receptors to a solid surface as in the case for SPA and flash plate formats.

However, one major disadvantage of the FP format for receptor binding assays is the difficulty in maintaining ligand affinity after labeling with fluorophores; specifically, the labeling can induce unwanted steric hindrance in receptor–ligand pairing. As mentioned earlier, not all ligands can be fluorescently tagged, particularly the small molecular weight antagonists and agonists commonly used in binding assays.

The receptor density should be relatively high (>1 pmol/mg) for a successful FP assay. Thus, nonrecombinant membrane receptor preparations are not feasible in this format. Also, in the case of green fluorescent tags, the cellular components present in the membrane preparation significantly increase the background, and colored or fluorescent compounds in the screening collection can contribute to the noise in the assay and/or false positives and negatives.

Labeling the ligand with red-shifted fluorophores such as rhodamine and red Alexa dyes can potentially alleviate this problem.[21] Far-red dyes with excitation and emission spectra near the infrared region have been introduced recently and may offer several advantages compared to green fluorescence for HTS.[22] Appropriate filters, dichroic mirrors and instrument settings should be checked before running any fluorescent assay; these are one of the most common factors for the failure in fluorescent assays and can be easily circumvented.

TABLE 9.1
Receptor Binding Assays in High Throughput Screening

Receptor Binding Assay	Radioactive	Homogeneous	Compound Interference	Throughput	Cells or Membranes?
Filtration	Y	N	Low	Low	C + M
SPA beads	Y	Y	Medium	Medium	M
FlashPlate	Y	Y	Medium	Medium	M
FlashBlue beads	Y	Y	Medium	Medium	M
LEADseeker beads	Y	Y	Low–medium	High	M
Imageplate	Y	Y	Low–medium	High	M
Fluorescence polarization	N	Y	Low–high	High	C + M
FMAT	N	Y	Low–high	Low	C + M
TRF	N	N	Low	Low	M
TR-FRET	N	Y	Low–medium	High	M

Y, yes; N, no; C, cells; M, membranes.

Finally, TR-FRET can be employed to measure receptor binding. TR-FRET utilizes time resolution to reduce background interference. It is based on long-lived fluorescence from lanthanides such as terbium (Tb) and europium (Eu) that can be measured after the short-lived background fluorescence has disappeared. The lanthanide-labeled ligand serves as the donor whereas a fluorophore-tagged WGA derivative binds to the membrane and serves as an acceptor for the emission of the lanthanide.

9.5 FUNCTIONAL GPCR ASSAYS

Whether inactive GPCRs are pre-coupled to G proteins and how GPCRs can activate heterotrimeric G proteins are issues subject to a lot of speculation. In general, activated GPCRs stimulate heterotrimeric G proteins in membranes in a manner that involves GDP release (a rate-limiting step) and then GTP binding. GTP binding to the α subunit of the G protein leads to a conformational change in its switch regions and promotes dissociation of the G protein complex into α- and $\beta\gamma$-subunits. Both entities then have the ability to stimulate or inhibit downstream effectors.[23,24]

The α-subunit defines the G protein subtype; the 16 different α-subunits are classified into four distinct families: G_s, $G_{i/o}$, G_q, and $G_{12/13}$.[24–26] $G\alpha_s$ activates adenylate cyclase (AC) and therefore leads to an increase in cAMP synthesis; $G\alpha_{i/o}$ inhibits AC. $G\alpha_q$ activates PLCβ and increases intracellular calcium levels and PKC activity. Heterotrimeric G_{12} and G_{13} have been shown to mediate signals from GPCRs to Rho GTPase activation and to proteins from the ERM family.[27]

In addition to recognizing effectors for specific subtypes of G proteins, great efforts made to identify novel effectors for G proteins led to discoveries of new effector proteins such as K^+ channels, Rap1GAP, src tyrosine kinase, Bruton tyrosine kinase, and others. However, their roles and receptor-mediated activation mechanisms of some of these effector proteins are still unclear.[28] Functional HTS assays mainly target well-characterized pathways such as GTP binding, cAMP, and Ca^{2+} assays. When GPCRs are overexpressed, they can display ligand-independent activities (constitutive or basal activities) that can be measured using second messenger or reporter gene assays. Different assay technologies now available allow ligand occupancy of GPCRs to be converted into robust functional assay signals as described in the next section.

9.5.1 GTP BINDING ASSAYS

A GTPγS binding assay detects the level of G protein activation following agonist stimulation. In the resting stage, G proteins exist as $G\alpha(GDP)_{\beta\gamma}$ heterotrimers with GDP bound to Gα; there is a low rate of GDP–GTP exchange. Subsequent to agonist stimulation of the receptor, GDP dissociates from Gα, which allows GTP to bind to Gα. Agonist-mediated GPCR activation increases the rate of guanine nucleotide exchange and allows the dissociated G protein subunits to interact with effector proteins to produce functional responses. In the GTPγS binding assay, occupancy of an agonist to a receptor can be measured as an increase in the binding of the nonhydrolyzable GTP analog [^{35}S] GTPγS to cell membranes. Generally, in the GTPγS binding assay, [^{35}S] GTPγS replaces the endogenous GTP and binds to the Gα subunit to form Gα[^{35}S] GTPγS. [^{35}S] GTPγS binding assays are best suited for G_i-coupled receptors because G_i proteins are expressed at higher levels and have higher GDP–GTP exchange rates than other G proteins.[29]

A critical aspect of the [^{35}S] GTPγS binding assays is the suppression of basal binding of the labeled nucleotide to unstimulated G proteins using high concentrations of GDP. The amount of GDP needed for optimal agonist stimulation of [^{35}S] GTPγS binding varies across several receptor systems. For the agonist-induced responses at muscarinic M1 and M3 receptors expressed in CHO or human embryonic kidney cells, low levels (0.1 µM) of GDP are required. In membranes from the same cell type, however, muscarinic M2 and M4 receptor-stimulated GTP binding requires 10-fold higher levels of GDP.[30,31] Therefore, a user should titrate concentrations of GDP ranging from 0.1 to 10 µM for each membrane receptor.

GTP binding is always performed in presence of magnesium and sodium ions and the concentrations of these ions should be optimized for individual assays. Mg^{2+} increases both basal and agonist-stimulated [^{35}S] GTPγS binding and has a preferential effect on the activated state that results in higher signals.[32,33] Typically Mg^{2+} ions are kept at a 1 to 10 mM range.

Na^+ ions decrease the basal [^{35}S] GTPγS binding and can improve the signal-to-noise ratio.[34] Na^+ ions exert their effects through two possible mechanisms: allosteric regulation of receptor properties and changing the affinity of G proteins for GDP. In some systems, the concentration of sodium ions is inversely related to agonist efficacy. An agonist can behave partially or fully at high and low sodium concentrations, respectively. Therefore, care should be taken in fixing the concentration of sodium in the assay. Typically, sodium concentrations are optimized in the range of 0 to 200 mM. Some membranes may require mild detergents such as saponin in order for the GTP to reach the G protein.

Optimization of membrane protein reactions should be tested to achieve maximal stimulation with the agonist, but this varies with the receptor expression level in the membrane preparation. The increased background observed with G_s- and G_q-coupled receptors could be reduced by utilizing C-terminal anti-$G_{\alpha s}$ or anti-$G_{\alpha q}$ antibodies to immunoprecipitate the [^{35}S] GTPγS-bound α-subunit or to bring it in close proximity to the SPA beads. Another approach is to fuse the G_s or G_q receptor of interest to G_i to couple the activation of G_i indirectly to make the assay feasible in GTPγS-binding format.[35] In homogeneous assay formats, the receptor preparation is immobilized onto WGA-coated SPA beads or onto the surfaces of FlashPlates.[36] Results from the conventional GTPγS binding assay correlated well with FlashPlate and SPA formats; therefore, the application of these homogeneous methods would increase throughput for HTS campaigns using the GTPγS binding assay.

A nonradioactive GTP binding assay with a europium-labeled GTP derivative has been developed for a variety of GPCRs. This format requires a filtration step to remove the unbound Eu-GTP, and therefore its use for primary screening may be rather limited. The amount of Eu-GTP bound to the receptor is retained on the filter and detected by time-resolved fluorometry.[37] GTPγS binding assays can be used to search for candidate compounds for known, orphan, and constitutively activated GPCRs. Because this method measures the signal earliest in receptor-mediated events, it is not susceptible to feedback or amplification mechanisms that occur in downstream signaling pathways. However, as [^{35}S] GTPγS does not cross the cell membrane, this assay cannot be used to evaluate receptor activation in intact cells.

9.5.2 cAMP Assays

Production of cAMP is regulated through AC — an effector enzyme whose activity can be modulated by different GTP binding proteins. Ten different isoforms of AC have been identified, based on primary sequence, tissue distribution, and regulation.[38] Forskolin, a diterpene, can be used to directly activate AC to increase cAMP production. AC enzymes can be regulated by a variety of input signals (Ca^{2+}, calmodulin, $G_{\alpha S}$, $G_{\alpha i}$, PKC) and should be considered as signal integrators. Subsequent to G_s-coupled receptor activation, an increase in AC activity converts ATP to cAMP and inorganic pyrophosphate. G_i-coupled receptor activation results in the inhibition of AC activity. G_s proteins activate all isoforms of AC, whereas G_i proteins only inhibit a subset of isoforms. Therefore, measurement of G_i-coupled receptor-mediated inhibition of AC can prove challenging.

Many methods are available to determine the concentrations of cAMP in cells or membranes; the easiest is the detection of cAMP formation. Alternatively, one could assess AC activity rather than cAMP production. The adenine nucleotides of whole cells can be labeled with [^{3}H] adenine and then the amount of radiolabeled cAMP present in cells can be determined. The same assay can be conducted using cell membranes or homogenates to assess the conversion of radiolabeled [^{32}P] ATP to cAMP. These methods employ column chromatography to isolate the radiolabeled cAMP product. They are tedious and not suitable for HTS applications. cAMP detection methods in HTS and its advantages and disadvantages have been extensively reviewed by Williams.[39] Although a large variety of cAMP detection technologies employ antibodies, a number of available assays do not rely on antibodies. The sensitivity of a kit that relies on antibody detection is determined by the detection technology and the quality of the antibody. Antibody-free cAMP technologies such as gene reporter assays and Ca^{2+}-based assays are extensively used in HTS labs.

9.5.2.1   General Principles of cAMP Assays Identifying Active Compounds

For $G_{\alpha s}$-coupled receptors, an agonist will stimulate production of cAMP, while an antagonist will inhibit agonist-induced cAMP production. For $G_{\alpha i}$-coupled receptors, the effect of an agonist whose binding results in the inhibition of AC cannot easily be measured because the basal cAMP is too low. Cells are first stimulated to produce cAMP with forskolin. An agonist will inhibit forskolin-induced cAMP production and this is measurable. $G_{\alpha i}$-coupled receptor antagonist: cells are stimulated to produce cAMP with forskolin. The antagonist will block the effect of an agonist acting on the receptor and therefore forskolin-induced cAMP formation is re-established.

9.5.2.2   Antibody-Dependent cAMP Detection Assays

These assays are based on the competition between the cellular cAMP and the labeled cAMP with the anti-cAMP antibody. To achieve the best results possible from cAMP assays, it is essential to determine the optimal number of cells per well. The optimal cell concentration should produce a relatively low basal level of cAMP, which will yield a good response to a relatively low concentration of stimulator. When evaluating the inhibitory compounds, the level of stimulator will be as low as possible to maximize the sensitivity of the assay. Homogeneous formats are preferred in HTS due to higher throughput and ease of automation. These kits can determine cAMP levels in cell supernatants in microtiter plate formats using adherent or suspension cells; some of them even work in cell membranes. The results for suspension cells are generally more reproducible because each well contains approximately the same number of cells. All the kits include cAMP standards. A concentration response curve for the cAMP standard is performed to calculate the amounts of cAMP in the unknown samples. Plotting the bound fraction (B/B0) as a function of log cAMP concentration generates a standard curve. There is no linear relationship between the raw data and the amount of cAMP produced. Therefore, one could anticipate a shift in the affinities of compounds based on the analysis criteria (whether the researcher is analyzing data based on raw signal or converted cAMP levels). cAMP assays are typically performed in the presence of a phosphodiesterase inhibitor, IBMX, which prevents the breakdown of cAMP. It also prevents the identification of PDE inhibitors as false positives in a screen for a G_s-coupled receptor agonist.

9.5.2.2.1   Homogeneous Radiometric Assays
Radioactive assays such as PerkinElmer FlashPlates[40,41] and Biotrack cAMP kits from Amersham[42,43] are widely used HTS formats (see Section 9.4.1). The FlashPlate

technology utilizes a FlashPlate coated with anti-cAMP antibodies that are blocked to avoid nonspecific binding. Cells from tissue culture flasks are directly placed in 96- or 384-well plates and treated with compound or buffer for 30 min. The cells are lysed with lysis buffer containing [^{125}I] cAMP tracer and incubated for 3 h. cAMP bound to the antibody can be detected in close proximity to the scintillation surface on the flash plate. The Biotrack cAMP kit is a scintillation proximity assay that involves the addition of SPA beads, anti-cAMP antibody, and [^{125}I] cAMP tracer to lysed cell samples. The sensitivity of the assay could be improved by acetylation of standards and unknowns prior to assay. Although the Biotrack is an easy method, it can generate false positives caused by color quenching.

9.5.2.2.2　Fluorescent Polarization Assays

Fluorescence polarization (FP) cAMP assays (PerkinElmer and Amersham Biosciences) rely on measurements of the parallel and perpendicular components of fluorescence emission from a fluorescently tagged cAMP molecule such as fluorescein to the plane of a polarized excitation source.[44–46] The rotation speed of the labeled cAMP is reduced and therefore possesses a higher polarization value when it is bound to antibody. In the presence of cellular cAMP, labeled cAMP is free in solution and a lower polarization value is observed. An advantage of the FP format is that it enables the utilization of membranes instead of whole cells for cAMP assay. One major disadvantage of the format is the potential interference from compounds as discussed earlier. Alternate probes such as BODIPY TMR, Alexa 647 (PerkinElmer), MR 121, Evoblue, Cy3, and Cy5 could eliminate these artifacts. For example, the Biotrak cAMP FP immunoassay system utilizes Cya3B, a bright version of the standard Cy3 fluorescent dye conjugated to cAMP, to generate a signal window twice that achievable with conventional dyes such as fluorescein.[46]

9.5.2.2.3　Time-Resolved Fluorescence Assays

The HTRF® cAMP assay (Cis-Bio International, France) is based on the interaction of europium cryptate anti-cAMP antibody and a modified allophyocyanin-labeled cAMP (XL665).[47,48] In the absence of cellular cAMP, these two fluorescent molecules are in close proximity to allow fluorescence resonance energy transfer (FRET) and the XL665 emits a long-lived fluorescence at 665 nm. In the presence of cellular cAMP, XL665 will be displaced and will not be in close proximity to the europium-labeled antibody to allow FRET; only emission from the europium will be detected. Because the signal is measured as a ratio of europium emission at 620 nm and XL665 emission at 665 nm (620/665), the assay is unaffected by colored compounds.

9.5.2.2.4　AlphaScreen® (Amplified Luminescent Proximity Assay)

The PerkinElmer AlphaScreen is an assay based on the competition between cellular cAMP and biotinylated cAMP.[49] Anti-cAMP antibody-coated acceptor beads capture biotinylated cAMP to form a sandwich with streptavidin-coated donor beads that bring donor and acceptor beads into close proximity. Upon laser excitation, a photosensitizer in the streptavidin-coated donor bead converts oxygen to a more excited singlet state. The singlet state oxygen molecules react with thioxene derivative in the anti-cAMP-conjugated acceptor beads, generating a highly amplified signal in the 520- to 620-nm range. The beads are sensitive to temperature and light fluctuations and the experiments should be performed with caution.

9.5.2.2.5 Enzyme Complementation Technology

The HitHunter® assay from DiscoveRx (Sunnyvale, California) utilizes β-galactosidase fragment complementation. Inactive fragments, enzyme acceptor (EA), and enzyme donor (ED) complement to form an active enzyme. In the EFC-cAMP (enzyme fragment complementation) assay, the β-galactosidase donor fragment-cAMP (ED-cAMP) conjugate binds with the EA fragment to form an active β-galactosidase enzyme. Binding of ED-cAMP conjugate to the anti-cAMP antibody prevents its complementation with the EA fragment to form the active enzyme. The amount of β-galactosidase formed is thus proportional to the cAMP concentration in the cell lysate. Enzyme activity is subsequently detected using a chemiluminescent or fluorescent substrate.[50,51] This technology is less prone to compound artifacts and easily amenable to miniaturization and automation.

9.5.2.2.6 Electrochemiluminescence Technology

Electrochemiluminescence technology available from MesoScale Discovery (Gaithersburg, Maryland) uses cAMP labeled with a ruthenium derivative. In the absence of cellular cAMP, labeled cAMP is bound by an anti-cAMP antibody on the surface of a disposable carbon electrode. When a potential is applied to the electrode, labeled cAMP in close proximity to the electrode emits light. One disadvantage of the MesoScale technology is that the plate cannot be read more than once due to significant reduction in signal. Also, the limited availability of specifically designed instruments to read this technology has hindered its acceptance in the HTS field.

All the homogeneous methods offer the advantages of easy automation and feasibility for a robust HTS screen. However, the sensitivity of the assay to detect cAMP can vary from one technology to another and a researcher must select a format based on the level of sensitivity needed and the automation and miniaturization capabilities available in his or her laboratory. The nonhomogeneous formats such as the DELFIA® assay of PerkinElmer, the chemiluminescence assay of PE Biosystem, and the Catchpoint cAMP kit from Molecular Devices (Sunnyvale, California) are also in limited use, primarily because of the washing steps that cannot be easily automated. Compounds that interact with the cAMP antibody can appear as false positives in all the cAMP detection assays. However, performing the assay in the absence of cells can easily identify these false positives.

9.5.2.3 Antibody-Free Methods for Measuring cAMP

9.5.2.3.1 Reporter Gene Assays

The most common reporter genes utilized in assays for cAMP are based on the expression of β-lactamase or luciferase. These assays are based on a reporter construct consisting of promoter and reporter genes that are stably incorporated into the genome of the cell (or transiently transfected). The activity of the reporter gene is controlled by the transcriptional regulation of the promoter element, which is activated directly by the receptor present on the cell membrane. For example, receptors coupling to G_s activation have usually been monitored with reporter constructs containing cAMP-responsive elements (CREs) Upon stimulation of the G_s-coupled receptors, the elevated levels of cAMP lead to activation of protein kinase A (PKA). The catalytic subunit of PKA translocates to the nucleus where it phosphorylates and activates the cAMP response element binding protein (CREB) bound to an upstream CRE.[52,53] Reporter gene assays are well suited for high-throughput screening approaches in various assay formats investigating full or partial agonists, antagonists, or inverse agonists.

Other reporter systems available include β-galactosidase, β-glucuronidase (GUS), secreted alkaline phosphatase (SEAP), human growth hormone (hGH), green fluorescent protein (GFP), and chloramphenicol acetyl transferase (CAT). All these reporter gene assays have inherent advantages and disadvantages.[54]

Overall, reporter gene assays are comparable to second messenger assays in sensitivity but are typically much more amenable for HTS. In order to analyze reporter gene activity at a single cell level, β-galactosidase and β-lactamase have been used.[55] The β-lactamase assay is simple and rapid and does not pose any background problems due to the lack of endogenous β-lactamase activity in mammalian cells. β-Lactamase assays based on the calcium-sensitive transcription factor, nuclear factor of activated T-cell (NFAT), are used to study $G\alpha_q$, $G\alpha_q$ chimeric proteins, and promiscuous $G\alpha_{15/16}$ G-proteins. β-Lactamase assays utilize the fluorescent probe CCF2/4, which is composed of two fluorescent dyes, 7-hydroxycoumarin-3-carboxamide and fluorescein. Once CCF2/4-AM enters the cell, endogenous esterases cleave the ester to CCF2/4, which produces a green fluorescence due to the FRET at an emission wavelength of fluorescein at 530 nm. This provides the additional benefit of detecting cytotoxic compounds that inhibit esterase activity in cells. A lower fluorescence will be observed because the ester groups quench the fluorescence of CCF2/4.

Cleavage of CCF2/4 by β-lactamase causes the loss of FRET, resulting in blue fluorescence ($\lambda = 460$ nm). The signal is detected by using the ratio of two fluorescence wavelengths, rather than absolute changes in the fluorescence signal. This significantly reduces assay variability. Other substrates for β-lactamase are being generated to provide additional flexibility.[56]

GPCRs transduce signals from cell surfaces to intracellular effectors through various G proteins. As with other G proteins, a reporter gene assay could be used to characterize the coupling of GPCRs to $G_{12/13}$. The GPCR under the control of a modified serum-responsive element (SRE) is transfected into a cell lacking $G_{q/11}$ and the agonist-induced activation of GPCR is measured using a luciferase assay.[57] A longer incubation period is generally required to produce transcriptional regulation of the gene, which will help the detection of weak agonists in the screen. On the other hand, longer incubation times may lead to unmasking the inherent toxicity of the compounds. A potential interaction with other steps in the signaling cascade can also result in the generation of false positives in the screen.

9.5.2.4 cAMP Biosensor

A cAMP biosensor HTS platform known as ACT:One® is available from Atto Biosciences (Rockville, Maryland) to detect cAMP in live cells. The GPCR of

TABLE 9.2A
Functional GPCR Assays in High Throughput Screening

GTP Binding	Radioactive	Homogeneous	Compound Interference	Throughput	Cells or Membranes?	Deorphanizing
[35S]-GTPγS	Y	Y/N	Medium	Medium	M	N
Eu-GTP	N	N	Low	Low	M	N
cAMP	Y	Y	Y	Y	C or M	Y
FlashPlate	Y	Y	Medium	Medium	C + M	N
BioTrack	Y	Y	Medium	Medium	C + M	N
Fluorescence polarization	N	Y	Low–high	High	C + M	N
TRF	N	N	Low	High	C + M	N
AlphaScreen	N	Y	Medium	Medium	C + M	N
HTRF	N	Y	Low–medium	High	C + M	N
HitHunter	N	Y	Low	High	C + M	N
ECL	N	Y	Medium	High	C + M	N
β-Lactamase reporter	N	Y	Low–high	High	N	Y
Luciferase reporter	N	Y	Low–medium	Medium–high	N	Y
cAMP sensor	N	Y	Medium	Low	N	?

Y, yes; N, no; C, cells; M, membranes.

TABLE 9.2B
Functional GPCR Assays in High Throughput Screening

Ca²⁺ Assay	Radioactive	Homogeneous	Compound Interference	Throughput	Cells or Membranes?	Deorphanizing
Fluorophore	N	Y/N	Medium	Medium	C	Y
β-Lactamase reporter	N	Y	Low–high	High	C	Y
Luciferase reporter	N	Y	Low–medium	High	C	Y
Aequorin	N	Y	Low	Medium	C	Y
IP3 assay	Y	Y	Y	Y	C or M	Y
AlphaScreen	N	Y	Medium	Medium	C	N
HitHunter	N	Y	Low	High	C	N
Amersham	Y	N	Medium	Low	C	N
Desensitization	Y	Y	Y	Y	C or M	Y
Transfluor	N	Y	High	Low	C	Y
BRET	N	Y	Medium	Medium	C	N
CypHer	N	Y/N	High	Medium	C	N
Other	Y	Y	Y	Y	C or M	Y
Microphysiometer	N	N	Low	Low	C	Y
Melanophore	N	Y	Medium	High	C	Y

Y, yes; N, no; C, cells; M, membranes.

interest is stably expressed in a cell line coexpressing cyclic nucleotide gated (CNG) channels that act as cAMP sensors targeted to plasma membranes and colocalized with AC to permit sensitive detection of a local cAMP rise. A change in cAMP level can be detected using either calcium-sensitive or membrane potential dyes.[58] This platform is applicable to G_s- and G_i-coupled receptors. This technology allows screening of receptors not coupled by chimeric and promiscuous G proteins.

9.5.3 INTRACELLULAR CALCIUM ASSAYS

The cell-based calcium flux assay is considered one of the most important screening techniques used in pharmaceutical drug discovery. G_q-coupled receptor-mediated increases in intracellular Ca^{2+} release can be measured by calcium-sensitive fluorescent dyes, bioluminescent indicators, or reporter gene assays. G_q-coupled receptor activation can elevate a transient increase in intracellular calcium to 400 to 1000 nM from a basal cytosolic level of 10 to 100 nM. Most common calcium-sensitive dyes such as Fluo-3 and Fluo-4 (Invitrogen) used in HTS are single wavelength compounds that result in a 100-fold increase in fluorescence when bound to calcium.

The increase in intracellular calcium is measured by FLIPR® technology (Molecular Devices Corporation, Sunnyvale, California). FLIPR was designed to perform homogeneous, kinetic, cell-based fluorometric assays for both adherent and nonadherent cell lines. FLIPR is equipped with an argon ion laser that excites a fluorescent dye. A cooled CCD camera images the entire plate and detects the emitted light. The device has the advantages of delivering compounds to 96- or 384-well plates and integrates data signals over a time interval.[59,60] Other readers available to measure calcium transients include the ImageTrack (PerkinElmer) and FDSS 6000 (Hamamatsu Photonics, Hamamatsu, Japan). The advantages and disadvantages of each instrument should be carefully evaluated before an investment is made in these expensive instruments.

Typically, cells are seeded in clear-bottom black plates overnight. It is often necessary to seed weakly adherent cells into poly-D-lysine-coated plates to improve adherence and minimize peeling off of cells during reagent addition. The cells expressing the GPCR of interest are loaded with the acetoxymethyl ester of Fluo-4 (or Fluo-3, its analog) followed by repeated washings to remove the extracellular dye. Upon cellular uptake, the acetoxymethyl ester will be cleaved by esterases to liberate Fluo-4. Upon binding of calcium, the fluorescence of Fluo-4 increases, which can be detected at excitation and emission wavelengths of 488 nm and 520 nm, respectively. The increase in fluorescent response is directly correlated with the increase in cytosolic calcium levels. In FLIPR, the emission wavelength is limited to 530. The availability of a next generation of FLIPR, FLIPR TETRA, with tunable wavelength, enables the utilization of dyes that emit in the red region for calcium assays. This new FLIPR is also equipped with liquid handlers for 1536-well plates.[61]

Typically in an agonist assay, data are recorded for every 1 sec for 1 min, followed by a 5-sec exposure for another 2 min. GPCR-induced calcium responses reach a peak in less than 20 sec and the duration of the response will be finished by 3 min. For an antagonist assay, cells are loaded with compounds followed by the addition of agonists in FLIPR. Some cells are serum sensitive, resulting in fluctuations of intracellular calcium that may interfere with the results. Similarly in an antagonist assay, if the agonist of choice is present in the serum, it could change the affinity of the agonist. For example, LPA is present in serum in micromolar range. Growing cells in low serum media or media containing heat-inactivated serum would help to achieve full efficacy and affinity of LPA to endothelial differentiation gene receptor (Gopalakrishnan et al., unpublished observations).

Cells containing anionic transporters (e.g., CHO cells) should be loaded with probenecid to prevent pumping out of the dye during and after loading. FLIPR kits such as calcium assay and the calcium 3 assay from Molecular Devices allow for a homogeneous format by eliminating washing steps associated with the Fluo-3/Fluo-4 format, resulting in easier automation and faster throughput.

In the new calcium kits, the background fluorescence is reduced by the addition of a dye that absorbs the excitation light of the fluorescent dye and/or its emission light without affecting the fluorescence readout from the cells. This eliminates background fluorescence, and the signal evoked from the cells can be detected with higher resolution. A researcher should carefully examine the utilities of different dyes for a particular target because these dyes could exhibit altered pharmacologic profiles with a target. For chemokine receptors and other growth factor receptor targets, calcium 3 assay kits resulted in robust signals and are preferred over Fluo-4 and calcium assay kits.

Intracellular calcium assays offer several advantages for HTS, including the affinity and efficacy of compounds. They can also distinguish full agonist, partial agonists, and antagonists in a single assay setup. However, the characterization of the potencies of agonists and antagonists could be complicated due to the transient nature of the functional response. Activation of a G_q protein-coupled receptor revealed transient kinetics and the maximal response was observed in fewer than 20 sec, which could lead to a nonequilibrium condition during assessment of the potency of an antagonist. Therefore, a competitive antagonist could exhibit a pseudo-non-competitive behavior if incubated with the cells before adding agonist. Kaler et al.[62] reported significant differences in the potencies of some of the serotonin antagonists in HEK-5HT2A cells and -5HT2C cells in co-addition of antagonists along with agonist versus preincubation of antagonists with the cells using an automated flow-through fluorescence analysis system (similar to FLIPR technology in many aspects). A fast binding antagonist displays similar affinity in both addition formats; a slow binding antagonist exhibits a lower potency in co-addition compared to pre-addition. Furthermore, spiperone exhibited a classical competitive behavior with serotonin receptors in a co-addition procedure versus a noncompetitive behavior in a preincubation format. Therefore, a detailed experimental setup is needed for mechanistic evaluation of agonist–antagonist interactions in intracellular calcium assays.

GPCRs coupled to $G_{\alpha s}$ or $G_{\alpha i/o}$ proteins can be linked to PLCβ to generate an increase in intracellular calcium when coexpressed with chimeric $G_{\alpha qi/o5}$ or $G_{\alpha qs5}$ proteins. These chimeric proteins can be generated by replacement of five C-terminal amino acids of $G_{\alpha q}$ with the corresponding $G_{\alpha i/o}$ or $G_{\alpha s}$ residues.[63] The receptors do not activate all the chimeric G proteins equally and therefore it is important to test multiple chimeras to determine the best system for the target of interest. Alternatively, receptors can be coexpressed with certain promiscuous G proteins (i.e., $G\alpha_{15}$,

$G\alpha_{16}$) that are capable of interacting with a variety of G_s-, $G_{i/o}$-, and G_q-coupled receptors and signaling through calcium release.[64]

The promiscuous $G\alpha_{16}$ G-protein and the chimeric G proteins are broadly used tools for setting up robust assays for HTS.[63–65] Many orphan receptors have been deorphanized through a $G\alpha_{16}$ approach.[66] The ability of $G\alpha_{16}$ to function as a universal adapter protein for GPCRs by coupling to PLCβ allows a large number of receptors to generate an increase in intracellular calcium independent of the second messenger pathway normally modulated by the receptor. Therefore, the receptor of interest and the other endogenous receptors present in the parental cell line could couple to $G\alpha_{16}$.

One difficulty with using a cell line containing $G\alpha_{16}$ for an HTS campaign is this artificial coupling of other nontarget endogenous receptors through $G\alpha_{16}$. All the compounds that interact with nontarget receptors in cells will be identified as positives in the primary screening. Thus, testing all the positives in the parental cell line with $G\alpha_{16}$ is critical for eliminating the false positives. Again, not all GPCRs can be forced to couple to $G\alpha_{15}$ or $G\alpha_{16}$. For example, it has been reported that receptors such as dopamine D3, angiotensin ATII, somatostatin SST1, muscarinic acetylcholine M1, tachykinin NK2, melatonin MT1C, CCR1 and CCR2, and adrenergic α1C and α1D do not couple efficiently to $G\alpha_{16}$.[68–70] Liu et al. reported the application of a $G\alpha_{16/z}$ chimera to G_i-coupled receptors for higher sensitivity and promiscuity compared to $G\alpha_{16}$. The $G\alpha_{16/z}$ chimera was constructed by replacing 25 or 44 amino acids of the C-terminal domain of $G\alpha_{16}$ with those of $G\alpha_z$. These chimeras were found to be more sensitive for G_i-coupled receptors, although their ability to link G_s proteins to PLCβ and calcium was diminished.[71] Again, researchers should carefully consider the chimeric/promiscuous $G\alpha_{16}$ approach because it may alter the pharmacology profiles of agonists and antagonists.

Recent reports demonstrate that the magnitudes of agonist responses can be improved dramatically by performing the assay at elevated temperature instead of at room temperature. Wong et al. reported the enhanced responses of many neuronal receptors such as α1A and α2A adrenergic, histamine H1, serotonin 5HT1A and 5HT2A, and dopamine D2 and D3 at higher temperatures when coexpressed with $G\alpha_{15}$ to make it amenable for calcium assay format.[72]

Other instruments beside FLIPR have fully integrated robotic capabilities with 96- or 384-well injectors for measuring kinetic readouts for calcium assays including the FDSS of Hamamatsu and the ImageTrack of PerkinElmer. During the last decade, the sizes of corporate screening libraries have increased tremendously, in essence forcing researchers to miniaturize their assays from 96- to 384- to 1536- or 3456-well plates.

Scientists at Abbott reported a microarrayed compound screening format (μARCS) for identifying agonists for a GPCR.[73] In this format, 8,640 discrete compounds are spotted and dried onto a polystyrene sheet that occupies the same footprint as a 96-well plate. The cells expressing the receptor of interest were pre-loaded with Fluo-4, cast into a 1% agarose gel, placed above the compound sheets, and imaged successively using a CCD system. In addition to the μARCS format, gene reporter assays can be employed to increase the throughputs of calcium assays, and β-lactamase assays have been miniaturized to 3456-well plates. One advantage of gene reporter assays over calcium fluorophores is their ability to detect weak calcium signals from GPCRs either because the signal is amplified or the reporter gene construct can be activated through the PKC pathway. Indeed, G_q-mediated activation of PLCβ leads to generation of two types of second messengers: inositol-triphosphate (IP3) will stimulate calcium release, whereas diacylglycerol will stimulate PKC activity which activates NFAT transcription through calcineurin.

Intracellular calcium can also be measured using a bioluminescent indicator known as aequorin. The aequorin complex consists of the 22,000-dalton apoaequorin protein, molecular oxygen, and a coelenterazine luminophore. When calcium binds to this complex, the coelenterazine is oxidized to coelenteramide resulting in the release of carbon dioxide and blue light at 468 nm. The utilization of aequorin for GPCRs has been characterized in detail by Ungrin et al.[74] Euroscreen has patented aequorin-based cell lines for GPCR screening (AequoScreen); the receptor of interest is expressed in a cell line coexpressed with apoaequorin. The promiscuity of $G\alpha_{16}$ is well utilized in the aequorin-based assays to improve the coupling to calcium for both adherent and nonadherent cells. Typically, cells are loaded with coelenterazine for at least 3 h. During the incubation step, coelenterazine enters the cells and conjugates with apoaequorin to form aequorin, which is the active form of the enzyme. Upon receptor activation, the light is emitted for 20 to 30 sec. Conventional luminometers or instruments such as the Lumax (CyBio AG, Jena, Germany) and FDSS 6000 can be used to measure the luminescent signal. This functional screening assay has been tested with several GPCRs. The potency of agonists is similar to those obtained from radioligand binding and/or other functional assays. Unlike fluorescent calcium indicators, Ca^{2+}-bound aequorin can be detected without illuminating the sample, thereby eliminating the interference due to autofluorescence from the screening compounds.

9.5.4 INOSITOL-TRIPHOSPHATE ASSAYS

Whereas most HTS labs utilize gene reporter or calcium fluorophore technologies to detect G_q- and/or G_i-coupled receptors, some technologies detect inositol-triphosphate (IP3) formation in cells. Historically, IP3 assays depended on [^{3}H]inositol loading into cells. Radioactivity was then detected in aqueous and organic fractions after column separation. This technology is not readily amenable for HTS. IP3 formation is determined by measuring the displacement of a labeled IP3 analog from an IP3 binding protein.

Kits are available from Amersham, PerkinElmer (AlphaScreen) and DiscoveRx (enzyme complementation). These assays may prove useful in secondary screening. One disadvantage of IP3 detection is the limited timeframe during which its production can be detected due to the kinetics of IP3 formation after receptor stimulation which is extremely fast. Therefore, care must be taken to ensure proper timing as well as fast and complete lysis of stimulated cells.

9.5.5 MELANOPHORE-BASED RECEPTOR FUNCTIONAL ACTIVITY

Melanophore cells derived from the neural crest of *Xenopus laevis* offer a highly sensitive system for the screening of ligands for GPCRs. The amphibian melanophores contain pigmented organelles called melanosomes, which are filled with the dark pigment melanin. Many G_s- and G_i-coupled receptors have been shown to influence pigment dispersion and aggregation after stimulation with the appropriate ligands.

The state of pigment distribution within a cell reflects activation of AC or phospholipase C.[75] When melanophores are stimulated by agonists such as αMSH that elevate cAMP above basal levels, melanophore dispersion is induced in the cytoplasm and the cells will appear dark. Inhibition of cAMP production results in melanophore aggregation into cells and they appear light. Therefore, the melanophore system can be used as an endogenous signaling technique to mediate cell darkening or lightening due to pigment dispersion and aggregation, respectively. The pigment dispersion and aggregation can be determined in a microtiter plate by measuring absorbance or by imaging the cell response. The main advantage of this format is its application to $G\alpha_i$-coupled receptors. DeCamp et al. reported the expression of more than 100 GPCRs in melanophores.[76] However, melanophores themselves express a range of endogenous GPCRs, and the specificity of compounds hitting the receptor of interest should be elucidated by other methods. This technology may have application for the study of orphan GPCRs for which functional expression in mammalian cells is difficult to achieve. The assay platform is simple, automatable, and miniaturizable to a 1536-well format, and one can efficiently screen for inverse agonists.[77]

9.6 INVERSE AGONISTS AND CONSTITUTIVE ACTIVITIES

Inverse agonists represent an established drug class possessing the property of reducing receptor-mediated constitutive activity of GPCRs. It is the ability of GPCRs to undergo an agonist-independent conversion from the inactive state of the receptor to the activated form that makes them capable of activating G proteins.[78,79] Arena pharmaceuticals (San Diego, California) developed a technology to identify inverse agonists. Its constitutively activated receptor technology (CART) is based on the generation of receptor mutations that induce constitutive signaling.[75] A receptor can be expressed constitutively by changing the amino acid sequences of the interior loops of the receptor so that it does not require a stimulus such as agonist binding to initiate the signaling cascade.

It is important to establish an acceptable level of basal activity by overexpressing the receptor target or altering the molecular structure of an intracellular loop or intracellular portion of the GPCR to generate a CART-activated form of the GPCR. When transfected into mammalian cell lines or *Xenopus* melanophores, such mutant receptors express the CART-activated form of these receptors at the cell surface and display constitutive activity. Compound screening can be done using this system to identify agonists that increase signaling and antagonists (inverse agonists), which results in a decrease in constitutive activity. Therefore, this technology is not limited to molecules that compete only with the receptor–ligand interactions. Thus, it can facilitate the identification of molecules that modulate the signaling cascade. Also, constitutive activity of receptors can result in an increase in the sensitivity of agonists and the same system could be used to identify agonists and inverse agonists during drug screening. However, its application to HTS is still unproven.

9.7 MICROPHYSIOMETER ASSAYS

The microphysiometer assay system is the most generic assay available for measuring the metabolic activation of cells expressing GPCRs following activation by an agonist. The cellular acidification rate response induced by agonists can be monitored by a cytosensor microphysiometer. The cytosensor system uses a silicon-based light addressable potential sensor (LAPS) to monitor changes in pH. Differences in pH under basal, stimulated conditions due to an agonist and inhibited stage caused by an antagonist can be monitored with this system. The major disadvantage of this format is very low throughput.[80]

9.8 GPCR INTERNALIZATION AND TRAFFICKING

Prolonged exposure of an agonist to a receptor leads to decreased sensitivity of the receptor to a subsequent agonist challenge. Several distinct mechanisms are involved in this process, for example, uncoupling of receptors from G proteins and reduction in the number of cell surface receptors by internalization or desensitization.[81] This phenomenon controls the initiation and termination of the signal and regulates the intensity of response. An early event in the desensitization process is the phosphorylation of the activated receptor by G protein-coupled receptor kinase (GRK) followed by binding of β-arrestin to the phosphorylated receptor. Binding of β-arrestin to the receptor sterically precludes coupling between the receptor and G protein leading to signal termination. β-arrestin acts as an adapter protein that targets GPCRs to clathrin-coated pits for endocytosis. Some GPCRs disappear from cell surfaces within minutes of ligand occupancy. Once internalized, they exhibit distinct patterns of interaction with β-arrestins.

Class A GPCRs, for example, the β2 adrenergic receptors, rapidly dissociate from β-arrestin and are immediately recycled to the plasma membrane. For some GPCRs, receptor occupancy for minutes to hours results in net loss of receptors from cells. For example, angiotensin II receptors form stable receptor–β-arrestin complexes that are accumulated in the endocytic vesicles for degradation or recycling to membranes.[82]

The Transfluor® technology from Norak Biosciences (Morrisville, North Carolina) is based on labeling β-arrestin with GFP so that the recycling of the receptor–arrestin complex can be monitored. Because desensitization occurs only with an activated receptor, monitoring of arrestin translocation within a cell provides a fluorescent bioassay to screen for GPCR ligands. The GFP–β-arrestin is evenly distributed throughout the cytoplasm in the absence of receptor activation. In response to stimulation, the translocation of β-arrestin to the membrane occurs within seconds followed by the movement to endocytic vesicles within minutes.

The application of high content screening to Transfluor technology allows one to screen agonists and antagonists in this format.[83–85] The Transfluor technology has been validated on various platforms such as the INCell Analyzer (Amersham Biosciences), Acumen Explorer (TTP Lab Tech, Royston, United Kingdom), ArrayScan (Cellomics, Pittsburgh, Pennsylvania), Discovery-1 (Molecular Devices) and Opera (Evotec OAI, Hamburg, Germany). The application of Transfluor technology has also been validated for a variety of GPCRs.

Bioluminescence resonance energy (BRET) assays have also been reported as translocation techniques for GPCRs.[86,87] The receptor is fused to renilla luciferase (Rluc) and β-arrestin is fused to GFP. The β-arrestin moiety interacts with the phosphorylated receptor such that Rluc is brought into close proximity with GFP. This results in efficient energy transfer between Rluc and GFP; BRET is measured at 515/410 nm wavelength.

Another generic format for receptor internalization is outlined below. CypHer® is a pH-sensitive dye for receptor internalization studies (CypHer 5 is nonfluorescent at neutral pH and fluorescent at acidic pH). When GPCRs internalize upon agonist binding and activation, the pH changes from neutral at the cell surface to acidic in the intracellular vesicles. Therefore, receptor internalization can be measured as an increase in the fluorescent signal from the cell. The CypHer dye can be coupled to any receptor antibody.[88] The antibody bound to the probe is internalized with the receptor and becomes fluorescent in the endocytotic vesicles. Receptor internalization studies are under development and not yet commonly used in HTS.

9.9 SCREENING FOR MODULATORS OF ORPHAN RECEPTORS

Seven hundred twenty genes belong to the GPCR super family in the human genome and approximately half of them encode sensory receptors. Of the remaining 360 receptors, ligands have been identified for 210; the other 150 are considered orphan receptors[89–92] (described in detail in Chapter 16). Because several known GPCR modulators have proven clinically useful drugs, the orphan GPCRs are viewed by the pharmaceutical and biotechnology industries as rich sources of novel drug targets. Thus, many companies are engaged in identifying the natural ligands and small molecule modulators for these orphan targets. Several strategies now used for identification of ligands for orphan receptors have been reviewed.[93–98]

The key to capitalizing on the potential therapeutic benefits of these orphan GPCRs lies in the ability to identify selective ligands for them; this will allow elucidation of the functions and clinical potentials of these receptors. The primary method for screening the receptors is based on function, using cell-based assays. The microphysiometer assay has been used to identify natural ligands for the human APJ receptor and apelin was identified as the natural ligand.[99] Utilization of FLIPR to measure intracellular Ca^{2+} mobilization identified ligands for several orphan GPCRs. Many orphan receptors have undergone deorphaning through $G\alpha_{16}$ coupling. Aequorin-based Ca^{2+} assays, cAMP assays, reporter gene assays, melanophore-based assays, electrophysiological measurements, and high-content receptor internalization assays have also served as tools for finding ligands for orphan receptors.

9.10 CONCLUSIONS

High throughput screening assays to identify modulators of GPCR function have evolved from filter binding to measuring GPCR internalization using high-content screening platforms. The decision to utilize a specific technology depends on the expected properties of a compound and the characteristics of the technology (sensitivity, quenching, cost, throughput, miniaturization, required instrumentation). For example, identification of ligands for orphan receptors will likely require functional cell-based assays (gene reporter systems, melanophores, Ca^{2+}, etc.) rather than ligand binding assays. Also, the receptor can sometimes define the choice of technology; screening for antagonists of G_q-coupled receptors will preferably be performed with a Ca^{2+} assay in the FLIPR whereas a G_i-coupled receptor will be assayed by cAMP or Ca^{2+} using promiscuous or chimeric G proteins. If a compound needs to function as an allosteric modulator on a specific GPCR, ligand binding assays will not entirely reveal such compounds and signal transduction assay would be a better choice.

Tremendous progress has been made in understanding the signal transduction and regulation of GPCRs (transactivation, homodimerization, heterodimerization, novel effectors). Although GPCR assay tools have improved in terms of sensitivity, throughput and ease of use, most current screening technologies rely on classic GPCR mechanisms such as receptor binding, GTP binding, and cAMP and Ca^{2+} signaling. Some high content screening technologies take advantage of newer paradigms such as desensitization or internalization of GPCRs (Transfluor and CypHer dyes). These high-content screening methodologies are becoming integral parts of the functional analysis required for the lead discovery and optimization processes.

REFERENCES

1. Beggs, M. HTS: where next? Drug Disc World. 2001, 2, 25–30.
2. Kell, D. Screensavers: trends in high-throughput analysis. Trends Biotechnol. 1999, 17, 89–91.
3. Lavery, P.; Brown, M.J.; Pope, A.J. Simple absorbance-based assays for ultra-high throughput screening. J Biomol Screen. 2001, 6, 3–9.
4. Burns, D.J; Kofron, J.L.; Warrior, U.; Beutel, B.A. Well-less, gel permeation formats for ultra-HTS. Drug Disc Today. 2001, 6, S40–S47.
5. Hodgson, J. Receptor screening and the search for new pharmaceuticals. Biotechnology. 1992, 10, 973–980.
6. Knight, P.J.; Pfeifer, T.A.; Grigliatti, T.A. A functional assay for G-protein-coupled receptors using stably transformed insect tissue culture cell lines. Anal Biochem. 2003, 320, 88–103.
7. Jamieson, A.C.; Miller, J.C.; Pabo, C.O. Drug discovery with engineered zinc finger proteins. Nat Rev. Drug Disc. 2003, 2,361–368.
8. Qureshi, S.A.; Sanders, P.; Zeh, K.; Whitney, M.; Pollok, B.; Desai, R.; Whitney. P.; Robers, M.; Hayes, S.A. A one-arm homologous recombination approach for developing nuclear receptor assays in somatic cells. Assay Drug Dev Technol. 2003, 1, 767–776.

9. Zeh, K.; Sanders, P.; Londo, P.; Crute, J.J.; Pollok, B.A.; Whitney, M.A. Gain-of-function somatic cell lines for drug discovery applications generated by homologous recombination. Assay Drug Dev Technol. 2003, 1, 755–765.

10. Vanhauwe, J.F.; Ercken, M.; van de Wiel, D.; Jurzak, M.; Leysen, J.E. Effects of recent and reference antipsychotic agents at human dopamine D2 and D3 receptor signaling in Chinese hamster ovary cells. Psychopharmacology. 2000, 150, 383–390.

11. Zhang, J-H.; Chung, T.D.Y.; Oldenburg, K.R.; A simple statistical parameter for use in evaluation and validation of high throughput screening. J Biomol Screen. 1999, 4, 67–73.

12. Cheng, X.; Hochlowski, J.; Tang, H.; Hepp, D.; Beckner, C.; Kantor, S.; Schmitt, R. Studies on repository compound stability in DMSO under various conditions. J Biomol Screen. 2003, 8, 292–304.

13. Kozikowski, B.A.; Burt, T.M.; Tirey, D.A.; Williams, L.E.; Kuzmak, B.R.; Stanton, D.T.; Morand, K.L.; Nelson, S.L. The effect of room temperature storage on the stability of compounds in DMSO. J Biomol Screen. 2003, 8, 205–209.

14. Kozikowski, B.A.; Burt, T.M.; Tirey, D.A; Williams, L.E.; Kuzmak, B.R.; Stanton, D.T.; Morand, K.L.; Nelson, S.L. The effect of freeze/thaw cycles on the stability of compounds in DMSO. J Biomol Screen. 2003, 8, 210–215.

15. Vanhauwe, J.F.M.; Fraeyman, N.; Francken, B.J.B; Luyten, W.H.M.L.; Leysen, J.E. Comparison of the ligand binding and signaling properties of human dopamine D2 and D3 receptors in Chinese hamster ovary cells. J Pharm Exp Ther. 1999, 290, 908–916.

16. Vanhauwe, J.F.M.; Josson, J.F.M; Luyten, W.H.M.L; Driessen, A.J.M.; Leysen, J.E. G-protein sensitivity of ligand binding to human dopamine D2 and D3 receptors expressed in *E. coli*: clues for a constrained D3 receptor structure. J Pharm Exp Ther. 2000, 295, 274–283.

17. Cook, N.D. Scintillation proximity assay: a versatile high-throughput screening technology. Drug Disc Today. 1996, 1, 287–294.

18. Lowitz, K.; Prescott, G.; Guyer, D.; Dunst, R. Comparison of human ghrelin receptor binding assays using SPA and the LEADseeker multimodality imaging system. Society for Biomolecular Screening 9th Annual Conference, Portland, OR, September 21–25, 2003. Danbury, CT, p. 189.

19. Miraglia, S.; Swartzman, E.E.; Mellentin-Michelotti, J.; Evangelista, L.; Smith, C.; Gunawan, I. I.; Lohman, K.; Goldberg, E.M.; Manian, B.; Yuan, P.M. Homogeneous cell- and bead-based assays for high throughput screening using fluorometric microvolume assay technology. J Biomol Screen. 1999, 4, 193–204.

20. Banks, P.; Harvey, M. Considerations for using fluorescence polarization in screening of G protein-coupled receptors. J Biomol Screen. 2002, 7, 111–117.

21. Banks, P.; Gosselin, M.; Prystay, L. Impact of a red-shifted dye label for high throughput fluorescence polarization assays of G protein-coupled receptors. J Biomol Screen. 2000, 5, 329–334.

22. Harris, A.; Cox, S.; Burns, D.; Norey, C. Miniaturization of fluorescence polarization receptor-binding assays using CyDye-labeled ligands. J Biomol Screen. 2003, 8, 410–420.

23. Sprang, S.R. G protein mechanisms: insights from structural analysis. Annu Rev Biochem. 1997, 66, 639–678.

24. Hamm, H.E. The many faces of G protein signaling. J Biol Chem. 1998, 273, 669–672.

25. Strathmann, M.P.; Simon, M.I. Gα12 and Gα13 subunits define a fourth class of G protein subunits. Proc Natl Acad Sci USA. 1991, 88, 5582–5586.

26. Simon, M.I.; Strathmann, M.P.; Gautam.N. Diversity of G proteins in signal transduction. Science. 1991, 252, 802–808.

27. Suzuki, N.; Nakamura, S.; Mano, H.; Kozasa, T. Gα12 activates Rho GTPase through tyrosine-phosphorylated leukemia-associated RhoGEF, Proc Natl Acad Sci USA. 2003, 100, 733–738.

28. Cabrera-Vera, T.M.; Vanhauwe, J.; Thomas, T.O.; Medkova, M.; Preininger, A.; Mazzoni, M.R; Hamm, H.E. Insights into G protein structure, function, and regulation. Endocr Rev. 2003, 24, 765–781.

29. Harrison, C.; Traynor, J.R. The [^{35}S]GTPγS binding assay: approaches and applications in pharmacology, Life Sci. 2003, 74, 489–508.

30. Lazareno, S.; Farries, T.; Birdsall, N.J. Pharmacological characterization of guanine nucleotide exchange reactions in membranes from CHO cells stably transfected with human muscarinic receptors M1–M4. Life Sci. 1993, 52, 449–456.

31. Offermanns, S.; Wieland, T.; Homann, D.; Sandmann, J.; Bombien, E.; Spicher, K.; Schultz, G.; Jakobs, K.H. Transfected muscarinic acetylcholine receptors selectively couple to G$_i$-type G proteins and G$_{q/11}$. Mol Pharmacol. 1994, 45, 890–898.

32. Birnbaumer, L.; Abramowitz, J.; Brown, A.M. Receptor–effector coupling by G proteins. Biochim Biophys Acta. 1990, 1031, 163–224.

33. Tian, W.N.; Duzic, E.; Lanier, S.M.; Deth, R.C. Determinants of β2-adrenergic receptor activation of G proteins: evidence for a precoupled receptor/G protein state. Mol Pharmacol. 1994, 45, 524–531.

34. Selley, D.E.; Cao, C.C.; Liu, Q.; Childers, S.R. Effects of sodium on agonist efficacy for G-protein activation in μ-opioid receptor-transfected CHO cells and rat thalamus. Br J Pharmacol. 2000, 130, 9879–9896.

35. Milligan, G. Principles: extending the utility of [^{35}S]GTPγS binding assays. Trends Pharmacol Sci. 2003, 24, 87–90.

36. Ferrer, M.; Kolodin, G.D.; Zuck, P.; Peltier, R.; Berry, K.; Mandala, S.M.; Rosen, H.; Ota, H.; Ozaki, S.; Inglese, J.; Strulovici, B. A fully automated [^{35}S]GTPγS scintillation proximity assay for the high throughput screening of G$_i$-linked G protein-coupled receptors. Assay Drug Dev Technol. 2003, 1, 261–273.

37. Frang. H.; Mukkala, V.M.; Syysto, R.; Ollikka, P.; Hurskainen, P.; Scheinin, M.; Hemmila, I. Nonradioactive GTP binding assay to monitor activation of G protein-coupled receptors, Assay Drug Dev Technol. 2003, 1, 275–280.

38. Hanoune, J.; Defer, N. Regulation and role of adenylyl cyclase isoforms. Annu Rev Pharmacol Toxicol. 2001, 41, 145–174.

39., Williams, C. cAMP detection methods in HTS: selecting the best from the rest. Nat Rev Drug Disc. 2004, 3, 125–135.

40. NEN Life Science Products. A novel adenylyl cyclase activation assay on FlashPlate (application note). Boston, MA, 1998.

41. Kariv, I.; I.; Stevens, M.E.; Behrens, D.L.; Oldenburg, K.R. High throughput quantitation of cAMP production mediated by activation of seven-transmembrane domain receptors. J Biomol Screen. 1999, 4, 27–32.

42. Amersham Biosciences. High throughput screening for cAMP formation by scintillation proximity radioimmunoassay. Proximity News, 23, 1996.

43. Amersham Biosciences. Miniaturization of the cAMP SPA direct screening assay to 384-well format. Proximity News, 57, 1998.

44. Prystay, L.; Gagne, A.; Kasila, P.; Yeh, L.A.; Banks, P. Homogeneous cell-based fluorescence polarization assay for the direct detection of cAMP. J Biomol Screen. 2001, 6, 75–82.

45. Allen, M.; Hall, D.; Collins, B.; Moore, K. A homogeneous high throughput nonradioactive method for measurement of functional activity of Gs-coupled receptors in membranes. J Biomol Screen. 2002, 7, 35–44.

46. Horton, J., Gardner, N.; Tinkler, S.; Cox, C.; Norey, M.; Briggs, M.; Baxendale, P. A simple, non-radioactive, functional assay for compound screening at G-protein coupled receptors: drug screening and cellular assays, Amersham Biosciences, Life Science News Online, 2003.

47. Degorce, F.; Cougouluegne, F.; Jacquemart, L.; Préaudat, M.; Achard, S.; Enomoto, K.; Ohta, H.; Nishigaki, A.; Dohi, K.; Takemoto, H.; Seguin, P.; Mathis G. Assessment of tumor necrosis factor alpha (TNF) in cell culture with a new HTRF assay. Society for Biomolecular Screening 6th Annual Conference, Vancouver, Canada, September 2000. Danbury, CT, p. 186.

48. Gabriel, D.; Vernier, M.; Pfeifer, M.J.; Dasen, B.; Tenaillon, L.; Bouhelal, R. High throughput screening technologies for direct cyclic AMP measurement. Assay Drug Dev Technol. 2003, 1, 291–303.

49. Bosse, R.; Bouchard, N.; Chelsky, D.; Howard, B. Alphascreen™: true format flexibility 96 to 384 to 1536 cAMP assays. Society for Biomolecular Screening 6th Annual Conference, Vancouver, Canada, September 2000. Danbury, CT, p. 176.

50. Golla, R.; Seethala, R. A homogeneous enzyme fragment complementation cyclic AMP screen for GPCR agonists. J Biomol Screen. 2002, 7, 515–525.

51. Weber, M.; Ferrer, M.; Zheng, W.; Inglese, J.; Strulovici, B.; Kunapuli, P. A. 1536-well cAMP assay for G_s- and G_i-coupled receptors using enzyme fragmentation complementation. Assay Drug Dev Technol. 2004, 2, 39–49.

52. De Cesare, D.; Fimia, G.M.; Sassone-Corsi, P. Signaling routes to CREM and CREB: plasticity in transcriptional activation. Trends Biochem Sci. 1999, 24, 281–285.

53. Hill, S.J.; Baker, J.G.; Rees, S. Reporter gene systems for the study of G-protein-coupled receptors. Curr Opin Pharmacol. 2001, 1, 526–532.

54. Naylor, L.H. Reporter gene technology: the future looks bright. Biochem Pharmacol 1999, 5, 749–757.

55. Zlokarnik, G.; Negulescu, P.A.; Knapp, T.E.; Mere, L.; Burres, N.; Feng, L.; Whitney, M.; Roemer,K.; Tsien, R.Y. Quantitation of transcription and clonal selection of single living cells with β-lactamase as reporter. Science. 1998, 279, 84–88.

56. Kunapuli, P.; Ransom, R.; Murphy, K.L.; Pettibone, D.; Kerby, J.; Grimwood, S.; Zuck, P.; Hodder, P.; Lacson, R.; Hoffman, I.; Inglese, J.; Strulovici, B. Development of an intact cell reporter gene β-lactamase assay for G protein-coupled receptors for high-throughput screening. Anal Biochem. 2003, 314, 16–29.

57. Liu, B.; Wu, D. Analysis of the coupling of G12/13 to G protein-coupled receptors using a luciferase reporter assay. Methods Mol Biol. 2004, 237, 145–149.

58. Langmead, C.; Patel, N.; Ratti, E.; Heath, J.; Wood, M. Act one: a novel real-time cAMP assay using the FLIPR. Eighth International Drug Discovery Conference, June 2004, Molecular Devices Corporation, Berkeley Hills, CA.

59. Schroeder, K.S.; Neagle, B.D. FLIPR: a new instrument for accurate, high throughput optical screening. J Biomol Screen. 1996, 1, 75–80.

60. Miller,T.R.; Witte, D.G.; Ireland, L.M.; Kang, C.H.; Roch, J.M.; Masters, J.N.; Esbenshade, T.A.; Hancock, A.A. Analysis of apparent non-competitive responses to competitive H1 histamine receptor antagonists in fluorescent imaging plate reader-based calcium assays. J Biomol Screen. 1999, 4, 249–258.

61. Stables, J. Evaluation of FLIPR Tetra instrument for 384- and 1536-well calcium mobilization assays. Eighth International Drug Discovery Conference, June 2004, Molecular Devices Corporation, Berkeley Hills, CA.

62. Kaler, G.; Otto, M.; Okun, A.; Okun, I. Serotonin antagonist profiling on 5HT2A and 5HT2C receptors by nonequilibrium intracellular calcium response using an automated flow-through fluorescence analysis system, HT-PS 100. J Biomol Screen. 2002, 7, 291–301.

63. Coward, P.; Chan, S.D.; Wada, H.G.; Humphries, G.M.; Conklin, B.R. Chimeric G proteins allow a high-throughput signaling assay of G_i-coupled receptors. Anal Biochem. 1999, 270, 242–248.

64. Milligan, G.; Marshall, F.; Rees, S. G16 as a universal G protein adapter: implications for agonist screening strategies. Trends Pharmacol Sci. 1996, 17, 235–237.

65. Mody, S.M.; Ho, M.K.; Joshi, S.A.; Wong, Y.H. Incorporation of $G\alpha(z)$-specific sequence at the carboxyl terminus increases the promiscuity of $G\alpha(16)$ toward G(i)-coupled receptors. Mol Pharmacol. 2000, 57, 13–23.

66. Elshourbagy, N.A.; Ames, R.S.; Fitzgerald, L.R.; Foley, J.J.; Chambers, J.K.; Szekeres, P.G.; Evans, N.A.; Schmidt, D.B.; Buckley, P.T.; Dytko, G.M.; Murdock, P.R.; Milligan, G.; Groarke, D.A.; Tan, K.B.; Shabon, U.; Nuthulaganti, P.; Wang, D.Y.; Wilson, S.; Bergsma, D.J.; Sarau, H.M. Receptor for the pain modulatory neuropeptides FF and AF is an orphan G protein-coupled receptor. J Biol Chem. 2000, 275, 25965–25971.

67. Chambers, J.K.; Macdonald, L.E.; Sarau, H.M.; Ames, R.S.; Freeman, K.; Foley, J.J.; Zhu, Y.; McLaughlin, M.M.; Murdock, P.; McMillan, L.; Trill, J.; Swift, A.; Aiyar, N.; Taylor, P.; Vawter, L.; Naheed, S.; Szekeres, P.; Hervieu, G.; Scott, C.; Watson, J.M.; Murphy, A.J.; Duzic, E.; Klein, C.; Bergsma, D.J.; Wilson, S.; Livi, G.P. A G protein-coupled receptor for UDP glucose. J Biol Chem. 2000, 275, 10767–10771.

68. Lai, F.P.; Mody, S.M.; Yung, L.Y.; Kam, J.Y.; Pang, C.S.; Pang, S.F.; Wong, Y.H Molecular determinants for the differential coupling of G16 to the melatonin MT1, MT2 and Xenopus Mel1c receptors. J Neurochem. 2002, 80, 736–745.

69. Marchese, A.; George, S.R.; Kolakowski, L.F. Jr.; Lynch, K.R.; O'Dowd, B.F. Novel GPCRs and their endogenous ligands: expanding the boundaries of physiology and pharmacology. Trends Pharmacol Sci. 1999. 20, 370–375.

70. Lee, C.H.; Shin, I.C.; Kang, J.S.; Koh, H.C.; Ha, J.H.; Min. C.K. Differential coupling of Gq family of G-proteins to muscarinic M1 receptor and neurokinin-2 receptor. Arch Pharm Res. 1998, 21, 423–428.

71. Liu, A.M.; Ho, M.K.; Wong, C.S.; Chan, J.H.; Pau, A.H.; Wong, Y.H. $G\alpha(16/z)$ chimeras efficiently link a wide range of G protein-coupled receptors to calcium mobilization. J Biomol Screen. 2003. 8, 39–49.

72. Wong, S.; Shirikhande, A. Functional assay for agonist activation of receptors, US2003.0104489A1, 2002.

73. Gopalakrishnan, S.M.; Moreland, R.B.; Kofron, J.L.; Helfrich, R.J.; Gubbins, E.; McGowen, J.; Masters, J.N.; Donnelly-Roberts, D.; Brioni, J.D.; Burns, D.J.; Warrior, U. A cell-based microarrayed compound screening format for identifying agonists of G-protein-coupled receptors. Anal Biochem. 2003, 321, 192–201.

74. Ungrin, M.D.; Singh, L.M.; Stocco, R.; Sas, D.E.; Abramovitz, M. An automated aequorin luminescence-based functional calcium assay for G-protein-coupled receptors. Anal Biochem. 1999. 272, 34–42.

75. Chen, W.J.; Jayawickreme, C.; Watson, C.; Wolfe, L.; Holmes, W.; Ferris, R.; Armour, S.; Dallas, W.; Chen, G.; Boone, L.; Luther, M.; Kenakin, T. Recombinant human CXC-chemokine receptor-4 in melanophores are linked to Gi protein: seven transmembrane coreceptors for human immunodeficiency virus entry into cells. Mol Pharmacol. 1998, 53, 177–181.

76. DeCamp, D.L.; Thompson, T.M.; de Sauvage, F.J.; Lerner, M.R. Smoothened activates Galphai-mediated signaling in frog melanophores. J Biol Chem. 2000. 275, 26322.
77. Wise, A.; Jupe, S.C.; Rees, S. The identification of ligands at orphan G-protein coupled receptors. Annu Rev Pharmacol Toxicol. 2004, 44, 43–66.
78. Milligan G. Constitutive activity and inverse agonists of G protein-coupled receptors: a current perspective. Mol Pharmacol. 2003, 64, 1271–1276.
79. Kenakin T. Inverse, protein, and ligand-selective agonism: matters of receptor conformation. FASEB J. 2001, 15, 598–611.
80. Patel, H.; Porter, R.H.; Palmer, A.M.; Croucher, M.J. Comparison of human recombinant adenosine A2B receptor function assessed by Fluo-3-AM fluorometry and microphysiometry. Br J Pharmacol. 2003. 138, 671–677.
81. Spampinato, S.; Di Toro, R.; Alessandri, M.; Murari, G. Agonist-induced internalization and desensitization of the human nociceptin receptor expressed in CHO cells. Cell Mol Life Sci. 2002. 59, 2172–2183.
82. Luttrell L.M.; Lefkowitz R.J. The role of β-arrestins in the termination and transduction of G-protein-coupled receptor signals. J Cell Sci. 2002. 115, 455–465.
83. Ghosh R.N.; Chen Y.T.; DeBiasio R.; DeBiasio R.L.; Conway B.R.; Minor L.K.; Demarest K.T. Cell-based, high-content screen for receptor internalization, recycling and intracellular trafficking. Biotechniques. 2000, 29, 170–175.
84. Oakley, R.H.; Hudson, C.C.; Cruickshank, R.D.; Meyers, D.M.; Payne, R.E. Jr.; Rhem, S.M.; Loomis, C.R. The cellular distribution of fluorescently labeled arrestins provides a robust, sensitive, and universal assay for screening G protein-coupled receptors. Assay Drug Dev Technol. 2002, 1, 21–30.
85. Milligan, G. High-content assays for ligand regulation of G-protein-coupled receptors. Drug Discov Today. 2003, 8, 579–585.
86. Bertrand, L.; Parent, S.; Caron, M.; Legault, M.; Joly, E.; Angers, S.; Bouvier, M.; Brown, M.; Houle, B.; Menard, L. The BRET2/arrestin assay in stable recombinant cells: a platform to screen for compounds that interact with G protein-coupled receptors (GPCRs). J Recept Signal Transduct Res. 2002, 22, 533–541.
87. Berglund, M.M.; Schober, D.A.; Statnick, M.A.; McDonald, P.H.; Gehlert, D.R. The use of bioluminescence resonance energy transfer 2 to study neuropeptide Y receptor agonist-induced β-arrestin 2 interaction. J Pharmacol Exp Ther. 2003, 306, 147–156.
88. Adie, E.J.; Francis, M.J.; Davies, J.; Smith, L.; Marenghi, A.; Hather, C.; Hadingham, K.; Michael, N.P.; Milligan, G.; Game, S. CypHer 5: a generic approach for measuring the activation and trafficking of G protein-coupled receptors in live cells. Assay Drug Dev Technol. 2003, 1, 251–259.
89. Lander, E.S.; Linton, L.M.; Birren, B.; Nusbaum, C.; Zody, M.C. et. al. Initial sequencing and analysis of human genome. Nature. 2001, 409, 860–921.
90. Pierce, K.L.; Premont, R.T.; Lefkowitz, R.J. Seven-transmembrane receptors. Nat Rev Mol Cell Biol. 2002, 3, 639–650.
91. Howard, A.D.; McAllister, G.; Feighner, S.D.; Liu, Q.; Nargund, R.P.; Van der Ploeg, L.H.; Patchett, A.A. Orphan G-protein-coupled receptors and natural ligand discovery. Trends Pharmacol Sci. 2001, 22, 132–140.
92. Wise, A.; Gearing, K.; Rees, S. Target validation of G-protein coupled receptors. Drug Disc Today. 2002. 7, 235–246.
93. Cacace, A.; Banks, M.; Spicer, T.; Civoli, F.; Watson, J. An ultra-HTS process for the identification of small molecule modulators of orphan G-protein coupled receptors. Drug Disc Today. 2003, 8, 785–792.

94. Howard, A.D.; McAllister, G.; Feighner, S.D.; Liu, Q.; Nargund, R.P.; Van der Ploeg, L.H.; Patchett, A.A.Orphan G-protein coupled receptors and natural ligand discovery. Trends Pharmacol Sci. 2001, 22, 132–140.

95. Civelli, O.; Nothacker, H.P.; Saito, Y.; Wang, Z.; Lin, S.H.; Reinscheid, R.K. Novel neurotransmitters as natural ligands of orphan G-protein-coupled receptors. Trends Neurosci. 2001, 24, 230–237.

96. Hinuma, S.; Onda, H.; Fujino, M. The quest for novel bioactive peptides utilizing orphan seven-transmembrane-domain receptors. J Mol Med. 1999. 77, 495–504.

97. Stadel, J.M.; Wilson, S.; Bergsma, D.J. Orphan G protein-coupled receptors: a neglected opportunity for pioneer drug discovery. Trends Pharmacol Sci, 1997, 18, 430–437.

98. Wilson, S.; Bergsma, D.J.; Chambers, J.K.; Muir, A.I. Fantom, K.G.; Ellis, C.; Murdock, P.R.; Herrity, N.C.; Stadel, J.M. Orphan G-protein-coupled receptors: the next generation of drug targets? Br J Pharmacol. 1998, 125, 1387–1392.

99. Tatemoto, K.; Hosoya, M.; Habata, Y.; Fujii, R.; Kakegawa, T.; Zou, M.X.; Kawamata, Y.; Fukusumi, S.; Hinuma, S.; Kitada, C.; Kurokawa, T.; Onda, H.; Fujino, M. Isolation and characterization of a novel endogenous peptide ligand for the human APJ receptor. Biochem Biophys Res Commun. 1998, 251, 471–476.

10 Molecular Bioinformatics of Receptor Binding and Activation

Kevan P. Willey, Heike Obermann,
and James B. Procter

CONTENTS

ABSTRACT

Analysis of 7-transmembrane receptors (7TMRs) determined that receptor classification according to ligand size conforms to the phylogenetic analysis of receptor sequences revealed in the human genome, allowing functional genomic interpretations to be made and the deorphanization of novel receptors to be attempted. The structural flexibility of membrane receptors and the disposition of cysteine residues suggest a mechanism of receptor activation through redox control that satisfies the broad range of binding and activation modalities of 7TMR ligands. Evidence from the mapping of ligand templates, the performance of disparate biochemical assays, and the realization that smell molecules are the "lowest common denominator" agonists for 7TMRs all support this molecular mechanism of ligand binding and receptor activation. If this activation step can be confirmed unequivocally, then understanding the physicochemical mechanisms of receptor activation will have profound consequences for drug discovery initiatives.

10.1 INTRODUCTION: NOT ALL 7TMRS ARE GPCRS

A common system of receptor activation in G protein-coupled receptors (GPCRs) has been contemplated often, primarily because of the conserved heptahelical structural motif and the shared signaling through G proteins. However, the great diversity in ligands has been used as a counter-argument against such considerations. The prevailing modern view is that no single "lock" exists for all agonist "keys" in GPCRs and that each ligand–receptor system has its own mechanistic binding and activation principle.[1] Subsequent to ligand binding, it is presumed that ligand-induced conformational changes in the receptor are reasonably conserved for all receptors, allosterically regulating signal transduction intracellularly, and empowering the heterotrimeric G proteins and other effector mechanisms to modulate and interfere with cellular responses.

Not all 7TMRs are physiological receptors and transduce specific ligand signals, although all are pharmacological receptors, even if the only known ligands are simply the bound lipid molecules of the cell membranes. A distinction between binding ability (ligand–receptor binding) and bioactivity (ligand-induced, receptor-mediated biological activation) is also emphasized and confusing terms such as binding "activity" or "actives" are to be avoided. This regrettable terminology is commonly used to designate "hits" or "positives" in early stage drug discovery screening assays, and the assays are too often indirect assessments of binding inhibition by another marker ligand and thus afflicted by a preponderance of false positives. As a result, few descriptive terms are truly descriptive and many scientific misconceptions and bio(un)informatic analyses can be traced to misleading annotation.

Examples of the 7TMR sequence and topology superfamily are to be found in all biological kingdoms. Whereas the predominant signaling pathway is G protein activation and hence the GPCR designation, not all of the heptahelical membrane proteins which are considered parts of this structural or sequence-related superfamily actually couple to G proteins. The heptahelical protein and 7TMR designations acknowledge their structural ubiquity without the confusion of inappropriate attribution of a GPCR signal-transduction mechanism. A pharmacological definition of a 7TMR will be used throughout this chapter; the GPCR designation is usually retained for those receptors which signal via G proteins.

10.2 RECEPTOR CLASSIFICATION BY LIGAND SIZE

The crystal structure of bovine rhodopsin[2] is considered to be applicable to all of the GPCR superfamily, with the α-helices traversing the lipid bilayer seven times. Bacteriorhodopsin and the many other prokaryotic opsins are not GPCRs, although their sequences and structures imply relatedness. A recent genomics survey of 16 species of Archaea and 96 species of bacteria identified two widespread families of novel 7TMRs that exhibit a bewildering array of amino and carboxyl terminal acquisitions, via gene fusion, characteristic of extracellular substrate-binding molecules and intracellular adenylate cyclases, histidine kinases, phosphatases, and other effectors typical of prokaryotic signal transduction pathways.[3]

10.2.1 ABC(DEF) CLASSIFICATION OF GPCRs

The conventional classification of the GPCR superfamily into six families (A–F) is maintained at the primary GPCR resource (GPCRDB, www.gpcr.org), which has moved to the University of Nijmegen, The Netherlands, with its cofounder Gerrit Vriend. The American "mirror" GPCRDB[4] has suffered the fate of so many comprehensive scientific databases and has been privatized. The 7TMRs and GPCRs are assigned to class 2 in the IUPHAR receptor database (www.iuphar.org), with individual receptors encoded in a manner similar to the EC nomenclature for enzymes (or even IP addresses), with individual receptor subcommittees publishing definitive reports in *Pharmacological Reviews*.

The A–F classification system is derived from a combination of multiple sequence alignment (MSA), endogenous ligand specificity, and species source. The two largest families are A and B. Family A includes rhodopsin-related members and is dominated by the biogenic amine receptors and many peptide hormones. Family B includes the remainder of the peptide hormones. Family C has a few receptors of the metabotropic glutamate type, whereas families D and E are pheromone receptors only to be found in yeasts; family F receptors are *Dictyostelium* cAMP receptors.

Rhodopsin is an undisputed member of the family A receptors and is also the best studied, truly archetypical GPCR. Visual transduction is mediated by activation and dissociation of both chromophore and G protein from the opsin, releasing the opsin constitutive activity which is desensitized and downregulated by specific phosphorylation. Thus, the dark-adapted form of rhodopsin with its covalently attached chromophore can be considered an inverse agonist or antagonist-blocked, constrained and inactive form of a constitutively active receptor.

Human GPCRs are of the family ABC types, although a diverse set of 7TM-like sequences related to the developmental products of the BOSS and frizzled genes in *Drosophila* are now included in the 7TMR superfamily. BOSS has a 500-amino acid amino-terminal domain that interacts with the *sevenless* gene-encoded receptor tyrosine kinase[5] and retains the characteristic disulfide bridge between the extracellular loops 1 and 2. The BOSS 7TMR and the sevenless RTK "ligand" coordinate development of the retinal photoreceptors of the fly ommatidia, with photoreceptor cells arranged in a group of seven around a central coordinator.

Family C metabotropic receptors are distinguished by huge amino and carboxyl acquisitions, whereas families A and B include some subfamilies with large extra-cellular extensions. Family A has the most representatives and interacts with the broadest range of ligands, from the opsins responsible for visual perception and the olfactory and some gustatory receptors responsible for smell and taste, to the receptors for monoamine, nucleotide, lipid, peptide and protein hormones, many chemokines, and the enzyme thrombin. Generally, family A 7TMRs are presumed to adopt the folding topology of rhodopsin, with a common 7TM core, extracellular amino and intracellular carboxyl termini, and a presumed dithiol bridge between the extracellular loops 1 and 2. For some of the receptors, TM1 can be the uncleaved signal peptide of the protein sequence and the extracellular sequences are generally glycosylated.

The great diversity in family A receptors has produced many attempts at classification that are almost as diverse in their results. Classification can be according to DNA or primary sequence, by structure of amino and carboxyl acquisitions, by protein–complex associations, by ligand, or by function (which necessarily is a combination of the above), such as olfactory or endocrine, or by pathology, such as mutations causing inactivity or constitutive activation and exogenous 7TMRs encoded and expressed by an infecting virus.

10.2.2 Four Classes of Ligands: Charged, Simple, Flexible, and Complex

Representative ligands of the human family A receptors fall into four major classes when listed according to their molecular weights (Figure 10.1). The small hydrophilic monoamines are charged and superseded by larger and more amphipathic molecules constituting an amorphous group, the more rigid members of which are hydrophobic and behave in a simple fashion according to Law of Mass Action kinetics. The extremely amphiphilic peptide hormones are flexible in everything they do and lead ultimately to the bulky globular proteins capable of complex formation and complex behavior. The descriptive terms — charged, simple, flexible and complex — define the physicochemical properties of the ligands along with their molecular behaviors, biomolecular interactions, receptor pharmacologies, ease of measurement, pharmacodynamics, pharmacokinetics, and whole body behavior.[6]

Ligand classes 1, 3 and 4 consist of agonists derived from amino acids and group 2 ligands consist of an amorphous mixture of amphipathic lipids, prostaglandins, and nucleotides. The small monoamines in class 1 are amino acids that became more polar and hydrophilic by the introduction of a positive charge or loss of the zwitterion. The proteins in class 4 are chains of amino acids that are long

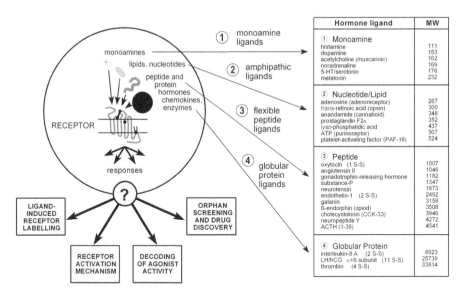

Hormone ligand	MW
① Monoamine	
histamine	111
dopamine	153
acetylcholine (muscarinic)	162
noradrenaline	169
5-HT/serotonin	176
melatonin	232
② Nucleotide/Lipid	
adenosine (adenoreceptor)	267
trans-retinoic acid (opsin)	300
anandamide (cannaboid)	348
prostaglandin F2α	352
lyso-phosphatidic acid	437
ATP (purinoceptor)	507
platelet-activating factor (PAF-16)	524
③ Peptide	
oxytocin (1 S-S)	1007
angiotensin II	1046
gonadotrophin-releasing hormone	1182
substance-P	1347
neurotensin	1673
endothelin-1 (2 S-S)	2492
galanin	3158
ß-endorphin (opiod)	3508
cholecystokinin (CCK-33)	3946
neuropeptide Y	4272
ACTH (1-39)	4541
④ Globular Protein	
interleukin-8 A (2 S-S)	8923
LH/hCG α+ß subunit (11 S-S)	25739
thrombin (4 S-S)	33814

FIGURE 10.1 Receptor classification according to ligand size and type. The diverse array of ligands which interact with rhodopsin family receptors can be classified according to their physicochemical properties[6] into four main classes, which are distinguishable in a table of representative ligands when listed according to molecular weight.

enough to adopt a globular molecular structure, largely driven by the most hydrophobic residues escaping from the solvent water molecules and burying themselves in the protein core. N- and O-glycans further reduce their surface hydrophobicity and enhance solubility.

Peptide chains with less than about 50 amino acids (MW <5000) cannot form the hydrophobic core that produces a stable globular protein. The patches of alternating hydrophobic and hydrophilic amino acids mean that the peptides in class 3 (as well as many class 2 nonpeptidaceous ligands) are therefore extremely amphiphilic and flexible molecules. They adopt different shapes and structures depending upon their environment, forming oligomeric complexes, aligning themselves at the water–lipid interfaces of cell surfaces, or insinuating themselves into proteins.

The placement of the retinal chromophore within the TM helical bundle is presumed to apply to the 7TMRs that bind smaller ligands, whether charged monoamines or amphipathic lipids. Flexible peptides and globular proteins are presumed to interact with the extracellular loops and variously sample the TM binding pockets directly or indirectly. Although flexible peptides and globular proteins are both amino acid polymers, the allometric relationship between sequence length and ligand physicochemistry is not simple. The property boundaries in behavioral characteristics are catastrophic when moving from a smaller flexible amphiphilic peptide to a larger peptide chain that can adopt the formal and restrained structure of a globular protein. These allometric relationships affect all aspects of the pharmacology and biology of the ligands and can be predicted from knowledge of the molecular size and constitution of the ligand alone.[6]

The great variety in ligands, of receptor subtypes, and the apparently haphazard utilization of G proteins, is confusing in its profligacy and has hindered the holistic comprehension of ligand-induced signal transduction in these receptors. The classification of receptors according to ligand size provides a perspective useful for biological comprehension, orphan allocation, and drug discovery initiatives.[7] As explained in the next section, this functional classification is supported by phylogenetic evidence.

10.3 LIGAND DIVERSITY CLASSIFICATION CONFORMS TO RECEPTOR PHYLOGENY

A recent phylogenetic analysis of the nearly 1000 human genome sequences related to GPCRs has reshuffled the ABC(DEF) classification. Incorporation of the latest sequence acquisitions to this superfamily led to the GRAFS classification system.[8]

10.3.1 GRAFS CLASSIFICATION OF 7TMRs

The five main families identified from multiple sequence alignment, with high bootstrap support, are designated glutamate, rhodopsin, adhesion, frizzled/taste2, and secretin. The first letters of the five family names constitute the GRAFS acronym designation. The rhodopsin and glutamate designations conform to the original A and C families, whereas family B has been split into the secretin GPCR family and the related but distinct non-GPCR adhesion family, with 7TMR sequences characterized by long amino terminal repeats, from 200 to 2800 residues, and with no signal transduction modalities as yet identified. Some of the recently identified taste receptors join either the glutamate group or the frizzled group, and all these peripheral members (except for the recent prokaryotic survey[3] examples) are now incorporated into the superfamily.

Chromosomal assignments of the phylogenetic groups conform to the known paralogy groups of the human genome, reinforcing the concepts of an ancient tetraploidic event and of duplication for chromosomes 4 and 5. The various paralogons are characterized by clusters of gene duplication and diversification, especially the more recently expanded chemokine cluster. Evidence of domain shuffling is observed for the more ancient adhesion receptors, with their large and repetitive extracellular domains consistent with the advent of metazoan complexity. (Within the discursive text,[8] certain receptor names have migrated confusingly into the wrong subfamily, including the receptors for TRH, cysteinyl-leukotrienes, and the immune formyl peptide.)

The rhodopsin segment of the GRAFS classification now includes the 460 or so olfactory receptors in the classical family A. Fredriksson and coauthors[8] identified four rhodopsin subfamilies designated α, β, γ, and δ (Figure 10.2). The α subfamily of rhodopsin receptors includes a prostaglandin, a monoamine, an opsin, and a melatonin receptor cluster. A fifth more diverse cluster incorporates the adenosine and cannabinoid receptors as well as the Edg receptors for charged lipids and the melanocortin receptors that bind a tridecapeptide. The β subfamily includes the major peptide hormones of the hypothalamus, pituitary, and digestive tract. The γ

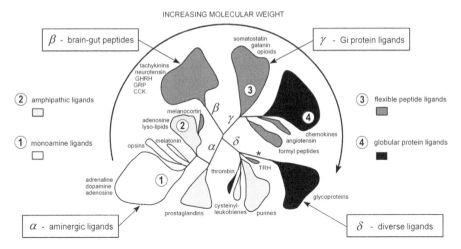

FIGURE 10.2 Concurrence of phylogenetic and functional classifications of 7TMRs. The GRAFS phylogenetic unrooted tree of rhodopsin receptors defines the αβγδ subfamily branches by a high bootstrap value.[8] The asterisk denotes the branching point for the 460 olfactory receptors. Areas incorporating numerous related receptor branches are identified by representative ligands and shaded according to the four ligand classes defined in Figure 10.1: ① monoamines, ② amphipathic ligands, ③ flexible peptides, and ④ complex globular proteins, in order of increasing molecular weight. The four ligand classes/receptor subfamilies can be designated α — aminergic, β — brain-gut peptide, γ — Gi protein ligand, and δ— diverse ligand class or receptor subfamily.

subfamily has three major branches: one for the galanin, somatostatin, and opioid receptors; another incorporating immune ligands such as the chemokines, formyl peptide, and leukotrienes with, unexpectedly, angiotensin/bradykinin; and a third branch populated by the two representatives of the 19-mer melanin-concentrating hormone. The δ subfamily includes an estimated 460 olfactory receptors, a few MAS1 oncogene-like receptors, the glycoprotein hormone receptor cluster, and a large purine receptor cluster that surprisingly brings receptors for the thrombins and cysteinyl-leukotrienes together.

10.3.2 RHODOPSIN SUBFAMILIES BIND DIFFERENT CLASSES OF LIGAND

Interestingly, the concordance of phylogenies (Figure 10.2) and the 7TMR classification into four classes according to ligand size (Figure 10.1) produces a biologically relevant interpretation for the designations of α, β, γ and δ: α for aminergic ligands, β for brain–gut peptides, γ ligands for their G_i preference, and δ ligands because they are the most diverse.

Thus, the α family of rhodopsin receptors interacts predominantly with aminergic-like ligands, typified by monoamine neurotransmitters, and it seems entirely appropriate to model their pharmacology on rhodopsin and to extrapolate from its three-dimensional (3D) structure. The β subfamily can be remembered for its homogeneous contingent of brain–gut peptides. The γ subfamily is dominated by ligands important for immune cell attraction or responses, and while it is tempting to make an association with that other important group of immune proteins, the γ-globulins, the most consistent GPCR feature for all members of this subfamily is their preference for G_i proteins. The δ subfamily is undoubtedly the most diverse in terms of ligand preference, ranging from the sublime olfactory stimuli provided by exotic small molecules to the ridiculous clustering of an extremely diverse ligand set, covering purine specificity to protease cleavage of a tethered ligand, together with cysteinyl-leukotrienes and the largest, most heterogeneous hormones, the glycoprotein cystine-knot growth factors, a subset that also includes the insulin-like relaxin molecule.

Demarcation of the receptor groups according to ligand size determines that the α, β, γ, and δ designations conform to the increasing molecular weights of the respective ligands. It is remarkable that the GRAFS classification of phylogenetic relationships supports the ligand-based classification of the rhodopsin receptor family shown in Figure 10.1, reinforcing the many clues that support biologically meaningful, functional, and evolutionary relationships. This could be the nearest that 7TMR analyses have approached to the ultimate goal of functional genomics.

10.4 STRUCTURAL FLEXIBILITIES OF MEMBRANE RECEPTORS

The seven individual helices for the five GRAFS families can be aligned (except for TM2 and TM7 from the G and F families) and each helix is self-consistent across the five families, providing no evidence of reordering and the production of a different folding topology.[8] Based on this remarkable degree of consistency, it is difficult to appreciate why 7TMRs are so resistant to structural elucidation attempts, assuming oligomeric complexing, coreceptors and/or other integral components are not necessary. One obvious hurdle is the purification and crystallization of a membrane protein while preserving its structural integrity. Analysis of the membrane-resident protein structures that have been elucidated suggests that membrane proteins are averse to the adoption of reasonably rigid structures and that they are notoriously flexible. Indeed, flexibility of structure can be expected to help many of these membrane proteins to perform their channel, transport, flux, enzymatic, and recognition functions at the cell surface.

10.4.1 LACK OF HYDROPHOBIC DRIVE IN 7TMRs

Water-soluble globular proteins adopt reasonably restrained formal structures during the folding process because of hydrophobic drive.[9] The generally hydrophobic nature of zwitterionic amino acids is worsened by the elimination of water during peptide bond formation in the production of extended chains of residues. Hydrophobic collapse ensures that the more hydrophilic residues are surface-exposed and many secondary modifications are exploited to increase aqueous solubility, including glycosylation and phosphorylation. Secondary and tertiary structures are constrained by this hydrophobic challenge, producing autonomous domains of a limited size range and resulting in the marked proclivity of shorter peptide sequences to coprecipitate or infiltrate membranes. The low dielectric constants of a lipid bilayer and of a protein interior, $\varepsilon = 2$ to 5, are indistinguishable in comparison with the high value of $\varepsilon = 80$ for water. This inability to discriminate between the interior physical properties of a protein and of a plasma membrane means that the orientation of transmembrane peptides cannot be derived reliably using predictive programs.[10] Thus, the membrane-resident portions of proteins are relieved of an aqueous imposition on their conformation.

The power of the hydrophobic drive is ineffectual in the lipophilic environment of the cell membrane and there is much less constraint on protein internal movement, resulting in an inevitable enhancement of flexibility. Nevertheless, the spate of prokaryotic membrane protein structure determinations in recent years (http://blanco.biomol.uci.edu/) was made possible by selective "tricks" that restrict protein flexibility. These include coprecipitation with antibody Fabs, site-directed mutagenesis (SDM), sequence truncation, chemical modifications, and "ligand rescue." For example, the 12-helix protein lactose permease, LacY, is a favored model for the multidrug transporters (MDRs or P-glycoproteins) responsible for small molecule metabolism and drug expulsion by cells. The broad specificity of MDRs for a wide range of molecules already implies some flexibility of recognition and response. Models of the permeases and MDRs were provided by distance constraints from many concerted approaches (e.g., a PubMed search lists 50 P-glycoprotein publications for Loo, T.W. and Clarke, D.M. in the past decade), including exhaustive tandem SCAM (scanning cysteine-accessibility mutagenesis) to generate a multitude of dithiol bridges. The ultimate elucidation of the crystal structure of LacY (1pv7) was only possible after the rational removal of a specific cysteine residue (C154G), to prevent nonspecific aggregation, and by the inclusion of substrate, which acted as a folding nucleus and preserved the structure during purification by ligand rescue. The water-filled channel is occupied by lactose and the binding site is provided by helices 1, 2 and 6, provided by both of the two autonomous bundles of six distorted helices.

The crystal form of rhodopsin (Figure 10.3B) loses its integrity on exposure to red light, as the chromophore dissociates after light activation and the crystals lose their diffractive ability.[11] Thus, even the reasonably rigid rhodopsin structure becomes more fluid on ligand loss, suggesting that flexibility is an intrinsic feature of 7TMRs. Given the low rates of successful crystallization for multispanning membrane proteins and the consequent paucity of structural information, it is all the more surprising that unrelated heptahelical membrane protein structures have been largely ignored by the 7TMR community.

10.4.2 Cytochrome C Oxidase, Subunit III

Cytochrome C oxidase is an integral membrane complex of up to 13 components in bacteria and mitochondria of which subunit III (designated chain C in 2occ, Figure 10.3A) is a 7TM "receptor" of unknown function. The helices of subunit III can be over 30 residues long, have no features in common with GPCRs, and form a topography similar to a Greek-key spiral when viewed from the amino terminal face. The void between TM2 and TM3 is partially filled by phosphatidylcholine aligned along TM3 in the bacterial structure. The quaternary ammonium and phosphate

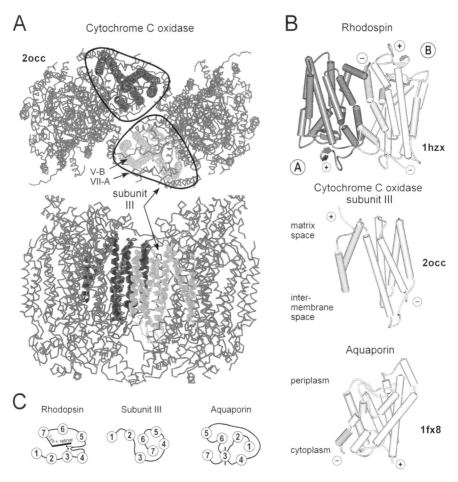

FIGURE 10.3 Broad diversity of known receptor 3D structures and topologies. A, a Cα ribbon trace of bovine mitochondrial cytochrome c oxidase (PDB code, 2occ) looking from the matrix space and from within the inner mitochondrial membrane to reveal the association of two 7TM subunit III molecules and their peptide 'ligands' provided by subunits V-B and VII-A. Comparative topologies of bovine rhodopsin (1hzx), subunit III (2occ) and human aquaporin (1fx8) from within the plane of the membrane, B, and diagrammatically from above, C. TM3 is produced from two independent half-helices in aquaporin.

groups of the lipid head group are attached noncovalently via ion pairs to Glu74 and Arg233 of the closely apposed helices TM2 and TM6, respectively.[12] The crystal structure of cytochrome C oxidase reveals the 13 component complexes have a dimer interface predominantly formed by apposition of two subunit III proteins. The amino terminal portion of subunit VII-A (chain J) forms a laterally oriented helix adjacent to the presumed membrane broaching of the helical bundle of subunit III, reminiscent of the disposition of helix 8 in the rhodopsin structure.

Intriguingly, the amino terminal portion of subunit V–B (chain F in 2occ) folds into two short helices that occupy the equivalent of the "ligand-binding surface" of a peptidic 7TMR. Only a few investigations have directly focused on the conformations of peptide ligands when bound to a 7TMR. As for most peptides up to ~40 residues long, little formal structure is evident in aqueous solution and the intrinsic amphiphilicity can only be assuaged by a hydrophobic folding focus. The conformations of PACAP(1–21)NH$_2$ bound both to micelles and to the PACAP receptor (of the family B or secretin type) were studied using NMR.[13] Residues 3–7 only produce a β-coil structure when bound to the receptor, whereas the 8–21 α-helix is evident in both bound conditions. The affinity of amphiphilic sequences for membranes offers the attractive prospect that flexible peptide ligands worm across cell surfaces in a search for their cognate serpentine receptors.

10.4.3 AQUAPORIN

The 3D structures of the aquaporin and glycerol facilitator family members do not seem related to 7TMRs at first glance,[14] but the particular arrangement of α-helices means that they should not be ignored. These water and substrate channel proteins consist of 6TM helices and two half-TM helices that are apposed amino-terminally in the 3D structure to produce a seventh helical TM span (like an extended arrangement of the antibiotic half-helix dimer, gramicidin A, 1jno). The final topography of 2.5:3.5:1 helices, in order along the primary sequence, is very reminiscent of a 7TMR (Figure 10.3). The conformations of the two half-TM helices determine that both the amino and carboxyl termini of the channel protein are on the same (cytoplasmic) side of the plasma membrane.

Domain swapping of the helices or incorporation of the periplasmic helical extensions can easily generate a multispanning protein with the amino and carboxyl termini on opposite sides of the bilayer, just like a 7TMR. Such contorted membrane orientations should not exclude the aquaporin-like structures from consideration as 7TMR models, as the recently discovered adiponectin Q subfamily has shown. This unusual subfamily consists of two major receptors for adiponectin, a 30-kDa ligand that forms higher order bioactive complexes, and three subtypes of membrane progestin receptors that bind a small hydrophobic steroid. From accessibility studies,[15] it appears that adiponectin 7TMRs lie upside-down in the lipid bilayer, in a manner similar to molecule (A) in Figure 10.3B. The many demonstrations of dimeric 7TMRs may need embellishing with the possibility that the two receptors may associate in the antiparallel manner observed in the crystal structure of rhodopsin, with one up and one down or in the exclusively upside-down orientation of the adiponectin (and presumably progestin) receptors.

Thus, there are many ways to produce a 7TMR structure and these three disparate examples will, hopefully, fire the imagination and allow many novel scenarios to be envisaged — and tested.

10.5 MOLECULAR BIOINFORMATICS IN DRUG DISCOVERY

Being the only representative structure, rhodopsin is the basis of homology modeling for the 7TMRs in many studies, of which some have spawned attempts at virtual screening. For many years, the molecular modeling community based their analyses on the low resolution electron microscopy (EM) structures of bacteriorhodopsin. Others attempted to derive models from the less well resolved bovine rhodopsin EM sources, using distance constraints from experimental data, MSAs, ligand specificities and presumed binding sites, or iterative procedures based on satisfying H-bond potentials. With the availability of the bovine rhodopsin crystal structure, and because of the limited superposition with bacteriorhodopsin, the *in silico* community has moved wholesale to rhodopsin-based models of GPCRs for virtual screening and drug discovery.

10.5.1 RHODOPSIN STRUCTURE REVELATIONS

The most conserved residue in each of the seven helices is the basis of a comparative numbering scheme, so that a particular receptor TM sequence can be compared with all others, e.g., rhodopsin Pro303[7.50] is the conserved residue in the NPxx(x)Y motif of TM7.[16]

The derivation of GPCR models from MSAs of the known sequences has a long tradition. A particularly comprehensive attempt,[17] performed prior to the elucidation of the rhodopsin crystal structure, used distance geometry derived from the satisfaction of H-bond potentials and residue packing constraints from MSAs of 410 known sequences. An approximate model of the helix bundle was used to derive an average 7TMR structure that provided a template to derive specific models for 26 representative receptor structures and delineated a conserved core of 43 residues. Many of the predicted packing arrangements were confirmed in the crystal structure, including a salt-bridge between Glu122[3.37] and His 211[5.46]. Whereas these models have been superseded by those derived from the rhodopsin crystal structure, many of the issues originally addressed were neglected by more recent analyses, especially contemplation of the physical interactions responsible for the structural stability of the transmembrane helical bundle. Apart from the obligatory network of interhelical H-bonds dictated by the modeling method, the involvement of sulfur-aromatic clusters organized as polarity gradients was highlighted, along with the colocalization of cysteine residues that may contribute to interhelical disulfide bonds. Subsequent resolution of the rhodopsin crystal structure revealed an extensive aromatic cage, supported by cysteine residues, that encloses the retinal chromophore (to be discussed in Section 10.6).

Although no biochemical evidence demonstrates the formation of endogenous disulfide bridges in the TM helices, their artificial construction has proven to be a powerful mechanism for establishing interhelical interfaces, as well as the constraints on and extent of relative helix movement upon photoactivation. Distance constraints to aid model prediction of the activated states of both rhodopsin and many of the receptors have been provided by the introduction of individual cysteine residues into the TM helices with subsequent monitoring of their effects. Endogenous and introduced cysteines provide sites for spin labels, fluorescent resonance energy transfer (FRET) analyses, and SCAM, revealing helical orientations, movement, and altered accessibility upon photon or ligand activation.[16,18] Several paired substitutions can induce disulfide bridges that prevent light activation of rhodopsin, with other bridges being permissive, although none was capable of producing constitutive activity.

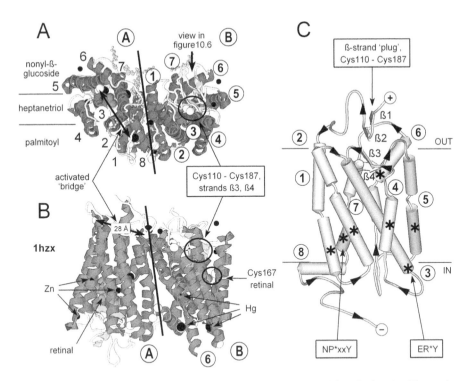

FIGURE 10.4 Essential structural features of 7TMRs illustrated by rhodopsin. The anti-parallel crystal dimer of rhodopsin (1hzx) viewed from the loops at the end of the helices, A, and from within the plane of the membrane, B, revealing the inhomogeneous distribution of participating lipophilic molecules (ball-and-stick) and metal ions (black Van der Waal's spheres). Retinal can be seen clearly from the cytoplasmic end in molecule Ⓐ, although it is nearer to the intradiscal surface where it is masked by the β-strand 'plug' and conserved cystine bridge, Cys110-Cys187, in molecule Ⓑ. The γ-sulfur of cysteines is shown. C, the molecular topology of rhodopsin reveals the low contiguity of the 8 helices (only TM3) and the position of the most conserved residues in each helix (asterisk). The highly conserved ER*Y and NP*xxY motifs are at the cytoplasmic ends of TM3 and TM7, respectively.

The $C\alpha$ distances for disulfide bridges in high-resolution protein structures vary from 3.8 to 6.8 Å[19] — a wide range that reflects the flexibility of the five torsional angles and explains the low success rates for placement of engineered disulfide bonds.[20] Only one novel disulfide is produced after photoactivation of rhodopsin, creating a bond that joins the endogenous cysteines $Cys145^{3.55}$ and $Cys316^{7.63}$ at the carboxyl ends of TM3 and TM7. Because these cysteines are 28 Å apart in the dark-adapted structure (Figure 10.4A and B), rhodopsin must undergo a drastic topological rearrangement upon activation.

The successful crystallization of bovine rhodopsin was dependent upon many critical factors: all experiments were performed under red light, including the x-ray diffraction experiments, otherwise the crystals deteriorated rapidly, the receptor was delipidated, and the solubilization detergent nonyl-β-glucoside was supplemented with the amphiphile heptanetriol and zinc, in order to stabilize the protein. All the

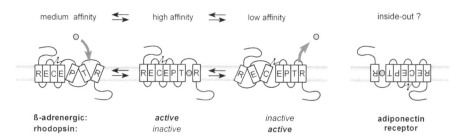

FIGURE 10.5 Receptor affinity cycle variations in signal transduction. The cycling of receptor affinity revealed by the association and dissociation of ligand and G protein with adrenergic receptors is not applicable to all 7TMRs.[7]

reported structures contained mercury ions, soaked into the crystals for phase determination of the diffraction patterns. The breakthrough publication reporting the 3D structure of rhodopsin (1f88) was recalculated to correct for twinning (1hzx), allowing the placement of the palmitoyl, detergent, and heptanetriol molecules. These lipid and detergent molecules were regionally distributed, reflecting the discrete distributions of amphipathy across the dimeric helical bundle (Figure 10.4A). Because polyunsaturated lipids are required for rhodopsin activation,[21] this emphasizes the different amphipathic demands of the structure. Similarly, the metal ions are not randomly distributed and, as usual, are mostly associated with cysteine residues (Figure 10.4B). The extent of infiltration by these stabilizing agents reveals the areas of accessible space within the protein, although it is not possible to distinguish between, say, the metal ions having gained access to the cysteines in the stably folded receptor structure or, by binding to the receptor, that they have contributed to stability and to the successful elucidation of the structure.

The 2.8-Å resolution rhodopsin structure has average temperature values around 50, although the intracellular loops have much higher values that are consistent with greater flexibility. A temperature factor >60 is usually cause for concern in structural assignment and the values for the cytoplasmic loops of the rhodopsin structures range up to 120. This flexibility is particularly evident for receptor chain Ⓑ, where the intracellular loops cannot be distinguished in the electron density (Figure 10.4B), which may be a reflection of the stabilizing influence of the missing interactive partner transducin.

The classical ternary complex model of ligand-induced, receptor-mediated allosteric activation of G protein envisages two receptor states: unactivated R and activated R*. More intermediate states have been elucidated, especially for rhodopsin, and the receptor affinity cycling of the β-AR paradigm depicts a G protein and ligand-dissociated, low-affinity state of the non-naive receptor (Figure 10.5).[7]

GDP–GTP exchange effects heterotrimeric G protein dissociation from the β-AR, producing a receptor conformation with a reduced affinity for ligand, which throws off the ligand and is no longer stimulated. However, neither ligands nor G proteins have to dissociate for effective coupling to adenylate cyclase,[22] and it is reasonable to assume that many receptors sample all of these affinity conformers even in the absence of ligand. The phenomenon of receptor "breathing" enables the entrapment of the agonist-induced conformer even in the absence of ligand, accommodating the concept of receptors with constitutively active mutations. As for allosterism in hemoglobin, the activated agonist-bound conformer may be the ligand-imposed "tense" form for some ligand–receptor combinations and the ligand-induced flexible or "relaxed" form for others. Evidence of receptor flexibility and conformational selection by ligands has been reported for AT1-R, using engineered mini-ligands of angiotensin.[23] The adiponectin receptor appears happy even when standing on its head.[15] Thus, some 7TMRs may be highly flexible; others may be relatively rigid; and yet others may require constant sampling of these states for signal transduction to be successful.

Parsimoniously, most experimental and modeling analyses of ligand binding in drug discovery are interested in mini-molecules like Lipinski-esque classical drugs[24] and assume ligand binding occurs in a rigid binding pocket within the helix bundle, like retinal in rhodopsin.

The post-crystal structure extrapolation to other 7TMRs inevitably provides a conserved core of residues,[16,26–29,31–33] with maintenance of the helical irregularity (although few were brave enough to model bent helices before the 3D structure became available). A review of exhaustive analyses using SCAM and receptor structure prediction[16] concentrates on the dopamine D2 receptor and is to be recommended as an example of the productive synergy that can be generated by combining experimental and computational biology. Extension of the general receptor model to other representatives of the rhodopsin family provided comparative information supporting the retention of a receptor core and binding pocket across the family — even for larger peptide and protein ligands that interact more with the amino terminal extensions or the receptor sequence and/or extracellular loops. The universal retention of a "binding pocket" could mean that it is always accessed, either directly or indirectly, by the remotely bound ligand. Ballesteros and co-authors[16] point out that the isosteres Cys and Ser, with Thr, have the ability to bend back upon themselves and bond with the sequentially adjacent backbone carbonyl of the preceding turn, encouraging the helix to bend and twist even in the absence of proline or glycine. Of course, the reactive chemistries of Cys and Ser are substantially different and subsequent publications only refer to Ser/Thr.[25]

10.5.2 Virtual Screening

Recent examples of molecular bioinformatic predictions of receptor structure are not content to use the concept of evolutionary entropy to map the conserved cores of global residues, but seek to delineate discrete areas that can be ascribed to different functionalities.[26,27] The conserved core of the global consensus receptor and the individual receptor signatures hint at discrete functionalities within the receptor, although again localized to predictable areas: around the "binding pocket" and that for G protein activation, with a communicating network between the two. Receptor signatures are then used for virtual screening (VS), the tuning of combinatorial libraries, and the generation of privileged scaffolds. The retention of the "binding pocket" imposes a practical boundary on any library, as molecules are flagged for their minimally aromatic qualities. A concept of privileged scaffolds, such as the spiro–piperidine–indane-based ligands,[27] is fulfilled before it is even tested. Inevitably, common skeletons produce a problem of cross-reactivity at other receptors, explaining why off-target effects are so often seen with prospective small molecule antagonists. The retention of the retinal "binding pocket" by most receptors, certainly *in virtuo* and perhaps also *in vivo*, does not enforce the endogenous cognate ligand to use this site, opening up other potential binding sites for possibly more precise pharmaceutical intervention. A conserved pocket may mean that an undiscovered ubiquitous allosteric modulator uses this pocket in many, if not all, receptors.

The generation of multiple comparative data from receptor models and VS places extraordinary demands on the analysis and interpretation of the results. The development of descriptors appropriate to clustering analysis is a fine art and an exacting science. Two recent examples of pattern analysis provide an interesting twist to receptor and pharmacophore fingerprinting, with one applied to cluster analysis of the receptors[28] and the other to the derivation of privileged scaffolds.[29] Süel and co-authors[28] considered that evolutionarily conserved residues in structural proximity will coassociate in clustering analyses, providing structural constraints and mutual dependency information. The scoring of these associations in a matrix of primary sequence and protein identity was converted to pseudocolor and analyzed using code originally developed for gene expression chips. Von Korff and Steger[29] used principal component analysis to generate self-organizing color maps to cluster pharmacophore information, providing distinctive patterns characteristic of ligand specificities.

In 2000, Caterina Bissantz and co-workers provided the original reference publication[30] for virtual screening and docking program assessment from known 3D structures (of soluble proteins, enzymes and a nuclear receptor), by which the performances of all subsequent programs are evaluated. The same group has now published a comparison of docking programs and scoring functions after a virtual screen of 1000-member chemical databases using homology models of five 7TMRs: dopamine D3, muscarinic M1, vasopressin V1a, β2-adrenergic, and δ-opioid receptors.[31,32] As may be expected from a homology model, with ~25% sequence identity for the TM regions, the performance of the VS was only mediocre, although better with antagonist-constrained models than when using agonists. They concluded that all antagonists stabilize a very similar ground state, simplifying the screen, but that agonists support several different bound states, complicating the screen because of receptor flexibility.

Structure prediction of the activated state of the receptors is absolutely necessary for any realistic attempts at VS and, in the absence of any surrogate structural information, it is extremely difficult and imprecise. Recently, an all-evidence prediction of the active and inactive opsin structures was extrapolated to the β2-AR and a VS run was performed on the two receptor states which exhibited marked preferences for agonists or antagonists, respectively.[33] Insufficient data coverage was identified for TM4, in particular, and it was suggested that the NPxxY motif enforced a kink in TM7 when the receptor was activated. The directions of helix movements upon accommodation of ligand were predicted to be similar to that reported for photoactivated rhodopsin, although the absolute extents were perceived to be less for the β-AR, which is probably an inevitable consequence of consensus modeling minimization.

Unfortunately, most analyses are quite conservative, often only reinforcing the current paradigm and not offering much in the way of novelty. The formulation of the query can constrain the results, as occurs when model parameterization favors retention of the rhodopsin template too assiduously. Most publications argue for a conserved network of core receptor residues that are responsible for allosteric communication from "out" to "in" and are predicted to be part of the machinery of the activation cascade. Modeling approaches break down if it is assumed that more than two molecules are involved in the formation of homodimers or heterodimers and higher order complexes,[34] or that the helices are free to move. This degree of flexibility is anathema to the practitioners of molecular modeling, docking routines, and virtual screening.[7]

A possible way out of this dilemma is to step back and take a fresh look at a comparison of sequence and structure, searching for major clues that could be interpreted as an activation mechanism via ligand-induced conformational alteration of the receptor.

10.6 SEQUENCE AND STRUCTURAL SEARCHES FOR AN ACTIVATION MECHANISM

The conventional view of ligand-induced allosteric regulation of G protein activation assumes some form of TM helix movement, involving some combination of translational or rotational movement, in whichever plane, and by bending, extension or uncoiling of a helix. Being able to capture this "activation" event would provide a desirable read-out of agonist activity and might also distinguish between other flavors of partial or inverse agonism, allosteric modulation, and antagonism.

The conserved sequence features of the rhodopsin family members are mapped onto the known rhodopsin structure to aid in the search of a putative activation mechanism. Apart from the obviously serpentine form and the S–S bridge, the highly conserved TM3 DRY and TM7 NPxxY motifs are the only other regions of particular interest (Figure 10.4C).

10.6.1 TM7 AND THE NPxxY MOTIF

The high conservation of the NP*xx(x)Y motif in the rhodopsin family receptors is reflected in the reference position for rhodopsin of Pro303$^{7.50}$, which is identified in the motif by an asterisk. Like all of the TM helices, except TM3, helix TM7 is not contiguous and is kinked. The location of the kink is maintained by Cys264$^{6.47}$, which acts like a peg upon which the TM7 helix is hung and closes a circle of interaction with retinal (Figure 10.6A). The γ-sulfur of Cys264$^{6.47}$ lies adjacent to Trp265$^{6.48}$, part of the highly conserved TM6 CWxP* motif, which provides an interactive surface for the cyclohexenyl ring of retinal. The polyene chain of the chromophore is attached to Lys296$^{7.43}$, one helical turn upstream from the kink in TM7 which is pegged down by Cys264$^{6.47}$. The stretch of helix between Lys296$^{7.43}$ and the kink adopts a 3_{10} configuration.

The importance of Asn302$^{7.49}$ is revealed by its interaction with Asp$^{2.50}$ in the core of the helical bundle. Intriguingly, the strict conservation of these residues is

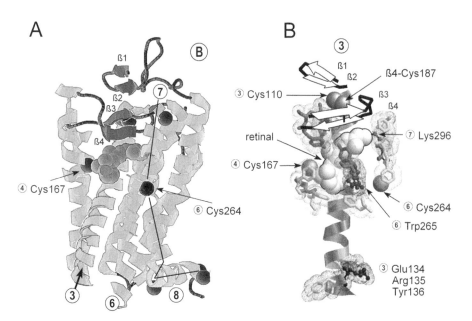

FIGURE 10.6 The conserved dithiol bridge and β4-strand fasten retinal to TM3. A, rhodopsin molecule B, sliced with a slabbing plane parallel to helix 3 (*cf.* figure 10.3A), reveals the polyene tail aligned with and extending the β-sheet. Lateral location of the ring directly by 3 Cys167 (γ-sulfur is shown) and through 6 Trp265 indirectly by 6 Cys264, acting as a pivot underneath which helix 7 (indicated by the thin line) is bent. B, the sulfur-aromatic cage fastening retinal to helix 3, with Tyr (darker atomic bonds) near the β-strands and mainly Phe (lighter atomic bonds) in the center of the bilayer, extends to incorporate 7 Lys296 (ball-and-stick). Helix 3 is contiguous from Cys110 to Arg135 (ball-and-stick).

violated by the GnRH receptor, in which the consensus aspartate Asp$^{2.50}$ is an asparagine and Asn$^{7.49}$ is an aspartate. The GnRH receptor is a minimalist 7TMR and is unusually short; helix TM1 is a retained signal peptide and sequence termination immediately at the end of the TM7 removes cysteine palmitoylation and desensitizing phosphorylation sites. Investigation had confirmed that this was a mutually compensating double substitution before the crystal structure of rhodopsin was known.[35] "Restoration" of the conserved Asp$^{2.50}$ was detrimental to binding and activation with GnRH, which was recoverable when the double substitution was made by "restoration" of Asn$^{7.49}$ as well.

The highly conserved Tyr296$^{7.53}$ is usually overlooked in most sequence–structure comparisons, although preoccupation with the non-conservation of the NPxxY motif in the GnRH receptor also led to SDM studies at this position. The importance of the phenyl ring was revealed by a Y332A receptor mutant that bound ligand but demonstrated no functional activity, whereas a Y332F mutant behaved like the wild-type receptor.[36]

It is not usually realized that the NPxx(x)Y motif is also a phosphorylation acceptor site.[37] Because it lies between helix 7 as it exits the membrane and helix 8 lying parallel to the membrane and is accessible to an antibody after photoactivation of rhodopsin,[38] then kinase interaction and phosphorylation of this interhelical loop is a feasible proposition. Surprisingly few studies have addressed the role of the tyrosine experimentally, although most recently an alanine substitution was shown to produce a biologically inactive B2 bradykinin receptor, which was constitutively phosphorylated and internalized.[39]

The substantial movement of the intracellular ends of the TM helices, moving 28 Å nearer together, was already evident in the colocalization of helices TM3 and TM7 after photoactivation, allowing a cystine bridge to form (Figure 10.4). The Tyr296[7.53] position between helices 7 and 8 is separated from Arg135[3.50] at the carboxyl terminus of TM3 by only one hydrophobic reside. The arginine of the TM3 DRY motif was one of the first residues implicated in signal transmission for both the rhodopsin and the secretin receptor families.

10.6.2 TM3 AND THE DRY MOTIF

In the rhodopsin family receptors, the highly conserved Arg[3.50] in the DR*Y sequence at the cytosolic end of TM3 has been proposed to operate as an arginine switch (Figure 10.4B). The highly positive, large side chain was hypothesized to swing away from the receptor upon ligand binding, possibly being released from the conserved TM2 aspartate, Asp[2.50], and to switch the G protein into an active conformation.[40] The motif equivalent to the DRY sequence (ERY in rhodopsin) is EGLY in the secretin family receptors, which suggests that a negative charge and a tyrosine residue may also have a role in a common signal transduction mechanism. Arginine (and tyrosine) residues are often present at the cytosolic ends of TM helices, although this may be due to the general use of cationic residues as helix terminators in most transmembrane proteins (or possibly potential phosphorylation sites; see above).

The arginine "switch" concept has evolved into the concept of an arginine "cage" by the incorporation of a hydrophobic residue Val139[3.54], which further restricts arginine movement.[41] Interestingly, the universal, constitutively activating mutation position[42] β2-AR Ala296[6.34] is occupied by Thr251 in rhodopsin and lies adjacent to Arg135[3.50], implying that its activating effect is by perturbation of the arginine switch. As mentioned above, Arg135[3.50] is separated from Tyr296[7.53] by only one hydrophobic residue, Leu72[2.39], suggesting that movement of the intracellular ends of TM helices 3 and 7 is coordinated during allosteric activation. This proposition has been condoned recently by the demonstration that constitutively activating mutations of the LH receptor involving Met389[2.43] correlated with the size of the mutated residue.[43] Tyr623[7.53] in the LH receptor was identified as the residue perturbed by the larger substitutions, encouraging proximity modifications at the DRY motif which promoted constitutive activity.

Not only is TM3 the only continuous and unbroken helix of the seven, but also it has highly conserved residues at both ends, with Arg135[3.50] intracellularly and Cys110[3.25] extracellularly (Figure 10.6B).

10.6.3 TM3 AND THE CONSERVED CYSTINE BRIDGE

The extracellular dithiol bridge between TM3 and the second outer loop is the only motif common to the 7TMR superfamily, apart from the ubiquitous presence of the

7TM helices. The second extracellular loop in rhodopsin produces a small region of antiparallel β-sheet, strands β3 and β4, which is extended by strands β1 and β2 from the amino terminal chain to produce a "plug" over retinal in the binding site (Figure 10.6). Retinal is located against TM3 and surrounded by a π-bonded cage of 12 aromatic residues, mainly tyrosine toward the solvent-exposed parts of the receptor and phenylalanine toward the center of the bilayer. The cyclohexenyl ring of retinal is wedged between Trp265$^{6.48}$ and an aromatic stack supported by Cys167$^{4.56}$. As noted above, the covalent attachment of retinal to Lys296$^{7.43}$ is reinforced by the involvement of Cys264$^{6.47}$, especially in fixing TM7 and in supporting the major π-bonded contribution from Trp265$^{6.48}$. Cys264$^{6.47}$ is also part of the retinal aromatic cage that isolates Lys296$^{7.43}$.

The β4 strand arises from a helical turn that is divorced from the shortest helix 4, broken prematurely by a PP, or PxP, sequence in most receptors. The β4 strand is locked against TM3 by the cystine bond contributed by β4-Cys187 and the conserved cysteine at the amino terminal end of TM3, Cys110$^{3.25}$. The β4 strand is H-bonded to the retinal polyene chain, almost as if it were an extension of the β sheet (Figure 10.6A). The integrity of the extensive aromatic sulfur cage is critically dependent upon the juxtaposition of residues imposed by the restraining cystine bond between TM3 and β4.

From the earliest cloning days, SDM of all of the individual cysteines in the β2-AR demonstrated that those in the second extracellular loop were important for agonist binding,[44] rationalizing the many biochemical reports showing that reduction generally increased receptor-mediated activity, and usually at the expense of receptor stability. These original observations were confirmed by innumerable reports using the same and other 7TMRs and almost invariably attributed to an indispensable structural role for the extracellular dithiol bridge (Figure 10.7).

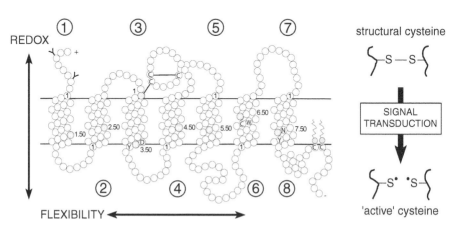

FIGURE 10.7 Flexibility and redox provide a broad range of binding and activity modalities. Membrane receptors sample the external oxidizing and internal reducing environments of the cell and are mediators of signal transduction between external ligands and internal second messenger systems, possibly by exploiting the redox potential gradient across the membrane.

The receptor sits in the plasma membrane and samples the external oxidizing and intracellular reducing environments simultaneously. Extracellular cysteines are presumed to be relatively inert structural cysteines forming molecular bridges, whereas intracellular cysteines are highly reactive and often enzymatically active. For a comprehensive overview of the chemical activities of cysteines and involvement in cellular processes, see the excellent reviews from Jacob and co-authors.[45,46]

Having a disulfide bridge that is structural does not preclude the same bridge from involvement in the functional activity of the receptor. We considered that signal transduction induced by ligand binding could involve the exterior disulfide-bonded cysteine residues interacting, directly or indirectly, with the reduced cysteine residues of the G protein cascade within the cell. This process can be modeled as a progressive extracellular-to-intracellular disulfide–thiol exchange initiated by external ligand binding that transduces the signal internally.

This proposition satisfied a number of prejudices and expectations: (1) reduction of the cystine bridge may be the liberating event for activation and allosteric or conformational changes in the receptor; (2) flexibility was already identified as a major characteristic of multispanning TMRs; (3) the two together would provide a broad range of binding and activation modalities, sufficient to accommodate such a diverse set of ligands, temporal responses, and second messenger specificities; (4) reduction of the cystine bridge would also liberate the participating cysteines for labeling by thiol-reactive reagents; and (5) what better way to communicate a signal from the outer oxidizing environment to the inner reducing environment than by the invocation of a redox-dependent switch?

10.7 THE CYSTEINE SHUFFLE

To test the feasibility of tagging receptor cysteines with a thiol-labeling reagent, the redox sensitivity of a radioligand receptor assay (RRA) for the 25-kDa glycoprotein hormone TSH was established. Unlike the 12 cysteines in the amino terminal extension of the related LH receptor, TSH (and FSH) receptors have only 11 extracellular cysteines, beyond the 2 involved in the conserved cystine bridge, suggesting that a free thiol may be available and influence ligand binding. Indeed, low concentrations of the thiol alkylating agent, *N*-ethyl maleimide (NEM), or thiol oxidizing agents promote the binding of TSH to its receptor, implying that reduced thiols hinder TSH binding. Dithiol reductants such as dithiothreitol (DTT) destroy binding and also reverse this effect of NEM, reinforcing the importance of intact cystine bridges to the maintenance of receptor structure.

The results in Figure 10.8A are from an early experiment designed to show that ligand binding to a receptor increases the sensitivity of the receptor to redox reagents — an indirect demonstration of ligand-induced conformational change in the receptor. The extent of radiolabeled-TSH binding in the control log–dose response curve of the TSH RRA is lowered when the receptor preparation is pretreated with a maximally tolerating dose of NEM (about 300 µM), to block already extant thiols. Addition of 300 µM NEM to the control curve or the pretreated receptor curve lowers binding to background levels as the receptor is destroyed. These results

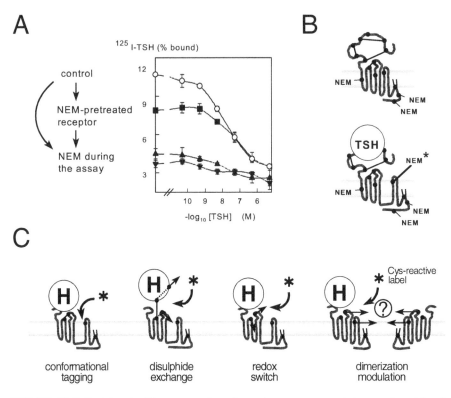

FIGURE 10.8 Cysteine shuffling as a marker of receptor rearrangement. A, experimental and B, diagrammatic demonstration of enhanced receptor cysteine accessibility during ligand binding.[47] C, various ways by which ligand-induced tagging of receptor cysteines may arise.

demonstrate unequivocally that the receptor preparation is more sensitive to the damaging effects of NEM in the presence of ligand than in its absence. By using an NEM that incorporates a label or marker molecule, it should be possible to tag the receptor cysteines that have become newly exposed, as a direct consequence of ligand binding and receptor conformational changes.

The procedure for ligand-induced tagging of receptor cysteines is depicted in Figure 10.8B. Readily accessible cysteines are blocked by pretreatment of the receptor by NEM. The pretreated receptor is then exposed to ligand, which binds and induces a conformational change in the receptor. Any newly exposed receptor cysteines, as a consequence of ligand binding, can be tagged by a subsequent or concomitant exposure of the ligand–receptor mixture to a labeled NEM preparation.

The principle of the assay allows the order and/or duration of reagent addition and exposure to be altered according to circumstance, e.g., the second pictogram in Figure 10.8B of NEM* with bound ligand can precede the first, and is an effective way of capturing receptor thiol groups in flux between hidden and exposed. By applying this technique to other receptors in the rhodopsin family, including adrenergic, peptide, and protein hormone receptors, it became rapidly clear that all tested receptors were redox sensitive and capable of ligand-induced labeling. Even peptide ligands with (endothelin, oxytocin) and without (GnRH, angiotensin) endogenous cystine bonds conformed and their receptors were labeled. Because the procedure was applicable to intact cells as well as membranes, it was eminently suitable for HTS, and the Cys-screen method of ligand-induced labeling of receptor cysteines as a measure of receptor conformational change has been patented.[47]

Ligand binding can influence the exposure of receptor cysteines in several ways and some of the most interesting are depicted in Figure 10.8C. The cysteine shuffle can involve straightforward conformational tagging or result from disulfide exchange between ligand and receptor, a ligand-induced redox switch, or even through the modulation of multicomponent complexes (as observed for the metabotropic 7TMR family and receptor tyrosine kinases). Each of these mechanisms is capable of producing a readout in a screening scenario.

The proposition that the transition from the inactive to the active form of the receptor is dependent upon or reflected in the disposition of the cysteine residues in the receptor is attractive for two major reasons: (1) the general placement of cysteine residues in proteins means that they are ideal monitors of conformational changes in protein molecules and (2) cysteines are the most chemically reactive amino acids, which provide a convenient group for specific labeling and also a reactive species with the potential for signal transmission. Therefore, an assay that measures ligand-induced thioreactive molecular tagging of a receptor could have the potential to discriminate between agonists and antagonists in a single screen.

Regardless of whether the cysteine residues are oxidized or reduced, their relative hydrophobicity means that they are usually to be found traversing the hydrophobic–hydrophilic boundary of proteins.[6] A survey of cysteine locations in protein structures reveals that the majority of cysteines reside just below the solvent-exposed surface of protein molecules (Scherzler, Procter, and Willey, unpublished observations).

In extracellular proteins, cysteines form S–S bonds that are conformational constrainers and also tie exterior and interior peptide chains together. Upon reduction of these bonds, a protein is no longer constrained and is free to adopt other conformations. Binding of ligand would be expected to induce a conformational change in the receptor, such that the cysteines lying at or near the surface of the molecule could become newly hidden or newly exposed. An exposed thiol group at or near the surface of the receptor can be detected readily using a multitude of marker molecules. The receptor is the pharmacological discriminator and it is highly appropriate for it to be tagged in the course of the Cys-screen. A major advantage of covalent cysteine labeling is its "permanent" nature, which allows signal detection to be measured as and when required.

The fundamental question is whether the cysteines are passively associated with or actively involved in hormone recognition and activation. Either activity may be useful in the development of a readout of hormone–receptor interaction, and it is conceivable that different cysteine residues exhibit differential activity profiles in one receptor compared with another.

10.8   REDOX CONTROL OF RECEPTOR ACTIVATION

Over the years, the sensitivity of ligand–receptor interactions to redox reagents has been demonstrated for most 7TMR systems. The roles of 7TMR and G protein cysteines have been extensively investigated using redox reagents and SDM, both alone and in combination. Ligand interactions with and activations of 7TMRs are sensitive to mild oxidation (generally beneficial) and reducing agents (generally destructive), as well as divalent metal ions and other typical cysteine-reactive entities. Conversely, G proteins are sensitive to NEM and reducing agents promote their activity. Interpretations from the use of redox reagents have always been unequivocal, both because of the different redox dosage sensitivities of the various components and because of the intricate feedback of, say, G protein release that downregulates receptor affinity and invites phosphorylation and internalization of the receptor. Anything that compromises one of the components may affect the performance of the others, directly or indirectly.

The ligand and G protein exist in different redox environments and communicate via a receptor that samples both environments. Are the confusing effects of redox reagents caused by different cellular backgrounds or do they reflect the microenvironment within the receptor? Substantial published evidence supports the role of the receptor, with its flexibility, as the mediator of redox-mediated signaling, and here are a few examples — some old and some new.

10.8.1   LIGANDS DEVOID OF SULFUR

Over 25 years ago, investigation of polyamines *in vivo* established that the tetramine disulfide, benextramine, was an α-adrenergic receptor (α-AR) antagonist. Further work established that inhibition by benextramine was irreversible, time-dependent, and essentially specific for α-ARs (muscarinic and nicotinic acetylcholine receptors were also affected, although these were considered ionic interactions). A hypothesis was proposed that a buried thiol group in the receptor was unmasked by benextramine binding; thiol–dithiol interchange resulted in covalent attachment to the receptor that could be alleviated using a cationic thiol such as cysteamine.[48]

More recently, an alkylating derivative of the α2-adrenergic agonist clonidine was found to bind irreversibly to α2-ARs. A covalent bond is formed with the sulfhydryl side chain of a cysteine residue in TM5 exposed in the binding cavity, leading to inactivation of the receptor. Subsequently, the recognition of inactive and active receptor conformations was claimed after differential labeling of receptor cysteines.[49] Thus, the cysteine residues of adrenergic receptors have both structural and functional roles in ligand activation, with the receptor demonstrating a range of flexibility.

Benextramine (but not other thiol-reactive agents such as cysteamine or DTNB) also blockades the 7TMR for neuropeptide Y.[50] Thiol-reducing agents reversed the effect of benextramine, freeing the receptor for subsequent NPY binding. Benextramine is an antagonist of α-adrenergic, and presumably NPY, hormone activity. Although covalent attachment of benextramine to these receptors was perceived, there was no evidence to suggest that other antagonists, or even agonist ligands, are capable of covalent linkage with the receptor.

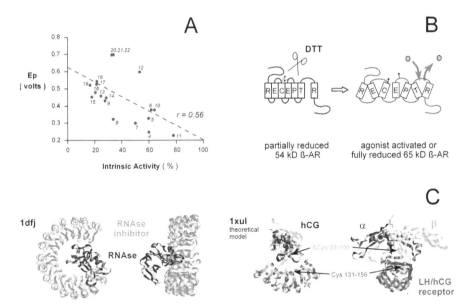

FIGURE 10.9 Evidence for redox control of 7TMR binding and activation. A, the relative adenylate cyclase activity of 19 β-adrenergic partial agonists is correlated with their reductive potential measured by cyclic voltammetry, E_p, reproduced with permission from Wong A, *et al*, in Molecular Pharmacology, © 1987 American Society for Pharmacology and Experimental Therapeutics.[51] B, exposure to either agonist or reducing agent produces a fully unfolded β-adrenergic receptor.[52] C, the retention of RNAse by its horseshoe-shaped inhibitor is through disulfide exchange under redox control (1dfj) and a theoretical model for the LH/hCG receptor (1xul) places a cystine bridge adjacent to the receptor-determinant loop of the hormone which has intrinsic thioredoxin activity.[53]

Based on the redox sensitivities of β-ARs and involvement of the attendant molecules of the four-state ternary complex of ligand, receptor, and G protein, a redox mechanism for the action of agonists at β-adrenergic receptors was proposed nearly 20 years ago.[51] The hypothesis was presented that agonists activate β-ARs by reducing them and this was examined by analyzing 41 agonists and antagonists. The structural features that determined binding affinity were shown to be distinct from those that determined agonist activity. Agonist activity was shown to be related to the oxidation–reduction properties of the ligands, which were determined primarily by the nature of the substituents on the phenyl ring. Of the 19 antagonists, all exhibited E_p values greater than 0.75 V (electrochemical peak potential for the first oxidative wave in cyclic voltammetry), suggesting that they were difficult to oxidize. Partial agonists, however, exhibited a wide range of E_p (0.25 to 0.7 V) with values lower than those of the antagonists (Figure 10.9A).

Also in the mid-1980s, Western blots with DTT-treated β-AR demonstrated the importance of receptor cysteines and their availability during ligand activation. Indeed, the behavior of the receptor in the presence and absence of thiol reduction led to a hypothesis of ligand binding controlled by disulfide bridge breakdown.[52] Pure β-AR migrated with a size of 54 kDa on SDS-PAGE, which increased to 65 kDa after chemical reduction with reagents that cleave disulfide bridges (Figure 10.9B). A possible explanation for this behavior is that the smaller molecule was compact because of intramolecular disulfide bridges. Cleavage of these bonds would allow true unfolding of an elongated molecule that would migrate nearer to its true molecular size. Agonist exposure before purification of the receptor increased the proportion of the larger, unfolded molecular species, implying that this was concomitant with receptor activation. Because the cytosol of a cell is a strongly reducing environment and most intracellular cysteines exist as free sulfhydryl groups, the authors concluded that all the cysteines in the extracellular loops and also in the TM helices 2, 3, 6 and 7 were disulfide-bonded in the vacant receptor configuration. Agonist binding would promote disulfide exchange and transform the receptor into its transducing configuration. The authors[52] consider "whether receptor disulfides are avenues to receptor activation or merely an unavoidable but provocative dead end." In the intervening years, this issue has still not been resolved — a problem mainly related to the advent of accessible molecular biology and the cloning boom of the late 1980s. Although many novel 7TMR sequences and ligands have since been discovered, biochemical pharmacology was seriously neglected in the 1990s and only recently underwent a rejuvenation, with SCAM and other mutation replacement and removal strategies for chemically reactive residues.

10.8.2 LIGANDS CONTAINING CYSTEINE

It has been known for many years that the glycosylated trophic hormones LH and FSH exert intrinsic thioredoxin-like activities.[53] Thioredoxin uses a localized disulfide bond to form mixed disulfides with cysteine-containing proteins, promoting either the cleavage of preformed disulfide bridges or the formation of cystine bridges from reduced thiols. Thioredoxin activity is usually measured by its ability to refold denatured RNAse, a property expressed by LH and FSH that is stronger than that expressed by thioredoxin itself. Both the α and β subunits of the hormones possess a similar localized disulfide bond that may be responsible for this catalytic property.

The thioredoxin activity of the glycotrophic hormones was considered an interesting biochemical anomaly until molecular cloning of the receptors revealed an extensive amino terminal sequence consisting of numerous leucine-rich repeats with 12 (LHR) or 11 (FSHR, TSHR) cysteines. A model for this domain relies upon another protein with characteristic leucine-rich repeats, a natural inhibitor of RNAse designated ribonuclease inhibitor (RNI). The 3D molecular structure of RNI complexed with RNAse (1dfj) reveals an annulus of a repetitive motif containing an α helix and a β strand. The inhibitory capacity of RNI depends upon the redox status of the many cysteine residues in RNI and the intermolecular disulfide bridges it forms with its captive RNAse.[54]

Subsequently, the 3D molecular structure of deglycosylated hCG (a paralog of LH) was solved and it revealed three unexpected features: (1) these hormones belong to a structural family of cystine-knot growth factors that also includes TGFα (and thereby activin/inhibin), NGF, and PDGF; (2) the α and β subunits have similar folds; and (3) the extra disulfide bond of the β subunit acts as a unique "seat-belt" loop, wrapping around the α subunit. This latter disulfide bond

(Cys26–Cys110) is the most labile of the 11 in the hormones, is part of the receptor-reactive surface, and is adjacent to the vicinal cysteine loop (Cys93–Cys100) crucial for bioactivity.[6]

The evident similarity of RNI and the receptors, the ability of RNI and RNAse to undergo disulfide exchange, and the abilities of hormones to interact with RNAse imply that cysteines and disulfide exchange may be features to be expected of glycoprotein hormone interactions with their receptors. The docking of hCG to a model of its receptor serendipitously places the labile ligand cystine adjacent to a cystine in the receptor. This may allow the thioredoxin activity of the hormones to be involved in some aspect of signal transduction (Figure 10.9C).

Mixed disulfide exchange can be expected from other peptide and protein ligands with free thiols or cystine bridges, providing a redox-dependent chemical reactivity with the capability of receptor activation or modulation. The defining characteristic for the large families of chemokines is the disposition of cysteines, splitting them into CC, CXC, or CX3C chemokines. Human interleukin-8 is a 72-residue CXC chemokine with two cystine bridges, Cys7–Cys34 and Cys9–Cys50, both of which are critical for structure and function. The ligand surface incorporating the 7–34 cystine is implicated in agonist activity and includes the residues preceding both Cys7 and Cys34 in the sequence.[55]

The demonstration that thioredoxin has chemokine activity[56] and that β-defensins interact with chemokine receptors[57] implies, in the absence of any sequence or structural similarities, that the cystine bridges in these proteins are responsible for the chemokine receptor effects.

Many peptide ligands have a constraining disulfide or two and at least one set is an absolute requirement for bioactivity. The most recently discovered peptide ligands are the vasoactive hormones, endothelin and urotensin, and they exhibit many of the characteristics to be expected from ligands that undergo disulfide exchange with a receptor, including protracted association rates that never attain equilibrium and the inability to dissociate from the receptor. Both of the endothelial endocrine peptides have receptor affinities (low pM) that are greater than those of most other peptide hormones and are markedly more potent than the *in vitro* biological responses engendered by that binding — the converse to most other systems. The protracted kinetics, the recalcitrance to wash-out from treated tissues,[58] and the better binding than bioactivity potencies are all indicative of ligand persistence at the receptor through disulfide exchange between ligand and receptor. Whereas the exaggerated binding potencies are artefacts of an assessment that assumes equilibrium binding and a dissociable system,[59] the "covalent" persistence at the receptor may well explain their remarkable efficacy.

Endothelin is the most studied of the two peptides and receptor binding was recognized long ago to be extremely sensitive to redox. Indeed, 2D-PAGE of endothelin–receptor complexes confirmed the covalent attachment of the ligand to its receptor, which was prevented by prior blockade of free thiols in the receptor preparation.[60] Preincubation of receptors with agonist (but not with antagonist) protected receptors from the alkylating reagents. Obviously, the receptor has a labile sulfhydryl group at or near the binding site that may undergo disulfide exchange with the ligand.

Thus, redox capabilities are relevant to 7TMR-ligand interactions and signal transduction for small and large ligands alike, especially those with cysteines, but the receptors also have interactive cysteines that are especially important for monoamine binding and agonist activities, which appear to be modulated by the reductive potential of the ligand molecule.

Cysteines are remarkable chameleon residues, but where is the specificity? Consider for a moment the serine residue. Nothing is remarkable about serines; they are considered to serve as the isostere for cysteine in SDM and VS (neglecting the redox switch potential of the –SH as opposed to the –OH group). Serine (and threonine) residues pepper the surfaces of proteins. However, put an innocuous serine at the edge of a hydrophobic pocket, adjacent to a histidine and an aspartate, and cleavage of a peptide chain can be achieved. Indeed, caspases use this catalytic triad, only substituting the serine with a cysteine and, of course, these intracellular effectors of apoptosis, in response to oxidative damage, are under redox control. The microenvironment of the receptor creates the redox specificity and the most likely "hot spot" is the highly conserved, indispensable cystine bridge that locks retinal into its aromatic cage.

10.9 SMELLS RANK AS THE SMALLEST 7TMR AGONISTS

The availability of a ligand or receptor cystine or cysteine thiol is a facile mechanism by which ligand-induced shuffling of receptor cysteines can be achieved (Cf. Figure 10.8C). However, as noted, not all agonist ligands and hormones possess cysteine residues and thus another means requires identification. The dependence of the bioactivity potency of adrenergic analogues upon their chemical reducing potential offers a substantial clue.[51]

10.9.1 SULFUR, CHARGES, RINGS

The presence of a positive charge appears to be a minimal requirement in monoamine agonist ligands and an aromatic structure is common. The long wavelength absorbance and intrinsic fluorescence of aromatic amino acids are much appreciated in the study of proteins, and the potent quenching ability of adjacent sulfhydryl and disulfide moieties is a frequent and annoying feature, although it is also evidence of a close affinity between aromatic and cysteine residues. The association of sulfur and π-bonded atoms in forming extended alternating chains was noted many years ago in a survey of protein structures[61] — an observation that has been rediscovered many times since then. Aromatic ring stacking is well recognized, but the incorporation of sulfur-containing amino acids into a stack received less attention until recently.[62] Cation–π interactions are also widespread in protein structures and extensive associations of all three groups, sulfur, ring and cation, especially in ligand binding, have been reported.[63] Of course, these favored interactions do not have to be contributed by different parts of a single molecule. Because sulfur-containing chemicals exhibit an affinity for aromatic rings, then the converse must be true and the aromatic rings usually to be found in ligands would be expected to seek out an association with sulfhydryl and cysteine groups in a receptor. The same interactions would be preferentially selected during ligand–receptor binding and in the generation of bioactive molecules such as hormones and neurotransmitters.[7] Certainly, the redox activation of the β-AR implies that amine, aromatic, and sulfhydryl groups are important for agonist activity. Assuming a scenario for ligand–receptor interaction in which important reactions can occur between amine, aromatic, and sulfhydryl groups, how is the redox activity of a monoamine manifest at the receptor and what is the result?

A recent theoretical publication developed a model for a δ-opioid binding site[64] in which the docked arrangement of the ligands suggested a reaction mechanism for the cleavage of the disulfide bond (Figure 10.10). Semi-empirical quantum chemical calculations determined that the interaction between the carboxyl side chain of an aspartic acid and the disulfide bond led to the polarization and withdrawal of a proton from the protonated nitrogen of the ligand to one of the sulfur atoms. A mixed sulfenic acid and carboxylic acid anhydrate is formed as an intermediate, as well as a thiol, which results in cleavage of the disulfide bond. The authors concluded that the suggested mechanism may explain the action of agonists and antagonists and it was assumed to be common to many GPCRs.

In a limited experiment, Brandt and colleagues[64] showed that agonist-stimulated cells took up more of a cysteine-reactive label than did unstimulated cells. This cannot be interpreted as a convincing demonstration of receptor tagging, especially because the molar excess of reactive thiols as opposed to receptors in a cell is of the order of 10^{12}, defying the generation of a specific signal. A case could be made for a general activation of the cell and an increased turnover in exposed thiol groups, such as those in the heterotrimeric G proteins, which occupy intersubunit interfaces and are exposed on dissociation.[65]

This chapter started with a sense of vision and it will end with a taste or smell of what is to come in drug discovery with 7TMRs by celebrating the award of the 2004 Nobel Prize for Medicine to Richard Axel and Linda Buck for their discoveries of odorant receptors and the organization of the olfactory system.[66]

10.9.2 THE ESSENCE OF AGONISM

The different categories of smells and aromatic substances can be readily distinguished. Pleasant and fruity odors are generally esters, ketones, and alcohols distinguished by the presence of an oxygen, often double-bonded. Distinctive and less pleasant smells are associated with nitrogen, such as the fishy smells of quaternary amines. The word *aromatic* conjures up a mental image of ring structures or π-orbitals and a wholesome smell or whiff of a volatile organic compound. The worst smell of all is methyl mercaptan, and sulfurous compounds in general are associated with death and decay, the curate's egg and Hades.

The distinctive chemistries of oxygen-containing -ols and -ones, basic amines, aromatic rings, and sulfurous compounds are responsible for the distinct categories of smells and are the essential features of templates that preferentially react with 7TMRs or GPCRs,[67,68] for example, piperidine with a basic and aromatic region, or a quaternary nitrogen, or cyclam groups. It is hardly surprising that the receptors responsible for smell discrimination are 7TMRs.

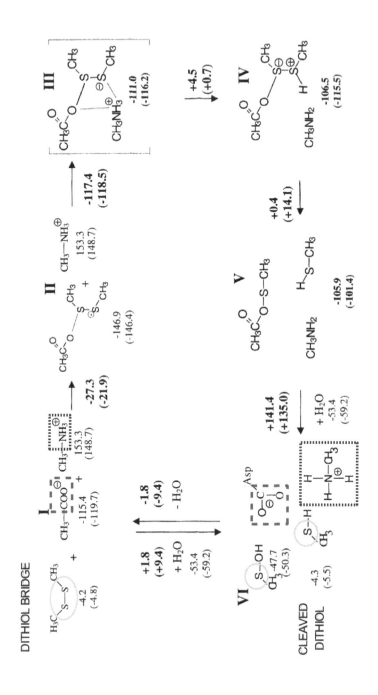

FIGURE 10.10 A mechanism of agonist-induced cleavage of the receptor dithiol bridge. Semi-empirical quantum chemical calculations for the cleavage of cystine by a carboxylic acid and protonated amine. Reproduced with permission from Brandt W, *et al*, in European Journal of Biochemistry, © 1999 Blackwell Publishing.[64]

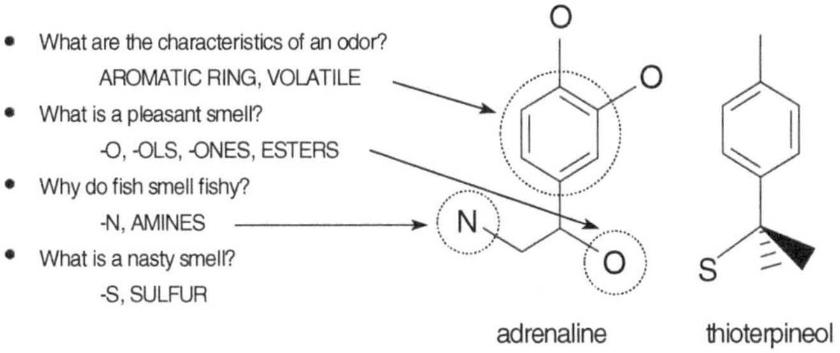

- What are the characteristics of an odor?
  AROMATIC RING, VOLATILE
- What is a pleasant smell?
  -O, -OLS, -ONES, ESTERS
- Why do fish smell fishy?
  -N, AMINES
- What is a nasty smell?
  -S, SULFUR

adrenaline   thioterpineol

FIGURE 10.11 Smell molecules are lowest common denominator agonist templates. The major categories of odor molecules have definitive chemistries that are related to the small molecule templates of ligands bioactive at 7TMRs.

The >1000 genes that encode the olfactory receptors have extremely diverse sequences and each olfactory epithelial cell expresses only one gene. Linda Buck of the Howard Hughes Medical Institute should know, for she developed a single-cell PCR technique to clone many of them. An olfactory chemical will stimulate an array of receptors and cells to different extents, depending upon its structure and their specificity determinants. Many different olfactory stimulants were tested and each was matched to a particular sequence by cloning the individual 7TMR from each of the neurons that fired.[69] It was found that only one olfactory 7TMR is expressed by a single neuron, although an array of receptors (and cells) will respond to a single olfactory chemical to different extents. The signals generated in the olfactory sensory neurons are mapped onto the olfactory cortex to produce a combinatorial code for sensing smell.

Some odorants are perceived as having different odors at different concentrations. A striking example is thioterpineol, whose odor is described as 'tropical fruit' at a low concentration, as 'grapefruit' at a higher concentration, and as 'stench' at a still higher concentration. Our studies indicate that an odorant at different concentrations can be recognized by different combinations of olfactory receptors. Thus, a change in the concentration of an odorant can change its receptor code, and this in turn may lead to a change in odor quality.[69]

Adrenaline is a biogenic amine derived from tyrosine and has a template typical for 7TMR reactivity (Figure 10.11). Certainly, the redox activation of the β-AR[51] and the proposed mechanism for disulfide bridge cleavage[64] imply that amine, aromatic, and sulfhydryl groups are important for agonist activity. The sulfur aromatic cage that locks retinal into rhodopsin demonstrates the intimacy and synergy of cysteine (with methionine) and tyrosine (with other aromatic rings) in ligand–receptor interactions. The active ingredient of grapefruit juice, 1-p-menthene-8-thiol, is one of the most powerful odorants known to science and is still detectable at 0.0001 parts per billion. Thioterpineol is probably an effective ligand at many 7TMRs, which raises the interesting and not only philosophical question of what sort of odor do hormones and neurotransmitters have.

The possession of a charge and an aromatic ring is a common feature of agonist activity, but not all ligands have a sulfur or cysteine moiety, unlike the receptors, although the most potent do. Smell molecules are small. The chemicals that smell and the receptors that respond provide the lowest common denominator for agonist activity at 7TMRs.[7]

10.10 BIOINFORMATICS WITH BIOLOGICAL SENSE

In conclusion, the molecular mechanism of binding and activation is reappraised by analysis of structure–function relationships in respect of receptor sequences and phylogenetic relationships. The wealth and superfluity of GPCR data bedevil holistic appreciation and contemplation of a universal activation mechanism for 7TMRs. The general lack of amenability of 7TMRs to manipulation and to structural determination efforts are recurrent themes throughout this volume on GPCRs in drug discovery, requiring highly concentrated effort and many ingenious approaches.

Bioinformatic analyses are less constrained and can evaluate virtual scenarios and exploit comparative information to glean new insights and comprehension. Despite this freedom, analyses can be too restrictive when developed from the safety of a particular model and may simply bolster the status quo, precluding the discovery of new models and failing to question established paradigms. Analyses may also be too rarified and fail to acknowledge the murkiness and non-ideal state of the biological condition, neglecting physiological context, the contributions of ligand pharmacokinetics and dynamics, and the relevance of general evolutionary (phylogeny) and individual developmental (ontogeny) constraints.

Just as chemoinformatics and molecular visualization depend on chemical sense, the execution of a computational biology approach that attempts to integrate all relevant information in a biologically holistic and comprehensive way, with biology and not computational tractability as the final arbiter, is bioinformatics with biological sense. Biological sense would suggest that the chameleon reactivities of cysteine residues, ranging from the textbook definitions of "simply" structural to the highly reactive chemistries of enzymatic activity and redox modulation of protein function, is a potential that nature cannot fail to exploit. By knowing the mechanism of receptor activation, drug discovery initiatives can also exploit this potential.

REFERENCES

1. Schwartz, T.W.; Rosenkilde, M.M. Is there a 'lock' for all agonist 'keys' in 7TM receptors? Trends Pharmacol Sci. 1996, 17, 213–216.
2. Palczewski, K.; Kumasaka, T.; Hori, T.; Behnke, C.A.; Motoshima, H.; Fox, B.A.; Le Trong, I.; Teller, D.C.; Okada, T.; Stenkamp, R.E.; Yamamoto, M.; Miyano, M. Crystal structure of rhodopsin: a G protein-coupled receptor. Science. 2000, 289, 739–745.
3. Anantharaman, V.; Aravind, L. Application of comparative genomics in the identification and analysis of novel families of membrane-associated receptors in bacteria. BMC Genomics. 2003, 4, 34.

4. Kolakowski, L.F., Jr. GCRDb: a G-protein-coupled receptor database. Receptors Channels. 1994, 2, 1–7.

5. Basler, K.; Christen, B.; Hafen, E. Ligand-independent activation of the sevenless receptor tyrosine kinase changes the fate of cells in the developing Drosophila eye. Cell. 1991, 64, 1069–1081.

6. Willey, K.P. An elusive role for glycosylation in the structure and function of reproductive hormones. Hum. Reprod. Update. 1999, 5, 330–355.

7. Willey, K.P. Structure-Guided Drug Design. D&MD Publications, Westborough, MA, 2004; 500 pp.

8. Fredriksson, R.; Lagerstrom, M.C.; Lundin, L.G, Schioth, H.B. The G-protein coupled receptors in the human genome form five main families: phylogenetic analysis, paralogon groups, and fingerprints. Mol Pharmacol. 2003, 63, 1256–1272.

9. Kauzmann, W. Some factors in the interpretation of protein denaturation. Adv Protein Chem. 1959, 14, 1–63.

10. Phoenix, D.A.; Harris, F.; Daman, O.A.; Wallace, J. The prediction of amphiphilic α-helices. Curr Protein Pept Sci. 2002, 3, 201–221.

11. Okada, T.; Le Trong, I.; Fox, B.A.; Behnke, C.A.; Stenkamp, R.E.; Palczewski, K. X-ray diffraction analysis of three-dimensional crystals of bovine rhodopsin obtained from mixed micelles. J Struct Biol. 2000, 130, 73–80.

12. Iwata, S.; Ostermeier, C.; Ludwig, B.; Michel, H. Structure at 2.8 Å resolution of cytochrome C oxidase from *Paracoccus denitrificans*. Nature. 1995, 376, 660–669.

13. Inooka, H.; Ohtaki, T.; Kitahara, O.; Ikegami, T.; Endo, S.; Kitada, C.; Ogi, K.; Onda, H.; Fujino, M.; Shirakawa, M. Conformation of a peptide ligand bound to its G-protein coupled receptor. Nat Struct Biol. 2001, 8, 161–165.

14. Fu, D.; Libson, A.; Miercke, L.J.; Weitzman, C.; Nollert, P.; Krucinski, J.; Stroud, R.M. Structure of a glycerol-conducting channel and the basis for its selectivity. Science. 2000, 290, 481–486.

15. Yamauchi, T.; Kamon, J.; Ito, Y.; Tsuchida, A.; Yokomizo, T.; Kita, S.; Sugiyama, T.; Miyagishi, M.; Hara, K.; Tsunoda, M.; Murakami, K.; Ohteki, T.; Uchida, S.; Takekawa, S.; Waki, H.; Tsuno, N.H.; Shibata, Y.; Terauchi, Y.; Froguel, P.; Tobe, K.; Koyasu, S.; Taira, K.; Kitamura, T.; Shimizu, T.; Nagai, R.; Kadowaki, T. Cloning of adiponectin receptors that mediate antidiabetic metabolic effects. Nature. 2003, 423, 762–769.

16. Ballesteros, J.A.; Shi, L.; Javitch, J.A. Structural mimicry in G protein-coupled receptors: implications of the high-resolution structure of rhodopsin for structure–function analysis of rhodopsin-like receptors. Mol Pharmacol. 2001, 60, 1–19.

17. Lomize, A.L.; Pogozheva, I.D.; Mosberg, H.I. Structural organization of G-protein-coupled receptors. J Comput Aided Mol Des. 1999, 13, 325–353.

18. Meng, E.C.; Bourne, H.R. Receptor activation: what does the rhodopsin structure tell us? Trends Pharmacol Sci. 2001, 22, 587–593.

19. Srinivasan, N.; Sowdhamini, R.; Ramakrishnan, C.; Balaram, P. Conformations of disulfide bridges in proteins. Int J Pept Protein Res. 1990, 36, 147–155.

20. Dani, V.S.; Ramakrishnan, C.; Varadarajan, R. MODIP revisited: re-evaluation and refinement of an automated procedure for modeling of disulfide bonds in proteins. Protein Eng. 2003, 16, 187–193.

21. Wang, Y.; Botelho, A.V.; Martinez, G.V.; Brown, M.F. Electrostatic properties of membrane lipids coupled to metarhodopsin II formation in visual transduction. J Am Chem Soc. 2002, 124, 7690–7701.

22. Bünemann, M.; Frank, M.; Lohse, M. Gi protein activation in intact cells involves subunit rearrangement rather than dissociation. Proc Natl Acad Sci USA. 2003, 100, 16077–16082.

23. Hunyady, L.; Vauquelin, G.; Vanderheyden, P. Agonist induction and conformational selection during activation of a G-protein-coupled receptor. Trends Pharmacol Sci. 2003, 24, 81–86.

24. Lipinski, C.A.; Lombardo, F.; Dominy, B.W; Feeney, P.J. Experimental and computational approaches to estimate solubility and permeability in drug discovery and development settings. Adv Drug Deliv Rev. 2001, 46, 3–26.

25. Deupi, X.; Olivella, M.; Govaerts, C.; Ballesteros, JA.; Campillo, M.; Pardo, L. Ser and Thr residues modulate the conformation of Pro-kinked transmembrane α-helices. Biophys J. 2004, 86 (1 Pt 1), 105–115.

26. Bondensgaard, K.; Ankersen, M.; Thogersen, H.; Hansen, B.S.; Wulff, B.S.; Bywater, R.P. Recognition of privileged structures by G-protein coupled receptors. J Med Chem. 2004, 47, 888–899.

27. Madabushi, S.; Gross, A.K.; Philippi, A.; Meng, E.C.; Wensel, T.G.; Lichtarge, O. Evolutionary trace of G protein-coupled receptors reveals clusters of residues that determine global and class-specific functions. J Biol Chem. 2004, 279, 8126–8132.

28. Suel, G.M.; Lockless, S.W.; Wall, M.A.; Ranganathan, R. Evolutionarily conserved networks of residues mediate allosteric communication in proteins. Nat Struct Biol. 2003, 10, 59–69.

29. von Korff, M.; Steger, M. GPCR-tailored pharmacophore pattern recognition of small molecular ligands. J Chem Inf Comput Sci. 2004, 44, 1137–1147.

30. Bissantz, C.; Folkers, G.; Rognan, D. Protein-based virtual screening of chemical databases. 1. Evaluation of different docking/scoring combinations. J Med Chem. 2000, 43, 4759–4767.

31. Bissantz, C.; Bernard, P.; Hibert, M.; Rognan, D. Protein-based virtual screening of chemical databases. II. Are homology models of G-Protein coupled receptors suitable targets? Proteins. 2003, 50, 5–25.

32. Bissantz, C.; Logean, A.; Rognan, D. High-throughput modeling of human G-protein coupled receptors: amino acid sequence alignment, three-dimensional model building, and receptor library screening. J Chem Inf Comput Sci. 2004, 44, 1162–1176.

33. Gouldson, P.R.; Kidley, N.J.; Bywater, R.P.; Psaroudakis, G.; Brooks, H.D.; Diaz, C.; Shire, D.; Reynolds, C.A. Toward the active conformations of rhodopsin and the β2-adrenergic receptor. Proteins. 2004, 56, 67–84.

34. George, S.R.; O'Dowd, B.F.; Lee, S.P. G-protein-coupled receptor oligomerization and its potential for drug discovery. Nat Rev Drug Discov. 2002, 1, 808–820.

35. Zhou, W.; Flanagan, C.; Ballesteros, J.A.; Konvicka, K.; Davidson, J.S.; Weinstein, H.; Millar, R.P.; Sealfon, S.C. A reciprocal mutation supports helix 2 and helix 7 proximity in the gonadotropin-releasing hormone receptor. Mol Pharmacol. 1994, 45, 165–170.

36. Arora, K.K.; Cheng, Z.; Catt, K.J. Dependence of agonist activation on an aromatic moiety in the DPLIY motif of the gonadotropin-releasing hormone receptor. Mol Endocrinol. 1996, 10, 979–986.

37. Mandiyan, V.; O'Brien, R.; Zhou, M.; Margolis, B.; Lemmon, M.A.; Sturtevant, J.M.; Schlessinger, J. Thermodynamic studies of SHC phosphotyrosine interaction domain recognition of the NPXpY motif. J Biol Chem. 1996, 271, 4770–4775.

38. Abdulaev, N.G.; Ridge, K.D. Light-induced exposure of the cytoplasmic end of transmembrane helix seven in rhodopsin. Proc Natl Acad Sci USA. 1998, 95, 10854–10859.

39. Kalatskaya, I.; Schussler, S.; Blaukat, A.; Muller-Esterl, W.; Jochum, M.; Proud, D.; Faussner, A. Mutation of tyrosine in the conserved NPXXY sequence leads to constitutive phosphorylation and internalization, but not signaling, of the human B2 bradykinin receptor. J Biol Chem. 2004, 279, 31268–31276.

40. Oliveira, L.; Paiva, A.C.; Sander, C.; Vriend, G. A common step for signal transduction in G protein-coupled receptors. Trends Pharmacol Sci. 1994, 15, 170–172.

41. Ballesteros, J.; Kitanovic, S.; Guarnieri, F.; Davies, P.; Fromme, B.J.; Konvicka, K.; Chi L.; Millar, R.P.; Davidson, J.S.; Weinstein, H.; Sealfon, S.C. Functional microdomains in G-protein-coupled receptors: the conserved arginine-cage motif in the gonadotropin-releasing hormone receptor. J Biol Chem. 1998, 273, 10445–10453.

42. Chalmers, D.T.; Behan, D.P. The use of constitutively active GPCRs in drug discovery and functional genomics. Nat Rev Drug Discov. 2002, 1, 599–608.

43. Fanelli, F.; Verhoef-Post, M.; Timmerman, M.; Zeilemaker, A.; Martens, J.W.; Themmen, A.P. Insight into mutation-induced activation of the luteinizing hormone receptor: molecular simulations predict the functional behavior of engineered mutants at M398. Mol Endocrinol. 2004, 18, 1499–1508.

44. Fraser, C.M. Site-directed mutagenesis of β-adrenergic receptors: identification of conserved cysteine residues that independently affect ligand binding and receptor activation. J Biol Chem. 1989, 264, 9266–9270.

45. Giles, G.I.; Jacob, C. Reactive sulfur species: an emerging concept in oxidative stress. Biol Chem. 2002, 383, 375–388.

46. Giles, N.M.; Giles, G.I.; Jacob, C. Multiple roles of cysteine in biocatalysis. Biochem Biophys Res Commun. 2003, 300, 1–4.

47. Willey, K.P.; Obermann, H.; Hunt, N.; Henco, K. A universal 2-step method for ligand-induced covalent labelling of cell surface and nuclear receptors by use of the same, single labelling reagent. U.S. Patent 6,156,529, December 5, 2000.

48. Melchiorre, C. Tetramine disulphides: a new tool in α-adrenergic pharmacology. Trends Pharmacol Sci. August 1981, 209–211.

49. Marjamaki, A.; Frang, H.; Pihlavisto, M.; Hoffren, A.M.; Salminen, T.; Johnson, M.S.; Kallio, J.; Javitch, J.A.; Scheinin, M. Chloroethylclonidine and 2-aminoethyl methanethiosulfonate recognize two different conformations of the human α(2A)-adrenergic receptor. J Biol Chem. 1999, 274, 21867–21872.

50. Li, W.; MacDonald, R.G.; Hexum, T.D. Role of sulfhydryl groups in Y2 neuropeptide Y receptor binding activity. J Biol Chem. 1992, 267, 7570–7575.

51. Wong, A.; Hwang, S.M.; Cheng, H.Y.; Crooke, S.T. Structure–activity relationships of β-adrenergic receptor-coupled adenylate cyclase: implications of a redox mechanism for the action of agonists at β-adrenergic receptors. Mol Pharmacol. 1987, 31, 368–376.

52. Malbon, C.C.; George, S.T.; Moxham, C.P. Intramolecular disulphide bridges: avenues to receptor activation? Trends Biochem Sci. 1987, 12, 172–175.

53. Boniface, J.J.; Reichert, L.E Jr. Evidence for a novel thioredoxin-like catalytic property of gonadotropic hormones. Science. 1990, 247, 61–64.

54. Ferreras, M.; Gavilanes, J.G.; Lopez-Otin, C.; Garcia-Segura, J.M. Thiol-disulfide exchange of ribonuclease inhibitor bound to ribonuclease A: evidence of active inhibitor-bound ribonuclease. J Biol Chem. 1995, 270, 28570–28578.

55. Rajarathnam, K.; Sykes, B.D.; Dewald, B.; Baggiolini, M.; Clark-Lewis, I. Disulfide bridges in interleukin-8 probed using non-natural disulfide analogues: dissociation of roles in structure from function. Biochemistry. 1999, 38, 7653–7658.

56. Bertini, R.; Howard, O.M.; Dong, H.F.; Oppenheim, J,J.; Bizzarri, C.; Sergi, R.; Caselli, G.; Pagliei, S.; Romines, B.; Wilshire, J.A.; Mengozzi, M.; Nakamura, H.; Yodoi, J.; Pekkari, K.; Gurunath, R.; Holmgren, A.; Herzenberg, L.A.; Herzenberg, L.A.; Ghezzi, P. Thioredoxin, a redox enzyme released in infection and inflammation, is a unique chemoattractant for neutrophils, monocytes, and T cells. J Exp Med. 1999, 189, 1783–1789.

57. Hoover, D.M.; Boulegue, C.; Yang, D.; Oppenheim, J.J.; Tucker, K.; Lu, W.; Lubkowski, J. The structure of human macrophage inflammatory protein-3α/CCL20: linking antimicrobial and CC chemokine receptor-6-binding activities with human β-defensins. J Biol Chem. 2002, 277, 37647–37654.

58. Hilal-Dandan, R.; Villegas, S.; Gonzalez, A.; Brunton, L.L. The quasi-irreversible nature of endothelin binding and G protein-linked signaling in cardiac myocytes. J Pharmacol Exp Ther. 1997, 281, 267–273.

59. Waggoner, W.G.; Genova, S.L.; Rash, V.A. Kinetic analyses demonstrate that the equilibrium assumption does not apply to [^{125}I]-endothelin-1 binding data. Life Sci. 1992, 51, 1869–1876.

60. Spinella, M.J.; Kottke, R.; Magazine, H.I.; Healy, M.S.; Catena, J.A.; Wilken, P.; Andersen, T.T. Endothelin-receptor interactions: role of a putative sulfhydryl on the endothelin receptor. FEBS Lett. 1993, 328, 82–88.

61. Morgan, R.S.; Tatsch, C.E.; Gushard, R.H.; McAdon, J.; Warme, P.K. Chains of alternating sulfur and π-bonded atoms in eight small proteins. Int J Pept Protein Res. 1978, 11, 209–217.

62. Meyer, E.A.; Castellano, R.K.; Diederich, F. Interactions with aromatic rings in chemical and biological recognition. Angew Chem Int Ed Engl. 2003, 42, 1210–1250.

63. Zacharias, N.; Dougherty, D.A. Cation-π interactions in ligand recognition and catalysis. Trends Pharmacol Sci. 2002, 23, 281–287.

64. Brandt, W.; Golbraikh, A.; Tager, M.; Lendeckel, U. A molecular mechanism for the cleavage of a disulfide bond as the primary function of agonist binding to G-protein-coupled receptors based on theoretical calculations supported by experiments. Eur J Biochem. 1999, 261, 89–97.

65. Garcia-Higuera, I.; Thomas, T.C.; Yi, F.; Neer, E.J. Intersubunit surfaces in G protein αβγ heterotrimers: analysis by cross-linking and mutagenesis of βγ. J Biol Chem. 1996, 271, 528–535.

66. http://nobelprize.org/medicine/laureates/2004/index.html

67. Onuffer, J.J.; Horuk, R. Chemokines, chemokine receptors and small-molecule antagonists: recent developments. Trends Pharmacol Sci. 2002, 23 (10), 459–467.

68. Müller, G. Medicinal chemistry of target family-directed masterkeys. Drug Discov Today. 2003, 8, 681–691.

69. Malnic, B.; Hirono, J.; Sato, T.; Buck, L.B. Combinatorial receptor codes for odors. Cell. 1999, 96, 713–723.

11 Structures and Dynamics of G Protein-Coupled Receptors

Judith Klein-Seetharaman and Michèle C. Loewen

CONTENTS

ABSTRACT

This chapter provides an overview of the field of structural biology of GPCRs. Rhodopsin is the most extensively studied GPCR to date and serves as the central theme of this chapter which describes some of the most important method developments and results that has led to our current understanding of the structures and dynamics of GPCRs. Nuclear magnetic resonance (NMR) spectroscopy has served as a primary tool in this effort, with some of the earliest studies carried out more than 30 years ago. Advances in NMR including solution and solid-state methods yielded results that, in combination with other biophysical methods (e.g., electron paramagnetic resonance, x-ray crystallography), applications of biochemical innovations (e.g., site-directed mutagenesis, recombinant protein expression), and computational modeling, revealed high-resolution structural details of the mechanisms of signal transduction for rhodopsin. The highly conserved seven-transmembrane (7TM) architectures of GPCRs allow this knowledge to be extended in light of studies on other receptors in this family (also reviewed here) to provide overall models of the structure and dynamics of GPCRs in general.

11.1   INTRODUCTION

G protein-coupled receptors (GPCRs) comprise the largest and most diverse family of integral membrane proteins responsible for signaling across cellular membranes. The initial signaling events are mediated by a wide variety of stimuli (including photons, Ca^{2+} ions, odorants, tasting molecules, amino acids, nucleotides, peptides, and proteins). The signal is transmitted to the cytoplasm, where downstream signaling is induced by receptor-mediated activation of G proteins. Based on pharmacological specificity and sequence conservation, GPCRs are divided into multiple classes.[1] The three main classes are (A) receptors related to rhodopsins, (B) secretin receptors, and (C) the metabotropic neurotransmitter receptors. Class A, the largest, contains more than 1200 distinct members listed in the GPCR database[1] and more than 7000 putative members in the Pfam family database.[2]

The GPCR family is characterized by a signature seven transmembrane (7TM) configuration that was first observed in bacteriorhodopsin by Henderson and Unwin[3] and was subsequently proposed for rhodopsin, the mammalian dim-light receptor.[4] High resolution crystal structures of bacteriorhodopsin confirmed the 7TM arrangement including extracellular (EC) loops and N-terminus, cytoplasmic (CP) loops and

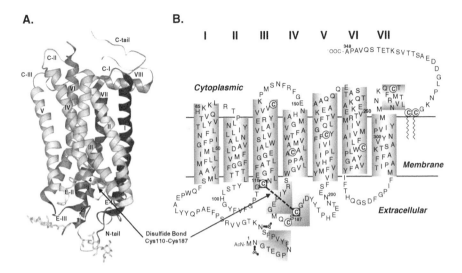

FIGURE 11.1 Structure of rhodopsin. A: Crystal structure model. Helices are labeled I–VIII. Loops connecting the helices are labeled C-I, C-II, C-III and C-tail for the cytoplasmic side and N-tail; E-I, E-II, and E-III for the extracellular side. Glycosylation, retinal and disulfide bond are represented by sticks. B: Secondary structure model. Helices and beta sheets are marked by gray boxes and arrows, respectively. The disulfide bond between Cys187 and Cys110 is indicated by a dotted line and labeled with an arrow. All ten cysteines in rhodopsin are highlighted with circles.

C-terminus, binding of the all-*trans* retinal ligand, and detailed conformational changes upon photo-activation.[5] Recent applications of atomic force microscopy to analysis of the conformational changes and folding dynamics of bacteriorhodopsin have also been reviewed elsewhere.[6] Although bacteriorhodopsin has provided extensive information about the 7TM structure, it is not a true GPCR, lacking the ability to mediate the signal transduction events with downstream G protein signaling pathways.

The only GPCR for which information about the 7TM structure is directly available is bovine rhodopsin. First, a low resolution structure was derived by electron microscopy,[7] followed by determination of the x-ray crystal structure for the inactive state[8] (Figure 11.1A). Thus, rhodopsin, has proven to be a model *par excellence* for structural and functional analyses of GPCRs. The ligand in rhodopsin, 11-*cis*-retinal, is covalently attached to the protein via a protonated Schiff base to the ε-amino group of Lys296 located in TM7 (Figure 11.1B). This gives rise to a characteristic absorption at 500 nm.

Activation of rhodopsin is achieved by light: capture of a photon by rhodopsin results in isomerization of *11-cis* retinal to the all-*trans* form that triggers a series of transient changes in the protein accompanied by characteristic absorption changes: dark rhodopsin (500 nm) → batho (548 nm) → lumi (497 nm) → meta I (480 nm) → meta II (380 nm) → meta III (450 nm) → opsin + free retinal (380 nm).[9] Only the meta II conformation can promote the activation of the G protein, the hallmark signaling event of the GPCR family.

This chapter will present the current state of knowledge of the structure and dynamics of rhodopsin in its dark conformation and in photo-intermediates, along with reviews of past and present state-of-the-art applied methodologies and in reference to other GPCRs in the ligand-free and ligand-bound states, where possible.

11.2 STRUCTURE, STABILITY, DYNAMICS AND CONFORMATIONAL CHANGES OF RHODOPSIN

11.2.1 EARLY MAGNETIC RESONANCE STUDIES OF CHROMOPHORE AND PROTEIN–LIPID INTERACTIONS

The earliest documented experiments applying nuclear magnetic resonance (NMR) to the analysis of rhodopsin were limited to studying the isolated solubilized visual pigment chromophores. Conformations of the polyene chains of all-*trans*, 9-*cis*, 11-*cis* and 13-*cis* retinal, as well as model forms of Schiff Base all-*trans* analogs, were solved by ^{1}H and natural abundance ^{13}C solution NMR providing details about the planarity, twisted irregularities, and motional freedom of various regions of the chromophore.[10,11] The production of the first isotopically enriched chromophore, designed to include ^{13}C at carbon 14 (C14), was reconstituted into rhodopsin and analyzed in a detergent-solubilized system.[12] Eventually a perturbed, protonated Schiff Base was confirmed, linking the 11-*cis* retinal to the protein in ground state rhodopsin.[11]

Some of the earliest studies probing interactions of rhodopsin with lipids and detergents were carried out by application of spin-label electron paramagnetic resonance (EPR) spectroscopy. Conformation and stability of rhodopsin in solubilized detergent-mixed micelles as well as the effect of rhodopsin on lipid bilayer mobility were probed by application of various spin-labeled system components.[13–15]

Other early magnetic resonance experiments focused on analysis of sonicated aqueous suspensions of rhodopsin in lipids and membranes for which rigid or motionally averaged NMR powder pattern spectra were collected.[16] Detailed studies of the dynamic properties of the rhodopsin–membrane system by molecular fluctuation sensitive spin-lattice (T_1) relaxation measurements as well as ^{13}C-^{1}H NOE showed small but significant decreases in T_1 relaxation times of the rod outer segment (ROS) membranes compared to ROS lipid vesicles.[16–18] These results, which were in agreement with EPR results obtained with spin-labeled lipids,[19–21] indicated that the average rate of phospholipid segmental motion was reduced four-fold by the presence of rhodopsin for lipids immediately adjacent to the receptor.

Moment analysis and spectral subtraction of deuterium (^{2}H) NMR data collected on rhodopsin in deuterated vesicles at low temperatures indicated two discrete lipid pools — one comprised of bulk gel lipids and the second comprised of phospholipids presumably in two-dimensional protein aggregates with rhodopsin.[22–24] Although probing head group mobility and lipid phases of rhodopsin membranes and extracted phospholipids by solid-state ^{31}P NMR[25–27] further supported a role for rhodopsin in organizing ROS phospholipids; subsequent measurement of ^{31}P T_1 relaxation indicated conformational head group changes in non-native high protein content membranes (probably arising from lipids trapped between adjacent proteins), but not when protein:lipid ratios were maintained at natural or lower levels.[28–30]

11.2.2 Advent of Recombinant Expression and Purification Technology

In contrast to other GPCRs, rhodopsin is readily available from native sources (bovine retina) in very high quantities and this source proved useful in limited early structural studies.[31] Significant structural information was obtained also from unpurified systems expressing GPCRs, and such studies have played important roles in studies of ligand binding pockets of GPCRs other than rhodopsin. However, high resolution structural studies require significant amounts of recombinant pure protein. A variety of efficient recombinant systems were tested and systematically evaluated for production of rhodopsin, including *Escherichia coli*, yeasts, oocytes, baculovirus insect cell systems, and mammalian COS, CHO, and HEK293 cells. In *E. coli*, negligible amounts of rhodopsin were expressed or the expressed protein was misfolded and not functional.[32]

Two yeast strains were tested for expression of rhodopsin. In *Saccharomyces cerevisiae*, 2.0 ± 0.5 mg/L were expressed, of which only 2 to 4% were correctly folded.[33] In *Pichia pastoris*, 0.3/L were produced, of which 4 to 15% were reported functional.[34] Rhodopsin is expressed in functional form in *Xenopus laevis* oocytes[35] and in Chinese hamster ovarian (CHO) cells,[36] although yields have not been reported. For small scale (microgram range) preparation of rhodopsin, transient transfection in the COS cell system[37] is the most widely used method. Typical wild-type expression yields are 25 µg protein/15-cm dish.

For production of larger quantities of functional rhodopsin, expression in baculovirus Sf9 and mammalian HEK293 cells has been successful. Whereas insect cell systems produced yields of ~1 mg/L rhodopsin, mammalian cell systems have been the mainstays in terms of providing large quantities of active rhodopsin for biophysical analyses. Stable mammalian cell lines expressing wild-type bovine rhodopsin were originally produced using a suspension-adapted HEK293 cell line and a vector system placing the opsin gene under the control of a constitutively enhanced cytomegalovirus promoter system.[38]

Yields of wild-type rhodopsin from this system produced in large-scale spinner flask preparations reached ~2 mg/L and this method has been extensively applied to the production of rhodopsin samples suitable for NMR analysis.[38–35] Recent optimizations of this stable cell line method include accommodation of a tetracycline-inducible promoter system, increasing yields five-fold, achieving levels of ~10 mg/L from bioreactors with enhanced nutrient mixtures, and permitting overexpression of a toxic, constitutively active rhodopsin mutant.[46,47]

Tetracycline induction also allowed production of nonglycosylated rhodopsin (important for the preparation of samples amenable to biophysical analyses such as protein crystallization and nuclear magnetic resonance) through addition of tunicamycin immediately prior to induction; although cell viability dropped after only 24 h, overall yields of nonglycosylated rhodopsin were ~3 mg/L.[47] Alternatively, a modified (N-acetylglucosamine transferase I-negative) HEK293S cell line with restricted glycosylation engineered and combined with the stable tetracycline induction system yielded more than ~6 mg/L of nonglycosylated rhodopsin.[47,48] Balancing cost and efficiency of functional expression, the COS and HEK293 cell expression

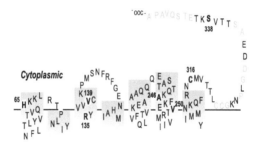

I II III IV V VI VII

FIGURE 11.2 Amino acids on the cytoplasmic surface of rhodopsin studied by cysteine mutagenesis. Highlighted in black letters are all residues that have been mutated to cysteines one at a time (see single cysteine mutants in Table 11.1, top). All other amino acids are omitted from this graph, except those in the C-terminal tail. Residues studied by cysteine mutagenesis are shown as black letters; those that have not been studied are shown as gray letters. Cysteines that have served as fixed reference points in di-cysteine mutants (see text) are marked in bold. These cysteines were fixed in mutants, where a second cysteine was systematically varied. The resulting pairs are listed in Table 11.1, bottom.

systems have proven to be the most useful in terms of providing microgram to hundreds of milligram quantities of active rhodopsin, respectively. A refined antibody affinity purification scheme allowed one-step purification of rhodopsin from native and recombinant systems.[37] Other alternative affinity chromatography methods include concanavalin A affinity purification of rhodopsin from natural sources and immobilized metal affinity chromatography in correlation with insect cell expression work.[12,49–51] An in-depth review of expression of GPCRs other than rhodopsin can be found in Chapter 8 of this book.

11.2.3 RHODOPSIN STRUCTURE VIA CYSTEINE MUTAGENESIS FOLLOWED BY BIOCHEMICAL AND BIOPHYSICAL ANALYSES

The structure and dynamics of the CP loop regions and CP ends of the helices of rhodopsin in solution were investigated using a combination of cysteine mutagenesis followed by biochemical and biophysical studies of the cysteine mutants. Figure 11.2 shows the residues on the CP side of the molecule that were replaced, one at a time, by cysteine. The top section of Table 11.1 lists the single cysteine mutants. In separate experiments, pairs of cysteines (di-cysteine mutants) were introduced, as also shown in Figure 11.2 and Table 11.1.

The unique chemistry of the cysteine sulfhydryl group allows specific derivatization of reactive and accessible cysteines with biophysical probes. Because dark-state, wild-type rhodopsin contains two reactive cysteines, Cys140 and Cys316, these cysteines were replaced by serines in the cysteine mutants to avoid ambiguity. The other cysteines in rhodopsin (palmitoylation sites, TM and EC cysteines, see Figure 11.1B) are not reactive in the dark and were therefore not replaced. Figure 11.3A

TABLE 11.1
Cysteine Mutants of Rhodopsin Studied via Biochemical and Biophysical Means

Mutant Type		Reference
Single Cysteine		
Asn55 – Ile75		53
Tyr136 – Met155		289
Gln225 – Ile256		290
Tyr306 – Leu321		291
Lys325, Asn326, Leu328, Asp331, Glu332, Thr335–Thr340		70, 131
Di-Cysteine		
Cysteine I (fixed)	Cysteine II (varied)	
Cys316	Cys60–Cys74	55
Cys65	Cys306–Cys321	56
Cys246 or Cys250	Cys311–Cys314	62
Cys139	Cys247–Cys252	119
Cys135	Cys250	62
Cys338	Cys240–Cys250	70

lists some of the biochemical and biophysical probes used to study the structure and dynamics of rhodopsin.

11.2.3.1 Cysteine Reactivity as Qualitative Indicator for Structure

Rhodopsin carrying a free sulfhydryl group reacts with 4,4′-dithiodipyridine to form the thiopyridinyl molecular entity derivative shown in Figure 11.3A, molecular entity 1.[52] The rate of this reaction is very sensitive to the accessibility of the cysteine. Cysteines buried in the micelle or protein interior are entirely unreactive, and cysteines exposed in the fully accessible aqueous portion of the CP domain react instantly.[53] This is shown for cysteines placed at positions 56 through 75 in Figure 11.3B, Panel I. The positions of these exposed cysteines correspond well with those identified by collisions of paramagnetic agents with spin-labeled derivatives (Figure 11.3A, molecular entity 3).[54]

The resulting thiopyridinyl derivative of rhodopsin (1) is reactive toward free sulfhydryl reagents by way of disulfide exchange. The rate of this exchange is an extremely sensitive probe for tertiary structure and has been used to prove the presence of light-induced conformational motions resulting in distinct tertiary structure changes,[53] even where EPR methods suggest the magnitude of conformational change to be small.[54] For example, changes in the rates of reactions in the dark as compared to the light are shown in Figure 11.3B, Panel II, for residues in the 60 through 74 sequence.

The disulfide exchange reaction of thiopyridinyl–rhodopsin (1) with sulfhydryl reagents has practical value in that the nature of the sulfhydryl reagent can be chosen relatively freely, thereby presenting a general scheme for derivatization of cysteine groups on proteins with reporter probes for which no reactive derivatives exist. For

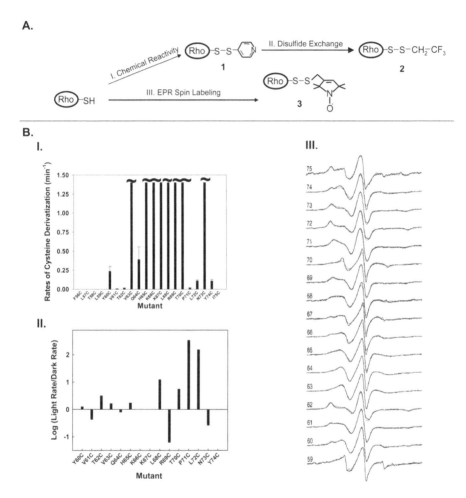

FIGURE 11.3 Biophysical probes used in studies of structure and dynamics of full-length rhodopsin. A: Molecular entities in various derivatization reactions of cysteines. The measurement of cysteine reactivity using 4,4′-dithiodipyridine yielding thiopyridinyl–cysteine derivative in process I, the exchange reaction with sulfhydryl reagents in process II, and the EPR spin-labeling procedure in process III yield molecular entities 1, 2, and 3, respectively. B: Examples for the results of processes I through III from A for the loop connecting helices I and II in rhodopsin. Panel I: Measurement of chemical reactivity of cysteines using 4,4′-dithiodipyridine.[53] Panel II: Release of thiopyridone via disulfide exchange using sulfhydryl reagents. Panel III: EPR spectra of spin-labeled cysteine derivatives.[54] (Reproduced with permission from Biochemistry. Copyright, American Chemical Society, Washington, D.C., 1999.)

example, it was possible to derivatize rhodopsin with the trifluoroethylthiol group resulting in molecular entity (2), Figure 11.3A, to enable ^{19}F-NMR applications.[42,43,45]

11.2.3.2   Distance Constraints by Rates of Disulfide Bond Formation

To further define tertiary structure in the CP domain of rhodopsin, disulfide bond formation was used to establish proximity between amino acids in rhodopsin,[55,56] a method developed by Falke and Koshland[57] in studies of the aspartate receptor.[58,59] The method is based on structural analysis of disulfides occurring in protein crystal structures that have shown preferred conformations for their formation.[60] The distance between α-carbon atoms across the disulfide ranges from about 4 to 9 Å in crystal structures, with 95% of all refined disulfides in the range of 4.4 to 6.8 Å. The average distances across left-handed and right-handed disulfides is 5.88 ± 0.49 Å and 5.07 ± 0.73 Å,[61] respectively.

Thus, the presence of a disulfide bond indicates that the α-carbons of the participating cysteines are about 5 to 6 Å apart. However, the geometry derived from crystallography may not hold in solution, especially for mobile regions at protein surfaces.[61] The formation of a disulfide bond between two cysteines does not imply a time-average proximity of the two residues in the protein structure. Once the disulfide bond is formed, the two cysteines are locked in a conformation that may not necessarily be favored.

The rates of disulfide bond formation were determined for sets of di-cysteine mutants of rhodopsin, in which a cysteine was kept constant at one site while the position of a second cysteine was varied at a proximal region (Table 11.1). The rates of disulfide bond formation were measured for different sets of di-cysteine mutants[62] and compared to proximities between amino acids deduced from a crystallographic model.[8] The reciprocal of the distances obtained as a function of residue position and the rates of disulfide bond formation are reproduced in Figure 11.4A for disulfide bonds between Cys316 in helix VIII at CP loop 4 and cysteines at positions 55 through 75 in CP loop 1.[55] The comparison showed excellent correlation between the rates of disulfide bond formation and the interthiol distances derived from the cysteine replacements in the crystal structure (Figure 11.4B). The three positions that most rapidly formed disulfide bonds with Cys316, H65C, L68C, and V61C, are 4 to 5 Å distant from Cys316, and facing it. In order for a disulfide bond to form, however, 3 to 4 Å translational movements would be necessary. This requires sufficient flexibility of the amino acids in this region of the rhodopsin CP face. The fact, however, that only those cysteines that faced Cys316 were able to bridge this small gap indicates that no unfolding of the ends of the helices occurs; instead, a movement of intact helices brings residues in CP loop 1 close to Cys316. These results presented the first direct evidence for substantive backbone motion in the CP domain of rhodopsin in solution.

11.2.3.3   Mobility, Accessibility, and Distance Constraints via Electron Paramagnetic Resonance Spectroscopy

Application of EPR spectroscopy to rhodopsin was reviewed earlier.[63,64] To enable application of EPR spectroscopy, nitroxide spin labels (molecular entity 3, Figure

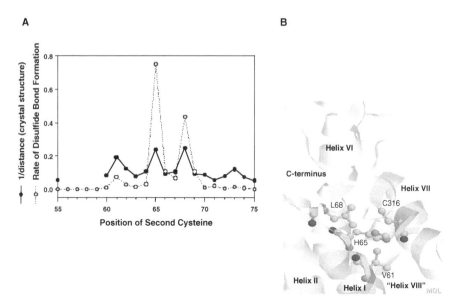

FIGURE 11.4 Disulfide bond formation in di-cysteine mutants Tyr60Cys-Tyr74Cys/Cys316.[55] A: Rates of disulfide bond formation at pH 7.7, 25°C (dotted lines, open circles). The reciprocals of distances between cysteines in the crystal structure of rhodopsin are shown as black lines, filled circles. (Reproduced with permission from Biochemistry. Copyright, American Chemical Society, Washington, D.C., 1999.) B: Visualization of positions of cysteines placed at sites where fastest disulfide bonds were formed with Cys316 (Leu68, His65 and Val61). All these positions are in closest proximity to Cys316 in the crystal structure model.

11.3A) are introduced either at single reactive cysteine sites (single cysteine mutants) or in pairs (di-cysteine mutants); see Section 11.2.3. EPR spectroscopy of spin-labeled single cysteine mutants is useful for estimating mobility and for obtaining qualitative indicators for tertiary structure contacts surrounding the nitroxides and the cysteines to which the nitroxides are attached. For example, EPR spectra of spin labels at positions 59 through 75 are shown in Figure 11.3B, Panel III. In conjunction with collision measurements, EPR spectroscopy can provide a measure of water accessibility. Applications of these methods to rhodopsin enabled derivation of the orientations of helices, protrusions of helices into the CP surface, relative flexibility of different loops, and qualitative assessment of conformational changes upon light activation.[54,65–67]

To obtain more quantitative information on conformational changes, di-cysteine mutants were studied. Application of EPR spectroscopy in the presence of two spin labels allowed detection of dipolar interactions among the labels. This resulted in derivation of distance constraints for the sets of di-cysteines listed in Table 11.1. Most significantly, changes in distances upon light activation allowed the derivation of a model for the mechanism of activation of rhodopsin.[67–70] The relative increase in the distances between the CP ends of helices suggested an opening of the helical bundle toward the CP face, exposing amino acids critical for interaction with the G protein. Although distance changes were observed in the majority of sites studied, the largest changes in relative distance were observed between helices III and VI, followed by smaller changes in relative distances between helices I and II with respect to helix VII and helix VIII.[63,64]

11.2.4 SOLID-STATE NUCLEAR MAGNETIC RESONANCE APPLICATIONS TO RHODOPSIN

11.2.4.1 Chromophore Structure and Schiff Base Linkage

The advent of ^{13}C magic angle spinning (MAS) solid-state NMR provided significant improvements over the original early suspension NMR experiments through increased line resolution by resolving broad chemical shift anisotropy powder patterns into sharp center bands at the isotropic chemical shift with side bands at the spinning frequency.[71–73] The first applications of ^{13}C cross polarization (CP) MAS to rhodopsin samples with the aim of observing protein–lipid interactions were limited by rotational diffusion of rhodopsin molecules in the bilayer on a time scale that interfered with decoupling and reduced cross-polarization efficiency.[74,75]

Spurred by success with a motionally restricted bacteriorhodopsin system,[76] two reports later demonstrated that restriction of rhodopsin rotational diffusion by application of low temperatures[75] or lyophilization[77] slowed motion sufficiently to permit effective application of CP MAS to analysis of chromophore structure. Interactions in rhodopsin provided evidence that the Schiff base C–N double bond linkage is protonated and in the *anti* conformation. Furthermore, these experiments supported the presence of negative counter ion(s) in close proximity to the retinyl moiety.

Ultimately, a combination of molecular orbital calculations, site-directed mutagenesis, and chemical shifts in [13]C -MAS spectra for 11-*cis* retinal carbons physically localized the counter ion to a carboxylate group situated above C12 of the chromophore in rhodopsin.[78,79] Specifically one of the carboxylate oxygens (O1) of Glu113 was suggested to be ~3Å from the C12 position.[80,81] Combining this with further analyses of the mechanism of the opsin shift,[82] the mechanism of energy storage in bathorhodopsin,[80,83] the unprotonated form of the Schiff base in the meta II rhodopsin intermediate[84] and a variety of site-directed mutagenesis studies[85–88] provided a coherent picture of the photoactivation mechanism of rhodopsin.[89]

Subsequently, a novel method for macroscopic alignment of bilayers on solid surfaces termed isopotential spin-dry ultracentrifugation (ISDU) was applied to rhodopsin preparations in native and reconstituted membranes.[90] Angles relative to the membrane normal for individual ^{2}H bonds in the chromophore were determined by this method. However, combining the ISDU-oriented bilayer method with ^{2}H MAS NMR[91] modified the experiment such that side band intensity became a function of size and orientation of the quadrupolar coupling tensor, allowing more precise measurement of the C–C(^{2}H)$_3$ bond vector orientation with respect to the membrane normal. Such magic angle-oriented sample spinning experiments were applied to labeled chromophores in reconstituted rhodopsin membranes and ultimately provided a three-dimensional structure and orientation of retinal within the binding pocket of rhodopsin in the ground state and upon photoactivation to the meta I intermediate.

Additional details about the chromophore structures were later solved by applications of one-dimensional CP MAS NMR rotational resonance measurement of distances between ^{13}C labeled carbons as well as a novel double quantum hetero-nuclear local field NMR method for determination of torsion angles about ^{13}C-labeled atoms.[92–95] Distance values and torsion angles confirmed the 11-*cis* confor-mation of retinal in the dark state of rhodopsin with a twisted C10–C13 unit. A relaxed planar all-*trans* conformation in the meta I rhodopsin intermediate was also confirmed. Most recently, these methods provided corrections to the magic angle oriented spinning model[91] of the 11-*cis* ground state chromophore such that a modest 28-degree twist is present in the C6–C7 torsion angle and the ring remains in the binding site in the meta I intermediate.[96,97]

As described earlier, dramatic progress was achieved in the field of high-yield recombinant expression of rhodopsin. This progress led to the first use of protein site-specific isotopically labeled recombinantly produced rhodopsin for MAS NMR experiments.

Homogeneously [α,ϵ-[15]N$_2$]-lysine labeled rhodopsin from a baculovirus Sf9 cell expression system was subjected to CP MAS ^{15}N NMR to investigate the Schiff base linkage.[98,99] The nitrogen of the lysine contributing to the Schiff base linkage was observed to resonate with an isotropic shift signal at 155.9 ppm relative to ammonium chloride.[99] This spectrum was correlated with the model of a protonated Schiff base stabilized by a complex counterion. Homogeneously α,ϵ-^{15}N-lysine and 2-^{13}C-glycine di-labeled rhodopsin produced by overexpression in a stably trans-fected mammalian suspension cell line (HEK293S) was also used to investigate Schiff base linkage.[39–40] The ^{15}N chemical shift of the lysine contributing to the Schiff base was observed at 156.8 ppm, a chemical shift characteristic of a protonated Schiff base nitrogen that has only weak counterion interactions.

The distance between the Schiff base nitrogen and the counterion was estimated at approximately 4.0 Å. This distance is consistent with a structural water molecule in the binding site hydrogen bonded with both the Schiff base nitrogen and the Glu113 counterion. These reports demonstrate the feasibility of applying solid-state MAS NMR to structural studies on isotopically labeled rhodopsin and potentially other GPCRs.

11.2.4.2   Other Tertiary Interactions

Noncovalent Protein–Chromophore Interactions — Aside from the obvious cova-lent linkage of the chromophore to the rhodopsin molecule by the protonated Schiff base and counterion interaction, the chromophore is expected to make other significant noncovalent interactions with the TM helices as alluded to in early ^{19}F studies.[100] Some of these interactions were identified through combinations of site-directed mutagenic analysis and ^{13}C MAS NMR of labeled chromophore and computational methods.[89] A model was proposed that highlights direct steric interactions between the retinal chromophore and TM helix III in the region of Gly121 that are modulated during the conversion of 11-*cis* to all-*trans* retinal. Subsequently, 11-*cis* chromophore labeled with ^{13}C at ten sites along most of the entire polyene chain or homogeneously was incorporated into native rhodopsin embedded in membranes.[101,102]

Application of 2D experiments was made possible by use of a 2D radio fre-quency-driven dipolar recoupling during the MAS experiments. Chemical shift differences between the free and bound ligand provided an NMR assay of the spatial and electronic structures of the chromophore for assessment of interactions between the chromophore and the binding site including protein-induced conformational restraints or electronic effects. Nonbonding interactions between the protons of the methyl groups of the β-ionone ring and the protein were observed. The interactions were attributed to aromatic residues Phe208, Phe212 (interacting with H16 and H17), and Trp265 (interacting with H18) determined to be in close proximity by crystallographic methods.[8,103]

Protein–Lipid Interactions — Rhodopsin is naturally packed and aligned in the ROS disk membranes in a regular parallel array of α-helices such that the overall intact ROS has an average diamagnetic anisotropy allowing orientation of intact ROS in solution with its long axis parallel to even very low magnetic fields.[104,105] Observation of ^{31}P spectra of such magnetically oriented intact ROS provided much more specific information than that obtained from earlier sonicated or reconstituted suspension samples (see Section 11.2.1), including resolution of the main phospholipid pools and metabolite phosphate groups present in these relatively complex biological systems.[106]

Rhodopsin Phosphorylation — Phosphorylation of activated rhodopsin is the first step in desensitization of the receptor. A comparison of ^{31}P spectra collected for phosphorylated and unphosphorylated ROS demonstrated the presence of a ^{31}P resonance specific for the phosphorylated sample.[107] Limited proteolysis of the C-terminus of rhodopsin, which removes the predicted phosphorylation sites, eliminated the resonance and determination of the p_{Ka} obtained from pH titration, indicating it was likely a serine phosphate.

Single pulse, CP, and rotational echo double resonance (REDOR) MAS techniques were applied to the analysis of phosphorylated rhodopsin in ROS membranes.[42] The single pulse and CP methods were used to probe the temperature dependence of the ^{31}P signal, which indicated that the C-terminus was highly mobile. REDOR analyses were then applied to measure internuclear distances between the ^{31}P label on the C-terminal tail and ^{19}F labels specifically incorporated (as described in Section 11.2.3.1) at Cys140 (helix III) and Cys316 (helix VIII) of the rhodopsin CP face of the native wild-type (WT) phosphorylated rhodopsin. ^{31}P-^{19}F distances were all estimated to be greater than 12 Å, suggesting that the rhodopsin C-terminus does not contact the areas of the CP face implicated in arrestin binding. This implies a change in the average position of the C-terminal tail upon phosphorylation from positions determined by other methods.[8,42]

11.2.5 SOLUTION NMR

While solution NMR was used in some of the earliest studies of rhodopsin–chromophore interactions (Section 11.2.1), overall resolution was very poor for the sonicated membrane suspensions and NMR efforts for the most part shifted to solid-state methods. However, with the advent of higher field instruments, novel pulse sequences, and recombinant expression technology, solution NMR can be applied to structural analyses of rhodopsin in a number of interesting ways.

11.2.5.1 Protein–Protein Interactions

Solution NMR was initially used to study the interaction between rhodopsin and the G protein α subunit known as transducin[108] via application of a transferred nuclear Overhauser effect (Tr-NOESY)[109] experiment to a system containing a fragment peptide of transducin (residues 340 through 350 with Lys341Arg) known to interact with rhodopsin.[108] The observed NOEs provided information on distances between various protons in the bound peptide, which along with distance geometry, NOE-constrained molecular dynamics, simulated annealing and grid searches produced a series of structures of the transducin peptide in association with and the excited meta II intermediate.

Although a number of problems including low signal-to-noise and assignment errors were later identified in this innovative work,[110] a similar analysis using Tr-NOESY and TOCSY (total correlation spectroscopy) experiments to compare free transducin peptide (340 to 350) to the meta II intermediate bound form was successful.[111] Disordered peptide was observed in solution which, upon light activation, underwent a dramatic shift in conformation to include a helical turn followed by an open reverse turn along the peptide chain. Comparison of this structure was made to the same segment identified in the crystal structure of transducin, which indicated a mechanism of signal transduction by which activated receptors control G proteins through reversible conformational changes.[111–115] This model was later confirmed by measurement of residual dipolar couplings.[116,117]

Magnetically oriented bilayer ^{31}P studies indicated that ROS disk membranes, which alone lack the cooperativity in alignment that exists in the intact ROS, actually have sufficiently large total magnetic susceptibility anisotropy to align with their membrane normals parallel to a magnetic field.[117] Residual dipolar contributions were observed (identified by smaller $^1J_{NH}$ splittings in ^{1}H-^{15}N HSQC spectra) between light-activated rhodopsin (meta II state) in oriented ROS disk membranes and a selectively ^{15}N di-labeled undecapeptide, closely resembling a fragment (residues 340 to 350) of the transducin G protein α subunit known to interact with activated rhodopsin.[117] Information on the conformation of the bound ligand and its orientation relative to the bilayer was determined from the measurement of N–H vector angles of 48 and 40 degrees with the disk normal for the labeled amino acids (Leu5 and Gly9, respectively). Transferred dipolar contributions were observed to return to their isotropic values at a rate determined by the decay of the meta II state of rhodopsin.

This magnetically oriented bilayer NMR system provided significantly more conformational information than in the transferred-NOE experiments.[108,111] Ultimately, this oriented bilayer-transferred dipolar coupling method (including observation of contributions to both $^1J_{NH}$ and $^1J_{CH}$ splittings), combined with determination of approximate distance restraints by transferred NOEs, was used to solve a high resolution structure of the G protein fragment.[116] The fragment was shown to take on a helical conformation upon binding to light-activated rhodopsin.

11.2.5.2 Rhodopsin Structure

The positions and widths of ^{19}F-NMR signals are indicators of chemical environment and mobility[118] and are therefore sensitive to light-induced conformational changes in rhodopsin.[43] The first solution NMR experiment analyzing the overall structure of full-length rhodopsin was carried out by application of ^{19}F NMR to detergent-solubilized recombinant rhodopsin.[43] A number of mutants (produced in HEK293S cells) containing single reactive cysteine residues in different regions of CP faces were labeled with ^{19}F as described above.[43] Each of these labeled derivatives specifically located at amino acid residue positions 67, 140, 245, 248, 311, and 316 of the CP face was purified in sufficient quantities for NMR analysis.

Recorded solution NMR spectra of the ground state rhodopsin resolved different chemical shifts for each of the ^{19}F labels with differing line widths representing the different environments and mobilities across the CP surface (Figure 11.5). Upon illumination to produce the meta II intermediate, significant upfield and downfield shifts of the ^{19}F resonances were observed, in particular those at amino acid positions 67 and 140 as well as at 248 and 316, respectively.[43] These shifts were entirely consistent with previous models of rhodopsin activation based on EPR analyses

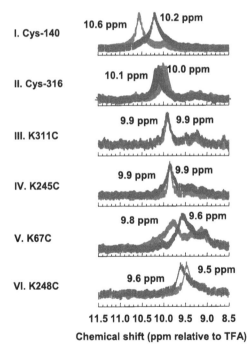

FIGURE 11.5 [^{19}F]NMR spectra of single cysteine mutants.[43] Single cysteine mutants of rhodopsin were derivatized with molecular entity 2, shown in Figure 11.3A. A minimum of 5 mg pure protein was analyzed in dodecyl maltoside micelles. [^{19}F]NMR was collected within 2-min acquisition time (average of 160 scans). Light gray lines indicate NMR spectra in the dark and dark gray lines indicate spectra after illumination.

(Section 11.2.3.3).[119] Subsequently, this method was expanded to report on measurements of short distances in the CP surfaces by determination of NOEs between ^{19}F labels in di-labeled rhodopsin mutants containing pairs of cysteines.[45] Specifically, labels were incorporated at the ends of helices III and VIII (WT Cys140 and Cys316), helices I and VIII (Cys65 and Cys316), and helices III and VI (Cys139 and Cys251), respectively. Distinct chemical shifts were observed for all the labels in the dark state of rhodopsin.

Analysis of homonuclear ^{19}F NOEs between the labeled cysteines indicated no evidence of proximity for the WT pair, but significant negative enhancements were observed for the other two pairs. The extent of the observed NOE enhancement correlated well with distances measured in the dark state crystal structure (29.8, 3.9, and 3.0 Å, respectively) and emphasized the advantage of this method over EPR distance measurement that has a lower distance limit of 8 Å, due in part to the size differences of the respective labels (molecular entities 2 and 3, Figure 11.3A).[8,45] A discrepancy between observed rates of relaxation and theoretical calculated values is acknowledged and concise quantitation of the distances was not attempted by this method.

Yeagle and co-workers set about solving the complete structure of bovine rhodopsin by independently analyzing the structures of each individual loop and TM segment by solution NMR and assembling the conformational information into a final overall structure.[120] The rationale for this method included observation of spontaneous assembly of the independent rhodopsin TM segments into a helical bundle in solution[121] and successful application of this method to solving the structure of bacteriorhodopsin in comparison to its previously derived crystal structure.[122–125] Structures of all the individual TM and loop segments were solved by solution two-dimensional homonuclear ^{1}H NMR (2DQ-COSY and NOESY) in DMSO.[126,127] Distance and dihedral angle constraints were obtained from the NMR data directly whereas long distance and interhelical constraints for the intact protein (for both ground and activated states) were obtained from previous structural studies of rhodopsin including methods such as electron microscopy, EPR, site-directed mutagenesis, cross-linking, and solid state NMR to produce the final model.[120]

The final calculated structure was in rough agreement with the crystal structure of ground state rhodopsin and included a more defined CP surface that was not well resolved in the crystal structure.[8,120] These experiments combined with modeling and information about the heterotrimeric G protein structure provided a calculated meta II intermediate structure.[112,116,117,120,128–130] Differences between the ground and activated states, in particular in the CP face, suggest how that surface is activated to interact with the G protein. Overall, the authors presented a model in which a conformational distortion driven by the energy of binding is induced in transducin when it binds to the activated form of rhodopsin.[129]

11.2.5.3 Conformational Dynamics in Rhodopsin

Quantitative evidence for conformational fluctuations in rhodopsin came from a solution NMR study of rhodopsin labeled with [α-^{15}N]-lysine (Figure 11.6A).[44] Despite the presence of 11 lysines in rhodopsin, only a single sharp signal (signal

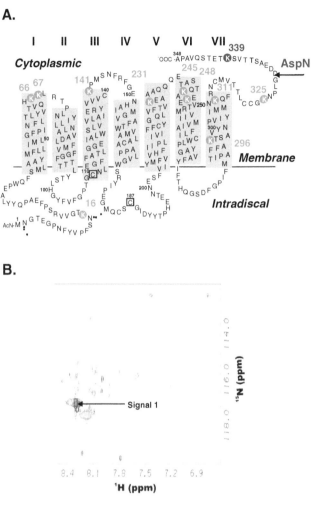

A.

B.

FIGURE 11.6 NMR spectroscopic study of α-^{15}N-lysine labeled rhodopsin.[44] A: Secondary structure model of bovine rhodopsin showing positions of the 11 lysine residues, with circles. Transmembrane helices are designated I to VII. The cleavage site for the AspN enzyme is indicated by an arrow. B: Conventional HSQC spectra at 37°C of α-^{15}N-lysine labeled rhodopsin. The predominant signal under all conditions observed (Signal 1) was assigned to Lys339 (shown in Figure 11.6A) based on the differences in spectra observed upon cleavage of a peptide sequence from the C-terminus of rhodopsin using AspN containing Lys339 as the only lysine. The spectrum of truncated rhodopsin after AspN digestion lacked signal 1, while that of peptide Asp330–Ala348 only contained signal 1.

1) was obtained under a variety of conditions. Signal 1 was unequivocally assigned to Lys339 in the C-terminal rhodopsin sequence (see Figure 11.6B and legend). The signal showed an unusually sharp line width and a ^{1}H chemical shift of ≈3.8 ppm. This observation demonstrated large mobility of the C-terminal peptide sequence of rhodopsin, a result in agreement with previous evidence. Spin labels attached to

engineered cysteine residues in the C-terminal sequence exhibited EPR spectra with mobilities similar to those of free spin labels in solution.[131]

Further, the amino acids in the C-terminus in rhodopsin are hardly visible in available x-ray diffraction data and have very high B factors (Figure 11.2A).[8] In addition to the predominant signal 1, a number of further resonances were observed at temperatures above 20°C that showed marked variation in intensity (Figure 11.6B). The heterogeneity in signal intensity of α-^{15}N-lysine resonance was also observed for α-^{15}N-tryptophan resonances, but not for ε-^{15}N-tryptophan resonances.[132] These results demonstrate the presence of exchange broadening, suggesting conformational fluctuations in the backbone of rhodopsin on a microsecond-to-millisecond time scale. Side chains, however, may be more restricted in their conformational dynamics.

Equilibrium fluctuations such as those observed in rhodopsin may also occur in other GPCRs. For example, single molecule spectroscopy of the β_2-adrenergic receptor revealed that ligand-bound and ligand-free conformations are in fact ensembles of multiple conformations[133] and functionally different agonists induce distinct conformations in its CP domain.[134] Thus, the widely accepted ternary complex model that accounts for the cooperative interactions between GPCRs, G proteins, and agonists[135] and the extended ternary model that distinguishes the inactive and the active conformations explicitly[136] have been modified to include a conformational stabilization component.[137] However, the extent to which these equilibrium dynamics affect folding, stability, and functions of GPCRs is not yet known.

11.2.6 CRYSTAL STRUCTURE AND STABILITY OF DARK STATE RHODOPSIN

Rhodopsin is the only GPCR whose three-dimensional structure has been determined by x-ray crystallography.[8,138] The crystal structure confirmed the vast majority of classical structure function studies using site-directed mutagenesis in conjunction with biochemical and biophysical studies carried out before the crystal structure was solved. Several reviews of this large body of previous studies in light of the crystal structure have been published.[63,64,103,139–148]

The implications of the crystal structure of rhodopsin for the structures and functions of other GPCRs have also been discussed extensively.[149–154] Recently, the crystal structure was used as a basis for the analysis of the structural elements that stabilize the rhodopsin dark state.[155] Computational analysis of the rhodopsin structure and comparison with data from previous *in vitro* mutational studies allowed identification of the most stable residues under thermal denaturation.[155] These residues form part of a core assumed to be important for the formation and stability of folded rhodopsin. The core includes the C110–C187 disulfide bond at the center of a residue forming the interface between the TM and the EC domains near the retinal binding pocket.

Fast peak mode analysis of a Gaussian network model of rhodopsin also identified the disulfide bond and the retinal ligand-binding pocket to be the most rigid regions in rhodopsin (Figure 11.7). This observation has significance for understand-

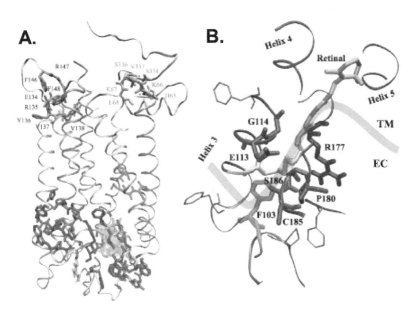

FIGURE 11.7 Core of structural stability in rhodopsin predicted by computational analysis of the rhodopsin crystal structure.[155] A: Overall view of the folding core within the rhodopsin structure. Critical folding residues in rhodopsin were identified using the FIRST and GNM computational methods (see text). Both methods select the cysteine residues that form the critical disulfide bonds, shown by space-filling spheres. The other critical residues are located in its vicinity. Together they form the largest mutually rigid cluster ("folding core"). The two next largest rigid clusters from FIRST analysis are shown at the CP end of the TM region. B: Detailed view of the most stable residues. At the center are the Cys110 and Cys187 residues that form the critical disulfide bond. Phe103, five of the common residues (Glu113, Gly114, Pro180, Cys185, and Ser186), and the retinal chromophore (retinal) are located within 4 Å of this disulfide bond and span the TM-EC interface (suggested by the thick curve). Arg177, demonstrates how this local stability is propagated across side chain interactions. Side chains for a few of the 45 FIRST folding core residues are shown by thin sticks to orient the reader.

ing a large number of previous mutagenesis studies. In particular, many naturally occurring point mutations have been characterized in rhodopsin.[156,157] Most of these mutations are associated with retinitis pigmentosa (RP), a disease that ultimately leads to photoreceptor degradation and consequent loss of vision. RP mutations are found in each of the three structural domains of rhodopsin. However, most of the mutations are found in the TM and EC domains.[156–160] Mutations in the EC domain cause partial to complete misfolding; misfolding is defined as the loss of ability to bind 11-*cis* retinal.[161–165]

Studies of both naturally occurring RP and designed mutations in the EC domain showed that misfolding involved the formation of a non-native disulfide bond between C185 and C187, instead of the naturally occurring C110–C187 bond.[166] Studies of RP mutations in the TM domain of rhodopsin showed that the mutations also cause misfolding by formation of an abnormal disulfide bond[167,168] identified by mass spectrometric analysis to be between C185 and C187.[169] The abnormal disulfide bond was the same regardless of whether the mutations were in the TM or EC domain. This suggests that packing of the helices in the TM domain and folding to a tertiary structure in the EC domain are coupled.[170] However, the molecular details of the relationship between the disulfide bond and the couplings of the structures in the TM and EC domains are not known.

The computational analysis of the crystal structure using simulated thermal denaturation now suggests a mechanism for this observation. The rhodopsin structure is most rigid and has the fewest degrees of steric freedom surrounding the interface between the TM and EC domains.[155] Indeed, comparison of the predicted core of stability with the existing mutagenesis experiments confirms that 90% of the amino acids predicted to form part of the core cause misfolding upon mutation. The observed high degree of conservation (78.9%) of this disulfide bond across all GPCR classes suggests that it is critical for the stabilities and functions of the vast majority of GPCRs.[155]

11.3 STRUCTURES, DYNAMICS, AND CONFORMATIONAL CHANGES OF GPCRs

11.3.1 HOMOLOGY MODELING AND GPCR FRAGMENT STUDIES

Because rhodopsin is the only GPCR member for which a three-dimensional structure is known, it forms the template for most current models of other GPCR structures based on homology modeling.[171–195] The use of rhodopsin in studies of other GPCRs was reviewed recently[154] and the reliability of such models has been discussed.[152] Recent improvements in the methodology achieved by combining homology modeling with molecular dynamics simulations in the lipid bilayer have been reported.[196]

The low sequence identities between many GPCRs and rhodopsin present difficulties for homology modeling approaches. Therefore, other computational approaches using helix packing moments,[197] hydrophobicity, and helix distortions[182] have been used successfully in the analysis of GPCR sequences for structure prediction. The predicted structures have been validated recently by drug discovery applications.[182] However, both modeling approaches, those using homology to rhodopsin and those based on sequence alone, have difficulties in exactly predicting helix orientation. Furthermore, the structures of the intracellular and extracellular sequences connecting the helices are not reliable.

A significant number of studies provided structural information on fragments corresponding to sequences of GPCR helices, loops, and termini. Most prominently, the x-ray crystal structure of the N-terminal fragment of the metabotropic glutamate receptor provided insight into the mechanisms of ligand binding in these GPCRs.[198] This study is particularly meaningful because the fragment and full-length receptor have identical ligand affinities.

Other studies have explored the use of NMR spectroscopy in the determination of structures of GPCR fragments. Thus, fragments corresponding to sequences derived from various receptors have been investigated structurally using NMR spectroscopy, for example, the *Saccharomyces cerevisiae* pheromone receptors,[199–203] tachykinin NK-1 receptor,[204] cannabinoid receptor CB2,[205] α2A adrenergic receptor,[206] interleukin-6 receptor,[207] thromboxane A2 receptor,[208] V(1A) vasopressin receptor,[209] and angiotensin II receptor.[210,211] A variety of solvent systems were used for these studies. Significant similarities between structures of fragments and the rhodopsin structure were observed. For example, the second intracellular loop of the V(1A) vasopressin receptor appears to be structurally related to the second intracellular loop of rhodopsin.[209] However, a comprehensive comparison of the entire structure of rhodopsin assembled from the individual structures of various fragments determined by NMR spectroscopy in solution (Section 11.2.6.2)[126,128,212–218] and the rhodopsin crystal structure showed significant differences between the structures derived from these two techniques.[128] Although some of the differences may be in part attributable to the differences between x-ray crystallography and NMR spectroscopy, they are largely due to the lack of constraints imposed by the tertiary structures of the whole molecules in the fragment studies. Two approaches have been used to simulate the constraints imposed by the helices experimentally.

In the first approach, the constraints were mimicked by including detergent micelles in the studies. NMR structures of loop fragments of several subtypes of cholecystokinin (CCK) receptors, parathyroid hormone (PTH) receptor, neurokinin–substance P receptor, and bradykinin 2 receptor were determined in the presence of flanking TM regions that anchored the loops to some extent in the micelles.[219–231] Along with models of the TM helices based on rhodopsin, the overlapping experimentally determined loop structures were grafted onto the homology models to generate new models that were validated experimentally[219–231] and reviewed by Mierke and Giragossian.[221]

Two major goals of these studies were determining the structures of the receptors and also investigating the structural bases for interactions between receptor and peptide hormone ligands and the induced conformational changes. Ligands often undergo structural transitions from an unstructured state in solution to a structured conformation upon binding to a receptor. In some cases, the intermolecular interactions between EC receptor domains and ligands were determined via NMR spectroscopic methods.[219,220] However, in other cases, the ligand would not bind to the receptor fragment, presumably because the entire tertiary structure had to be present to provide sufficient contacts to stabilize the receptor–ligand complex.

For example, the third EC loop of the PTH receptor was shown to be a major determinant in the binding of PTH, but the fragment alone does not bind PTH, thus demonstrating the cooperative natures of binding events involving more than a single loop. Similar observations were made for the intracellular loops of rhodopsin. Whereas peptide fragments corresponding to the intracellular loops of rhodopsin can compete with full-length rhodopsin for the binding to the transducin G protein, all loops together compete cooperatively better than individual loops.[232]

In the second approach, GPCR loops were grafted onto other proteins that were more suitable for structural studies. Peptide sequences corresponding to the CP loops of rhodopsin alone or together were inserted into thioredoxin.[233,234] The constructs were shown to be functional by their ability to activate the G protein. NMR spectroscopic investigation in solution provided structural insights into the loop regions of rhodopsin that are not well resolved in crystal structures.[233] Another useful scaffold is bacteriorhodopsin, due to the ease with which structural studies are possible (see Section 11.1, Introduction). The third CP loop of rhodopsin was inserted into bacteriorhodopsin and analyzed by atomic force microscopy (AFM).[235] Similarly, sequences corresponding to predicted loop regions of the CCR5 receptor were also grafted on bacteriorhodopsin for the purposes of future structural studies.[236]

11.3.2 STRUCTURE AND DYNAMICS OF LIGANDS IN CONTEXT OF FULL-LENGTH RECEPTORS

In order to fully understand the effect of a receptor structure on its cognate ligand, the ligand must be investigated in the presence of the full-length receptor. The large body of studies describing the conformations of retinal bound to opsin using solid-state NMR and Raman spectroscopy was reviewed elsewhere.[91,99,101,237–240]

NMR spectroscopic approaches to the study of drugs and ligands bound to membrane receptors were reviewed by Watts[241] and are discussed by Lange and Baldus in Chapter 13 of this book. The ability to introduce isotope labels into full-length rhodopsin by recombinant methods has recently allowed the study of the Schiff base in rhodopsin directly by solid-state NMR (Section 11.2.4.1).[40,99] Furthermore, the recent availability of large quantities of recombinant GPCRs other than rhodopsin also enabled studies of their respective ligands by structural methods. For example, acetylcholine analogs bound to M2 muscarinic acetylcholine receptor were investigated by NMR spectroscopy in solution.[242] Transferred nuclear Overhauser effects indicated that the receptor selectively recognized only certain conformers that were distinct from the predominant conformations of these ligands in solution.[242]

The pituitary adenylate cyclase activating polypeptide bound to its receptor was studied by NMR spectroscopy.[243] [13]C,[15]N-labeled neurotensin was studied recently in the presence of full-length receptors expressed in *Escherichia coli*, purified in milligram amounts, and reconstituted into lipid vesicles. The ligand was disordered in the absence of receptor and formed a β sheet when bound to the receptor.[244]

Because ligands can usually be easily derivatized and used as molecular probes, many biophysical techniques exist to study ligand–receptor interactions.[245,246] For example, fluorescent energy transfer has proven useful in the study of the tachykinin neurokinin-2[247] and other receptors.[248] EPR spectroscopy of a spin-labeled peptide corresponding to an EC loop of mass oncogene receptor has been described.[249] Photoaffinity cross-linking has proven very useful in studying the interactions of retinal and opsin[250–252] and neuropeptide Y and its receptor.[253]

11.3.3 CONFORMATIONAL CHANGES

The openings of the helical bundles described above for rhodopsin, and in particular the CP ends of helices III and IV, were demonstrated subsequent to their first discovery in rhodopsin and other GPCRs. In the β2-adrenergic receptor, fluorescence spectroscopy and cysteine reactivity changes upon ligand binding have shown that agonists and inverse agonists induce conformational changes in TM domains III and VI similar to light activation in rhodopsin.[254–259] These motions appear to be conserved across Class A receptors and also extend to Class B receptors, as shown by Zn-His studies carried out in parallel on the β2-adrenergic and the PTH receptors.[260] Further studies on a constitutively active mutant at position Phe282 at the CP end of helix VI in the

β_2-adrenergic receptor and corresponding position Phe303 in the α_{1B}-adrenergic receptor support a model in which rigid body motion of helix VI acts as a pivot because of its kink that can transduce and amplify agonist-induced conformation change in the EC–TM domain to result in a change in the TM–CP domain.[261,262]

Similar models have been proposed for the muscarinic receptor,[263] the thyroid-stimulating hormone (TSH) receptor,[264] and the C5a receptor.[265] Furthermore, studies of the dopamine D2 receptor confirmed the opening of the helical bundle toward the cytoplasmic side upon ligand binding, supporting a model in which TM I, VII, and H8 make contacts that stabilize the inactive state.[266] Cysteine accessibility studies of the μ-, δ-, and κ-opioid receptors revealed patterns very similar to those observed with the dopamine D2 receptor.[267,268]

11.3.4 CONSTRAINTS LOCKING RECEPTORS IN THEIR INACTIVE CONFORMATIONS

What prevents the conformational changes observed in many GPCRs upon ligand binding or captivation of a photon in the absence of stimulus? Overwhelming evidence suggests that this is achieved through specific contacts and that the release of the resulting constraints is a key event in the activation of rhodopsin,[250,269] β_2-adrenergic,[136,259,270,271] α_{1B}-adrenergic,[272,273] muscarinic acetylcholine,[274] gonadotropin releasing hormone,[275] μ opioid,[276] dopamine D2,[277] cannabinoid,[278] histamine H2,[279] serotonin 5HT$_{2A}$, and C5a[265] receptors, and probably GPCRs in general.[280]

The constraints cluster in several conserved "microdomains": (1) an ionic interaction between charges of ligand and receptor corresponding to the retinal Schiff base and the Glu113 counterion in rhodopsin,[269] (2) an Asp/Glu Arg Tyr motif at the cytoplasmic end of helix III and the $X_1BBX_2X_3B$ motif at the cytoplasmic end of helix VI (B = basic; X = nonbasic),[271,281] (3) Asn/Asp pairs in helices I and II, respectively, and the Asn Pro XX Tyr motif in helix VII,[138] and (4) the aromatic clusters surrounding the ligand binding pockets.[87,250,251,270,281,282] In rhodopsin, an important interaction within this aromatic cluster is between the highly conserved Trp265 and the retinal, as demonstrated by cross-linking[250] and x-ray crystallography.[8]

Upon light activation, Trp265 no longer cross-links to retinal, suggesting that the close contact between Trp265 and the ionone ring of the retinal has been broken. Instead, the ionone ring of retinal can be cross-linked with Ala169,[252] a residue located >10 Å away from the ionone ring in the dark state crystal structure.[8] The enhanced conformational flexibility of Trp265 would allow rearrangement of the indole side chain, proposed to be an important event in activation of rhodopsin.[283] The important role of the Trp265 side chain has also been studied in the β_2-adrenergic receptor.[270] Biased Monte Carlo techniques of conformational memories demonstrated that rotamer changes among the equivalent of Tryp205 and a nearby cysteine and phenylalanine residue are highly correlated and represent a "rotamer toggle switch" from Cyst/Trpg+/Pheg+ in the inactive state to Cysg+/Trpt/Phet in the active state. This switch may modulate the helix kink induced by the conserved Pro in helix VI that has been proposed to provide the necessary flexibility to allow movements of the CP ends of helices.

A change in Trp side chain from g+ to t upon activation is consistent with the observations in rhodopsin, in which the indole side chain of Trp265 changes orientation upon activation[283] and a single tryptophan was shown to tilt toward the membrane during activation.[284] In addition to Trp265, Trp161, another tryptophan, also highly conserved and located two helical turns below Ala169, appears to play a critical role in the light activation process of rhodopsin.[252,283]

11.3.5 CONSEQUENCES OF CONFORMATIONAL CHANGES FOR GPCR FUNCTION

The conformational changes enable the receptors to activate two opposing signaling cascades, one leading to sensitization, the other leading to desensitization. Sensitization is initiated by binding and activation of the G protein, whereas desensitization is followed by activation of a kinase that phosphorylates the C-termini of the receptors. The phosphorylated receptor binds to arrestin, which effectively switches the signal off. How important are the conformational changes for these events?

The relative changes in distances between helices have been probed by individual cross-links,[62,119] combined cross-links,[285] and metal chelation.[260,286] Whereas preventing the relative displacement between helices I/II and helix VII did not produce any effects on interactions of rhodopsin with G proteins or rhodopsin kinase, preventing the movement between helices III and IV abolished both G protein activation and phosphorylation by rhodopsin kinase. Interestingly, tethering the C-terminal tail to the third CP loop resulted in enhanced G protein activation and abolished phosphorylation by rhodopsin kinase as expected. These results show that the conformational changes in the CP face that are necessary for protein–protein interactions need not be cooperative and may be segmental.

This hypothesis is supported by recent studies of the parathyroid hormone (PTH) receptor.[287,288] Receptor activation and internalization can be selectively dissociated by using different PTH peptide fragments.[287] Furthermore, activation-independent PTH receptor internalization was found to be regulated by NHERF1 (EBP50) that appears to act as a molecular switch that legislates the conditional efficacy of PTH fragments.[288] Very little is known about the complexes of receptors with any of these proteins, and structural studies to date have been restricted to NMR spectroscopic investigations of a peptide derived from the C-terminus of the transducin G protein in the presence of rhodopsin (Section 11.2.6.1).[111,116,117] Ideally, one would need information on the structure and dynamics of all of the different conformations in the presence and absence of other proteins of the signaling cascades.

11.4 CONCLUSIONS

Achieving the current understanding of the structures and dynamics of GPCRs required more than 30 years of biophysical studies by hundreds of researchers around the world. The breadth of methodologies applied is extremely wide, ranging from EPR and NMR, to electron and atomic force microscopy, cysteine scanning mutagenesis, cross-linking, and most recently x-ray crystallography and computational modeling. All this effort has provided us with a huge amount of knowledge including an atomic resolution three-dimensional structure of an inactive state GPCR, demonstration of specific aspects of conformational changes within the transmembrane segments that occur upon activation, and indications of the unique dynamics that make signal transduction from the extracellular to the intracellular possible.

Many challenges remain to be addressed. The loop domains differentiate one GPCR from another and provide GPCRs with ligand and signaling specificities. To date, we have only the most limited ideas of the structures and mechanisms of protein–ligand and protein–protein interactions mediated by these loop domains that comprise the primary drug targets of the pharmaceutical industry. Unfortunately, these types of knowledge are not as generally applicable across GPCRs and will likely require case-by-case analyses. Additionally, the overall structures of the activated states of the receptors remain to be elucidated. These areas are now and will continue to be the foci of structural biological studies of GPCRs for the foreseeable future.

ACKNOWLEDGMENTS

The authors gratefully acknowledge support from the Sofya Kovalevskaya Prize of the Humboldt Foundation/Zukunftsinvestitionsprogramm der Bundesregierung Deutschland, National Science Foundation ITR grants 0225636, 0225656, and CAREER CC044917 NIH grant NLM1ROILM007994, and the National Research Council of Canada GHI grant 49/64-GH2P6.

REFERENCES

1. Horn, F., E. Bettler, L. Oliveira, F. Campagne, F.E. Cohen and G. Vriend. GPCRDB information system for G protein-coupled receptors. Nucleic Acids Res 2003, 31, 294–207.

2. Bateman, A., E. Birney, L. Cerruti, R. Durbin, L. Etwiller, S.R. Eddy, S. Griffiths-Jones, K.L. Howe, M. Marshall and E.L. Sonnhammer. The Pfam protein families database. Nucleic Acids Res 2002, 30, 276–280.

3. Henderson, R. and P.N. Unwin. Three-dimensional model of purple membrane obtained by electron microscopy. Nature 1975, 257, 28–32.

4. Hargrave, P.A., J.H. McDowell, D.R. Curtis, J.K. Wang, E. Jusczak, S.L. Fong, J.K.M. Rao and P. Argos. The structure of bovine rhodopsin. Biophys Struct Mech 1983, 9, 235–244.

5. Neutze, R., E. Pebay-Peyroula, K. Edman, A. Royant, J. Navarro and E.M. Landau. Bacteriorhodopsin: a high-resolution structural view of vectorial proton transport. Biochim Biophys Acta 2002, 1565, 144–167.

6. Fotiadis, D. and A. Engel. High-resolution imaging of bacteriorhodopsin by atomic force microscopy. Methods Mol Biol 2004, 242, 291–303.

7. Unger, V.M. and G.F. Schertler. Low resolution structure of bovine rhodopsin determined by electron cryomicroscopy. Biophys J 1995, 68, 1776–1786.

8. Palczewski, K., T. Kumasaka, T. Hori, C.A. Behnke, H. Motoshima, B.A. Fox, I. LeTrong, D.C. Teller, T. Okada, R.E. Stenkamp, M. Yamamoto and M. Miyano. Crystal structure of rhodopsin: a G protein-coupled receptor. Science 2000, 289, 739–745.

9. Wald, G. Molecular basis of visual excitation. Science 1968, 162, 230–239.

10. Rowan, R., 3rd, A. Warshel, B.D. Sykes and M. Karplus. Conformation of retinal isomers. Biochemistry 1974, 13, 970–981.

11. Shriver, J.W., G.D. Mateescu and E.W. Abrahamson. A proton and carbon-13 nuclear magnetic resonance spectroscopy study of the conformation of a protonated 11-cis-retinal Schiff base. Biochemistry 1979, 18, 4785–4792.

12. Shriver, J., G. Mateescu, R. Fager, D. Toricha and E.W. Abrahamson. Unprotonated chromophore–protein bond in visual pigments from ^{13}C-NMR spectra. Nature 1977, 2705, 271–274.

13. Hong, K. and W.L. Hubbell. Lipid requirements for rhodopsin regenerability. Biochemistry 1973, 12, 4517–4523.

14. Knudsen, P. and W.L. Hubbell. Stability of rhodopsin in detergent solutions. Membr Biochem 1978, 1, 297–322.

15. Delmelle, M. and M. Pontus. Magnetic resonance study of spin-labelled rhodopsin. Biochim Biophys Acta 1974, 365, 47–56.

16. Brown, M.F., A.J. Deese and E.A. Dratz. Proton, carbon-13, and phosphorus-31 NMR methods for the investigation of rhodopsin–lipid interactions in retinal rod outer segment membranes. Methods Enzymol 1982, 81, 709–728.

17. Brown, M.F., G.P. Miljanich and E.A. Dratz. Proton spin-lattice relaxation of retinal rod outer segment membranes and liposomes of extracted phospholipids. Proc Natl Acad Sci USA 1977, 74, 1978–1982.

18. Deese, A.J., E.A. Dratz, L. Hymel and S. Fleischer. Proton NMR T1, T2, and T1 rho relaxation studies of native and reconstituted sarcoplasmic reticulum and phospholipid vesicles. Biophys J 1982, 37, 207–216.

19. Watts, A., I.D. Volotovski and D. Marsh. Rhodopsin-lipid associations in bovine rod outer segment membranes: identification of immobilized lipid by spin labels. Biochemistry 1979, 18, 5006–5013.

20. Cornell, B.A., M.M. Sacre and D. Chapman. The modulation of lipid bilayer fluidity by intrinsic polypeptides and proteins. FEBS Lett 1978, 90, 29–35.

21. Favre, E., A. Baroin, A. Bienvenue and P.F. Devaux. Spin-label studies of lipid–protein interactions in retinal rod outer segment membranes: fluidity of the boundary layer. Biochemistry 1979, 18, 1156–1162.

22. Bienvenue, A., M. Bloom, J.H. Davis and P.F. Devaux. Evidence for protein-associated lipids from deuterium nuclear magnetic resonance studies of rhodopsin–dimyristoyl phosphatidylcholine recombinants. J Biol Chem 1982, 257, 3032–3038.

23. Ryba, N.J., C.E. Dempsey and A. Watts. Protein–lipid interactions at membrane surfaces: a deuterium and phosphorus nuclear magnetic resonance study of the interaction between bovine rhodopsin and the bilayer head groups of dimyristoyl phosphatidylcholine. Biochemistry 1986, 25, 4818–4825.

24. Deese, A.J., E.A. Dratz, F.W. Dahlquist and M.R. Paddy. Interaction of rhodopsin with two unsaturated phosphatidylcholines: a deuterium nuclear magnetic resonance study. Biochemistry 1981, 20, 6420–6427.

25. De Grip, W.J., E.H. Drenthe, C.J. Van Echteld, B. De Kruijff and A.J. Verkleij. A possible role of rhodopsin in maintaining bilayer structure in the photoreceptor membrane. Biochim Biophys Acta 1979, 558, 330–337.

26. Deese, A.J., E.A. Dratz and M.F. Brown. Retinal rod outer segment lipids form bilayers in the presence and absence of rhodopsin: a ^{31}P NMR study. FEBS Lett 1981, 124, 93–99.

27. Ellena, J.F., R.D. Pates and M.F. Brown. ^{31}P NMR spectra of rod outer segment and sarcoplasmic reticulum membranes show no evidence of immobilized components due to lipid–protein interactions. Biochemistry 1986, 25, 3742–3748.

28. Yeagle, P.L., B.S. Selinsky and A.D. Albert. Perturbations of phospholipid head groups by membrane proteins in biological membranes and recombinants. Biophys J 1984, 45, 1085–1089.

29. Albert, A.D. and P.L. Yeagle. Phospholipid domains in bovine retinal rod outer segment disk membranes. Proc Natl Acad Sci USA 1983, 80, 7188–7191.

30. Yeagle, P.L. 31P nuclear magnetic resonance studies of the phospholipid–protein interface in cell membranes. Biophys J 1982, 37, 227–239.

31. Molday, R.S. and D. MacKenzie. Monoclonal antibodies to rhodopsin: characterization, cross-reactivity, and application as structural probes. Biochemistry 1983, 22, 653–660.

32. Harada, Y., T. Senda, T. Sakamoto, K. Takamoto and T. Ishibashi. Expression of octopus rhodopsin in *Escherichia coli*. J Biochem (Tokyo) 1994, 115, 66–75.

33. Mollaaghababa, R., F.F. Davidson, C. Kaiser and H.G. Khorana. Structure and function in rhodopsin: expression of functional mammalian opsin in *Saccharomyces cerevisiae*. Proc Natl Acad Sci USA 1996, 93, 11482–11486.

34. Abdulaev, N.G. and K.D. Ridge. Heterologous expression of bovine opsin in *Pichia pastoris*. Methods Enzymol 2000, 315, 3–11.

35. Khorana, H.G., B.E. Knox, E. Nasi, R. Swanson and D.A. Thompson. Expression of a bovine rhodopsin gene in *Xenopus* oocytes: demonstration of light-dependent ionic currents. Proc Natl Acad Sci USA 1988, 85, 7917–7921.

36. Weiss, E.R., R.A. Heller-Harrison, E. Diez, M. Crasnier, C.C. Malbon and G.L. Johnson. Rhodopsin expressed in Chinese hamster ovary cells regulates adenylyl cyclase activity. J Mol Endocrinol 1990, 4, 71–79.

37. Oprian, D.D., R.S. Molday, R.J. Kaufman and H.G. Khorana. Expression of a synthetic bovine rhodopsin gene in monkey kidney cells. Proc Natl Acad Sci USA 1987, 84, 8874–8878.

38. Reeves, P.J., R.L. Thurmond and H.G. Khorana. Structure and function in rhodopsin: high level expression of a synthetic bovine opsin gene and its mutants in stable mammalian cell lines. Proc Natl Acad Sci USA 1996, 93, 11487–11492.

39. Reeves, P.J., J. Klein-Seetharaman, E.V. Getmanova, M. Eilers, M.C. Loewen, S.O. Smith and H.G. Khorana. Expression and purification of rhodopsin and its mutants from stable mammalian cell lines: application to NMR studies. Biochem Soc Trans 1999, 27, 950–955.

40. Eilers, M., P.J. Reeves, W. Ying, H.G. Khorana and S.O. Smith. Magic angle spinning NMR of the protonated retinylidene Schiff base nitrogen in rhodopsin: expression of 15N-lysine- and ^{13}C-glycine-labeled opsin in a stable cell line. Proc Natl Acad Sci USA 1999, 96, 487–492.

41. Eilers, M., W. Ying, P.J. Reeves, H.G. Khorana and S.O. Smith. Magic angle spinning nuclear magnetic resonance of isotopically labeled rhodopsin. Methods Enzymol 2002, 343, 212–222.

42. Getmanova, E., A.B. Patel, J. Klein-Seetharaman, M.C. Loewen, P.J. Reeves, N. Friedman, M. Sheves, S.O. Smith and H.G. Khorana. NMR spectroscopy of phosphorylated wild-type rhodopsin: mobility of the phosphorylated C-terminus of rhodopsin in the dark and upon light activation. Biochemistry 2004, 43, 1126–1133.

43. Klein-Seetharaman, J., E.V. Getmanova, M.C. Loewen, P.J. Reeves and H.G. Khorana. NMR spectroscopy in studies of light-induced structural changes in mammalian rhodopsin: applicability of solution ^{19}F NMR. Proc Natl Acad Sci USA 1999, 96, 13744–13749.

44. Klein-Seetharaman, J., P.J. Reeves, M.C. Loewen, E.V. Getmanova, J. Chung, H. Schwalbe, P.E. Wright and H.G. Khorana. Solution NMR spectroscopy of [α-5N]lysine-labeled rhodopsin: the single peak observed in both conventional and TROSY-type HSQC spectra is ascribed to Lys339 in the carboxyl-terminal peptide sequence. Proc Natl Acad Sci USA 2002, 99, 3452–3457.

45. Loewen, M.C., J. Klein-Seetharaman, E.V. Getmanova, P.J. Reeves, H. Schwalbe and H.G. Khorana. Solution ^{19}F nuclear Overhauser effects in structural studies of the cytoplasmic domain of mammalian rhodopsin. Proc Natl Acad Sci USA 2001, 98, 4888–4892.

46. Bruel, C., K. Cha, P.J. Reeves, E. Getmanova and H.G. Khorana. Rhodopsin kinase: expression in mammalian cells and a two-step purification. Proc Natl Acad Sci USA 2000, 97, 3004–3009.

47. Reeves, P.J., J.M. Kim and H.G. Khorana. Structure and function in rhodopsin: a tetracycline-inducible system in stable mammalian cell lines for high-level expression of opsin mutants. Proc Natl Acad Sci USA 2002, 99, 13413–13418.

48. Reeves, P.J., N. Callewaert, R. Contreras and H.G. Khorana. Structure and function in rhodopsin: high-level expression of rhodopsin with restricted and homogeneous N-glycosylation by a tetracycline-inducible N-acetylglucosaminyl-transferase I-negative HEK293S stable mammalian cell line. Proc Natl Acad Sci USA 2002, 99, 13419–13424.

49. Janssen, J.J., P.H. Bovee-Geurts, M. Merkx and W.J. DeGrip. Histidine tagging both allows convenient single-step purification of bovine rhodopsin and exerts ionic strength-dependent effects on its photochemistry. J Biol Chem 1995, 270, 11222–11229.

50. Steinemann, A. and L. Stryer. Accessibility of the carbohydrate moiety of rhodopsin. Biochemistry 1973, 12, 1499–1502.

51. DeGrip, W.J. Purification of bovine rhodopsin over concanavalin A-sepharose. Methods Enzymol 1982, 81, 197–207.

52. Grassetti, D.R. and J.F.J. Murray. Determination of sulfhydryl groups with 2,2'- or 4,4'-dithiodipyridine. Arch Biophys 1967, 119, 41–49.

53. Klein-Seetharaman, J., J. Hwa, K. Cai, C. Altenbach, W.L. Hubbell and H.G. Khorana. Single-cysteine substitution mutants at amino acid positions 55–75, the sequence connecting the cytoplasmic ends of helices I and II in rhodopsin: reactivity of the sulfhydryl groups and their derivatives identifies a tertiary structure that changes upon light-activation. Biochemistry 1999, 38, 7938–7944.

54. Altenbach, C., J. Klein-Seetharaman, J. Hwa, H.G. Khorana and W.L. Hubbell. Structural features and light-dependent changes in the sequence 59–75 connecting helices I and II in rhodopsin: a site-directed spin labeling study. Biochemistry 1999, 38, 7945–7949.

55. Klein-Seetharaman, J., J. Hwa, K. Cai, C. Altenbach, W.L. Hubbell and H.G. Khorana. Probing the dark state tertiary structure in the cytoplasmic domain of rhodopsin: proximities between amino acids deduced from spontaneous disulfide bond formation between Cys316 and engineered cysteines in cytoplasmic loop 1. Biochemistry 2001, 40, 12472–12478.

56. Cai, K., J. Klein-Seetharaman, C. Altenbach, W.L. Hubbell and H.G. Khorana. Probing the dark state tertiary structure in the cytoplasmic domain of rhodopsin: proximities between amino acids deduced from spontaneous disulfide bond formation between cysteine pairs engineered in cytoplasmic loops 1, 3, and 4. Biochemistry 2001, 40, 12479–12485.

57. Falke, J.J. and D.E. Koshland, Jr. Global flexibility in a sensory receptor: a site-directed cross-linking approach. Science 1987, 237, 1596–1600.

58. Falke, J.J., A.F. Dernburg, D.A. Sternberg, N. Zalkin, D.L. Milligan and D.E. Koshland, Jr. Structure of a bacterial sensory receptor: a site-directed sulfhydryl study. J Biol Chem 1988, 263, 14850–14858.

59. Milligan, D.L. and D.E. Koshland, Jr. Site-directed cross-linking: establishing the dimeric structure of the aspartate receptor of bacterial chemotaxis. J Biol Chem 1988, 263, 6268–6275.

60. Richardson, J.S. The anatomy and taxonomy of protein structure. Adv Prot Chem 1981, 34, 167–339.

61. Kosen, P.A. Stability of protein pharmaceuticals, part A. In: Chemical and Physical Pathways of Protein Degradation M.C. Manning, Ed., 1992, Plenum Press, New York.

62. Cai, K., J. Klein-Seetharaman, J. Hwa, W.L. Hubbell and H.G. Khorana. Structure and function in rhodopsin: effects of disulfide cross-links in the cytoplasmic face of rhodopsin on transducin activation and phosphorylation by rhodopsin kinase. Biochemistry 1999, 38, 12893–12898.

63. Columbus, L. and W.L. Hubbell. A new spin on protein dynamics. Trends Biochem Sci 2002, 27, 288–295.

64. Hubbell, W.L., C. Altenbach, C.M. Hubbell and H.G. Khorana. Rhodopsin structure, dynamics, and activation: a perspective from crystallography, site-directed spin labeling, sulfhydryl reactivity, and disulfide cross-linking. Adv Protein Chem 2003, 63, 243–290.

65. Altenbach, C., K. Yang, D.L. Farrens, Z.T. Farahbakhsh, H.G. Khorana and W.L. Hubbell. Structural features and light-dependent changes in the cytoplasmic interhelical E–F loop region of rhodopsin: a site-directed spin-labeling study. Biochemistry 1996, 35, 12470–12478.

66. Altenbach, C., K. Cai, H.G. Khorana and W.L. Hubbell. Structural features and light-dependent changes in the sequence 306–322 extending from helix VII to the palmitoylation sites in rhodopsin: a site-directed spin-labeling study. Biochemistry 1999, 38, 7931–7937.

67. Farahbakhsh, Z.T., K.D. Ridge, H.G. Khorana and W.L. Hubbell. Mapping light-dependent structural changes in the cytoplasmic loop connecting helices C and D in rhodopsin: a site-directed spin labeling study. Biochemistry 1995, 34, 8812–8819.

68. Altenbach, C., K. Cai, J. Klein-Seetharaman, H.G. Khorana and W.L. Hubbell. Structure and function in rhodopsin: mapping light-dependent changes in distance between residue 65 in helix TM1 and residues in the sequence 306–319 at the cytoplasmic end of helix TM7 and in helix H8. Biochemistry 2001, 40, 15483–15492.

69. Altenbach, C., J. Klein-Seetharaman, K. Cai, H.G. Khorana and W.L. Hubbell. Structure and function in rhodopsin: mapping light-dependent changes in distance between residue 316 in helix 8 and residues in the sequence 60–75 covering the cytoplasmic end of helices TM1 and TM2 and their connection loop CL1. Biochemistry 2001, 40, 15493–15500.

70. Cai, K., R. Langen, W.L. Hubbell and H.G. Khorana. Structure and function in rhodopsin: topology of the C-terminal polypeptide chain in relation to the cytoplasmic loops. Proc Natl Acad Sci USA 1997, 94, 14267–14272.

71. Pines, A., M.G. Gibby and J.S. Waugh. Proton-enhanced NMR of dilute spins in solids. J Chem Phys 1973, 59, 569–590.

72. Haberkorn, R.A., J. Herzfeld and R.G. Griffin. High resolution phosphorus-31 and carbon-13 nuclear magnetic resonance spectra of unsonicated model membranes. J Am Chem Soc 1978, 100, 1296–1298.

73. Maricq, M.M. and J.S. Waugh. NMR in rotating solids. J Chem Phys 1979, 70, 3300–3316.

74. de Groot, H.J., G.S. Harbison, J. Herzfeld and R.G. Griffin. Nuclear magnetic resonance study of the Schiff base in bacteriorhodopsin: counterion effects on the ^{15}N shift anisotropy. Biochemistry 1989, 28, 3346–3353.

75. Smith, S.O., I. Palings, V. Copie, D.P. Raleigh, J. Courtin, J.A. Pardoen, J. Lugtenburg, R.A. Mathies and R.G. Griffin. Low-temperature solid-state ^{13}C NMR studies of the retinal chromophore in rhodopsin. Biochemistry 1987, 26, 1606–1611.

76. Harbison, G.S., S.O. Smith, J.A. Pardoen, J.M. Courtin, J. Lugtenburg, J. Herzfeld, R.A. Mathies and R.G. Griffin. Solid-state ^{13}C NMR detection of a perturbed 6-s-*trans* chromophore in bacteriorhodopsin. Biochemistry 1985, 24, 6955–6962.

77. Mollevanger, L.C., A.P. Kentgens, J.A. Pardoen, J.M. Courtin, W.S. Veeman, J. Lugtenburg and W.J. de Grip. High-resolution solid-state ^{13}C-NMR study of carbons C-5 and C-12 of the chromophore of bovine rhodopsin: evidence for a 6-S-*cis* conformation with negative-charge perturbation near C-12. Eur J Biochem 1987, 163, 9–14.

78. Han, M., B.S. DeDecker and S.O. Smith. Localization of the retinal protonated Schiff base counterion in rhodopsin. Biophys J 1993, 65, 899–906.

79. Sakmar, T.P., R.R. Franke and H.G. Khorana. Glutamic acid-113 serves as the retinylidene Schiff base counterion in bovine rhodopsin. Proc Natl Acad Sci USA 1989, 86, 8309–8313.

80. Han, M. and S.O. Smith. NMR constraints on the location of the retinal chromophore in rhodopsin and bathorhodopsin. Biochemistry 1995, 34, 1425–1432.

81. Han, M. and S.O. Smith. High-resolution structural studies of the retinal–Glu113 interaction in rhodopsin. Biophys Chem 1995, 56, 23–29.

82. Smith, S.O., I. Palings, M.E. Miley, J. Courtin, H. de Groot, J. Lugtenburg, R.A. Mathies and R.G. Griffin. Solid-state NMR studies of the mechanism of the opsin shift in the visual pigment rhodopsin. Biochemistry 1990, 29, 8158–8164.

83. Smith, S.O., J. Courtin, H. de Groot, R. Gebhard and J. Lugtenburg. ^{13}C magic-angle spinning NMR studies of bathorhodopsin, the primary photoproduct of rhodopsin. Biochemistry 1991, 30, 7409–7415.

84. Smith, S.O., H. de Groot, R. Gebhard and J. Lugtenburg. Magic angle spinning NMR studies on the metarhodopsin II intermediate of bovine rhodopsin: evidence for an unprotonated Schiff base. Photochem Photobiol 1992, 56, 1035–1039.

85. Han, M., S.W. Lin, S.O. Smith and T.P. Sakmar. The effects of amino acid replacements of glycine 121 on transmembrane helix 3 of rhodopsin. J Biol Chem 1996, 271, 32330–32336.

86. Han, M., S.W. Lin, M. Minkova, S.O. Smith and T.P. Sakmar. Functional interaction of transmembrane helices 3 and 6 in rhodopsin: replacement of phenylalanine 261 by alanine causes reversion of phenotype of a glycine 121 replacement mutant. J Biol Chem 1996, 271, 32337–32342.

87. Han, M., J. Lou, K. Nakanishi, T.P. Sakmar and S.O. Smith. Partial agonist activity of 11-cis-retinal in rhodopsin mutants. J Biol Chem 1997, 272, 23081–23085.

88. Han, M., M. Groesbeek, T.P. Sakmar and S.O. Smith. The C9 methyl group of retinal interacts with glycine-121 in rhodopsin. Proc Natl Acad Sci USA 1997, 94, 13442–13447.

89. Shieh, T., M. Han, T.P. Sakmar and S.O. Smith. The steric trigger in rhodopsin activation. J Mol Biol 1997, 269, 373–384.

90. Grobner, G., G. Choi, I.J. Burnett, C. Glaubitz, P.J. Verdegem, J. Lugtenburg and A. Watts. Photoreceptor rhodopsin: structural and conformational study of its chromophore 11-cis retinal in oriented membranes by deuterium solid state NMR. FEBS Lett 1998, 422, 201–204.

91. Grobner, G., I.J. Burnett, C. Glaubitz, G. Choi, A.J. Mason and A. Watts. Observations of light-induced structural changes of retinal within rhodopsin. Nature 2000, 405, 810–813.

92. Verdegem, P.J., P.H. Bovee-Geurts, W.J. de Grip, J. Lugtenburg and H.J. de Groot. Internuclear distance measurements up to 0.44 nm for retinals in the solid state with 1-D rotational resonance ^{13}C MAS NMR spectroscopy. J Am Chem Soc 1997, 119, 169.

93. Verdegem, P.J., P.H. Bovee-Geurts, W.J. de Grip, J. Lugtenburg and H.J. de Groot. Retinylidene ligand structure in bovine rhodopsin, metarhodopsin-I, and 10-methyl-rhodopsin from internuclear distance measurements using ^{13}C-labeling and 1-D rotational resonance MAS NMR. Biochemistry 1999, 38, 11316–11324.

94. Feng, X., P.J. Verdegem, M. Eden, D. Sandstrom, Y.K. Lee, P.H. Bovee-Geurts, W.J. de Grip, J. Lugtenburg, H.J. de Groot and M.H. Levitt. Determination of a molecular torsional angle in the metarhodopsin-I photointermediate of rhodopsin by double-quantum solid-state NMR. J Biomol NMR 2000, 16, 1–8.

95. Feng, X., P.J. Verdegem, Y.K. Lee, M. Helmle, S.C. Shekar, H.J. de Groot, J. Lugtenburg and M.H. Levitt. Rotational resonance NMR of ^{13}C2-labelled retinal: quantitative internuclear distance determination. Solid State Nucl Magn Reson 1999, 14, 81–90.

96. Spooner, P.J., J.M. Sharples, M.A. Verhoeven, J. Lugtenburg, C. Glaubitz and A. Watts. Relative orientation between the beta-ionone ring and the polyene chain for the chromophore of rhodopsin in native membranes. Biochemistry 2002, 41, 7549–7555.

97. Spooner, P.J., J.M. Sharples, S.C. Goodall, H. Seedorf, M.A. Verhoeven, J. Lugtenburg, P.H. Bovee-Geurts, W.J. DeGrip and A. Watts. Conformational similarities in the beta-ionone ring region of the rhodopsin chromophore in its ground state and after photoactivation to the metarhodopsin-I intermediate. Biochemistry 2003, 42, 13371–13378.

98. DeGrip, W.J., C.H. Klaassen and P.H. Bovee-Geurts. Large-scale functional expression of visual pigments: towards high-resolution structural and mechanistic insight. Biochem Soc Trans 1999, 27, 937–944.

99. Creemers, A.F., C.H. Klaassen, P.H. Bovee-Geurts, R. Kelle, U. Kragl, J. Raap, W.J. de Grip, J. Lugtenburg and H.J. de Groot. Solid state 15N NMR evidence for a complex Schiff base counterion in the visual G-protein-coupled receptor rhodopsin. Biochemistry 1999, 38, 7195–7199.

100. Mirzadegan, T., C. Humblet, W.C. Ripka, L.U. Colmenares and R.S. Liu. Modeling rhodopsin, a member of G-protein coupled receptors, by computer graphics: interpretation of chemical shifts of fluorinated rhodopsins. Photochem Photobiol 1992, 56, 883–893.

101. Verhoeven, M.A., A.F. Creemers, P.H. Bovee-Geurts, W.J. De Grip, J. Lugtenburg and H.J. de Groot. Ultra-high-field MAS NMR assay of a multispin labeled ligand bound to its G-protein receptor target in the natural membrane environment: electronic structure of the retinylidene chromophore in rhodopsin. Biochemistry 2001, 40, 3282–3288.

102. Creemers, A.F., S. Kiihne, P.H. Bovee-Geurts, W.J. DeGrip, J. Lugtenburg and H.J. de Groot. $^{(1)}$H and $^{(13)}$C MAS NMR evidence for pronounced ligand–protein interactions involving the ionone ring of the retinylidene chromophore in rhodopsin. Proc Natl Acad Sci USA 2002, 99, 9101–9106.

103. Teller, D.C., T. Okada, C.A. Behnke, K. Palczewski and R.E. Stenkamp. Advances in determination of a high-resolution three-dimensional structure of rhodopsin, a model of G-protein-coupled receptors (GPCRs). Biochemistry 2001, 40, 7761–7772.

104. Worcester, D.L. Structural origins of diamagnetic anisotropy in proteins. Proc Natl Acad Sci USA 1978, 75, 5475–5477.

105. Chalazonitis, N., R. Chagneux and A. Arvanitaki. Rotation of external segments of photoreceptors in constant magnetic field. C R Acad Sci Hebd Seances Acad Sci D 1970, 271, 130–133.

106. Mollevanger, L.C., E.A. Dratz, B. De Kruijff, C.W. Hilbers and W.J. De Grip. ^{31}P-NMR investigation of magnetically oriented rod outer segments: spectral analysis and identification of individual phospholipids. Eur J Biochem 1986, 156, 383–390.

107. Albert, A.D., J.S. Frye and P.L. Yeagle. Light-induced membrane protein phosphorylation in the bovine rod outer segment. A magic angle spinning ^{31}P-NMR study. Biophys Chem 1990, 36, 27–31.

108. Dratz, E.A., J.E. Furstenau, C.G. Lambert, D.L. Thireault, H. Rarick, T. Schepers, S. Pakhlevaniants and H.E. Hamm. NMR structure of a receptor-bound G-protein peptide. Nature 1993, 363, 276–281.

109. Campbell, A.P. and B.D. Sykes. Theoretical evaluation of the two-dimensional transferred nuclear Overhauser effect. J Magn Reson 1991, 93, 77–92.

110. Dratz, E.A., J.E. Furstenau, C.G. Lambert, D.L. Thireault, H. Rarick, T. Schepers, S. Pakhlevaniants and H.E. Hamm. NMR Structure of a receptor bound G-protein peptide. Nature 1997, 390, 424.

111. Kisselev, O.G., J. Kao, J.W. Ponder, Y.C. Fann, N. Gautam and G.R. Marshall. Light-activated rhodopsin induces structural binding motif in G protein alpha subunit. Proc Natl Acad Sci USA 1998, 95, 4270–4275.

112. Lambright, D.G., J. Sondek, A. Bohm, N.P. Skiba, H.E. Hamm and P.B. Sigler. The 2.0 Å crystal structure of a heterotrimeric G protein. Nature 1996, 379, 311–319.

113. Sondek, J., D.G. Lambright, J.P. Noel, H.E. Hamm and P.B. Sigler. GTPase: mechanisms of G proteins from the 1.7- crystal structure of transducin α-GDP-AIF-4. Nature 1994, 372, 276–279.

114. Lambright, D.G., J.P. Noel, H.E. Hamm and P.B. Sigler. Structural determinants for activation of the α subunit of a heterotrimeric G protein. Nature 1994, 369, 621–628.

115. Noel, J.P., H.E. Hamm and P.B. Sigler. The 2.2 Å crystal structure of transducin-α complexed with GTP-γ-S. Nature 1993, 366, 654–663.

116. Koenig, B.W., G. Kontaxis, D.C. Mitchell, J.M. Louis, B.J. Litman and A. Bax. Structure and orientation of a G protein fragment in the receptor bound state from residual dipolar couplings. J Mol Biol 2002, 322, 441–461.

117. Koenig, B.W., D.C. Mitchell, S. Konig, S. Grzesiek, B.J. Litman and A. Bax. Measurement of dipolar couplings in a transducin peptide fragment weakly bound to oriented photo-activated rhodopsin. J Biomol NMR 2000, 16, 121–125.

118. Gerig, J.T. Fluorine nuclear magnetic resonance of fluorinated ligands. Methods Enzymol 1989, 177, 3–23.

119. Farrens, D.L., C. Altenbach, K. Yang, W.L. Hubbell and H.G. Khorana. Requirement of rigid-body motion of transmembrane helices for light activation of rhodopsin. Science 1996, 274, 768–770.

120. Yeagle, P.L., G. Choi and A.D. Albert. Studies on the structure of the G-protein-coupled receptor rhodopsin including the putative G-protein binding site in unactivated and activated forms. Biochemistry 2001, 40, 11932–11937.

121. Yu, H., M. Kono, T.D. McKee and D.D. Oprian. A general method for mapping tertiary contacts between amino acid residues in membrane-embedded proteins. Biochemistry 1995, 34, 14963–14969.

122. Katragadda, M., J.L. Alderfer and P.L. Yeagle. Assembly of a polytopic membrane protein structure from the solution structures of overlapping peptide fragments of bacteriorhodopsin. Biophys J 2001, 81, 1029–1036.

123. Katragadda, M., J.L. Alderfer and P.L. Yeagle. Solution structure of the loops of bacteriorhodopsin closely resembles the crystal structure. Biochim Biophys Acta 2000, 1466, 1–6.

124. Luecke, H., B. Schobert, H.T. Richter, J.P. Cartailler and J.K. Lanyi. Structure of bacteriorhodopsin at 1.55 Å resolution. J Mol Biol 1999, 291, 899–911.

125. Pebay-Peyroula, E., G. Rummel, J.P. Rosenbusch and E.M. Landau. X-ray structure of bacteriorhodopsin at 2.5 angstroms from microcrystals grown in lipidic cubic phases. Science 1997, 277, 1676–1681.

126. Katragadda, M., A. Chopra, M. Bennett, J.L. Alderfer, P.L. Yeagle and A.D. Albert. Structures of the transmembrane helices of the G-protein coupled receptor, rhodopsin. J Peptide Res 2001, 58, 79–89.

127. Yeagle, P.L., A. Salloum, A. Chopra, N. Bhawsar, L. Ali, G. Kuzmanovski, J.L. Alderfer and A.D. Albert. Structures of the intradiskal loops and amino terminus of the G-protein receptor, rhodopsin. J Peptide Res 2000, 55, 455–465.

128. Albert, A.D. and P.L. Yeagle. Structural studies on rhodopsin. Biochim Biophys Acta 2002, 1565, 183–195.

129. Yeagle, P.L. and A.D. Albert. A conformational trigger for activation of a G protein by a G protein-coupled receptor. Biochemistry 2003, 42, 1365–1368.

130. Choi, G., J. Landin, J.F. Galan, R.R. Birge, A.D. Albert and P.L. Yeagle. Structural studies of metarhodopsin II, the activated form of the G-protein coupled receptor, rhodopsin. Biochemistry 2002, 41, 7318–7324.

131. Langen, R., K. Cai, C. Altenbach, H.G. Khorana and W.L. Hubbell. Structural features of the C-terminal domain of bovine rhodopsin: a site-directed spin-labeling study. Biochemistry 1999, 38, 7918–7924.

132. Klein-Seetharaman, J., N.V.K. Yanamala, F. Javeed, E. Getmanova, P.J. Reeves, H. Schwalbe and H.G. Khorana. Differential dynamics in the G protein-coupled receptor rhodopsin revealed by solution NMR. Proc Natl Acad Sci USA 2004, 101, 3409–3413.

133. Peleg, G., P. Ghanouni, B.K. Kobilka and R.N. Zare. Single-molecule spectroscopy of the beta(2) adrenergic receptor: observation of conformational substates in a membrane protein. Proc Natl Acad Sci USA 2001, 98, 8469–8474.

134. Ghanouni, P., Z. Gryczynski, J.J. Steenhuis, T.W. Lee, D.L. Farrens, J.R. Lakowicz and B.K. Kobilka. Functionally different agonists induce distinct conformations in the G protein coupling domain of the beta 2 adrenergic receptor. J Biol Chem 2001, 276, 24433–24436.

135. De Lean, A., J.M. Stadel and R.J. Lefkowitz. A ternary complex model explains the agonist-specific binding properties of the adenylate cyclase-coupled beta-adrenergic receptor. J Biol Chem 1980, 255, 7108–71017.

136. Samama, P., S. Cotecchia, T. Costa and R.J. Lefkowitz. A mutation-induced activated state of the beta 2-adrenergic receptor: extending the ternary complex model. J Biol Chem 1993, 268, 4625–4636.

137. Gether, U. and B.K. Kobilka. G protein-coupled receptors. II. Mechanism of agonist activation. J Biol Chem 1998, 273, 17979–17982.

138. Okada, T., Y. Fujiyoshi, M. Silow, J. Navarro, E.M. Landau and Y. Shichida. Functional role of internal water molecules in rhodopsin revealed by x-ray crystallography. Proc Natl Acad Sci USA 2002, 99, 5982–5987.

139. Klein-Seetharaman, J. Dynamics in rhodopsin. Chembiochem 2002, 3, 981–986.

140. Okada, T., O.P. Ernst, K. Palczewski and K.P. Hofmann. Activation of rhodopsin: new insights from structural and biochemical studies. Trends Biochem Sci 2001, 26, 318–324.

141. Dohlman, H.G., J. Thorner, M.G. Caron and R.J. Lefkowitz. Model systems for the study of seven-transmembrane-segment receptors. Annu Rev Biochem 1991, 60, 653–588.

142. Mirzadegan, T., G. Benko, S. Filipek and K. Palczewski. Sequence analyses of G-protein-coupled receptors: similarities to rhodopsin. Biochemistry 2003, 42, 2759–2767.

143. Ridge, K.D., N.G. Abdulaev, M. Sousa and K. Palczewski. Phototransduction: crystal clear. Trends Biochem Sci 2003, 28, 479–487.

144. Stojanovic, A. and J. Hwa. Rhodopsin and retinitis pigmentosa: shedding light on structure and function. Receptors Channels 2002, 8, 33–50.

145. Stenkamp, R.E., D.C. Teller and K. Palczewski. Crystal structure of rhodopsin: a G-protein-coupled receptor. Chembiochem 2002, 3, 963–967.

146. Sakmar, T.P., S.T. Menon, E.P. Marin and E.S. Awad. Rhodopsin: insights from recent structural studies. Annu Rev Biophys Biomol Struct 2002, 31, 443–484.

147. Meng, E.C. and H.R. Bourne. Receptor activation: what does the rhodopsin structure tell us? Trends Pharmacol Sci 2001, 22, 587–593.

148. Filipek, S., R.E. Stenkamp, D.C. Teller and K. Palczewski. G protein-coupled receptor rhodopsin: a prospectus. Annu Rev Physiol 2003, 65, 851–879.

149. Ballesteros, J.A., L. Shi and J.A. Javitch. Structural mimicry in G protein-coupled receptors: implications of the high-resolution structure of rhodopsin for structure-function analysis of rhodopsin-like receptors. Mol Pharmacol 2001, 60, 1–19.

150. Sakmar, T.P. Structure of rhodopsin and the superfamily of seven-helical receptors: the same and not the same. Curr Opin Cell Biol 2002, 14, 189–195.

151. Ballesteros, J. and K. Palczewski. G protein-coupled receptor drug discovery: implications from the crystal structure of rhodopsin. Curr Opin Drug Discov Dev 2001, 4, 561–574.

152. Archer, E., B. Maigret, C. Escrieut, L. Pradayrol and D. Fourmy. Rhodopsin crystal: new template yielding realistic models of G-protein-coupled receptors? Trends Pharmacol Sci 2003, 24, 36–40.

153. Klabunde, T. and G. Hessler. Drug design strategies for targeting G-protein-coupled receptors. Chembiochem 2002, 3, 928–944.

154. Filipek, S., D.C. Teller, K. Palczewski and R. Stenkamp. The crystallographic model of rhodopsin and its use in studies of other G protein-coupled receptors. Annu Rev Biophys Biomol Struct 2003, 32, 375–397.

155. Rader, A.J., G. Anderson, B. Isin, H.G. Khorana, I. Bahar and J. Klein-Seetharaman. Identification of core amino acids stabilizing rhodopsin. Proc Natl Acad Sci USA 2004, 101, 7246–7251.

156. Berson, E.L. Retinitis pigmentosa: the Friedenwald lecture. Invest Ophthalmol Vis Sci 1993, 34, 1659–1676.

157. Dryja, T.P., E.L. Berson, V.R. Rao and D.D. Oprian. Heterozygous missense mutation in the rhodopsin gene as a cause of congenital stationary night blindness. Nat Genet 1993, 4, 280–283.

158. Macke, J.P., C.M. Davenport, S.G. Jacobson, J.C. Hennessey, F. Gonzalez-Fernandez, B.P. Conway, J. Heckenlively, R. Palmer, I.H. Maumenee, P. Sieving et al. Identification of novel rhodopsin mutations responsible for retinitis pigmentosa: implications for the structure and function of rhodopsin. Am J Hum Genet 1993, 53, 80–89.

159. Inglehearn, C.F., T.J. Keen, R. Bashir, M. Jay, F. Fitzke, A.C. Bird, A. Crombie and S. Bhattacharya. A completed screen for mutations of the rhodopsin gene in a panel of patients with autosomal dominant retinitis pigmentosa. Hum Mol Genet 1992, 1, 41–45.

160. Sung, C.H., C.M. Davenport and J. Nathans. Rhodopsin mutations responsible for autosomal dominant retinitis pigmentosa: clustering of functional classes along the polypeptide chain. J Biol Chem 1993, 268, 26645–26649.

161. Anukanth, A. and H.G. Khorana. Structure and function in rhodopsin: requirements of a specific structure for the intradiscal domain. J. Biol. Chem. 1994, 269, 19738–19744.

162. Doi, T., R.S. Molday and H.G. Khorana. Role of the intradiscal domain in rhodopsin assembly and function. Proc Natl Acad Sci USA 1990, 87, 4991–4995.

163. Kaushal, S. and H.G. Khorana. Structure and function in rhodopsin. 7. Point mutations associated with autosomal-dominant retinitis pigmentosa. Biochemistry 1994, 33, 6121–6128.

164. Liu, X., P. Garriga and H.G. Khorana. Structure and function in rhodopsin: correct folding and misfolding in two-point mutations in the intradiscal domain of rhodopsin identified in retinitis pigmentosa. Proc Natl Acad Sci USA 1996, 93, 4554–4559.

165. Ridge, K.D., Z. Lu, X. Liu and H.G. Khorana. Structure and function in rhodopsin: separation and characterization of the correctly folded and misfolded opsins produced on expression of an opsin mutant gene containing only the native intradiscal cysteine codons. Biochemistry 1995, 34, 3261–3267.

166. Hwa, J., P.J. Reeves, J. Klein-Seetharaman, F. Davidson and H.G. Khorana. Structure and function in rhodopsin: further elucidation of the role of the intradiscal cysteines, Cys-110, -185, and -187, in rhodopsin folding and function. Proc Natl Acad Sci USA 1999, 96, 1932–1935.

167. Garriga, P., X. Liu and H.G. Khorana. Structure and function in rhodopsin: correct folding and misfolding in point mutants at and in proximity to the site of the retinitis pigmentosa mutation Leu-125→Arg in the transmembrane helix C. Proc Natl Acad Sci USA 1996, 93, 4560–4564.

168. Hwa, J., P. Garriga, X. Liu and H.G. Khorana. Structure and function in rhodopsin: packing of the helices in the transmembrane domain and folding to a tertiary structure in the intradiscal domain are coupled. Proc Natl Acad Sci USA 1997, 94, 10571–10576.

169. Hwa, J., J. Klein-Seetharaman and H.G. Khorana. Structure and function in rhodopsin: mass spectrometric identification of the abnormal intradiscal disulfide bond in misfolded retinitis pigmentosa mutants. Proc Natl Acad Sci USA 2001, 98, 4872–4876.

170. Khorana, H.G. Molecular biology of light transduction by the mammalian photoreceptor, rhodopsin. J Biomol Struct Dynamics 2000, 11, 1–16.

171. Man, O., Y. Gilad and D. Lancet. Prediction of the odorant binding site of olfactory receptor proteins by human-mouse comparisons. Protein Sci 2004, 13, 240–254.

172. Miedlich, S.U., L. Gama, K. Seuwen, R.M. Wolf and G.E. Breitwieser. Homology modeling of the transmembrane domain of the human calcium sensing receptor and localization of an allosteric binding site. J Biol Chem 2004, 279, 7254–7263.

173. Kim, S.K., Z.G. Gao, P. Van Rompaey, A.S. Gross, A. Chen, S. Van Calenbergh and K.A. Jacobson. Modeling the adenosine receptors: comparison of the binding domains of A2A agonists and antagonists. J Med Chem 2003, 46, 4847–4859.

174. Petrel, C., A. Kessler, F. Maslah, P. Dauban, R.H. Dodd, D. Rognan and M. Ruat. Modeling and mutagenesis of the binding site of Calhex 231, a novel negative allosteric modulator of the extracellular $Ca^{(2+)}$-sensing receptor. J Biol Chem 2003, 278, 49487–49494.

175. Malherbe, P., N. Kratochwil, M.T. Zenner, J. Piussi, C. Diener, C. Kratzeisen, C. Fischer and R.H. Porter. Mutational analysis and molecular modeling of the binding pocket of the metabotropic glutamate 5 receptor negative modulator 2-methyl-6-(phenylethynyl)-pyridine. Mol Pharmacol 2003, 64, 823–832.

176. Ruan, K.H., J. Wu, S.P. So and L.A. Jenkins. Evidence of the residues involved in ligand recognition in the second extracellular loop of the prostacyclin receptor characterized by high resolution 2D NMR techniques. Arch Biochem Biophys 2003, 418, 25–33.

177. Trent, J.O., Z.X. Wang, J.L. Murray, W. Shao, H. Tamamura, N. Fujii and S.C. Peiper. Lipid bilayer simulations of CXCR4 with inverse agonists and weak partial agonists. J Biol Chem 2003, 278, 47136–47144.

178. Berkhout, T.A., F.E. Blaney, A.M. Bridges, D.G. Cooper, I.T. Forbes, A.D. Gribble, P.H. Groot, A. Hardy, R.J. Ife, R. Kaur, K.E. Moores, H. Shillito, J. Willetts and J. Witherington. CCR2: characterization of the antagonist binding site from a combined receptor modeling/mutagenesis approach. J Med Chem 2003, 46, 4070–4086.

179. Anzini, M., L. Canullo, C. Braile, A. Cappelli, A. Gallelli, S. Vomero, M.C. Menziani, P.G. De Benedetti, M. Rizzo, S. Collina, O. Azzolina, M. Sbacchi, C. Ghelardini and N. Galeotti. Synthesis, biological evaluation, and receptor docking simulations of 2-[(acylamino)ethyl]-1,4-benzodiazepines as kappa-opioid receptor agonists endowed with antinociceptive and antiamnesic activity. J Med Chem 2003, 46, 3853–3864.

180. Hulme, E.C., Z.L. Lu and M.S. Bee. Scanning mutagenesis studies of the M1 muscarinic acetylcholine receptor. Receptors Channels 2003, 9, 215–228.

181. Schulz, A. and T. Schoneberg. The structural evolution of a P2Y-like G-protein-coupled receptor. J Biol Chem 2003, 278, 35531–35541.

182. Becker, O.M., S. Shacham, Y. Marantz and S. Noiman. Modeling the 3D structure of GPCRs: advances and application to drug discovery. Curr Opin Drug Discov Dev 2003, 6, 353–361.

183. Mehler, E.L., X. Periole, S.A. Hassan and H. Weinstein. Key issues in the computational simulation of GPCR function: representation of loop domains. J Comput Aided Mol Des 2002, 16, 841–853.

184. Johren, K. and H.D. Holtje. A model of the human M2 muscarinic acetylcholine receptor. J Comput Aided Mol Des 2002, 16, 795–801.

185. Shim, J.Y., W.J. Welsh and A.C. Howlett. Homology model of the CB1 cannabinoid receptor: sites critical for nonclassical cannabinoid agonist interaction. Biopolymers 2003, 71, 169–189.

186. Bhave, G., B.M. Nadin, D.J. Brasier, K.S. Glauner, R.D. Shah, S.F. Heinemann, F. Karim and R.W. Gereau. Membrane topology of a metabotropic glutamate receptor. J Biol Chem 2003, 278, 30294–30301.

187. Chambers, J.J. and D.E. Nichols. A homology-based model of the human 5-HT2A receptor derived from an in silico activated G-protein coupled receptor. J Comput Aided Mol Des 2002, 16, 511–520.

188. Malherbe, P., N. Kratochwil, F. Knoflach, M.T. Zenner, J.N. Kew, C. Kratzeisen, H.P. Maerki, G. Adam and V. Mutel. Mutational analysis and molecular modeling of the allosteric binding site of a novel, selective, noncompetitive antagonist of the metabotropic glutamate 1 receptor. J Biol Chem 2003, 278, 8340–8347.

189. Stenkamp, R.E., S. Filipek, C.A. Driessen, D.C. Teller and K. Palczewski. Crystal structure of rhodopsin: a template for cone visual pigments and other G protein-coupled receptors. Biochim Biophys Acta 2002, 1565, 168–182.

190. Deraet, M., L. Rihakova, A. Boucard, J. Perodin, S. Sauve, A.P. Mathieu, G. Guillemette, R. Leduc, P. Lavigne and E. Escher. Angiotensin II is bound to both receptors AT1 and AT2, parallel to the transmembrane domains and in an extended form. Can J Physiol Pharmacol 2002, 80, 418–425.

191. Lopez-Rodriguez, M.L., M. Murcia, B. Benhamu, M. Olivella, M. Campillo and L. Pardo. Computational model of the complex between GR113808 and the 5-HT4 receptor guided by site-directed mutagenesis and the crystal structure of rhodopsin. J Comput Aided Mol Des 2001, 15, 1025–1033.

192. Gao, Z.G., A. Chen, D. Barak, S.K. Kim, C.E. Muller and K.A. Jacobson. Identification by site-directed mutagenesis of residues involved in ligand recognition and activation of the human A3 adenosine receptor. J Biol Chem 2002, 277, 19056–19063.

193. Kassack, M.U., P. Hogger, D.A. Gschwend, K. Kameyama, T. Haga, R.C. Graul and W. Sadee. Molecular modeling of G-protein coupled receptor kinase 2: docking and biochemical evaluation of inhibitors. AAPS PharmSci 2000, 2, E2.

194. Greasley, P.J., F. Fanelli, A. Scheer, L. Abuin, M. Nenniger-Tosato, P.G. DeBenedetti and S. Cotecchia. Mutational and computational analysis of the α(1b)-adrenergic receptor: involvement of basic and hydrophobic residues in receptor activation and G protein coupling. J Biol Chem 2001, 276, 46485–46494.

195. Nikiforovich, G.V. and G.R. Marshall. 3D model for TM region of the AT-1 receptor in complex with angiotensin II independently validated by site-directed mutagenesis data. Biochem Biophys Res Commun 2001, 286, 1204–1211.

196. Aburi, M. and Smith, P.E. Modeling and simulation of the human delta opioid receptor. Prot Science 2004, 13, 1997–2008.

197. Liu, W., M. Eilers, A.B. Patel and S.O. Smith. Helix packing moments reveal diversity and conservation in membrane protein structure. J Mol Biol 2004, 337, 713–729.

198. Kunishima, N., Y. Shimada, Y. Tsuji, T. Sato, M. Yamamoto, T. Kumasaka, S. Nakanishi, H. Jingami and K. Morikawa. Structural basis of glutamate recognition by a dimeric metabotropic glutamate receptor. Nature 2000, 407, 971–977.

199. Arevalo, E., R. Estephan, J. Madeo, B. Arshava, M. Dumont, J.M. Becker and F. Naider. Biosynthesis and biophysical analysis of domains of a yeast G protein-coupled receptor. Biopolymers 2003, 71, 516–531.

200. Arshava, B., S.F. Liu, H. Jiang, M. Breslav, J.M. Becker and F. Naider. Structure of segments of a G protein-coupled receptor: CD and NMR analysis of the *Saccharomyces cerevisiae* tridecapeptide pheromone receptor. Biopolymers 1998, 46, 343–357.

201. Naider, F., F.X. Ding, N.C. VerBerkmoes, B. Arshava and J.M. Becker. Synthesis and biophysical characterization of a multidomain peptide from a *Saccharomyces cerevisiae* G protein-coupled receptor. J Biol Chem 2003, 278, 52537–52545.

202. Naider, F., B. Arshava, F.X. Ding, E. Arevalo and J.M. Becker. Peptide fragments as models to study the structure of a G-protein coupled receptor: the alpha-factor receptor of *Saccharomyces cerevisiae*. Biopolymers 2001, 60, 334–350.

203. Valentine, K.G., S.F. Liu, F.M. Marassi, G. Veglia, S.J. Opella, F.X. Ding, S.H. Wang, B. Arshava, J.M. Becker and F. Naider. Structure and topology of a peptide segment of the sixth transmembrane domain of the *Saccharomyces cerevisae* alpha-factor receptor in phospholipid bilayers. Biopolymers 2001, 59, 243–256.

204. Berlose, J.P., O. Convert, A. Brunissen, G. Chassaing and S. Lavielle. Three-dimensional structure of the highly conserved seventh transmembrane domain of G-protein-coupled receptors. Eur J Biochem 1994, 225, 827–843.

205. Choi, G., J. Landin and X.Q. Xie. The cytoplasmic helix of cannabinoid receptor CB2, a conformational study by circular dichroism and ^{1}H NMR spectroscopy in aqueous and membrane-like environments. J Peptide Res 2002, 60, 169–177.

206. Chung, D.A., E.R. Zuiderweg, C.B. Fowler, O.S. Soyer, H.I. Mosberg and R.R. Neubig. NMR structure of the second intracellular loop of the α-2A adrenergic receptor: evidence for a novel cytoplasmic helix. Biochemistry 2002, 41, 3596–3604.

207. Schwantner, A., A.J. Dingley, S. Ozbek, S. Rose-John and J. Grotzinger. Direct determination of the interleukin-6 binding epitope of the interleukin-6 receptor by NMR spectroscopy. J Biol Chem 2004, 279, 571–576.

208. Ruan, K.H., S.P. So, J. Wu, D. Li, A. Huang and J. Kung. Solution structure of the second extracellular loop of human thromboxane A2 receptor. Biochemistry 2001, 40, 275–280.

209. Demene, H., S. Granier, D. Muller, G. Guillon, M.N. Dufour, M.A. Delsuc, M. Hibert, R. Pascal and C. Mendre, eds. Active peptidic mimics of the second intracellular loop of the V(1A) vasopressin receptor are structurally related to the second intracellular rhodopsin loop: a combined ^{1}H NMR and biochemical study. Biochemistry 2003, 42, 8204–8213.

210. Franzoni, L., G. Nicastro, T.A. Pertinhez, M. Tato, C.R. Nakaie, A.C. Paiva, S. Schreier and A. Spisni. Structure of the C-terminal fragment 300–320 of the rat angiotensin II AT1A receptor and its relevance with respect to G-protein coupling. J Biol Chem 1997, 272, 9734–9741.

211. Nicastro, G., F. Peri, L. Franzoni, C. de Chiara, G. Sartor and A. Spisni. Conformational features of a synthetic model of the first extracellular loop of the angiotensin II AT1A receptor. J Peptide Sci 2003, 9, 229–243.

212. Chopra, A., P.L. Yeagle, J.A. Alderfer and A.D. Albert. Solution structure of the sixth transmembrane helix of the G-protein-coupled receptor, rhodopsin. Biochim Biophys Acta 2000, 1463, 1–5.

213. Yeagle, P.L., J.L. Alderfer and A.D. Albert. Three-dimensional structure of the cytoplasmic face of the G protein receptor rhodopsin. Biochemistry 1997, 36, 9649–9654.

214. Yeagle, P.L., J.L. Alderfer and A.D. Albert. Structure determination of the fourth cytoplasmic loop and carboxyl terminal domain of bovine rhodopsin. Mol Vis 1996, 2, 12.

215. Yeagle, P.L., J.L. Alderfer and A.D. Albert. Structure of the carboxy-terminal domain of bovine rhodopsin. Nat Struct Biol 1995, 2, 832–834.

216. Yeagle, P.L., J.L. Alderfer and A.D. Albert. Structure of the third cytoplasmic loop of bovine rhodopsin. Biochemistry 1995, 34, 14621–14625.

217. Yeagle, P.L., J.L. Alderfer, A.C. Salloum, L. Ali and A.D. Albert. The first and second cytoplasmic loops of the G-protein receptor, rhodopsin, independently form beta-turns. Biochemistry 1997, 36, 3864–3869.

218. Yeagle, P.L., C. Danis, G. Choi, J.L. Alderfer and A.D. Albert. Three dimensional structure of the seventh transmembrane helical domain of the G-protein receptor, rhodopsin. Mol Vis 2000, 6, 125–131.

219. Giragossian, C., M. Pellegrini and D.F. Mierke. NMR studies of CCK-8/CCK1 complex support membrane-associated pathway for ligand-receptor interaction. Can J Physiol Pharmacol 2002, 80, 383–387.

220. Giragossian, C. and D.F. Mierke. Determination of ligand-receptor interactions of cholecystokinin by nuclear magnetic resonance. Life Sci 2003, 73, 705–713.

221. Mierke, D.F. and C. Giragossian. Peptide hormone binding to G-protein-coupled receptors: structural characterization via NMR techniques. Med Res Rev 2001, 21, 450–471.

222. Monticelli, L., S. Mammi and D.F. Mierke. Molecular characterization of a ligand-tethered parathyroid hormone receptor. Biophys Chem 2002, 95, 165–172.

223. Pellegrini, M., M. Royo, M. Chorev and D.F. Mierke. Conformational characterization of a peptide mimetic of the third cytoplasmic loop of the G-protein coupled parathyroid hormone/parathyroid hormone related protein receptor. Biopolymers 1996, 40, 653–666.

224. Pellegrini, M., A. Bisello, M. Rosenblatt, M. Chorev and D.F. Mierke. Binding domain of human parathyroid hormone receptor: from conformation to function. Biochemistry 1998, 37, 12737–12743.

225. Pellegrini, M., M. Royo, M. Rosenblatt, M. Chorev and D.F. Mierke. Addressing the tertiary structure of human parathyroid hormone-(1–34). J Biol Chem 1998, 273, 10420–10427.

226. Pellegrini, M. and D.F. Mierke. Structural characterization of peptide hormone/receptor interactions by NMR spectroscopy. Biopolymers 1999, 51, 208–220.

227. Pellegrini, M., A.A. Bremer, A.L. Ulfers, N.D. Boyd and D.F. Mierke. Molecular characterization of the substance P*neurokinin-1 receptor complex: development of an experimentally based model. J Biol Chem 2001, 276, 22862–22867.

228. Piserchio, A., A. Bisello, M. Rosenblatt, M. Chorev and D.F. Mierke. Characterization of parathyroid hormone/receptor interactions: structure of the first extracellular loop. Biochemistry 2000, 39, 8153–8160.

229. Piserchio, A., G.N. Prado, R. Zhang, J. Yu, L. Taylor, P. Polgar and D.F. Mierke. Structural insight into the role of the second intracellular loop of the bradykinin 2 receptor in signaling and internalization. Biopolymers 2002, 63, 239–246.

230. Ulfers, A.L., A. Piserchio and D.F. Mierke. Extracellular domains of the neurokinin-1 receptor: structural characterization and interactions with substance P. Biopolymers 2002, 66, 339–349.

231. Ulfers, A.L., J.L. McMurry, D.A. Kendall and D.F. Mierke. Structure of the third intracellular loop of the human cannabinoid 1 receptor. Biochemistry 2002, 41, 11344–11350.

232. König, B., A. Arendt, J.H. McDowell, M. Kahlert, P.A. Hargrave and K.P. Hofmann. Three cytoplasmic loops of rhodopsin interact with transducin. Proc Natl Acad Sci USA 1989, 86, 6878–6882.

233. Brabazon, D.M., N.G. Abdulaev, J.P. Marino and K.D. Ridge. Evidence for structural changes in carboxyl-terminal peptides of transducin-α subunit upon binding a soluble mimic of light-activated rhodopsin. Biochemistry 2003, 42, 302–311.

234. Abdulaev, N.G., T. Ngo, R. Chen, Z. Lu and K.D. Ridge. Functionally discrete mimics of light-activated rhodopsin identified through expression of soluble cytoplasmic domains. J Biol Chem 2000, 275, 39354–39363.

235. Heymann, J.B., M. Pfeiffer, V. Hildebrandt, H.R. Kaback, D. Fotiadis, B. Groot, A. Engel, D. Oesterhelt and D.J. Muller. Conformations of the rhodopsin third cytoplasmic loop grafted onto bacteriorhodopsin. Structure Fold Des 2000, 8, 643–653.

236. Abdulaev, N.G., T.T. Strassmaier, T. Ngo, R. Chen, H. Luecke, D.D. Oprian and K.D. Ridge. Grafting segments from the extracellular surface of CCR5 onto a bacteriorhodopsin transmembrane scaffold confers HIV-1 coreceptor activity. Structure (Camb) 2002, 10, 515–525.

237. Mathies, R.A. Photons, femtoseconds and dipolar interactions: a molecular picture of the primary events in vision. Novartis Found Symp 1999, 224, 70–101.

238. Lugtenburg, J., R.A. Mathies, R.G. Griffin and J. Herzfeld. Structure and function of rhodopsins from solid state NMR and resonance Raman spectroscopy of isotopic retinal derivatives. Trends Biochem Sci 1988, 13, 388–393.

239. Vogel, R. and F. Siebert. Fourier transform IR spectroscopy study for new insights into molecular properties and activation mechanisms of visual pigment rhodopsin. Biopolymers 2003, 72, 133–148.

240. Zheng, L. and J. Herzfeld. NMR studies of retinal proteins. J Bioenerg Biomembr 1992, 24, 139–146.

241. Watts, A. NMR of drugs and ligands bound to membrane receptors. Curr Opin Biotechnol 1999, 10, 48–53.

242. Furukawa, H., T. Hamada, M.K. Hayashi, T. Haga, Y. Muto, H. Hirota, S. Yokoyama, K. Nagasawa and M. Ishiguro. Conformation of ligands bound to the muscarinic acetylcholine receptor. Mol Pharmacol 2002, 62, 778–787.

243. Inooka, H., T. Ohtaki, O. Kitahara, T. Ikegami, S. Endo, C. Kitada, K. Ogi, H. Onda, M. Fujino and M. Shirakawa. Conformation of a peptide ligand bound to its G-protein coupled receptor. Nat Struct Biol 2001, 8, 161–165.

244. Luca, S., J.F. White, A.K. Sohal, D.V. Filippov, J.H. van Boom, R. Grisshammer and M. Baldus, The conformation of neurotensin bound to its G protein-coupled receptor, Proc Natl Acad Sci USA 2003, 100, 10706–10711.

245. Chollet, A. and G. Turcatti. Biophysical approaches to G protein-coupled receptors: structure, function and dynamics. J Comput Aided Mol Des 1999, 13, 209–219.

246. Khanolkar, A.D., S.L. Palmer and A. Makriyannis. Molecular probes for the cannabinoid receptors. Chem Phys Lipids 2000, 108, 37–52.

247. Turcatti, G., K. Nemeth, M.D. Edgerton, U. Meseth, F. Talabot, M. Peitsch, J. Knowles, H. Vogel and A. Chollet. Probing the structure and function of the tachykinin neurokinin-2 receptor through biosynthetic incorporation of fluorescent amino acids at specific sites. J Biol Chem 1996, 271, 19991–19998.

248. Eidne, K.A., K.M. Kroeger and A.C. Hanyaloglu. Applications of novel resonance energy transfer techniques to study dynamic hormone receptor interactions in living cells. Trends Endocrinol Metab 2002, 13, 415–421.

249. Pertinhez, T.A., C.R. Nakaie, A.C. Paiva and S. Schreier. Spin-labeled extracellular loop from a seven-transmembrane helix receptor: studies in solution and interaction with model membranes. Biopolymers 1997, 42, 821–829.

250. Nakayama, T.A. and H.G. Khorana. Orientation of retinal in bovine rhodopsin determined by cross-linking using a photoactivatable analog of 11-cis-retinal. J Biol Chem 1990, 265, 15762–15769.

251. Nakayama, T.A. and H.G. Khorana. Mapping of the amino acids in membrane-embedded helices that interact with the retinal chromophore in bovine rhodopsin. J Biol Chem 1991, 266, 4269–4275.

252. Borhan, B., M.L. Souto, H. Imai, Y. Shichida and K. Nakanishi. Movement of retinal along the visual transduction path. Science 2000, 288, 2209–2212.

253. Bettio, A. and A.G. Beck-Sickinger. Biophysical methods to study ligand-receptor interactions of neuropeptide Y. Biopolymers 2001, 60, 420–437.

254. Gether, U., S. Lin, P. Ghanouni, J.A. Ballesteros, H. Weinstein and B.K. Kobilka. Agonists induce conformational changes in transmembrane domains III and VI of the β-2 adrenoceptor. EMBO J 1997, 16, 6737–6747.

255. Gether, U., S. Lin and B.K. Kobilka. Fluorescent labeling of purified β-2 adrenergic receptor: evidence for ligand-specific conformational changes. J Biol Chem 1995, 270, 28268–28275.

256. Gether, U., J.A. Ballesteros, R. Seifert, E. Sanders-Bush, H. Weinstein and B.K. Kobilka. Structural instability of a constitutively active G protein-coupled receptor: agonist-independent activation due to conformational flexibility. J Biol Chem 1997, 272, 2587–2590.

257. Javitch, J.A., D. Fu, G. Liapakis and J. Chen. Constitutive activation of the β-2 adrenergic receptor alters the orientation of its sixth membrane-spanning segment. J Biol Chem 1997, 272, 18546–18549.

258. Jensen, A.D., F. Guarnieri, S.G. Rasmussen, F. Asmar, J.A. Ballesteros and U. Gether. Agonist-induced conformational changes at the cytoplasmic side of transmembrane segment 6 in the β-2 adrenergic receptor mapped by site-selective fluorescent labeling. J Biol Chem 2001, 276, 9279–9290.

259. Rasmussen, S.G., A.D. Jensen, G. Liapakis, P. Ghanouni, J.A. Javitch and U. Gether. Mutation of a highly conserved aspartic acid in the β-2 adrenergic receptor: constitutive activation, structural instability, and conformational rearrangement of transmembrane segment. Mol Pharmacol 1999, 56, 175–184.

260. Sheikh, S.P., J.P. Vilardarga, T.J. Baranski, O. Lichtarge, T. Iiri, E.C. Meng, R.A. Nissenson and H.R. Bourne, Similar structures and shared switch mechanisms of the β-2 adrenoceptor and the parathyroid hormone receptor: Zn(II) bridges between helices III and VI block activation, J Biol Chem 1999, 274, 17033–17041.

261. Chen, S., F. Lin, M. Xu, R.P. Riek, J. Novotny and R.M. Graham. Mutation of a single TMVI residue, Phe(282), in the beta(2)-adrenergic receptor results in structurally distinct activated receptor conformations. Biochemistry 2002, 41, 6045–6053.

262. Chen, S., F. Lin, M. Xu, J. Hwa and R.M. Graham. Dominant-negative activity of an alpha(1B)-adrenergic receptor signal-inactivating point mutation. EMBO J 2000, 19, 4265–4271.

263. Spalding, T.A., E.S. Burstein, S.C. Henderson, K.R. Ducote and M.R. Brann. Identification of a ligand-dependent switch within a muscarinic receptor. J Biol Chem 1998, 273, 21563–21568.

264. Govaerts, C., A. Lefort, S. Costagliola, S.J. Wodak, J.A. Ballesteros, J. Van Sande, L. Pardo and G. Vassart. A conserved Asn in transmembrane helix 7 is an on/off switch in the activation of the thyrotropin receptor. J Biol Chem 2001, 276, 22991–22999.

265. Baranski, T.J., P. Herzmark, O. Lichtarge, B.O. Gerber, J. Trueheart, E.C. Meng, T. Iiri, S.P. Sheikh and H.R. Bourne. C5a receptor activation: genetic identification of critical residues in four transmembrane helices. J Biol Chem 1999, 274, 15757–15765.

266. Shi, L., M.M. Simpson, J.A. Ballesteros and J.A. Javitch. The first transmembrane segment of the dopamine D2 receptor: accessibility in the binding-site crevice and position in the transmembrane bundle. Biochemistry 2001, 40, 12339–12348.

267. Xu, W., J. Li, C. Chen, P. Huang, H. Weinstein, J.A. Javitch, L. Shi, J.K. de Riel and L.Y. Liu-Chen. Comparison of the amino acid residues in the sixth transmembrane domains accessible in the binding-site crevices of mu, delta, and kappa opioid receptors. Biochemistry 2001, 40, 8018–8029.

268. Xu, W., C. Chen, P. Huang, J. Li, J.K. de Riel, J.A. Javitch and L.Y. Liu-Chen. The conserved cysteine 7.38 residue is differentially accessible in the binding-site crevices of the mu, delta, and kappa opioid receptors. Biochemistry 2000, 39, 13904–13915.

269. Cohen, G.B., T. Yang, P.R. Robinson and D.D. Oprian. Constitutive activation of opsin: influence of charge at position 134 and size at position 296. Biochemistry 1993, 32, 6111–6115.

270. Shi, L., G. Liapakis, R. Xu, F. Guarnieri, J.A. Ballesteros and J.A. Javitch. Beta-2 adrenergic receptor activation: modulation of the proline kink in transmembrane 6 by a rotamer toggle switch. J Biol Chem 2002, 277, 40989–40996.

271. Ballesteros, J.A., A.D. Jensen, G. Liapakis, S.G. Rasmussen, L. Shi, U. Gether and J.A. Javitch. Activation of the beta 2-adrenergic receptor involves disruption of an ionic lock between the cytoplasmic ends of transmembrane segments 3 and 6. J Biol Chem 2001, 276, 29171–29177.

272. Scheer, A., F. Fanelli, T. Costa, P.G. De Benedetti and S. Cotecchia. Constitutively active mutants of the alpha 1B-adrenergic receptor: role of highly conserved polar amino acids in receptor activation. EMBO J 1996, 15, 3566–3578.

273. Greasley, P.J., F. Fanelli, O. Rossier, L. Abuin and S. Cotecchia. Mutagenesis and modelling of the alpha(1b)-adrenergic receptor highlight the role of the helix 3/helix 6 interface in receptor activation. Mol Pharmacol 2002, 61, 1025–1032.

274. Hogger, P., M.S. Shockley, J. Lameh and W. Sadee. Activating and inactivating mutations in N- and C-terminal i3 loop junctions of muscarinic acetylcholine Hm1 receptors. J Biol Chem 1995, 270, 7405–7410.

275. Ballesteros, J., S. Kitanovic, F. Guarnieri, P. Davies, B.J. Fromme, K. Konvicka, L. Chi, R.P. Millar, J.S. Davidson, H. Weinstein and S.C. Sealfon. Functional microdomains in G-protein-coupled receptors: conserved arginine-cage motif in the gonadotropin-releasing hormone receptor. J Biol Chem 1998, 273, 10445–10453.

276. Huang, P., J. Li, C. Chen, I. Visiers, H. Weinstein and L.Y. Liu-Chen. Functional role of a conserved motif in TM6 of the rat mu opioid receptor: constitutively active and inactive receptors result from substitutions of Thr6.34(279) with Lys and Asp. Biochemistry 2001, 40, 13501–13509.

277. Wilson, J., H. Lin, D. Fu, J.A. Javitch and P.G. Strange. Mechanisms of inverse agonism of antipsychotic drugs at the D(2) dopamine receptor: use of a mutant D(2) dopamine receptor that adopts the activated conformation. J Neurochem 2001, 77, 493–504.

278. Abadji, V., J.M. Lucas-Lenard, C. Chin and D.A. Kendall. Involvement of the carboxyl terminus of the third intracellular loop of the cannabinoid CB1 receptor in constitutive activation of Gs. J Neurochem 1999, 72, 2032–2038.

279. Alewijnse, A.E., H. Timmerman, E.H. Jacobs, M.J. Smit, E. Roovers, S. Cotecchia and R. Leurs. The effect of mutations in the DRY motif on the constitutive activity and structural instability of the histamine H(2) receptor. Mol Pharmacol 2000, 57, 890–898.

280. Gether, U. Uncovering molecular mechanisms involved in activation of G protein-coupled receptors. Endocr Rev 2000, 21, 90–113.

281. Visiers, I., J.A. Ballesteros and H. Weinstein. Three-dimensional representations of G protein-coupled receptor structures and mechanisms. Methods Enzymol 2002, 343, 329–371.

282. Ridge, K.D., S. Bhattacharya, T.A. Nakayama and H.G. Khorana. Light-stable rhodopsin. II. An opsin mutant (TRP-265----Phe) and a retinal analog with a nonisomerizable 11-cis configuration form a photostable chromophore. J Biol Chem 1992, 267, 6770–6775.

283. Lin, S.W. and T.P. Sakmar. Specific tryptophan UV-absorbance changes are probes of the transition of rhodopsin to its active state. Biochemistry 1996, 35, 11149–11159.

284. Chabre, M. and J. Breton. Orientation of aromatic residues in rhodopsin. Rotation of one tryptophan upon the meta I to meta II transition after illumination. Photochem Photobiol 1979, 30, 295–299.

285. Struthers, M., H. Yu and D.D. Oprian. G protein-coupled receptor activation: analysis of a highly constrained, "straitjacketed" rhodopsin. Biochemistry 2000, 39, 7938–7942.

286. Sheikh, S.P., T.A. Zvyaga, O. Lichtarge, T.P. Sakmar and H.R. Bourne. Rhodopsin activation blocked by metal-ion-binding sites linking transmembrane helices C and F. Nature 1996, 383, 347–350.

287. Sneddon, W.B., C.E. Magyar, G.E. Willick, C.A. Syme, F. Galbiati, A. Bisello and P.A. Friedman. Ligand-selective dissociation of activation and internalization of the parathyroid hormone receptor: conditional efficacy of PTH peptide fragments. Endocrinology, 2004, 145, 2815–2823.

288. Sneddon, W.B., C.A. Syme, A. Bisello, C.E. Magyar, M.D. Rochdi, J.L. Parent, E.J. Weinman, A.B. Abou-Samra and P.A. Friedman. Activation-independent parathyroid hormone receptor internalization is regulated by NHERF1 (EBP50). J Biol Chem 2003, 278, 43787–43796.

289. Ridge, K.D., C. Zhang and H.G. Khorana. Mapping of amino acids in the cytoplasmic loop connecting helices C and D in rhodopsin: chemical reactivity in the dark state following single cysteine replacements. Biochemistry 1995, 34, 8804–8811.

290. Yang, K., D.L. Farrens, W.L. Hubbell and H.G. Khorana. Structure and function in rhodopsin: single cysteine substitution mutants in the cytoplasmic interhelical E–F loop region show position-specific effects in transducin activation. Biochemistry 1996, 35, 12464–12469.

291. Cai, K., J. Klein-Seetharaman, D. Farrens, C. Zhang, C. Altenbach, W.L. Hubbell and H.G. Khorana. Single-cysteine substitution mutants at amino acid positions 306–321 in rhodopsin, the sequence between the cytoplasmic end of helix VII and the palmitoylation sites: sulfhydryl reactivity and transducin activation reveal a tertiary structure. Biochemistry 1999, 38, 7925–7930.

12 Toward Crystallization of G Protein-Coupled Receptors

Mark L. Chiu and Maria P. MacWilliams

CONTENTS

ABSTRACT

Although the high-resolution crystal structure of bovine rhodopsin has been determined, many structure-function relationships remain to be learned from other GPCRs. Many pharmaceutically interesting GPCRs cannot be modeled because of their amino acid sequence divergences from bovine rhodopsin and its related family members. Structure determination via x-ray crystallography of GPCRs can provide new avenues of engineering drugs with greater potencies and higher specificities. Since most membrane protein structures have been solved using x-ray crystallography, this method is most likely to be used for other GPCR structure elucidations. However, several obstacles must be overcome before this method becomes routine: overexpression of active GPCRs; solubilization and purification of active and stable GPCRs; and preparation of large, well-ordered crystals. This chapter outlines conditions for solubilization and purification using detergents and additives along with assessments of different crystallization formats of membrane proteins. It concludes with a case study of the crystallization of rhodopsin and some of the challenges that remain with GPCR structures.

12.1 INTRODUCTION

G protein-coupled receptors (GPCRs) comprise a family of seven-transmembrane (7TM) receptors that mediate most cell–cell communications in humans via a wide variety of extracellular activators such as hormones, light, neurotransmitters, ions, odorants, and amino acids.[1,2] Because of the physiological importance of GPCRs in metabolic, endocrine, neuromuscular, and CNS systems, many pharmaceutical companies have expended significant efforts to develop drug therapies that act on this family of proteins.[3,4] As shown in Chapter 3, more than 50% of marketed drugs treat diseases by targeting only 20 GPCRs.[5] Surveys of genomic data suggest that about 1000 GPCRs exist in the human genome: 614 annotated entries and the remainder classified as orphan receptors.[6–8] These as yet untargeted GPCRs may have important physiological roles and unique mechanisms of interaction and this drug discovery potential makes GPCRs one of the most important classes of proteins.[9]

One of the great challenges in the GPCR field is to elucidate the conformational changes associated with the binding of agonists, antagonists, and allosteric modulators.[10] Little is known about the structure–function relationships that govern the specificity of ligand binding and activation of the GPCR/G protein systems and the GPCR/G protein effector interactions.[11] This ignorance is due in part to the complex and multifactorial nature of signal transduction regulation. Notwithstanding, structural descriptions of GPCR ligand binding and activation could provide early diagnostic tools that could prevent progression of receptor-mediated diseases and serve as new strategies for more specific and effective therapeutic intervention.[12]

Extensive efforts have been made to model GPCRs based on the high-resolution structures of the seven-transmembrane domain (7TM) proteins known as bovine rhodopsin and bacteriorhodopsin.[6,13,14] Although current GPCR models attempt to accommodate related mutagenesis data in a three-dimensional (3D) context, much can be learned from authentic 3D structures of specific GPCRs. Unfortunately, many GPCRs cannot be modeled because of their divergence in sequence similarities and functional properties from bovine rhodopsin.[15] Hence, structures of representative family members could help open a new era in drug design. Knowledge of the structural determinants would elucidate the molecular bases of ligand binding and receptor conformation in normal and abnormal conditions.[16]

All members of the GPCR superfamily share a common structural framework: 7TM or membrane-spanning domains connected by three intracellular loops and three extracellular loops with an extracellular amino terminus and an intracellular carboxy terminus.[17,18] Superimposed on the basic structures of GPCRs are a number of variations that provide specificity in ligand binding, G protein coupling, and interactions with other proteins.

Sequence alignment of the 7TM domains divides the superfamily into several subfamilies (Figure 12.1). Family 1 (Class A) is the largest group and includes the rhodopsins and biogenic amine GPCRs. Family 2 (Class B) shows essentially little sequence homology to family 1 even within the 7TM segments. Members of this

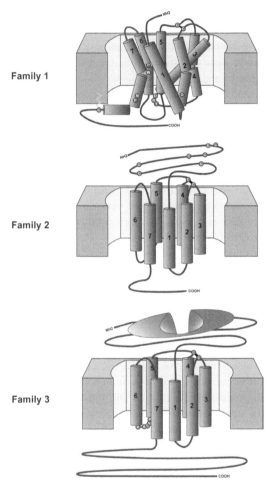

Family 1

Family 2

Family 3

FIGURE 12.1 GPCR families. Monomeric GPCRs are depicted with structural aspects of the three main GPCR families. Family 1 (Class A) GPCRs are characterized by highly conserved amino acids indicated by circles and a disulfide bridge that connects the first and second extracellular loops. Most of these receptors have palmitoylated cysteines in their carboxy terminal tails. The crystal structure of rhodopsin indicates that the transmembrane domains of family 1 receptors are tilted and kinked. Family 2 (Class B) GPCRs are characterized by relatively long amino terminal domains containing several cysteines that form a network of disulfide bridges. Little is known about the orientation of the transmembrane domains because of the divergence in amino acid sequences from family 1 GPCRs. Family 3 (Class C) GPCRs are characterized by long amino and carboxy termini. The ligand binding domain is located in the amino terminal domain — shown by the crystal structure of the ligand binding domain of the metabotropic glutamate receptor to form a disulfide linker. Except for two cysteines in extracellular loops 1 and 2 that form a putative disulfide bridge, the family 3 receptors do not have any features that characterize family 1 and 2 receptors. A unique characteristic of these receptors is that the third intracellular loop is short and highly conserved. Little is known about the orientation of the transmembrane domains. (Reproduced with permission from Nat. Rev. Drug Disc., 2002. 1: 808–820.)

family include receptors for peptide hormones such as secretin, parathyroid hormone, parathyroid hormone-related protein, and calcitonin. Family 3 (Class C) members include the GABA, metabotropic glutamate, and extracellular Ca^{2+} sensing receptors along with the putative taste and pheromone receptors. Family 3 GPCRs dimerize via interactions of the ligand binding domains.

Although the ligand binding properties of these domains in monomers are similar in dimers, the regulation of ligand binding and G protein activation in family 3 GPCRs can be affected.[19] Homodimerization and heterodimerization are possible with other GPCR families as well. The consequences of heterodimerization may extend the repertoire of GPCR pharmacology.[20] Each GPCR family has different post-translational characteristics that could be important for activity. When considering the aspects of purification and structural characterization of GPCRs, the interactions of lipid and detergents will be very different with monomers, various dimers, and higher-order oligomers. A more detailed description of GPCR oligomerization is covered in Chapter 15.

To advance our understanding of the structure–function relationships of GPCR pharmacology, we need the ability to obtain high-resolution structures of these membrane proteins. Four different methods are currently used for the elucidation of high-resolution structures of membrane proteins: x-ray crystallography,[21–23] electron crystallography,[22,24,25] atomic force microscopy (AFM),[26] and nuclear magnetic resonance (NMR) spectroscopy.[27–30] Of the high resolution integral membrane protein structures posted on two websites compiled by Stephen White[31] and Hartmut Michel,[32] the only listed member of the GPCR family is bovine rhodopsin. Most membrane protein structures have been determined via x-ray crystallography because data collection and processing are routine and the technique allows rapid solution of a large number of structures to high resolution with little concern for protein size. For detailed mechanistic insights, resolution below 2.3 Å is required, particularly if small conformational changes and the positions of structurally relevant water molecules are to be observed. Whereas this chapter addresses the steps toward the preparation membrane protein crystals, Chapter 13 covers a novel membrane protein structure determination method via solid-state NMR spectroscopy.

Structure determination via x-ray crystallography requires preparation of well ordered 3D crystals that diffract at resolutions below 3 Å. The main limitations of the technique include overexpression of active protein; purification of stable homogenous active protein; and development of reproducible crystallization methods[33] (Figure 12.2). This chapter presents ideas for meeting criteria for the crystallization of GPCRs.

12.2 OVEREXPRESSION AND PURIFICATION

The first bottleneck encountered in membrane protein crystallization is the availability of milligram quantities of purified homogeneous protein. Most GPCRs are intrinsically present in low amounts in cells — sometimes for short segments of the cell cycle; thus, high level overexpression is required to obtain milligram quantities of GPCRs. Production of GPCRs in mammalian cell overexpression systems often provides enough material for the screening of ligand interactions as noted in Chapter

Express in membranes

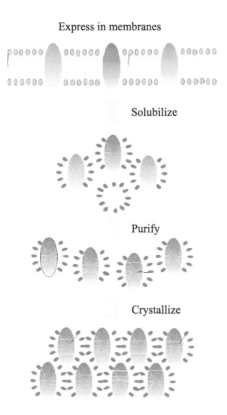

Solubilize

Purify

Crystallize

FIGURE 12.2 Membrane protein crystallization. The challenging steps for determining GPCR structures via x-ray crystallography include overexpression of active membrane protein; extraction and purification of active protein; and crystallization of membrane proteins. Upon concentration, the purified GPCRs can be incorporated into different reconstitution systems (micelles, bicelles, lipidic mesophases, and nanodiscs) that are conducive for crystallization in the presence of precipitants.

9. On the other hand, production of milligram quantities from such expression systems would require cost-prohibitive fermentation of hundreds of liters. In addition, the copurification of related endogenous GPCRs can be problematic.

Standard heterologous overexpression systems often yield misfolded proteins and the process of refolding GPCRs often does not produce high yields.[34–38] It is worthwhile to develop technology to overexpress GPCRs into membrane bilayers to produce a protein in a more active state. Differences in post-translational modifications such as glycosylation, phosphorylation, and acylation can result in decreased specific activity of a recombinantly expressed GPCR compared to a native GPCR.[39] This can be remedied by using recombinant site-directed mutagenesis methods to replace the respective amino acids that form such modifications. Because little is understood about which expression system should be used, empirical screening of overexpression of GPCRs in heterologous hosts is required.[40–42] Several higher-throughput heterologous expression systems have been described.[43,44] Chapter 8 discusses a variety of methods for GPCR overexpression.

In rare instances, membrane proteins exist in cell membranes in a crystalline form. Bacteriorhodopsin occurs in two-dimensional crystalline arrays in purple membrane patches in halobacteria; halorhodopsin and gap junctions can form crystalline arrays in cells when they are overexpressed. Density gradient centrifugation can be utilized to purify such membranes based on the fact that their high protein content gives them a unique density compared to densities of other cellular membranes.

However, most overexpression systems do not produce GPCRs to the high levels found for rhodopsin and bacteriorhodopsin and protein purification is required to separate the GPCR from the heterogeneous milieu of lipids and proteins in the membrane bilayer. Starting from a good recombinant source of protein, the purification procedure must yield pure and homogeneous protein preparations in sufficient yield. Because membrane proteins are purified as membrane protein–detergent complexes, standard protein purification methods are complicated by the presence of lipids and detergents.

GPCRs can be engineered to have affinity tags (FLAG, polyhistidine, strep-tag, V5, c-myc, and others) that permit interaction with an affinity matrix in the presence of detergent.[45–48] Although these tags are short peptides that are inserted either onto the amino, carboxy terminus, or extra-membranous loops, ligand binding properties and G protein activation of these recombinant GPCR constructs must be tested to ascertain their biological activities. In addition, the crystallizability of the protein might be perturbed and thus the ability to remove the affinity tags is desirable. The choice of protease cleavage site depends on the GPCR amino acid sequence.

12.3 DETERGENT SELECTION

The second bottleneck of GPCR crystallization is the purification of active and homogeneous GPCR. As with any membrane protein purification, GPCR molecules must be extracted from their native or recombinant lipid bilayer environments that comprise heterogeneous mixtures of proteins and lipids. The internal membrane spanning domains are highly hydrophobic and unprotected exposure to aqueous environment will cause unspecific protein aggregation and denaturation. Detergents are often used in membrane protein purification as a means to shield the hydrophobic surfaces of proteins while at the same time disrupting the biological membrane.[49] The resulting solubilized membrane proteins are present in aqueous solutions as protein–detergent complexes. The subsequent purification steps should yield homogeneous GPCR–detergent complexes.

The choice of detergent is a critical issue to consider for membrane protein stability.[49] Dozens of detergents are in common use in biochemistry, dozens more are less well characterized but still probably useful, and many novel detergents are under development.[50–53] Detergent-solubilized GPCRs are essentially protein–detergent complexes with the detergent hydrophobic tails oriented toward the hydrophobic surfaces of GPCRs and the polar detergent head groups facing the aqueous phase (Figure 12.2). The choices of detergents and lipids for characterizing purified receptors are important because the interfacial structures and chemistries involving the membrane spanning domains can affect the specificity of protein signaling and targeting.[54] Because membrane proteins are idiosyncratic in their interactions with detergents, it is impossible to identify the best detergents to use *a priori*. Detergent extraction of a GPCR with quantitative comparisons of yield and specific activity should be determined to judge which detergent would function best. Because little is known about the similarities and stabilities of GPCR surfaces, detergent extraction is likely to remain an empirical affair.

The ideal detergent for GPCR crystallization should possess the following properties: solubility in a variety of aqueous media over a large temperature range; inertness in terms of GPCR conformation and activity; ability to be easily removed or exchanged during vesicle reconstitution; compatibility with protein chromatography matrices; large-scale availability (100 to 1000 g) at reasonable cost; and well-defined homogeneity.[55,56]

The last criterion often excludes many common polymeric detergents with polydisperse populations such as Triton, Nonidet, Tween, amphipols, and Lubrol. Detergents that penetrate a protein may help denature the protein by disrupting the native hydrophobic and electrostatic interactions relevant to the stabilization of the native protein structure. The ideal detergent would have a large head group to prevent its complete insertion into the protein but would still be able to form a compact micelle structure.[57] Table 12.1 lists detergents used for membrane protein crystallization.

Several nonionic detergents fit the aforementioned criteria and are often preferred for the extraction of proteins. Nonionic detergents form more compact micelles because the absence of an intrinsic charge at the head group allows the detergent molecules in the micelles to approach each other more closely. Typically, the effective head group area is usually much larger as compared to those of ionic detergents.[58] Membrane proteins are typically extracted by detergent concentrations of 0.5 to 2% w/v using detergent:protein ratios of 1:1 to 3:1. Sufficient detergent must be present to remove the protein from its membrane bilayer environment and prevent the formation of nonspecific protein–protein aggregates or precipitates.

The capacity of a detergent to associate with GPCRs depends on the physical parameters of the detergent molecules: detergent tail length; head group size; and head group hydrophobicity. The use of nonionic and zwitterionic detergents with high critical micellar concentration values facilitates the detergent removal and renders them very suitable for reconstitution studies. In addition to detergents, additives are commonly used as stabilizers in crystallization procedures.[59] See Table 12.1. The presence of osmolytes (glycerol, sorbitol, glucose, trehalose, betaine, glycine, proline, and others) at high concentrations (10 to 20% w/v) during detergent solubilization prevents the inactivations that normally occur when proteins are extracted from natural membranes.[60] It is important to stress that membrane protein–detergent complexes are dynamic with the detergent molecules in equilibrium between the monomer, detergent micelle, and protein–detergent micelle states.[61] The dynamic aspect could explain why protein–detergent complexes are notoriously unstable and often result in aggregation or denaturation of the protein.

Receptors may require particular lipids for full retention of their biological properties in a micellar environment. Thus the choice of detergent may be linked to the differential solubilization of proteins and lipids by the detergents. Coextracted lipids can be essential for maintaining the activities of receptors.[62] Dramatic differences were observed in the ratios of individual lipids extracted by the same con-

TABLE 12.1
Components of Membrane Protein Crystallization

	Number of Membrane Proteins[a]	
	Prokaryotic	Eukaryotic
Detergent Phase Crystallization[32]		
1,2-Heptanoyl-*sn*-glycerol 3 phosphocholine		1
3-Laurylamido-N,N'-dimethylpropylamine oxide		1
Decanoyl-N-methyl glucamide		1
Decyl pentaoxyethylene	2	
Decyl-β-D-maltopyranoside	5	1
Dodecyl nonaoxyethylene	4	
Dodecyl octaoxyethylene	2	
Dodecyl-β-D-maltopyranoside	10	3
Foscholine 12	1	
Heptyl-thio-β-D-glucopyranoside		1
N,N-dimethyldecylamine-N-oxide	2	
N,N-dimethyldodecylamine-N-oxide	12	2
N,N-dimethylhexylamine-N-oxide	3	1
N,N-dimethylundecylamine-N-oxide	1	
Nonyl-β-D-glucopyranoside	1	2
Nonyl-thio-β-D-glucopyranoside		1
Octyl-2-hydroxy ethyl sulfoxide	3	
Octyl pentaoxyethylene	1	
Octyl tetraoxyethylene	6	
Octyl-β-D-glucopyranoside	12	4
Octyl-β-D-maltopyranoside	2	
Tridecyl-β-D-maltopyranoside	2	
Triton X100		1
Undecyl-β-D-maltopyranoside		3

Additives in Detergent Phase Crystallization

1,6-Hexanediol
2-Methyl-2,4-pentanediol
2-Propanol
Benzamidine
CHAPS
Digalactosyldiacyl glycerol
Dioleoyl phosphatidylcholine
Dioxane
Dodecyl octaoxyethylene
Ethanol
Ethylene glycol
Glycerol
Heptane-1,2,3-triol
Heptyl-β-D-glucopyranoside
Inositol

TABLE 12.2
Components of Membrane Protein Crystallization

Methyl-6-O-(N-heptylcarbamoyl)-α–D-glucopyranoside)
N,N-bis-(3-D-glyconamidopropyl deoxycholamide)
t-Butanol

Bicelle-Based Crystallization[146,147]

Dimyristoyl-*sn*-glycero-phosphocholine and 3[(3-cholamidopropyl)-dimethyl-ammonio]-
  2-hydroxy-1-propane sulfonate) (CHAPSO)
Dihexanoyl-*sn*-glycero-phosphocholine and dimyristoyl-*sn*-glycero-phosphocholine

Lipidic Mesophase-Based Crystallization[148]

1-Δ9-*cis*-octadecenoyl-*rac*-glycerol or monoolein
1-Δ9-*cis*-hexadecenoyl-*rac*-glycerol or monopalmitolein
1-Δ11-*cis*-hexadecenoyl-*rac*-glycerol or monovaccenin

[a] Most of the reported membrane protein structures came from prokaryotic membrane proteins.[31,146]

centrations of different detergents. Alternatively, purified membrane proteins extracted in detergents can be stabilized by the addition of lipids that are relevant modulators of receptors.[63] Table 12.1 lists components that can be used for membrane protein purification and stabilization. Table 12.2 is a compilation of resources for lipids and detergents.

The buoyant density of the lipid–detergent micelle is usually lower than that of the detergent–protein complex, which in turn is lower than that of the membrane vesicle. Such differences have been utilized for successful separation of these species by density gradient centrifugation.[64] The mixed lipid micelles formed by nonionic detergents can be separated conveniently by ion exchange chromatography.[65] In certain cases, affinity chromatography methods can be used with proteins solubilized with nonionic and zwitterionic detergents.

Changing the ionic strength, changing the hydrophile–lipophile balance value of the detergent, or inducing detergent phase separation[66,67] can bring about selective GPCR extraction. Evaluation of detergent extraction of GPCRs from membrane preparations requires the ability to distinguish the GPCR from other proteins coextracted with the detergent. Assessment of protein solubilization often is monitored best using Western blot analysis. Detection of recombinant protein with distinct amino or carboxy terminal antigen tags can be used to monitor the level of protein solubilization and for subsequent purification steps. In addition, membrane solubilization can be monitored using a wide variety of spectroscopic methods.[68]

Excessive detergent, however, can lead to both protein denaturation and the formation of protein-free micelles that may interfere with crystal formation. Rhodopsin is less thermally stable as the detergent concentration increases.[69] Each GPCR

TABLE 12.3
Lipid and Detergent Resources

www.ldb.chemistry.ohio-state.edu

The web site contains relational databases on the phase properties, miscibilities, and molecular structures of lipids, tables of critical micellar concentrations, aggregation numbers for an assortment of detergents, and links to other membrane-, lipid-, and detergent-related web sites.

www.embl-heidelberg.de/~gohlke/properties.html

www.anatrace.com

This web site lists many properties of commonly used detergents

www.avantilipids.com

This web site lists native and synthetic lipids with protocols and references on biophysical characterizations of different lipid phases.

Commercial Lipid Sources

Avanti Polar Lipids, Alabaster, AL: www.avantilipids.com

Cayman Chemical Company, Ann Arbor, MI

Nu-Chek-Prep, Inc., Elysian, MN

Commercial Detergent Sources

Amphotech Ltd., Beverly, MA

Anatrace, Maumee, OH

Bachem, San Carlos, CA

Calbiochem/Novabiochem/EMD Biosciences, San Diego, CA

Carbomer Inc., San Diego, CA

Dojindo Molecular Technologies, Gaithersburg, MD

ICN Biomedicals, Costa Mesa, CA

Research Organic, Inc., Cleveland, OH

Research Products International Corporation, Mt. Prospect, IL

Sigma Aldrich Chemical Co., St. Louis, MO

Soltec Ventures, Beverly, MA

Toronto Research Chemicals, North York, Ontario, Canada

is likely to have a select detergent and detergent concentration range under which crystallization is favorable.

The level of endogenous lipid in a solubilized GPCR sample can also influence the structural integrity and thus the crystallizability of the protein. Excess lipid can form protein-free lipid–detergent micelles that interfere with crystal growth and may also increase the effect size of the detergent belt surrounding the transmembrane domain of the protein, thus making crystal contacts sterically unfavorable. As with detergents, membrane protein crystallization is likely to be optimal at a select range of lipid–protein ratios[70] and it is important to monitor detergent and lipid concentrations during the protein purification process.[71]

Because GPCRs may be solubilized best with different mixtures of detergents, lipids, and additives, the respective purification protocols must be tailored to suit particular GPCRs. Purification protocols can be more standardized in the preliminary steps by utilizing interactions of affinity tags cloned onto GPCRs. However, these affinity tags may have to be removed to enhance crystallization. Proteases that remove affinity tags can be inhibited by detergents or by the availability of the recognition site in an overexpressed GPCR.[72,73]

Implicit in all purification procedures is the ability to assess the integrity of the protein. This is not a trivial issue because the state of a GPCR in a detergent phase may not be the same as that of a GPCR in membrane bilayer phase. Consideration must be given to the physicochemical properties of the matrix used to purify membrane proteins. Because it is difficult to quantify such properties, ligand binding is often the best method to measure the integrity of a GPCR during purification. A good sign of GPCR integrity in the detergent micelle is the retention of equilibrium dissociation constant (K_D) values obtained for the GPCR in native membranes.

Determination of rank order binding of ligands can also be helpful. However, the presence of non-native post-translational modifications and the absence of G proteins and other effector molecules could exert drastic effects on K_D values. During the last decade, significant progress has been made in developing analytical techniques for the screening of ligand binding to membranes and membrane receptors. Some of these assays include sedimentation, chromatographic, acoustic, optical surface plasmon resonance, biosensing, analytical centrifugation, chromatographic, isothermal titration calorimetry, and differential scanning calorimetry assays on a variety of model membrane systems.[2] The ability to optimize such assays for detergent-solubilized proteins holds much promise.

12.4 SAMPLE PREPARATION

GPCR purification procedures for crystallization must reach the goal of preparing reproducible, pure, and homogeneous solutions devoid of heterogeneity caused by proteolysis, different oligomerization states, and post-translational modifications. However, the presence of post-translational or oligomeric states should not be deterrents for protein crystallization as long as a monodispersity can be established. Crystallization of GPCRs can be hampered by the presence of flexible termini and extra-membranous loops. Removal of these flexible regions without grave effect on protein activity can be accomplished by using limited proteolytic cleavage followed by recombinant re-engineering of the protein.

GPCRs change their conformations considerably during ligand binding and G protein activation.[74] A general model of GPCR activation postulates that GPCRs are in equilibrium between active and inactive states. These states presumptively differ in the disposition of the transmembrane helices and, in turn, the cytoplasmic domains that determine G protein coupling. Agonists, according to this model, stabilize the activated state. For some family 1 GPCRs, an agonist activates the receptor by binding directly within the 7TM domains and altering the disposition of these helices. For GPCRs in families 2 and 3, agonist binding involves linking portions of the extracellular domains to the transmembrane domains of the receptor. A detailed mechanism whereby this signal is transmitted to the 7TM domain remains to be clarified. Hence, successful crystallization of GPCRs may require fixing the protein into a certain conformation by the use of specific ligands or inhibitors. If no conditions can be found to keep the protein stably homogeneous for crystallization, a selection of stable mutants could help.

TABLE 12.4
Criteria for Selection of Reconstitution Matrices for Crystallization of GPCRs[a]

Biological
    a. Membrane-like environment of reconstituted membrane protein
    b. Presence of native and active configuration
    c. Stability of membrane protein in reconstitution environment for more than 2 weeks
    d. Presence of monodisperse post-translational modification, otherwise use site-directed mutagenesis
        to replace amino acids that participate in post-translational modifications
Physical
    a. Monodispersity of pure component and protein–detergent/lipid complex
    b. Phase permitting 3D diffusion of precipitating solutions and proteins from 4 to 30°C
    c. Conditions that allow optical spectroscopic characterizations of membrane proteins
    d. Phase diagram with rheological properties that permit incorporation of membrane proteins of
        different sizes
Chemical
    a. Homogeneity
    b. Chemical inertness
    c. Ease of synthesis and preparation
    d. Capacity to be inert to precipitating solutions that induce protein crystallization

[a] Adapted from References 58 and 150.

For soluble proteins, purification to homogeneity generally provides monodispersity. For membrane proteins, the situation is significantly different. A purified GPCR sample is a complex mixture due to the presence of detergents, additives that associate with the detergents, and other solution components that maintain the integrity of the protein–detergent complex. Establishing how these components affect the monodispersity of the protein requires comparison of data obtained by several methods. Monodispersity of a GPCR preparation can be characterized via analytical ultracentrifugation,[75] light scattering,[76–78] self-interaction chromatography,[79,80] and fluorescence spectroscopy.[81] The criteria for membrane protein crystallization matrix selection are noted in Table 12.3. Likewise, some assessment of the membrane protein incorporated into the crystallization matrices should be made to determine GPCR integrity. Establishing thorough biochemical and biophysical characterizations of a purified GPCR preparation in combination with comprehensive screening for the most suitable detergent may be the most efficient strategy to cope with the difficulties of GPCR crystallization.

12.5 CRYSTALLIZATION METHODS

The next bottleneck for GPCR structure determination via X-ray crystallography is the ability to generate well-ordered 3D crystals. The GPCR–detergent complex should be concentrated to values greater than 5 mg/mL. This is not trivial because the process of protein concentration may also concentrate other solution components (such as detergents) that can be deleterious to GPCR activity. Careful assessment of preservation of homogeneity and protein activity at higher concentrations is required.

The amphipathic surfaces of membrane proteins allow several ways of forming 3D crystals. Membrane protein crystals can be ordered stacks of 2D crystals that contain ordered protein molecules within the membrane plane. Membrane protein crystals can also be formed by membrane protein–detergent complexes that form crystal contacts mediated by interactions between the polar surfaces of the protein that protrude from the lipid–detergent areas. With the exception of detergent phase properties, crystallization conditions in this situation are very similar to those for soluble proteins.

The detergent micelle requires space in the crystal lattice and may not contribute to rigid crystal contacts. Unless they contribute to crystal packing, lipids and detergents must be dynamic to permit reorganization during GPCR crystallization. Proteins with small hydrophilic domains can be especially difficult to crystallize. It is possible to enlarge the polar surfaces of a membrane protein by the attachment of polar domains of specifically binding antibody fragments.[82,82] Alternatively, antibody fragments can be useful to fix dynamic regions of GPCRs. Several excellent reviews discuss membrane protein crystallization.[23,84–87] Membrane protein crystallization methods have employed a variety of matrices that include detergents,[59,83] bicelles,[88] and lipid phases for 2D[85] and 3D crystallization.[89,90] Table 12.3 lists the criteria for using these matrices. Biophysical and biochemical methods should be developed to characterize membrane proteins that have been incorporated into these matrices. We will now discuss how these methods can be pertinent to GPCRs.

12.6 MICELLES AND BICELLES

Most of the known structures of membrane proteins have been determined via crystallization of membrane proteins in the detergent phase. Membrane proteins have been successfully crystallized in various nonionic and zwitterionic detergents mainly belonging to the alkyl dimethyl amine oxide, alkyl glycopyranoside, alkyl maltopyranoside, and alkyl polyoxyethylene groups as listed in Table 12.1. Detergents exert moderately complex phase behavior that can be perturbed in the presence of precipitants commonly used to induce protein crystallization.[91,92] Additional detergent mixtures in the protein solution or by the addition of crystallization trials can be useful. New generations of detergents, for example, lipopeptides,[50,93] amphipols,[94,95] and tripod amphiphiles,[96] may also be helpful.

Detergents are used typically at concentrations above the critical micellar concentration.[49,58] The clustering of micelles close to the critical temperature of phase separation provides high local concentrations of mixed micelles and thus high local protein concentrations. This encourages the incorporation of proteins in large and growing micelles, rather than mixed micelles consisting of proteins covered with layers of detergents.[97]

Prior to conducting crystallization trials, a researcher may need to replace the initial solubilization detergent with a detergent that is more compatible with protein crystallization. Dialysis of protein can be slow and incomplete, even with detergent binding resins such as BioBeads and Chelex.[98] Detergent exchange is more efficient by binding the protein onto an affinity or ion exchange column followed by repeated washes of buffer containing the detergent of interest. After a high level of membrane protein concentration is achieved, it is advantageous to determine the final detergent concentration accurately, thereby allowing calculation of the final detergent-to-protein ratio.[68] High-quality crystals may be obtained when excess detergent is removed to reach a controlled detergent-to-protein ratio[99] Crystallization of protein–detergent complexes is enhanced under conditions that approach the cloud point of the detergent used.[23,84] The cloud point or consolute boundary represents a phase boundary at which intermicellar attractive forces become sufficiently strong that the micelles coalesce into separate phases.[78]

12.7   LIPIDIC MESOPHASES

Despite the numerous successes associated with crystallization of protein–detergent complexes, it is not clear whether detergent micelles can fully and accurately reproduce the native environment of the protein, the lipid bilayer. The lipidic cubic phase has been employed to crystallize membrane proteins in a bilayer environment.[100] The lipidic cubic phases form gel-like materials containing 3D lipid bilayer structures arranged so as to form topologically distinct lipid and aqueous regions.[101,102] Proteins are added to lipids and then to precipitating agents that induce crystallization.[89,103] During crystallization, a mixture of phases exists that directly feeds the growing crystal face. This process is now called lipidic mesophase crystallization because of the collection of lipid phases present during crystallization. Crystallization in the lipidic cubic phases has been successful for archeael 7TM helix proteins[100,104] and other classes of membrane proteins such as light harvesting complexes and photosynthetic reaction centers.[105]

Lipidic cubic phase methods have primarily used hydrated *cis*-monounsaturated monoacyl glycerols to form the mesophase. Most commonly used lipidic cubic phases used for crystallization include monoolein, monopalmitolein, and monovaccenin.[106] By varying the lipids, the cubic phase will provide different host properties such as bilayer curvatures and aqueous channel diameters that can affect the manner of crystallization.

It is possible that the hosting cubic phase created by monoacyl glycerides alone will limit the range of membrane protein crystallizability. With a view to expanding the range of applicability of the method and make the hosting cubic phase more familiar to its guest proteins, it is possible to make lipidic mesophases that include other lipid types such as phosphatidylcholine, phosphatidylethanolamine, phosphatidylserine, cardiolipins, lyso-phosphatidylcholine, polyethylene glycol lipid, 2-monoolein, oleamide, and cholesterol.[107] Such lipids can have a role in maintaining particular affinity states of the receptors.[108] Phase boundaries related to the use of precipitating salts at various pH values[109] and effects of different alkyl glucopyranosides at different temperatures[110] have been reported. The presence of salt solutions does not affect the overall symmetry of lipidic mesophases.[109] Compilations of detergents and precipitants which are compatible with lipid matrices have been made.[102,111]

Lipidic cubic phase samples are highly viscous and have traditionally been prepared in batch mode; a new apparatus allows for the preparation and dispensing of small volumes[112] and a high throughput robotic system has been developed for crystallizing membrane proteins.[113] The robot can dispense submicroliter quantities of the lipidic mesophase onto specially designed 96-well glass plates. Quantitative evaluation of crystallization progress can be performed in these plates. A novel lipid that permits the formation of a bicontinuous lipidic cubic phase that is stable at 6°C has been synthesized recently.[114] With the availability of automation to test more conditions and the use of lipids with lower temperature profiles to enhance GPCR stability, this method offers great promise for the crystallization of membrane proteins.

12.8 BICELLES, LIPOPEPTIDES, AND NANODISCS

Membrane protein incorporation into bicelles is a crystallization method that incorporates aspects of both the lipidic cubic phase and protein–detergent complex methods.[88] Bicelles are small bilayered discs that dissolve in aqueous solution much as micelles do, but offer an environment more like a native biological membrane.[115] Bicelles can be viewed as flattened or discoidal structures whose perimeters are coated with detergent. Bacteriorhodopsin crystallization was conducted at temperatures at and above room temperature.[88] Improving the method will require finding conditions under which gelation occurs at lower temperatures because not all proteins will tolerate temperatures at or above room temperature.

Lipopeptides with amphipathic helices that are long enough to span lipid bilayers have been employed as detergent surrogates.[50] These peptides appear to solubilize membrane proteins by clustering concentrically around the transmembrane domains of the resident protein. The manner in which lipopeptides assemble around membrane proteins may allow retention of membrane protein function and stability when they are used at very low concentrations. Little is known yet about the relative stability of these protein–detergent complexes.

Nanodiscs have been developed recently to form discoidal nanoparticles that self-assemble with lipids and amphipathic α-helical proteins called membrane scaffold proteins (MSPs).[116–118] Nanodiscs of desired size and composition are results of directed self-assembly of components solubilized in aqueous detergent solutions. Their sizes can be controlled by protein engineering of the MSP.[119] Nanodiscs can be prepared in sizes ranging from 9.5 to 12.8 nm.[119] Because stabilization of the rims of the phospholipid bilayers of these self-assembled discoidal bilayer particles is accomplished by a protein instead of a detergent, overall stability is dramatically increased. Thus, the technique of using GPCRs encapsulated into nanodiscs for crystallization shows promise.

12.9 CRYSTALLIZATION

To determine the protein concentration range relevant for crystallization, it is worthwhile to test stability and solubility of the protein in solutions with varying concentrations of precipitants. Preparation of a solubility diagram could narrow the choice of detergent concentration and ensure suitable precipitant concentration in crystallization experiments.[120] To reach super-saturation of the membrane protein and thereby the nucleation for crystal growth, commonly used precipitants include poly-ethylene glycols, polyethylene glycol monomethyl ether, ammonium sulfate, and other inorganic salts.

A variety of crystallization formats are possible including hanging drop, sitting drop, microdialysis, free interface, and batch crystallization modes. Because of the significant number of variables to be screened for GPCR crystallization (buffer conditions, ligands, effector molecules, and detergent contents), the number of crystallization conditions required for achieving successful crystals can be signifi-cantly higher than the number required for finding soluble proteins. Each crystalli-zation method should be modified to accommodate the rheological properties of GPCRs incorporated into micelles, bicelles, lipidic mesophases, and nanodiscs.

Several vendors provide robotic support for crystallization methods using sitting drop and batch crystallization modes. In microbatch crystallizations, a small volume of protein solution was placed in contact with a large volume of oil. Although expected to partition to the oil phase, detergents in the aqueous phase showed no significant losses in concentration during the crystallization process.[121]

Little is known about how the purified protein–detergent complexes will respond to crystallization screening solutions that have been developed primarily for soluble proteins. There is some concern about the presence of detergent or lipid phase changes that can mask the presence of GPCR crystals. A map of detergent phase behavior can allow design of screens that could yield good protein crystals.[122] Methods using dyes can aid in identifying detergent phase separation.[123,124] Pro-tein–detergent complexes are known to crystallize preferentially near the cloud point and the location of the cloud point must be carefully determined. The main problem with membrane protein crystallization using the different micelle matrices is the need to distinguish protein crystals from detergent and buffer crystals.

Membrane protein crystals are often small; hence, diffraction analysis using synchrotron radiation may be required. To maximize the potential for structure determination, the ability to flash-cool crystals at near-liquid nitrogen temperatures is extremely important for successfully transporting small crystals and obtaining complete datasets using fewer crystals. Soaking the crystals in cryoprotectants will be required to minimize radiation damage. Because different matrices are used for protein crystallization, a variety of cryoprotectants are required for rapid vitrification.

12.10 CASE STUDY: RHODOPSIN

Rhodopsin is a highly specialized GPCR that detects photons in rod photoreceptor cells. It is a unique member of the GPCR family in that it contains an intrinsic inverse agonist, 11-*cis* retinal. The retinal forms a covalent bond to form a protonated Schiff base with a lysine at position 296 in helix VII. Photon absorption results in retinal isomerization to the all-*trans* configuration that drives the protein to the active metarhodopsin form. Pigments mediating color vision in cone cells share a common mechanism and evoke a cellular signaling cascade[125] through interactions between the meta II state and the corresponding G protein known as transducin.

Bovine rhodopsin is the first eukaryotic GPCR to be crystallized[126] and yield a crystal structure.[127] Rhodopsin also is part of the class A GPCR family whose members share amino acid sequence similarity.[128,129] Rhodopsin can be obtained from bovine retinae (0.5 to 1 mg/retina) by a sucrose density gradient centrifugation preparation of the rod outer segment disc membranes.[130,131] The screen for detergent solubilization is simplified by using a rapid spectroscopic assay to determine protein stability and viability in the presence of different detergents and additives.[132] Rhodopsin is stable enough in the dark to be purified further by various chromatographic procedures, and it remains stable in solution in a variety of detergents.[133]

Because of obstacles in obtaining 3D crystals of eukaryotic rhodopsins, structural data were derived only from 2D crystals formed from intact membranes of rod outer segments. An initial view of the seven helices of bovine rhodopsin was achieved via electron crystallography.[134–136] A subsequent x-ray crystallographic study determined its structure at 2.8 Å resolution.[127] This structure provided the template model at high resolution for the rhodopsin-like GPCRs and has been further refined at higher resolutions.[137,138]

The breakthrough in the 3D crystallization of bovine rhodopsin was not due to advances in the crystallization methodology of membrane proteins that were crucial for the successful structure determination of archeael rhodopsin. Rather, the discovery of a new protocol for the selective extraction of bovine rhodopsin from rod outer segments led to preparations yielding 3D crystals useful for x-ray crystallographic analyses. One unusual component of the extracting agents, the zinc ion, was used in high concentrations and observed in two positions in the refined x-ray crystallographic structure of rhodopsin where it stabilized the overall protein fold. A recent high-resolution structure of bovine rhodopsin to 2.2 Å confirmed the presence of the detergent and lipids in the structure.[139]

In bovine rhodopsin, the regions that protrude from the lipid bilayer are much larger and more organized than those of archeael rhodopsins. Bovine rhodopsin structure has the functionally important G protein-binding site on the cytosolic side and parts of the retinal binding pocket on the extracellular side. Unfortunately, the G protein-binding site is still poorly defined in the x-ray crystal structure of rhodopsin because of high temperatures and incomplete tracing of polypeptide chains in the loops between transmembrane helices V and VI and the carboxy terminal stretch. Because cytosolic loops are not well defined in the x-ray crystal structure, it is not immediately obvious how GPCRs bind to G proteins.[140,141] Even if further crystal refinement of this crystal form improves identification of the G protein binding site, the conformation may be of limited value for explaining in structural terms how GPCRs catalyze the guanine nucleotide exchange on G protein complexes because various studies demonstrate significant conformational changes of the binding site upon G protein binding and GPCR activation. Extra-membranous domains of other GPCRs are structurally divergent from rhodopsin and hence the rhodopsin structure cannot be used for modeling of these GPCRs.

The bovine rhodopsin structure is in the inactive (dark) state; the receptor changes its conformation during light activation. Nevertheless, the structure provides very useful information because the ligand binding site is clearly defined. The structure corresponds rather well with earlier proposals concerning which GPCR residues such as dopamine D2, muscarinic M1, and histamine H4 receptors are involved in the binding of biogenic amines. However, the orientation and position of transmembrane helices III and VI relative to the protein core may be different in the active conformations.

12.11 CONCLUSIONS

The suggestion has been made to rename GPCRs and call them 7TM proteins because they interact with many regulatory elements other than G proteins.[18] These interactions are involved in long-term desensitization due to prolonged time of agonist exposure; mechanisms of endocytosis with β-arrestin interactions of kinases for phosphorylation; expression of receptor splice variants exhibiting different sensitivities to ligands and G proteins; receptor coupling to more than one G protein class; and cross-talk between different GPCRs.[142] There is also the possibility of receptor oligomerization that may play a role in subtype specific receptor regulation.[12] Access to GPCR structures alone may not be sufficient to develop a mechanistic view of all of these interactions.

In suggesting that GPCRs be overexpressed in heterologous expression systems, it is possible that the purified GPCR will have non-native post-translational modifications. Ascertaining the role of such modifications may not be trivial because of the difficulties in comparing *in vivo* receptor states to different *in vitro* formats. Chimeric receptors can offer insights into the domains of GPCRs.[143] Chimeras involving swapping of the exofacial, transmembrane, and cytoplasmic segments that couple agonist binding into activation of effectors could be generated. They have been used to determine which regions of a GPCR are responsible for agonist binding and for activation of the receptor. They could be used to determine the minimal structural basis for ligand binding or effector coupling.[144,145]

Although fewer than 60 integral membrane proteins have been recorded in the protein data bank (compared to more than 25,000 soluble protein structures), the possibility of determining additional membrane protein structures within the next few years holds great promise. Systematic efforts to solve aspects of the three bottlenecks of membrane protein structure determination (overexpression, purification, and crystallization) via x-ray crystallography are in progress. These daunting technical challenges often require the multidisciplinary resources, leading to the establishment of several consortia such as MePNet, E-MeP, and other structural genomics initiatives sponsored in the United States, Canada, Japan, Switzerland, and Europe (see Chapter 14). These are exciting times that offer promise of developing structure-based drug designs based on GPCRs.

REFERENCES

1. Marinissen, M.J. and J.S. Gutkind, G protein-coupled receptors and signalling networks: emerging paradigms. Trends Pharmacol. Sci., 2001. 22: 368–376.
2. Cooper, M.A., Advances in membrane receptor screening and analysis. J. Mol. Recognit., 2004. 17: 286–315.
3. Sautel, M. and G. Milligan, Molecular manipulation of G-protein coupled receptors: a new avenue into drug discovery. Curr. Med. Chem., 2000. 7: 889–896.

4. Chalmers, D.T. and D.P. Behan, The use of constitutively active GPCRs in drug discovery and functional genomics. Nat. Rev. Drug Disc., 2002. 1: 599–608.

5. Drews, J., Drug discovery. Science, 2000. 287: 1960–1964.

6. Bywater, R.P., Location and nature of the residues important for ligand recognition in G-protein coupled receptors. J. Mol. Recognit., 2005. 18, 60–72.

7. Horn, F., E. Bettler, L. Oliveira, F. Campagne, F.E. Cohen, and G. Vriend, GPCRDB information system for G protein-coupled receptors. Nucleic Acids Res., 2003. 31: 294–297.

8. Civelli, O., R.K. Reinscheid, and H.P. Nothacker, Orphan receptors, novel neuropeptides and reverse pharmaceutical research. Brain Res., 1999. 848: 63–65.

9. Scussa, F., World's best selling drugs. Med. Ad News, 2002. 21: 1–46.

10. Behan, D.P. and D.T. Chalmers, The use of constitutively active receptors for drug discovery at the G protein-coupled receptor gene pool. Curr. Opin. Drug Disc. Dev., 2001. 4: 548–560.

11. Ellis, C.R., State of GPCR Research in 2004. Nat. Rev. Drug Disc., 2004. 3: 577–626.

12. Ulloa-Aguirre, A., D. Stanislaus, J.A. Janovick, and P.M. Conn, Structure–activity relationships of G-protein coupled receptors. Arch. Med. Res., 1999. 30: 420–435.

13. Müller, G., Towards 3D structures of GPCRs: a multidisciplinary approach. Curr. Med. Chem., 2000. 7: 861–888.

14. Bondensgaard, K., M. Ankersen, H. Thogersen, B.S. Hansen, B.S. Wulff, and R.P. Bywater, Recognition of privileged structures by G-protein coupled receptors. J. Med. Chem., 2004. 47: 888–899.

15. Foord, S.M., Receptor classification: post genome. Curr. Opin. Pharmacol., 2002. 2: 561–566.

16. Spiegel, A.M. and L.S. Weinstein, Inherited diseases involving G proteins and G protein-coupled receptors. Annu. Rev. Med., 2004. 55: 27–39.

17. Pierce, K.L., R.T. Premont, and R.J. Lefkowitz, Seven-transmembrane receptors. Nat. Rev. Mol. Cell Biol., 2002. 3: 639–650.

18. Vassilatis, D.K., J.G. Hohmann, H. Zeng, F. Li, J.E. Ranchalis, M.T. Mortrud, A. Brown, S.S. Rodriguez, J.R. Weller, A.C. Wright, J.E. Bergmann, and G.A. Gaitanaris, The G protein-coupled receptor repertoires of human and mouse. Proc. Natl. Acad. Sci. USA, 2003. 100: 4903–4908.

19. Bywater, R.P., A. Sorensen, P. Rogen, and P.G. Hjorth, Construction of the simplest model to explain complex receptor activation kinetics. J. Theoret. Biol., 2002. 218: 139–147.

20. Gouldson, P.R., C.R. Snell, R.P. Bywater, C. Higgs, and C.A. Reynolds, Domain swapping: a mechanism for functional rescue and activation in G-protein coupled receptors. Protein Eng., 1998. 11: 1181–1193.

21. Michel, H., Three-dimensional crystals of a membrane protein complex: photosynthetic reaction centre from *Rhodopseudomonas viridis*. J. Mol. Biol., 1982. 158: 567–572.

22. Kühlbrandt, W. and E. Gouaux, Membrane proteins. Curr. Opin. Struct. Biol., 1999. 9: 445–447.

23. Loll, P.J., Membrane protein structural biology: the high throughput challenge. J. Struct. Biol., 2003. 142: 144–153.

24. Werten, P.J., H.W. Remigy, B.L. de Groot, D. Fotiadis, A. Philippsen, H. Stahlberg, H. Grubmuller, and A. Engel, Progress in the analysis of membrane protein structure and function. FEBS Lett., 2002. 529: 65–72.

25. Engel, A., A. Hoenger, A. Hefti, C. Henn, R.C. Ford, J. Kistler, and M. Zulauf, Assembly of 2-D membrane protein crystals: dynamics, crystal order, and fidelity of structure analysis by electron microscopy. J. Struct. Biol., 1992. 109: 219–234.

26. Liang, Y., D. Fotiadis, S. Filipek, D.A. Saperstein, K. Palczewski, and A. Engel, Organization of the G protein-coupled receptors rhodopsin and opsin in native membranes. J. Biol. Chem., 2003. 278: 21655–21662.

27. Watts, A., S.K. Straus, S.L. Grage, M. Kamihira, Y.H. Lam, and X. Zhao, Membrane protein structure determination using solid-state NMR. Methods Mol. Biol., 2004. 278: 403–473.

28. Nielsen, N., A. Malmendal, and T. Vosegaard, Techniques and applications of NMR to membrane proteins. Mol. Membr. Biol., 2004. 21: 129–141.

29. Fernandez, C. and K. Wuthrich, NMR solution structure determination of membrane proteins reconstituted in detergent micelles. FEBS Lett., 2003. 555: 144–150.

30. Opella, S.J., Membrane protein NMR studies. Methods Mol. Biol., 2003. 227: 307–320.

31. White, S.H., Blanco.biomol.uci.edu/Membrane_Proteins_xtal.html. 2005.

32. Michel, H., www.mpibp-frankfurt.mpg.de/michel/public/memprotstruct.html. 2005.

33. McPherson, A., Crystallization of Biological Macromolecules. 1999, Cold Spring Harbor, NY: Cold Spring Harbor Laboratory Press.

34. Kiefer, H., J. Krieger, J.D. Olszewski, G. Von Heijne, G.D. Prestwich, and H. Breer, Expression of an olfactory receptor in *Escherichia coli*: purification, reconstitution, and ligand binding. Biochemistry, 1996. 35: 16077–16084.

35. Kiefer, H., *In vitro* folding of alpha-helical membrane proteins. Biochim. Biophys. Acta, 2003. 1610: 57–62.

36. Bazarsuren, A., U. Grauschopf, M. Wozny, D. Reusch, E. Hoffmann, W. Schaefer, S. Panzner, and R. Rudolph, *In vitro* folding, functional characterization, and disulfide pattern of the extracellular domain of human GLP-1 receptor. Biophys. Chem., 2002. 96: 305–318.

37. Grauschopf, U., H. Lilie, K. Honold, M. Wozny, D. Reusch, A. Esswein, W. Schafer, K.P. Rucknagel, and R. Rudolph, The N-terminal fragment of human parathyroid hormone receptor 1 constitutes a hormone binding domain and reveals a distinct disulfide pattern. Biochemistry, 2000. 39: 8878–8887.

38. Schrattenholz, A., S. Pfeiffer, V. Pejovic, R. Rudolph, J. Godovac-Zimmermann, and A. Maelicke, Expression and renaturation of the N-terminal extracellular domain of torpedo nicotinic acetylcholine receptor alpha-subunit. J. Biol. Chem., 1998. 273: 32393–32399.

39. Papoucheva, E., A. Dumuis, M. Sebben, D.W. Richter, and E.G. Ponimaskin, The 5-Hydroxytryptamine (1A) receptor is stably palmitoylated, and acylation is critical for communication of receptor with G_i protein. J. Biol. Chem., 2004. 279: 3280–3291.

40. Sarramegna, V., F. Talmont, P. Demange, and A. Milon, Heterologous expression of G-protein-coupled receptors: comparison of expression systems from the standpoint of large-scale production and purification. Cell Mol. Life Sci., 2003. 60: 1529–1546.

41. Grisshammer, R. and C.G. Tate, Overexpression of integral membrane proteins for structural studies. Q. Rev. Biophys., 1995. 28: 315–422.

42. Tate, C.G., Overexpression of mammalian integral membrane proteins for structural studies. FEBS Lett., 2001. 504: 94–98.

43. Laible, P.D., H.N. Scott, L. Henry, and D.K. Hanson, Towards higher-throughput membrane protein production for structural genomics initiatives. J. Struct. Funct. Genomics, 2004. 5: 167–172.

44. Lundstrom, K., Structural genomics on membrane proteins: the MePNet approach. Curr. Opin. Drug Disc. Dev., 2004. 7: 342–346.

45. Weiss, H.M. and R. Grisshammer, Purification and characterization of the human adenosine A(2a) receptor functionally expressed in *Escherichia coli*. Eur. J. Biochem., 2002. 269: 82–92.

46. White, J.F., L.B. Trinh, J. Shiloach, and R. Grisshammer, Automated large-scale purification of a G protein-coupled receptor for neurotensin. FEBS Lett., 2004. 564: 289–293.

47. Grisshammer, R., P. Averbeck, and A.K. Sohal, Improved purification of a rat neurotensin receptor expressed in *Escherichia coli*. Biochem. Soc. Trans., 1999. 27: 899–903.

48. Ratnala, V.R., H.G. Swarts, J. VanOostrum, R. Leurs, H.J. DeGroot, R.A. Bakker, and W.J. DeGrip, Large-scale overproduction, functional purification and ligand affinities of the His-tagged human histamine H1 receptor. Eur. J. Biochem., 2004. 271: 2636–2646.

49. Helenius, A. and K. Simons, Solubilization of membrane proteins by detergents. Biochim. Biophys. Acta, 1975. 415: 29–79.

50. McGregor, C.L., L. Chen, N.C. Pomroy, P. Hwang, S. Go, A. Chakrabartty, and G.G. Prive, Lipopeptide detergents designed for the structural study of membrane proteins. Nat. Biotechnol., 2003. 21: 171–176.

51. Popot, J.L., E.A. Berry, D. Charvolin, C. Creuzenet, C. Ebel, D.M. Engelman, M. Flotenmeyer, F. Giusti, Y. Gohon, Q. Hong, J.H. Lakey, K. Leonard, H.A. Shuman, P. Timmins, D.E. Warschawski, F. Zito, M. Zoonens, B. Pucci, and C. Tribet, Amphipols: polymeric surfactants for membrane biology research. Cell Mol. Life Sci., 2003. 60: 1559–1574.

52. McQuade, D.T., M.A. Quinn, S.M. Yu, A.S. Polaus, M.P. Krebs, and S.H. Gellman. Rigid Amphiphiles for Membrane Protein Manipulation. Angew. Chem. Int. Ed. Engl., 2000. 39, 758–761.

53. Sanders, C.R., A.K. Hoffman, D.N. Gray, M.H. Keyes, and C.R. Ellis, French swimwear for membrane proteins. Chembiochem, 2004. 5: 423–426.

54. Haltia, T. and E. Freire, Forces and factors that contribute to the structural stability of membrane proteins. Biochim. Biophys. Acta, 1995. 1241: 295–322.

55. Garavito, R.M. and S. Ferguson-Miller, Detergents as tools in membrane biochemistry. J. Biol. Chem., 2001. 276: 32403–32406.

56. Zulauf, M., Detergent phenomena in membrane protein crystallization. In: Crystallization of Membrane Proteins, H. Michel, Ed., 1991. Boca Raton, FL: CRC Press, pp. 54–71.

57. Tan, E.H.L. and R.R. Birge, Correlation between surfactant/micelle structure and the stability of bacteriorhodopsin in solution. Biophys. J., 1996. 70: 2385–2395.

58. Helenius, A., D.R. McCaslin, E. Fries, and C. Tanford, Properties of detergents. Methods Enzymol., 1979. 56, 734–749.

59. Michel, H., Crystallization of Membrane Proteins, 1990. Boca Raton, FL: CRC Press.

60. Maloney, P.C. and S.V. Ambudkar, Functional reconstitution of prokaryote and eukaryote membrane proteins. Arch. Biochem. Biophys., 1989. 269: 1–10.

61. Moroi, Y., Micelles: Theoretical and Applied Aspects, 1992. New York: Plenum Press.

62. Banerjee, P., J.B. Joo, J.T. Buse, and G. Dawson, Differential solubilization of lipids along with membrane proteins by different classes of detergents. Chem. Phys. Lipids, 1995. 77: 65–78.

63. Huidobro-Toro, J.P. and R.A. Harris, Brain lipids that induce sleep are novel modulators of 5-hydroxytryptamine receptors. Proc. Natl. Acad. Sci. USA, 1996. 93: 8078–8082.

64. Engelman, D.M., D.M. Terry, and H.J. Morowitz, Biochim. Biophys. Acta, 1967. 135: 381–390.

65. Jacobs, E.E., F.H. Kirkpatrick, B.C. Andrews, W. Cunningham, and F.L. Crane, Biochem. Biophys. Res. Commun., 1966. 25: 96–103.

66. Albertsson, P.A., Application of the phase partition method to a hydrophobic membrane protein, phospholipase A1 from Escherichia coli. Biochemistry, 1973. 12: 2525–2530.

67. Hjerten, S. and K.E. Johansson, Biochim. Biophys. Acta, 1972. 288: 312–325.

68. Goni, F.M. and A. Alonso, Spectroscopic techniques in the study of membrane solubilization, reconstitution and permeabilization by detergents. Biochim. Biophys. Acta, 2000. 1508: 51–68.

69. de Grip, W.J., Thermal stability of rhodopsin and opsin in some novel detergents. Methods Enzymol., 1982. 81: 256–265.

70. daCosta, C.J.B. and J.E. Baenziger, A rapid method for assessing lipid:protein and detergent:protein rations in membrane protein crystallization. Acta Crystallogr., 2003. D59: 77–83.

71. Eriks, L.R., J.A. Mayor, and R.S. Kaplan, A strategy for identification and quantification of detergents frequently used in the purification of membrane proteins. Anal. Biochem., 2003. 323: 234–241.

72. Mohanty, A.K., C.R. Simmons, and M.C. Wiener, Inhibition of tobacco etch virus protease activity by detergents. Protein Exp. Purif., 2003. 27: 109–114.

73. Mohanty, A.K. and M.C. Wiener, Membrane protein expression and production: effects of polyhistidine tag length and position. Protein Exp. Purif., 2004. 33: 311–325.

74. Kobilka, B., U. Gether, R. Seifert, S. Lin, and P. Ghanouni, Characterization of ligand-induced conformational states in the beta 2 adrenergic receptor. J. Receptor Signal Transduct. Res., 1999. 19: 293–300.

75. Reynolds, J.A. and C. Tanford, Determination of molecular weight of protein moiety in protein–detergent complexes without direct knowledge of detergent binding. Proc. Natl. Acad. Sci. USA, 1976. 73: 4467–4470.

76. Durbin, S.D. and G. Feher, Protein crystallization. Annu. Rev. Phys. Chem., 1996. 47: 171–204.

77. Ferre-D'Ámare, A.R. and S.K. Burely, Dynamic light scattering in evaluating crystallizability of macromolecules. Methods Enzymol., 1997. 276: 157–166.

78. Hitscherich, C.J., B. Aseyev, J. Wiencek, and P.J. Loll, Effects of PEG on detergent micelles: implications for the crystallization of integral membrane proteins. Acta Crystallogr., 2001. D57: 1020–1029.

79. Garcia, C.D., D.J. Hadley, W.W. Wilson, and C.S. Henry, Measuring protein interactions by microchip self-interaction chromatography. Biotechnol. Progr., 2003. 19: 1006–1010.

80. Tessier, P.M., A.M. Lenhoff, and S.I. Sandler, Rapid measurement of protein osmotic second virial coefficients by self-interaction chromatography. Biophys. J., 2002. 82: 1620–1631.

81. Waka, Y., K. Hamamoto, and N. Mataga, Heteroeximer systems in aqueous micellar solutions. Photochem. Photobiol., 1980. 32: 27–35.

82. Ostermeier, C., S. Iwata, B. Ludwig, and H. Michel, Fv fragment-mediated crystallization of the membrane protein bacterial cytochrome C oxidase. Nat. Struct. Biol., 1995. 2: 842–846.

83. Hunte, C. and H. Michel, Crystallisation of membrane proteins mediated by antibody fragments. Curr. Opin. Struct. Biol., 2002. 12: 503–508.

84. Garavito, R.M., D. Picot, and P.J. Loll, Strategies for crystallizing membrane proteins. J. Bioenerg. Biomembr, 1996. 28: 13–27.

85. Kühlbrandt, W., Three-dimensional crystallization of membrane proteins. Q. Rev. Biophys., 1988. 21: 429–477.

86. Ostermeier, C. and H. Michel, Crystallization of membrane proteins. Curr. Opin. Struct. Biol., 1997. 7: 697–701.

87. Reiss-Husson, F., Crystallization of membrane proteins. In: Crystallization of Nucleic Acids and Proteins, A. DuCruix and R. Giege, Eds., 1992. Oxford: IRL Press, pp. 175–193.

88. Faham, S. and J.U. Bowie, Bicelle crystallization: a new method for crystallizing membrane proteins yields a monomeric bacteriorhodopsin structure. J. Mol. Biol., 2002. 316: 1–6.

89. Caffrey, M., A lipid's eye view of membrane protein crystallization in mesophases. Curr. Opin. Struct. Biol., 2000. 10: 486–497.

90. Caffrey, M., Membrane protein crystallization. J. Struct. Biol., 2003. 142: 108–132.

91. Rosenow, M.A., J.C. Williams, and J.P. Allen, Amphiphiles modify the properties of detergent solutions used in crystallization of membrane proteins. Acta Crystallogr., 2001. D57: 925–927.

92. Wiener, M.C., Existing and emergent roles for surfactants in the three-dimensional crystallization of integral membrane proteins. Curr. Opin. Colloid Interface Sci., 2001. 6: 412–419.

93. Schafmeister, C.E., L.J. Miercke, and R.M. Stroud, Structure at 2.5 Å of a designed peptide that maintains solubility of membrane proteins. Science, 1993. 262: 734–738.

94. Gorzelle, B.M., A.K. Hoffman, M.H. Keyes, D.N. Gray, D.G. Ray, and C.R. Sanders, Amphipols can support the activity of a membrane enzyme. J. Am. Chem. Soc., 2002. 124: 11594–11595.

95. Tribet, C., R. Audebert, and J.L. Popot, Amphipols: polymers that keep membrane proteins soluble in aqueous solutions. Proc. Natl. Acad. Sci. USA, 1996. 93: 15047–15050.

96. Yu, S.M., D.T. McQuade, M.A. Quinn, C.P. Hackenberger, M.P. Krebs, A.S. Polans, and S.H. Gellman, An improved tripod amphiphile for membrane protein solubilization. Protein Sci., 2000. 9: 2518–2527.

97. Roth, M., A. Lewit-Bentley, H. Michel, J. Deisenhofer, R. Huber, and D. Oesterhelt, Detergent structure in crystals of a bacterial photosynthetic reaction centre. Nature (London), 1989. 340: 659–662.

98. Seddon, A.M., P. Curnow, and P.J. Booth, Membrane proteins, lipids and detergents: not just a soap opera. Biochim. Biophys. Acta, 2004. 1666: 105–117.

99. Dahout-Gonzalez, C., G. Brandolin, and E. Pebay-Peyroula. Crystalization of the bovine ADP/ATP carrier is critically dependent upon the detergent-to-protein ratio. Acta Crystallogr. D. Biol. Crystallogr., 2003. 59, 2353–2359.

100. Landau, E.M. and J.P. Rosenbusch, Lipidic cubic phases: a novel concept for the crystallization of membrane proteins. Proc. Natl. Acad. Sci. USA, 1996. 93: 14532–14535.

101. Larsson, K., Aqueous dispersions of cubic lipid–water phases. Curr. Opin. Colloid Interface Sci., 2000. 5: 64–69.

102. Rummel, G., A. Hardmeyer, C. Widmer, M.L. Chiu, P. Nollert, K.P. Locher, I.I. Pedruzzi, E.M. Landau, and J.P. Rosenbusch, Lipidic cubic phases: new matrices for the three-dimensional crystallization of membrane proteins. J. Struct. Biol., 1998. 121: 82–91.

103. Nollert, P., J. Novarro, and E.M. Landau, Crystallization of membrane proteins in cubo. Methods Enzymol., 2002. 343: 183–199.

104. Gordeliy, V.I., R. Schlesinger, R. Efremov, G. Buldt, and J. Heberle, Crystallization. In: Lipidic Cubic Phase, in Membrane Protein Protocols: Expression, Purification, and Characterization, B.S. Selinsky, Ed., 2003. Totowa, NJ: Humana Press, pp. 305–316.

105. Chiu, M.L., P. Nollert, M.C. Loewen, H. Belrhali, E. Pebay-Peyroula, and J.P. Rosenbusch, Crystallization *in cubo*: general applicability to membrane proteins. Acta Crystallogr., 2000. D56: 781–784.

106. Gordeliy, V.I., Molecular basis of transmembrane signalling by sensory rhodopsin II-transducer complex. Nature (London), 2002. 419: 484–487.

107. Cherezov, V., J. Clogston, Y. Misquitta, W. Abdel-Gawad, and M. Caffrey, Membrane protein crystallization *in meso*: lipid tailoring of the cubic phase. Biophys. J., 2002. 81: 225–242.

108. Fahrenholz, F., U. Klein, and G. Gimpl, Conversion of the myometrial oxytocin receptor from low to high affinity state by cholesterol. In: Oxytocin, R. Ivell and J. Russell, Eds., 1995. New York: Plenum Press.

109. Vargas, R., L. Mateu, and A. Romero, The effect of increasing concentrations of precipitating salts used to crystallize proteins on the structure of the lipidic Q224 cubic phase. Chem. Phys. Lipids, 2004. 127: 103–111.

110. Misquitta, Y. and M. Caffrey, Detergents destabilize the cubic phase of monoolein: implications for membrane protein crystallization. Biophys. J., 2003. 85: 3084–3096.

111. Cherezov, V., J. Clogston, Y. Misquitta, W. Abdel-Gawad, and M. Caffrey, Membrane protein crystallization *in meso*: lipid type-tailoring of the cubic phase. Biophys. J., 2002. 83: 3393–3407.

112. Cheng, A., B. Hummel, H. Qiu, and M. Caffrey, A simple mechanical mixer for small viscous lipid-containing samples. Chem. Phys. Lipids, 1998. 95: 11–21.

113. Cherezov, V., A. Peddi, L. Muthusubramaniam, Y.F. Zheng, and M. Caffrey, A robotic system for crystallizing membrane and soluble proteins in lipidic mesophases. Acta Crystallogr. D Biol. Crystallogr., 2004. 60: 1795–1807.

114. Misquitta, Y., V. Cherezov, F. Havas, S. Patterson, J.M. Mohan, A.J. Wells, D.J. Hart, and M. Caffrey, Rational design of lipid for membrane protein crystallization. J. Struct. Biol., 2004. 148: 169–175.

115. Sanders, C.R. and R.S. Prosser, Bicelles: a model membrane system for all seasons? Structure, 1998. 6: 1227–1234.

116. Bayburt, T.H., J.W. Carlson, and S.G. Sligar, Reconstitution and imaging of a membrane protein in a nanometer-size phospholipid bilayer. J. Struct. Biol., 1998. 123: 37–44.

117. Bayburt, T.H. and S.G. Sligar, Single-molecule height measurements on microsomal cytochrome P450 in nanometer-scale phospholipid bilayer disks. Proc. Natl. Acad. Sci. USA, 2002. 99: 6725–6730.

118. Bayburt, T.H. and S.G. Sligar, Self-assembly of single integral membrane proteins into soluble nanoscale phospholipid bilayers. Protein Sci., 2003. 12: 2476–2481.

119. Denisov, I.G., Y.V. Grinkova, A.A. Lazarides, and S.G. Sligar, Directed self-assembly of monodisperse phospholipid bilayer nanodiscs with controlled size. J. Am. Chem. Soc., 2004. 126: 3477–3487.

120. Odahara, T., Stability and solubility of integral membrane proteins from photosynthetic bacteria solubilized in different detergents. Biochim. Biophys. Acta, 2004. 1660: 80–92.

121. Barends, T.R. and B.W. Dijkstra, Oils used in microbatch crystallization do not remove a detergent from the drops they cover. Acta Crystallogr., 2003. D59: 2345–2347.

122. Song, L. and E. Gouaux, Membrane protein crystallization: application of sparse matrices to the α-hemolysin heptamer. Methods Enzymol., 1997. 276: 60–74.

123. Loll, P.J., M. Allaman, and J. Wiencek, Assessing the role of detergent–detergent interactions in membrane protein crystallization. J. Cryst. Growth, 2001. 232: 432–438.

124. Wiener, M.C. and C.F. Snook, The development of membrane protein crystallization screens based upon detergent solution properties. J. Cryst. Growth, 2001. 232: 426–431.

125. Kawata, A., T. Oishi, Y. Fukada, Y. Shichida, and T. Yoshizawa, Photoreceptor cell types in the retina of various vertebrate species: immunocytochemistry with antibodies against rhodopsin and iodopsin. Photochem. Photobiol., 1992. 56: 1157–1166.

126. Okada, T., I. Le Trong, B.A. Fox, C.A. Behnke, R.E. Stenkamp, and K. Palczewski, X-ray diffraction analysis of three-dimensional crystals of bovine rhodopsin obtained from mixed micelles. J. Struct. Biol., 2000. 130: 73–80.

127. Palczewski, K., T. Kumasaka, T. Hori, C.A. Behnke, H. Motoshima, B.A. Fox, I. Le Trong, D.C. Teller, T. Okada, R.E. Stenkamp, M. Yamamoto, and M. Miyano, Crystal structure of rhodopsin: a G protein-coupled receptor. Science, 2000. 289: 739–745.

128. Sakmar, T.P., Opsins. In: Handbook of Receptors and Channels: G Protein Coupled Receptors, S.J. Peroutka, Ed., 1994. Boca Raton, FL: CRC Press.

129. Strader, C.D., T.M. Foing, M.R. Tota, D. Underwood, and R.A. Dixon, Structure and function of G protein-coupled receptors. Annu. Rev. Biochem., 1994. 63: 101–132.

130. Papermaster, D.S., Preparation of retinal rod outer segments. Methods Enzymol., 1982. 81: 48–52.

131. Nemes, P.P. and E.A. Dratz, Preparation of isolated osmotically intact bovine rod outer segment disk membranes. Methods Enzymol., 1982. 81: 116–123.

132. Fong, S.L., A.T.C. Tsin, C.D.B. Bridges, and G.I. Lian, Detergents for extraction of visual pigments: types, solubilization, and stability. Methods Enzymol., 1982. 81: 133–140.

133. Okada, T., K. Takeda, and T. Kouyama, Highly selective separation of rhodopsin from bovine rod outer segment membranes using combination of divalent cation and alkyl(thio)glucoside. Photochem. Photobiol., 1998. 67: 495–499.

134. Schertler, G.F., C. Villa, and R. Henderson, Projection structure of rhodopsin. Nature, 1993. 362: 770–772.

135. Mielke, T., C. Villa, P.C. Edwards, G.F. Schertler, and M.P. Heyn, X-ray diffraction of heavy-atom labelled two-dimensional crystals of rhodopsin identifies the position of cysteine 140 in helix 3 and cysteine 316 in helix 8. J. Mol. Biol., 2002. 316: 693–709.

136. Mielke, T., U. Alexiev, M. Glasel, H. Otto, and M.P. Heyn, Light-induced changes in the structure and accessibility of the cytoplasmic loops of rhodopsin in the activated MII state. Biochemistry, 2002. 41: 7875–7884.

137. Teller, D.C., T. Okada, C.A. Behnke, K. Palczewski, and R.E. Stenkamp, Advances in determination of a high-resolution three-dimensional structure of rhodopsin, a model of G-protein-coupled receptors (GPCRs). Biochemistry, 2001. 40: 7761–7772.

138. Okada, T., A. Terakita, and Y. Shichida, Structure–function relationship in G protein-coupled receptors deduced from crystal structure of rhodopsin, Tanpak. Kakusan Koso, 2002. 47: 1123–1130.

139. Okada, T., M. Sugihara, A.N. Bondar, M. Elstner, P. Entel, and V. Buss, The retinal conformation and its environment in rhodopsin in light of a new 2.2 Å crystal structure. J. Mol. Biol., 2004. 342: 571–583.

140. Seifert, R., K. Wenzel-Seifert, U. Gether, and B.K. Kobilka, Functional differences between full and partial agonists: evidence for ligand-specific receptor conformations. J. Pharmacol. Exp. Ther., 2001. 297: 1218–1226.

141. Gether, U. and B.K. Kobilka, G protein-coupled receptors. II. Mechanism of agonist activation. J. Biol. Chem., 1998. 273: 17979–17982.

142. Bockaert, J. and J.P. Pin, Molecular tinkering of G protein-coupled receptors: an evolutionary success. EMBO J., 1999. 18: 1723–1729.

143. Yin, D., S. Gavi, H.Y. Wang, and C.C. Malbon, Probing receptor structure/function with chimeric GPCRs. Mol. Pharmacol., 2004. 65: 1323–1332.

144. Kobilka, B.K., T.S. Kobilka, K. Daniel, J.W. Regan, M.G. Caron, and R.J. Lefkowitz, Chimeric alpha 2–beta 2 adrenergic receptors: delineation of domains involved in effector coupling and ligand binding specificity. Science, 1988. 240: 1310–1316.

145. Kubo, T., H. Bujo, I. Akiba, J. Nakai, M. Mishina, and S. Numa, Location of a region of the muscarinic acetylcholine receptor involved in selective effector coupling. FEBS Lett., 1988. 241: 119–125.

146. Bowie, J.U., www.doe-mbi.ucla.edu/~salem/bicelle_method_faq.html. 2005.

147. Avanti Lipids, www.avantilipids.com/BicellePreparation.asp. 2005.

148. Caffrey, M., www.ldb.chemistry.ohio-state.edu. 2005.

149. Rosenbusch, J.P., A. Lustig, M. Grabo, M. Zulauf, and M. Regenass, Approaches to determining membrane protein structures to high resolution: do selections of subpopulations occur? Micron, 2001. 32: 75–90.

13 Novel Solid-State NMR Methods for Structural Studies on G Protein-Coupled Receptors

Adam Lange and Marc Baldus

CONTENTS

ABSTRACT

The structural characterization of G protein-coupled receptors (GPCRs) and their ligands in a natural membrane environment can be challenging to achieve with standard techniques such as x-ray crystallography and solution state nuclear magnetic resonance (NMR). Instead, NMR techniques specifically designed for the study of slowly tumbling or solid-phase systems (solid-state NMR) can offer unique possibilities for elucidating structural or dynamic parameters at atomic resolution. A series of experiments using technical advancements and tailored to retrieving multiple structural parameters and even complete three-dimensional molecular structures from a single isotope-labeled sample are described. Currently available methods for obtaining such samples in the context of ssNMR-based structural studies on GPCRs will be reviewed. Even if the receptor protein cannot be obtained in isotope-labeled form, high resolution solid-state NMR methods can offer unique possibilities to study ligand binding to GPCRs. These methods are applicable to the neurotensin–NTS-1 system studied in detail and recent developments in the area of high resolution ssNMR to study structural aspects of GPCRs will be presented.

13.1 INTRODUCTION

Many chemical and biophysical processes involve molecules that are insoluble or noncrystalline. Consider, for instance, the perception of light and smell mediated by G protein-coupled receptors (GPCRs) or the formation of protein aggregates in the context of Alzheimer's disease. In both cases, the concerned molecules are partially or fully immobilized and often resist structural analysis by standard techniques such as x-ray crystallography and solution-state nuclear magnetic resonance (NMR). Instead, NMR techniques specifically designed for the study of slowly tumbling or solid phase systems (solid-state NMR or ssNMR) can offer unique possibilities to elucidate structural or dynamic parameters at atomic resolution.

For more than three decades, ssNMR has been applied to the structural studies of membranes and associated proteins. Unlike in solution, the spectral resolution and overall sensitivity of ssNMR are influenced by the size and orientation dependence of the nuclear spin interactions, i.e., the chemical shielding and the homo- and heteronuclear dipolar spin–spin couplings. As a result, spectral resolution and overall sensitivity can be compromised. Earlier biophysical applications of solid-state NMR concentrated on the determination of *individual* structural and dynamical parameters in insoluble and noncrystalline systems. Combined with site-specific single or pairwise isotope labeling, powerful applications ranging from the determination of intermolecular distances and torsion angles in membrane proteins and their natural ligands[1–3] to the determination of a complete three-dimensional (3D) structure of a membrane peptide[4] were reported. In addition to rhodopsin and other retinal proteins, a variety of membrane-embedded systems have been the subjects of ssNMR-based structural studies, often involving macroscopically oriented systems.[5,6] Interested readers are referred to a series of review articles[5–9] for further information.

Recent progress in NMR instrumentation, in particular the combination of ultrahigh magnetic fields (i.e., above 14 Tesla) with rapid sample rotation about the "magic" angle (magic angle spinning, MAS),[10] where the size and orientation dependence of the nuclear spin interactions is minimized, have considerably improved the use of ssNMR in biophysical applications. Such conditions have not permitted widespread application of high-resolution ^{1}H NMR spectroscopy — the most sensitive detection method in solution — but have greatly extended the possibilities for studying multiply or uniformly ^{13}C,^{15}N-isotope-labeled biomolecules at high sensitivity and adequate spectral resolution.

In this chapter, we will discuss a series of experiments that make use of these technical advancements and are tailored to retrieving *multiple* structural parameters or even the complete 3D molecular structure from a single, isotope-labeled sample. Currently available methods for obtaining such samples in the context of ssNMR-based structural studies on GPCRs will be reviewed. Even if the receptor protein cannot be obtained in isotope-labeled form, high resolution solid-state NMR methods can offer unique possibilities to study ligand binding to GPCRs. Some of these methods will be applied to the neurotensin–NTS-1 system that we have been investigating in collaboration with Dr. Grisshammer and co-workers at the National Institutes of Health (NIH). If membrane proteins can be obtained in isotope-labeled forms, such studies can also be extended to infer information about the protein structure itself. Finally, recent developments that could further expand the utility of high-resolution ssNMR for studying structural aspects of GPCRs will be discussed.

13.2 HIGH-RESOLUTION SOLID-STATE NMR

The spectral resolution and overall sensitivity of solid-state NMR spectra are influenced by the size and orientational dependence of the nuclear spin interactions. As a result, the NMR spectra of randomly oriented molecules are usually broadened extensively, making the application of MAS mandatory. Combined with ultrahigh magnetic fields and radio frequency (r.f.)-based NMR decoupling schemes, high-resolution ssNMR (HR-ssNMR) conditions can be established (Figure 13.1). Much of our current research[11] involves the design of multidimensional correlation experiments that employ through-space (dipolar, D), through-bond (J), or relaxation-mediated (R) spin–spin transfer to study molecular structures under HR-ssNMR conditions.

For example, we can exploit the exquisite sensitivity of the NMR resonance frequency to local (protein backbone or hydrogen bonding) or overall structure (in the case of macroscopically oriented samples). See Figure 13.2. As shown in Sections 13.4.1 and 13.4.2, these conformation-dependent chemical shifts allow inference of the secondary structure of a peptide ligand bound in high affinity to a GPCR and also provide new routes to study conformation ensembles at the levels of individual residues.

We can make use of a variety of spin–spin interactions that are directly related to an internuclear distance (Figure 13.2). In contrast to selectively ($^{13}C,^{13}C$ or $^{13}C,^{15}N$) labeled molecules, the spin dynamics in a uniformly labeled protein are dominated by one-bond interactions that can provide the bases for spectral assignments[12] and information about local dihedral angles,[13–16] but are of limited use for establishing 3D distance constraints. Such interactions can be probed by employing chemical shift selective transfer methods[17–20] or chemical dilution.[21–23] In fact, studying different "block labeled" protein variants can lead to the 3D structure of a solid-phase protein,[23] but requires the use of several protein variants, in particular a uniformly labeled sample that most easily provides a route to obtaining $^{13}C,^{15}N$ resonance assignments.[24]

The derivation of 3D structures by solution-state NMR has relied greatly on the detection of $^1H,^1H$ contacts that provide short-, medium-, and long-range constraints for a sample of interest.[25,26] Because of the limited spectral resolution of ^1H ssNMR data, we have begun designing correlation methods that *indirectly* encode $^1H,^1H$ or $^1H,X$ (where X may designate ^{13}C or ^{15}N nuclei) contacts in high spectral resolution. Because evolution and detection periods involve ^{13}C or ^{15}N nuclei, such pulse schemes may be discriminated using the acronyms NHC, CHC,[27] NHHC, and CHHC.[28,29] Recent results obtained on globular proteins indicate that such methods may become of similar value in solid-state NMR to determine 3D structures[28,29] or probe intermolecular interactions[30–32] in high resolution.

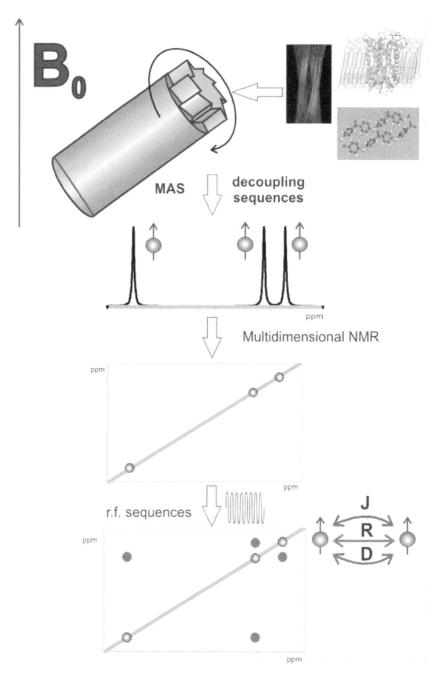

FIGURE 13.1 Basic tools to study molecular structures using high-resolution solid-state NMR. The sample of interest is measured under MAS conditions using multidimensional correlation spectroscopy and radio-frequency pulse schemes that enhance resolution (decoupling schemes) or establish structurally relevant correlations.

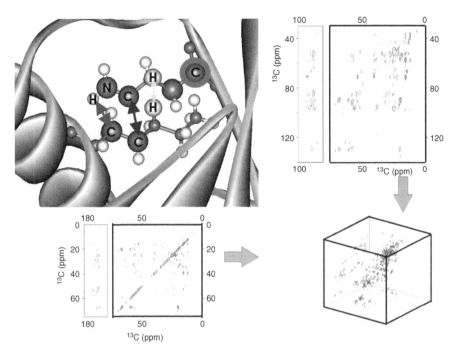

FIGURE 13.2 Structural parameters of an HR-ssNMR experiment: resonance frequencies (such as encircled for the carboxyl carbon) are sensitive to local and overall structure. Two-spin interactions (indicated by arrows) permit determination of $(^{13}C,^{13}C)$, $(^{1}H,^{1}H)$ and $(^{1}H,^{13}C/^{15}N)$ contacts in two or three spectral dimensions.

In general, two-spin interactions are most easily probed by 2D correlation spectroscopy.[33] In the case of spectral overlap, extending the experiment to three or more spectral dimensions is possible at the expense of a longer experimental time. For 3D spectroscopy (see Figure 13.2), various solution-state NMR methods for faster acquisition of multidimensional spectra that circumvent the sampling of the full t_1,t_2 space have been developed.[34] Reduced dimensionality experiments[35] have been applied in solid-state MAS NMR spectroscopy[36–38] and nonlinear sampling.[39] However, due to the low sensitivity of many ssNMR experiments, the main requirement for a multidimensional sampling scheme is not the reduction of the total number of scans, but rather the maximization of the average signal-to-noise ratio per scan without sacrificing resolution. This requirement can be fulfilled conveniently by a triangular sampling protocol suggested earlier for solution-state NMR[40] that reduces the 3D NMR measurement by a factor 1–2 and can be readily incorporated into CCHHC or NHHCC experiments that provide spectral assignment and $^{1}H,^{1}H$ distance constraints within one 3D experiment.[41]

13.3 SAMPLE PREPARATION

In order to investigate structural or dynamical properties of GPCRs and GPCR–ligand complexes by ssNMR spectroscopy, isotopic labeling is usually mandatory. For example, partial or uniform $^{13}C, ^{15}N$ labeling of a GPCR agonist or antagonist allows for the detection of the ligand signal in high sensitivity and spectral resolution while the background signal resulting from natural abundance signal contributions (GPCRs, detergents, lipids, buffer components) can be strongly suppressed. The natural abundance of the NMR active nucleus ^{13}C is about 1%. Thus, 100 nonlabeled sites and one labeled site contribute to comparable signal intensity in a 1D cross-polarization spectrum.[42,43]

In detergent-solubilized or reconstituted GPCR samples, the molar ratio between detergents and lipids, respectively, on the one hand and the GPCR–ligand on the other hand is often larger than 100. Hence, 2D or double-quantum filtering techniques (2QF)[44,45] must be applied. In the resulting spectra, one labeled site contributes to as much signal intensity as 10,000 (^{13}C-^{13}C correlation spectroscopy, ^{13}C double-quantum filtering), or even 26,500 (^{13}C-^{15}N correlation spectroscopy) nonlabeled sites. These multiple-pulse experiments are based on an efficient reintroduction of the dipolar coupling that is averaged to zero under high-speed MAS conditions.

Through-space polarization transfer is not possible if the molecules are subject to isotropic tumbling or exhibit high mobility, as is the case in aqueous solutions at room temperature. Similar to the solution state, through-bond transfer schemes could be applied.[46,47] Another way to circumvent this problem is to conduct experiments at low temperatures (≤ 200 K) well below the freezing point of the solution, making a low-temperature NMR set-up necessary.

Working at low temperatures has two advantages. First, it leads to a higher sensitivity due to the changed Boltzmann distribution of the spins among the two possible spin states. Second, biological samples are electrically lossy and interact with the r.f. electric field inside the sample coil. As a result, overheating and finally degradation of the sample can occur. Such problems are minimized by conducting measurements below the glass transition of the sample. In addition, working at low temperatures is of paramount importance if loss of receptor activity and change in the aggregation state (e.g., dissociation of the bound ligand) over an extended period is to be prevented.

It has been demonstrated[48] (and will be discussed in detail below) that the application of state-of-the-art ssNMR technology permits the spectroscopic study of 25 nmol of labeled material, for example, microgram quantities of a labeled GPCR agonist and milligram quantities of the corresponding nonlabeled purified receptor. Expression and purification of milligram quantities of functional GPCRs for large-scale crystallization experiments have been demonstrated in several cases.[49–52] Examples include the N-terminally truncated form of rat NTS-1 (levocabastine-insensitive neurotensin type I receptor)[53–55] expressed as a maltose-binding protein fusion in *Escherichia coli*,[56] the truncated, nonglycosylated turkey β-adrenergic receptor expressed in insect cells using recombinant baculoviruses,[57] and the D6 chemokine receptor obtained from mammalian cells.[58]

In all cases the GPCRs must be solubilized from cell membranes and purified in the presence of detergents. Exceptions are rhodopsin and bacteriorhodopsin, which are highly abundant in membranes isolated from fresh cattle eyes and in the purple patches of the plasma membranes of *Halobacterium salinarium*, respectively. In these cases, the isolated and purified membranes can be directly centrifuged into zirconia MAS rotors, frozen, and stored at 190 K.[59,60] Furthermore,

GPCRs can be overexpressed in inclusion bodies in *E. coli*, purified, and refolded into active conformations.[61]

A straightforward approach to studying solubilized GPCRs is the preparation of frozen solutions. The receptor is highly concentrated (>10 mg/ml) and, if necessary, incubated with its agonist or antagonist. Subsequently, the solution is transferred to a rotor, and NMR measurements are performed on frozen samples. Unlike the rhodopsin preparation, this approach is limited by the small volume of a typical 4-mm zirconia MAS rotor, i.e., 50 to 100 µL. Sophisticated ssNMR experiments that can be employed to determine the 3D conformation of a protein at atomic resolution are by a factor of 1.5 to 3 less sensitive than experiments used for NMR resonance assignments[28,32] at present. Hence, their performance requires samples containing at least 60 nmol of labeled material. This can be challenging to achieve with detergent-solubilized GPCR samples, where an increase in concentration of the solubilized receptor leads to an excess of detergent micelles. The resulting high-detergent concentration can seriously destabilize the GPCRs because the detergents compete with both lipid–lipid and lipid–protein and also with protein–protein interactions.[62]

Several groups have shown that solubilized and purified GPCRs can be functionally reconstituted into lipid vesicles. A frequently used strategy for proteoliposome preparation involves saturation of preformed liposomal vesicles with detergents, incubation with the solubilized membrane protein, and detergent removal by various methods, depending on the physicochemical properties of the detergent.[63,64] Applications to GPCRs include the functional reconstitution of the N-terminally truncated form of rat NTS-1[48] and the co-reconstitution of the HIV receptor CCR5 with its co-receptor CD4.[65] The detergent was removed with hydrophobic polystyrene beads. A different approach is based on selective extraction of detergents from mixed detergent–lipid–protein micelles using cyclodextrin inclusion compounds. This method has been successfully applied to the functional reconstitution of bovine rhodopsin[66] and the human histamine H_1 receptor.[67] For ssNMR studies, the recovered proteoliposomes are incubated with ligand and can subsequently be centrifuged into zirconia MAS rotors for ssNMR experiments.

13.4 APPLICATIONS

13.4.1 Receptor-Bound Conformation of Neurotensin

Neurotensin (NT) is a 13-amino acid peptide[68] involved in a variety of neuromodulatory functions in the central and peripheral nervous systems.[69] NT binds to its NTS-1 and NTS-2 GPCRs.[53–55,70] NTS-1 interacts with its agonist NT with high (sub-nanomolar) affinity. Similar observations were made for the N-terminally truncated form of rat NTS-1 when expressed as a maltose-binding protein fusion in *E. coli*[71,71] and purified in the presence of detergents.[56,71] Notably, both the full-length peptide and also the C-terminal part of NT(8–13) have been found to interact with NTS-1 with high affinity (see References 54 and 73).

As described in Section 13.3, HR-ssNMR 2QF permits the study of NT(8–13) (1 kDa)–NTS-1 (101 kDa) binding in a detergent–lipid environment. As shown previously by ssNMR experiments and *ab initio* calculations,[74] the resulting resonance frequencies are both diagnostic for each individual peptide residue[75,76] and are also very sensitive to polypeptide backbone conformation.[75,77–82]

To elucidate the backbone conformation of neurotensin in complex with its receptor, we[48] conducted a series of 2QF 2D correlation experiments to identify the peptide-bound signal and correlate the ssNMR spectral assignments with backbone structures. The resulting fold appears in Figure 13.3 and suggests a β-strand conformation of the agonist bound to NTS-1. The current analysis does not yet permit the refinement of the side chain conformations of NT(8–13). For this reason, all side chains shown in Figure 13.3 are shaded and only included for reference. As noted

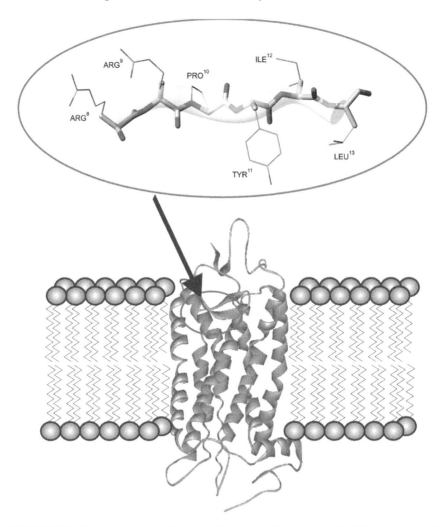

FIGURE 13.3 Representative backbone conformation of neurotensin(8–13) bound to its G protein coupled receptor NTS-1 (of currently unknown structure, model: rhodopsin[99]) as determined by solid state NMR.[48] The indicated binding site results from mutagenesis experiments.

earlier, ssNMR methods are now available to determine entire 3D structures and will be applied in future ssNMR studies on the NT-NTS1 system.

Published molecular modeling studies[83–85] of NT bound to the NTS-1 receptor revealed conflicting information about the presumed structure of the bound agonist. Pang et al.[83] predicted a compact conformation of NT(8–13), with a proline type I turn for the segment Arg^9–Pro^{10}–Tyr^{11}–Ile^{12}. The corresponding backbone angles (φ,ψ) for Pro^{10} and Tyr^{11} would be given by –60°, –30° and –90°, 0°, respectively, contradicting the ssNMR data presented here. More recently, mutagenesis and structure–activity studies combined with modeling techniques were utilized to predict the receptor binding site and conformation of bound NT(8–13).[84,85] In this model, NT(8-13) adopted a linear backbone conformation in qualitative agreement with our ssNMR results.

Further evidence that the ssNMR-derived structure represents the bioactive conformation of neurotensin comes from a recent report by Bittermann and coworkers[86] who rigidized NT by chemical synthesis (using 4,4 spirolactam) around the Pro^{10} backbone angle to (A) $\psi = 120°$ and (B) $\psi = -120°$. Whereas compound A exhibiting a dihedral angle close to the value predicted from the 2D ssNMR data revealed a nanomolar binding affinity to the NT receptor, compound B is characterized by a binding affinity three orders of magnitude lower.

13.4.2 ELUCIDATING BINDING PATHWAY OF NEUROTENSIN

For ligand-based drug design, the elucidation of the GPCR-bound conformation and also the optimization of the ligand membrane permeability and its stability against proteolytic degradation are of critical relevance.[87] Additional information relating the receptor-bound ligand structure to the conformation of free ligand in solution can hence be highly desirable. Previous solution-state NMR studies of neurotensin in aqueous, methanol, and SDS solutions[88] indicated an inherent conformational flexibility with no discernible elements of secondary structure in water and methanol. In the following section, we explain how ssNMR can be used to infer the conformational distribution of free NT(8-13) in different chemical solvents at the levels of individual residues.[89]

We again make use of the inherent sensitivity of the isotropic chemical shift to molecular structures detected under MAS. Because the chemical shift depends on residue type[75,76] and backbone conformation[75,80,90] and is also influenced by the natures and structural topologies of the neighboring residues, we developed a strategy that adequately samples the allowed conformational space for a given polypeptide, predicts the backbone chemical shifts $C\alpha$ and $C\beta$ using standard routines, and finally applies a selection criterion based on results of a 2D ssNMR experiment. Each peptide conformation gives rise to one distinct cross-peak position for each residue that reflects a particular backbone conformation. Conformational heterogeneity hence translates into 2D cross-peak patterns that provide spectroscopic snapshots of all backbone conformations present in the polypeptide of interest.

The validity of this concept was successfully tested[89] on a globular protein (ubiquitin) for which solid-state NMR resonance assignments and high-resolution x-ray and NMR structures were available. The concept was applied subsequently to

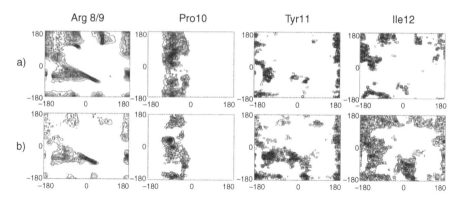

FIGURE 13.4 Ramachandran plot revealing backbone angles predicted from a 2D ssNMR analysis[89] of NT(8–13) in (a) a flash frozen water solution or (b) in model lipids. Backbone angle pairs (ψ,ϕ) consistent with the ssNMR data lead to a nonzero entry in the (ψ,ϕ) data matrix.

studying the conformational space sampled by NT(8–13) in an aqueous solution by analysis of ssNMR 2D data obtained after rapid freezing to –50°C and comparison to molecular dynamics (MD) simulations. Although discrepancies between both approaches and among MD simulations using different force fields were detected, a remarkable overall agreement observed suggests that the conformational ensemble characterizing NT(8–13) in H_2O contains β-strand dihedral angle distributions for Arg[8], Arg[9], Ile[12], and Pro[10] residues and can adopt helical torsion angles for Tyr[11] (Figure 13.4, Ramachandran plots consistent with 2D ssNMR data).

The same ssNMR-based analysis was subsequently employed to monitor the torsion angle space adopted by NT(8–13) in model lipids in which the solubilized receptor was shown to be functional. Compared to the situation in H_2O, residue types characterized by a positive free energy difference for a transition from an aqueous to a lipid environment exhibit remarkable reductions of the available (ϕ,ψ) space. On the other hand, hydrophobic residues sample a larger segment of the Ramachandran backbone angle maps in lipids. For reasons mentioned earlier, the corresponding torsion angle distributions may provide valuable structural information during structure-based drug design.

13.4.3 ADDITIONAL APPLICATIONS TO MEMBRANE PEPTIDES AND PROTEINS

As noted earlier, isotope labeling generally allows extension of an HR-ssNMR study to the receptor, as demonstrated in selectively labeled versions of bacteriorhodopsin[91] or bovine rhodopsin.[92] As part of an initial test application, we applied the $^{13}C,^{15}N$ resonance assignment methods described in Section 13.2 to uniformly labeled $^{13}C,^{15}N$ samples of the LH2 light harvesting complex.[93]

These HR-ssNMR experiments were conducted at –10°C and confirmed[80] the α-helical arrangements of the α and β apoproteins observed earlier in the x-ray structure.[94] Such studies are not restricted to low-temperature experiments (Figure

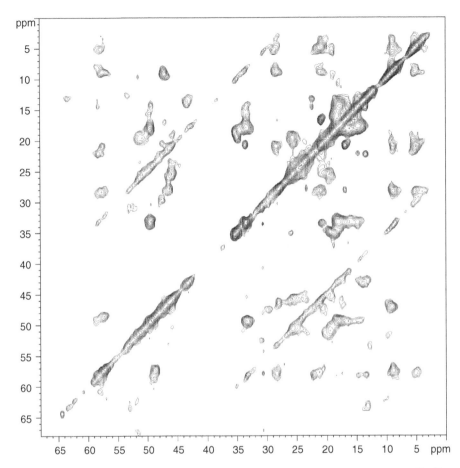

FIGURE 13.5 ^{13}C,[13]C correlation experiment on proteoliposomes containing a U-^{13}C,[15]N labeled mutant of phospholamban. The spectrum was recorded on a narrow-bore 800-MHz instrument at a sample temperature of 35°C using an MAS rate of 11 kHz (unpublished results). The protein sequence differs from the wild-type version of phospholamban by the three trans-membrane cysteines that have been replaced by A36, F41, and A46.

13.5). 2D correlation data were recorded above the liquid–crystalline to gel-phase transition for a lipid-reconstituted mutant of phospholamban, a 52-aa protein known to interact with Ca[2+]-ATPase.[95] Notably, the 2D experiment was conducted on a narrow-bore 800-MHz instrument using a custom-designed triple channel (^{1}H,^{13}C,[15]N) MAS NMR probe head (Bruker Biospin, Rheinstetten, Germany).

13.5 CONCLUSIONS AND OUTLOOK

As demonstrated here for the case of the neurotensin-NTS-1 system, considerable progress has been made in the use of HR-ssNMR to elucidate the bioactive confor-mation of a GPCR ligand. Similar approaches could be used to obtain key structure restraints and the complete 3D structure of (non)-peptidic ligands and study conformational ensembles at atomic resolution. A suite of ssNMR methods is now available to probe entire 3D molecular structures and the applicability of these methods is determined primarily by protein availability. As shown in the neurotensin case, functional protein quantities of a few milligrams will be sufficient for many such structural studies.

Ligand–GPCR interactions can be analyzed further by approaches tailored to probe molecular interfaces.[32] Such methods can be applied readily to studies of heterogeneously labeled receptor-ligand complexes. In combination with mutagenesis data, the 3D arrangement within the ligand binding pocket could be probed in detail without the necessity for a complete NMR spectral assignment of the receptor signals.

As demonstrated for globular proteins,[24,96,97] spectral resolution may be enhanced by studying nano- or microcrystalline material that may not allow for crystallographic analysis but is sufficient for HR-ssNMR studies as described. In general, HR-ssNMR investigations are not limited to highly purified protein samples. In addition, the methods presented are readily applicable to protein samples precipitated in functional form. Advances related to sample preparation (for example, including modular labeling, *in vitro* expression, and intein technology[98]) and improvements in NMR instrumentation will further expand the tractable protein size from recent studies involving protein dimers of about 2*85 aa[32,97] to the complete 3D characterization of larger (>300 aa) membrane proteins.

Solid-state NMR studies of multiply labeled biomolecules will benefit from improved procedures for calculating 3D structures, in particular in the presence of ambiguous or limited numbers of structural constraints. Unlike x-ray crystallography, protein motion does not hinder solid-state NMR methods. In fact, complementarily to solution-state NMR, it may provide a very efficient means to study protein folding, flexibility, and function under biologically relevant conditions. Hand in hand with solution-state techniques and crystallographic methods, solid-state NMR may provide insight into protein function and the chemistry of life with unprecedented accuracy and flexibility.

ACKNOWLEDGMENTS

We thank our collaborators and group members who contributed to the studies described in this chapter. Work on the NTS-1 system was conducted in close collaboration with Dr. Reinhard Grisshammer and co-workers at the National Institutes of Health. Financial support by the Max-Planck-Gesellschaft and through a Fonds der Chemischen Industrie fellowship to A.L. is gratefully acknowledged.

REFERENCES

1. F. Creuzet, A. McDermott, R. Gebhard, K. Vanderhoef, M.B. Spijkerassink, J. Herzfeld, J. Lugtenburg, M.H. Levitt, R.G. Griffin, Determination of membrane protein structure by rotational resonance NMR: bacteriorhodopsin. Science 251 (1991) 783–786.

2. M. Han, S.O. Smith, NMR constraints on the location of the retinal chromophore in rhodopsin and barthorhodopsin. Biochemistry 34 (1995) 1425–1432.

3. X. Feng, P.J.E. Verdegem, Y.K. Lee, D. Sandstrom, M. Eden, P. Bovee-Geurts, W.J. deGrip, J. Lugtenburg, H.J.M. deGroot, M.H. Levitt, Direct determination of a molecular torsional angle in the membrane protein rhodopsin by solid-state NMR. J. Am. Chem. Soc. 119 (1997) 6853–6857.

4. R.R. Ketchem, W. Hu, T.A. Cross, High-resolution conformation of gramicidin A in a lipid bilayer by solid-state NMR. Science 261 (1993) 1457–1460.

5. T.A. Cross, S.J. Opella, Solid-state NNR structural studies of peptides and proteins in membranes. Curr. Opin. Struct. Biol. 4 (1994) 574–581.

6. F.M. Marassi, S.J. Opella, NMR structural studies of membrane proteins. Curr. Opin. Struct. Biol. 8 (1998) 640–648.

7. M. Engelhard, B. Bechinger, Application of NMR spectroscopy to retinal proteins. Isr. J. Chem. 35 (1995) 273–288.

8. M. Eilers, W.W. Ying, P.J. Reeves, H.G. Khorana, S.O. Smith, G protein pathways, part A. Receptors 343 (2002), 212–222.

9. J.H. Davis, M. Auger, Static and magic angle spinning NMR of membrane peptides and proteins. Prog. Nucl. Magn. Reson. Spectrosc. 35 (1999) 1–84.

10. E.R. Andrew, A. Bradbury, R.G. Eades, Nuclear magnetic resonance spectra from a crystal rotated at high speed. Nature 182 (1958) 1659.

11. S. Luca, H. Heise, M. Baldus, High-resolution solid-state NMR applied to polypeptides and membrane proteins. Accounts Chem. Res. 36 (2003) 858–865.

12. M. Baldus, Correlation experiments for assignment and structure elucidation of immobilized polypeptides under magic angle spinning. Prog. Nucl. Magn. Reson. Spectrosc. 41 (2002) 1–47.

13. Y. Ishii, T. Terao, M. Kainosho, Relayed anisotropy correlation NMR: determination of dihedral angles in solids. Chem. Phys. Lett. 256 (1996) 133–140.

14. X. Feng, Y.K. Lee, D. Sandstrom, M. Eden, H. Maisel, A. Sebald, M.H. Levitt, Direct determination of a molecular torsional angle by solid-state NMR. Chem. Phys. Lett. 257 (1996) 314–320.

15. P.R. Costa, J.D. Gross, M. Hong, R.G. Griffin, Solid-state NMR measurement of psi in peptides: a NCCN 2Q-heteronuclear local field experiment. Chem. Phys. Lett. 280 (1997) 95–103.

16. M. Hong, J.D. Gross, R.G. Griffin, Site-resolved determination of peptide torsion angle phi from the relative orientations of backbone N-H and C-H bonds by solid-state NMR. J. Phys. Chem. B 101 (1997) 5869–5874.

17. D.P. Raleigh, M.H. Levitt, R.G. Griffin, Rotational resonance in solid state NMR. Chem. Phys. Lett. 146 (1988) 71–76.

18. K. Takegoshi, K. Nomura, T. Terao, Rotational resonance in the tilted rotating frame. Chem. Phys. Lett. 232 (1995) 424–428.

19. P.R. Costa, B.Q. Sun, R.G. Griffin, Rotational resonance tickling: accurate internuclear distance measurement in solids. J. Am. Chem. Soc. 119 (1997) 10821–10830.

20. L. Sonnenberg, S. Luca, M. Baldus, Multiple-spin analysis of (^{13}C,^{13}C) chemical-shift selective transfer in uniformly labeled biomolecules. J. Magn. Reson. 166 (2004) 100–110.

21. D.M. LeMaster, D.M. Kushlan, Dynamical mapping of E. coli thioredoxin via C-13 NMR relaxation analysis. J. Am. Chem. Soc. 118 (1996) 9255–9264.

22. M. Hong, K. Jakes, Selective and extensive C-13 labeling of a membrane protein for solid state NMR investigations. J. Biomol. NMR 14 (1999) 71–74.

23. F. Castellani, B. van Rossum, A. Diehl, M. Schubert, K. Rehbein, H. Oschkinat, Structure of a protein determined by solid-state magic-angle-spinning NMR spectroscopy. Nature 420 (2002) 98–102.

24. J. Pauli, M. Baldus, B. van Rossum, H. de Groot, H. Oschkinat, Backbone and side-chain C-13 and N-15 signal assignments of the α-spectrin SH3 domain by magic angle spinning solid-state NMR at 17.6 Tesla. Chembiochem 2 (2001) 272–281.

25. K. Wüthrich, NMR of Proteins and Nucleic Acids, Wiley Interscience, New York (1986).

26. K. Wüthrich, The way to NMR structures of proteins. Nat. Struct. Biol. 8 (2001) 923–925.

27. A. Lange, ^{1}H NMR spectroscopy in the solid-state: methods and biomolecular applications, Diploma thesis, University of Göttingen, 2002.

28. A. Lange, S. Luca, M. Baldus, Structural constraints from proton-mediated rare-spin correlation spectroscopy in rotating solids. J. Am. Chem. Soc. 124 (2002) 9704–9705.

29. A. Lange, K. Seidel, L. Verdier, S. Luca, M. Baldus, Analysis of proton–proton transfer dynamics in rotating solids and their use for 3D structure determination. J. Am. Chem. Soc. 125 (2003) 12640–12648.

30. I. de Boer, L. Bosman, J. Raap, H. Oschkinat, H.J.M. de Groot, 2D(13)C-C-13 MAS NMR correlation spectroscopy with mixing by true H-1 spin diffusion reveals long-range intermolecular distance restraints in ultra high magnetic field. J. Magn. Reson. 157 (2002) 286–291.

31. R. Tycko, Y. Ishii, Constraints on supramolecular structure in amyloid fibrils from two-dimensional solid-state NMR spectroscopy with uniform isotopic labeling. J. Am. Chem. Soc. 125 (2003) 6606–6607.

32. M. Etzkorn, A. Böckmann, A. Lange, M. Baldus, Probing molecular interfaces using 2D magic-angle-spinning NMR on protein mixtures with different uniform labeling. J. Am. Chem. Soc. 126 (2004) 14746–14751.

33. R.R. Ernst, G. Bodenhausen, A. Wokaun, Principles of Nuclear Magnetic Resonance in One and Two Dimensions, Clarendon Press, Oxford (1987).

34. E. Kupce, R. Freeman, Projection reconstruction of three-dimensional NMR spectra. J. Am. Chem. Soc. 125 (2003) 13958–13959.

35. T. Szyperski, G. Wider, J.H. Bushweller, K. Wuthrich, Reduced dimensionality in triple resonance NMR experiments. J. Am. Chem. Soc. 115 (1993) 9307–9308.

36. N.S. Astrof, C.E. Lyon, R.G. Griffin, Triple resonance solid state NMR experiments with reduced dimensionality evolution periods. J. Magn. Reson. 152 (2001) 303–307.

37. S. Luca, M. Baldus, Enhanced spectral resolution in immobilized peptides and proteins by combining chemical shift sum and difference spectroscopy. J. Magn. Reson. 159 (2002) 243–249.

38. J. Leppert, B. Heise, O. Ohlenschlager, M. Gorlach, R. Ramachandran, Triple resonance MAS NMR with (C-13, N-15) labelled molecules: reduced dimensionality data acquisition via C-13-N-15 heteronuclear two-spin coherence transfer pathways. J. Biomol. NMR 28 (2004) 185–190.

39. D. Rovnyak, C. Filip, B. Itin, A.S. Stern, G. Wagner, R.G. Griffin, J.C. Hoch, Multiple quantum magic angle spinning spectroscopy using nonlinear sampling. J. Magn. Reson. 161 (2003) 43–55.

40. K. Aggarwal, M.A. Delsuc, Triangular sampling of multidimensional NMR data sets. Magn. Reson. Chem. 35 (1997) 593–596.

41. H. Heise, K. Seidel, M. Etzkorn, S. Becker, M. Baldus, 3D Spectroscopy for resonance assignment and structure elucidation of proteins under MAS: novel pulse schemes and sensitivity considerations. 173 (2005) 64–74.

42. S.R. Hartmann, E.L. Hahn, Nuclear double resonance in rotating frame. Phys. Rev. 128 (1962) 2042–2053.

43. A. Pines, M.G. Gibby, J.S. Waugh, Proton-enhanced NMR of dilute spins in solids. J. Chem. Phys. 59 (1973) 569–590.

44. A. Bax, R. Freeman, S.P. Kempsell, Natural abundance C-13–C-13 coupling observed via double-quantum coherence. J. Am. Chem. Soc. 102 (1980) 4849–4851.

45. M. Munowitz, A. Pines, Principles and applications of multiple quantum NMR. Adv. Chem. Phys. 66 (1987) 1–152.

46. M. Baldus, B.H. Meier, Total correlation spectroscopy in the solid state: use of scalar couplings to determine the through-bond connectivity. J. Magn. Reson. Ser. A 121 (1996) 65–69.

47. M. Baldus, R.J. Iuliucci, B.H. Meier, Probing through-bond connectivities and through-space distances in solids by magic-angle-spinning nuclear magnetic resonance. J. Am. Chem. Soc. 119 (1997) 1121–1124.

48. S. Luca, J.F. White, A.K. Sohal, D.V. Filippov, J.H. van Boom, R. Grisshammer, M. Baldus, The conformation of neurotensin bound to its G protein-coupled receptor. Proc. Natl. Acad. Sci. USA 100 (2003) 10706–10711.

49. C.G. Tate, R. Grisshammer, Heterologous expression of G-protein-coupled receptors. Trends Biotechnol. 14 (1996) 426–430.

50. M. Bouvier, L. Menard, M. Dennis, S. Marullo, Expression and recovery of functional G-protein-coupled receptors using baculovirus expression systems. Curr. Opin. Biotechnol. 9 (1998) 522–527.

51. V. Sarramegna, R. Talmont, P. Demange, A. Milon, Heterologous expression of G protein-coupled receptors: comparison of expression systems from the standpoint of large-scale production and purification. Cell. Mol. Life Sci. 60 (2003) 1529–1546.

52. K. Lundstrom, Semliki Forest virus vectors for rapid and high-level expression of integral membrane proteins. Biochim. Biophys. Acta Biomembr. 1610 (2003) 90–96.

53. N. Vita, P. Laurent, S. Lefort, P. Chalon, X. Dumont, M. Kaghad, D. Gully, G. Lefur, P. Ferrara, D. Caput, Cloning and expression of a complementary DNA-encoding a high affinity human neurotensin receptor. FEBS Lett. 317 (1993) 139–142.

54. K. Tanaka, M. Masu, S. Nakanishi, Structure and functional expression of the cloned rat neurotensin receptor. Neuron 4 (1990) 847–854.

55. M. Watson, P.J. Isackson, M. Makker, M.S. Yamada, M. Yamada, B. Cusack, E. Richelson, Identification of a polymorphism in the human neurotensin receptor gene. Mayo Clin. Proc. 68 (1993) 1043–1048.

56. R. Grisshammer, P. Averbeck, A.K. Sohal, Improved purification of a rat neurotensin receptor expressed in *Escherichia coli*. Biochem. Soc. Trans. 27 (1999) 899–903.

57. T. Warne, J. Chirnside, G.F.X. Schertler, Expression and purification of truncated, non-glycosylated turkey beta-adrenergic receptors for crystallization. Biochim. Biophys. Acta Biomembr. 1610 (2003) 133–140.

58. P.E. Blackburn, C.V. Simpson, R.J.B. Nibbs, M. O'Hara, R. Booth, J. Poulos, N.W. Isaacs, G.J. Graham, Purification and biochemical characterization of the D6 chemokine receptor. Biochem. J. 379 (2004) 263–272.

59. M. Carravetta, X. Zhao, O.G. Johannessen, W.C. Lai, M.A. Verhoeven, P.H.M. Bovee-Geurts, P.J.E. Verdegem, S. Kiihne, H. Luthman, H.J.M. de Groot, W.J. deGrip, J. Lugtenburg, M.H. Levitt, Protein-induced bonding perturbation of the rhodopsin chromophore detected by double-quantum solid-state NMR. J. Am. Chem. Soc. 126 (2004) 3948–3953.

60. A.T. Petkova, J.G.G. Hu, M. Bizounok, M. Simpson, R.G. Griffin, J. Herzfeld, Arginine activity in the proton-motive photocycle of bacteriorhodopsin: solid state NMR studies of the wild-type and D85N proteins. Biochemistry 38 (1999) 1562–1572.

61. J.L. Baneres, A. Martin, P. Hullot, J.P. Girard, J.C. Rossi, J. Parello, Structure-based analysis of GPCR function: conformational adaptation of both agonist and receptor upon leukotriene B-4 binding to recombinant BLT1. J. Mol. Biol. 329 (2003) 801–814.

62. Y. Gohon, J.L. Popot, Membrane protein-surfactant complexes. Curr. Opin. Colloid Interface Sci. 8 (2003) 15–22.

63. J.L. Rigaud, B. Pitard, D. Levy, Reconstitution of membrane proteins into liposomes: application to energy-transducing membrane proteins. Biochim. Biophys. Acta Bioenerg. 1231 (1995) 223–246.

64. J.L. Rigaud, D. Levy, G. Mosser, O. Lambert, Detergent removal by non-polar polystyrene beads: applications to membrane protein reconstitution and two-dimensional crystallization. Eur. Biophys. J. Biophys. Lett. 27 (1998) 305–319.

65. F. Devesa, V. Chams, P. Dinadayala, A. Stella, A. Ragas, H. Auboiroux, T. Stegmann, Y. Poquet, Functional reconstitution of the HIV receptors CCR5 and CD4 in liposomes. Eur. J. Biochem. 269 (2002) 5163–5174.

66. W.J. DeGrip, J. VanOostrum, P.H.M. Bovee-Geurts, Selective detergent extraction from mixed detergent/lipid/protein micelles, using cyclodextrin inclusion compounds: a novel generic approach for the preparation of proteoliposomes. Biochem. J. 330 (1998) 667–674.

67. V.R.P. Ratnala, R.B. Hulsbergen, H.J.M. de Groot, W.J. de Grip, Analysis of histamine and modeling of ligand-receptor interactions in the histamine H-1 receptor for magic angle spinning NMR studies. Inflammation Res. 52 (2003) 417–423.

68. R. Carraway, S.E. Leeman, Isolation of a new hypotensive peptide, neurotensin, from bovine hypothalami. J. Biol. Chem. 248 (1973) 6854–6861.

69. S. Martin, J.M. Botto, J.P. Vincent, J. Mazella, Pivotal role of an aspartate residue in sodium sensitivity and coupling to G proteins of neurotensin receptors. Mol. Pharmacol. 55 (1999) 210–215.

70. P. Chalon, N. Vita, M. Kaghad, M. Guillemot, J. Bonnin, B. Delpech, G. LeFur, P. Ferrara, D. Caput, Molecular cloning of a levocabastine-sensitive neurotensin binding site. FEBS Lett. 386 (1996) 91–94.

71. J. Tucker, R. Grisshammer, Purification of a rat neurotensin receptor expressed in *Escherichia coli*. Biochem. J. 317 (1996) 891–899.

72. R. Grisshammer, R. Duckworth, R. Henderson, Expression of a rat neurotensin receptor in *Escherichia coli*. Biochem. J. 295 (1993) 571–576.

73. M. Goedert, Radioligand binding assays for study of neurotensin receptors. Methods Enzymol. 168 (1989) 462–481.

74. A.C. Dedios, J.G. Pearson, E. Oldfield, Secondary and tertiary structural effects on protein NMR chemical-shifts: an *ab initio* approach. Science 260 (1993) 1491–1496.

75. H. Saito, Conformation-dependent C-13 chemical shifts: a new means of conformational characterization as obtained by high resolution solid state C-13 NMR. Magn. Reson. Chem. 24 (1986) 835–852.
76. J. Cavanagh, W.J. Fairbrother, A.G. Palmer, N.J. Skelton, Protein NMR Spectroscopy: Principles and Practice, Academic Press, San Diego (1996).
77. J. Heller, D.D. Laws, M. Tomaselli, D.S. King, D.E. Wemmer, A. Pines, R.H. Havlin, E. Oldfield, Determination of dihedral angles in peptides through experimental and theoretical studies of alpha-carbon chemical shielding tensors. J. Am. Chem. Soc. 119 (1997) 7827–7831.
78. O.N. Antzutkin, J.J. Balbach, R.D. Leapman, N.W. Rizzo, J. Reed, R. Tycko, Multiple quantum solid-state NMR indicates a parallel, not antiparallel, organization of β sheets in Alzheimer's β-amyloid fibrils. Proc. Natl. Acad. Sci. USA 97 (2000) 13045–13050.
79. R.H. Havlin, D.D. Laws, H.M.L. Bitter, L.K. Sanders, H.H. Sun, J.S. Grimley, D.E. Wemmer, A. Pines, E. Oldfield, An experimental and theoretical investigation of the chemical shielding tensors of C-13(alpha) of alanine, valine, and leucine residues in solid peptides and in proteins in solution. J. Am. Chem. Soc. 123 (2001) 10362–10369.
80. S. Luca, D.V. Filippov, J.H. van Boom, H. Oschkinat, H.J.M. de Groot, M. Baldus, Secondary chemical shifts in immobilized peptides and proteins: qualitative basis for structure refinement under magic angle spinning. J. Biomol. NMR 20 (2001) 325–331.
81. A.T. Petkova, Y. Ishii, J.J. Balbach, O.N. Antzutkin, R.D. Leapman, F. Delaglio, R. Tycko, A structural model for Alzheimer's β-amyloid fibrils based on experimental constraints from solid state NMR. Proc. Natl. Acad. Sci. USA 99 (2002) 16742–16747.
82. C.P. Jaroniec, C.E. MacPhee, N.S. Astrof, C.M. Dobson, R.G. Griffin, Molecular conformation of a peptide fragment of transthryretin in an amyloid fibril. Proc. Natl. Acad. Sci. USA 99 (2002) 16748–16753.
83. Y.P. Pang, B. Cusack, K. Groshan, E. Richelson, Proposed ligand binding site of the transmembrane receptor for neurotensin(8–13). J. Biol. Chem. 271 (1996) 15060–15068.
84. S. Barroso, F. Richard, D. Nicolas-Etheve, J.L. Reversat, J.M. Bernassau, P. Kitabgi, C. Labbe-Jullie, Identification of residues involved in neurotensin binding and modeling of the agonist binding site in neurotensin receptor 1. J. Biol. Chem. 275 (2000) 328–336.
85. F. Richard, S. Barroso, D. Nicolas-Etheve, P. Kitabgi, C. Labbe-Jullie, Impaired G protein coupling of the neurotensin receptor 1 by mutations in extracellular loop 3. Eur. J. Pharmacol. 433 (2001) 63–71.
86. H. Bittermann, J. Einsiedel, J. Hubner, P. Gmeiner, Evaluation of lactam-bridged neurotensin-analogous adjusting psi (Pro10) close to the experimentally derived bioactive conformation of NT(8–13). J. Med. Chem. 47 (2004) 5587–5590.
87. V.J. Hruby, Designing peptide receptor agonists and antagonists. Nat. Rev. Drug Discov. 1 (2002) 847–858.
88. G.Y. Xu, C.M. Deber, Conformations of neurotensin in solution and in membrane environments studied by 2D NMR spectroscopy. Int. J. Peptide Protein Res. 37 (1991) 528–535.
89. H. Heise, S. Luca, B. de Groot, H. Grubmüller, M. Baldus, Probing conformational distributions by high-resolution solid-state NMR and MD simulations: application to neurotensin. (2005), submitted.

90. S. Spera, A. Bax, Empirical correlation between protein backbone conformation and C-alpha and C-beta C-13 nuclear magnetic resonance chemical shifts. J. Am. Chem. Soc. 113 (1991) 5490–5492.

91. A.T. Petkova, M. Baldus, M. Belenky, M. Hong, R.G. Griffin, J. Herzfeld, Backbone and side chain assignment strategies for multiply labeled membrane peptides and proteins in the solid state. J. Magn. Reson. 160 (2003) 1–12.

92. E. Getmanova, A.B. Patel, J. Klein-Seetharaman, M.C. Loewen, P.J. Reeves, N. Friedman, M. Sheves, S.O. Smith, H.G. Khorana, NMR spectroscopy of phosphory-lated wild-type rhodopsin: mobility of the phosphorylated C terminus of rhodopsin in the dark and upon light activation. Biochemistry 43 (2004) 1126–1133.

93. T.A. Egorova-Zachernyuk, J. Hollander, N. Fraser, P. Gast, A.J. Hoff, R. Cogdell, H.J.M. de Groot, M. Baldus, Heteronuclear 2D-correlations in a uniformly C-13, N-15 labeled membrane-protein complex at ultra-high magnetic fields. J. Biomol. NMR 19 (2001) 243–253.

94. G. McDermott, S.M. Prince, A.A. Freer, A.M. Hawthornthwaite-Lawless, M.Z. Papiz, R.J. Cogdell, N.W. Isaacs, Crystal structure of an integral membrane light harvesting complex from photosynthetic bacteria. Nature 374 (1995) 517–521.

95. C. Toyoshima, M. Asahi, Y. Sugita, R. Khanna, T. Tsuda, D.H. MacLennan, Modeling of the inhibitory interaction of phospholamban with the Ca^{2+} ATPase. Proc. Natl. Acad. Sci. USA 100 (2003) 467–472.

96. A. McDermott, T. Polenova, A. Bockmann, K.W. Zilm, E.K. Paulsen, R.W. Martin, G.T. Montelione, Partial NMR assignments for uniformly (C-13, N-15) enriched BPTI in the solid state. J. Biomol. NMR 16 (2000) 209–219.

97. A. Böckmann, A. Lange, A. Galinier, S. Luca, N. Giraud, H. Heise, M. Juy, R. Montserret, F. Penin, M. Baldus, Solid-state NMR sequential resonance assignments and conformational analysis of the 2*10.4 kDa dimeric form of the *Bacillus subtilis* protein crh. J. Biomol. NMR 27 (2003) 323–339.

98. D. Staunton, J. Owen, I.D. Campbell, NMR and structural genomics. Accounts Chem. Res. 36 (2003) 207–214.

99. K. Palczewski, T. Kumasaka, T. Hori, C.A. Behnke, H. Motoshima, B.A. Fox, I. Le Trong, D.C. Teller, T. Okada, R.E. Stenkamp, M. Yamamoto, M. Miyano, Crystal structure of rhodopsin: a G protein-coupled receptor. Science 289 (2000) 739–745.

14 Structural Genomics Initiatives

Kenneth Lundstrom

CONTENTS

ABSTRACT

Among the almost 30,000 crystal protein structures deposited in public databases, fewer than 1% represent membrane proteins. Only a single high-resolution structure — for the bovine rhodopsin receptor — is available for a GPCR. The low success rate is mainly due to the complexity of GPCR topology, making their overexpression, purification, and crystallization more demanding than is the case with soluble proteins. In this context, a broad-base platform in a high throughput format for a large number of targets is the key to technology improvements. Several national and international networks focusing on structural genomics of membrane proteins have been established. Most of the consortia include both prokaryotic and eukaryotic membrane proteins and not exclusively GPCRs. The Europe-based Membrane Protein Network (MePNet) exceptionally concentrates uniquely on GPCRs as targets.

14.1　INTRODUCTION

Structure-based design has played an important role in drug development. At least seven approved drugs were discovered based exclusively on crystal structure information.[1] The strength of structure-based drug design has been related to the possibility of rapid compound evaluation and has provided relevant indicators for identification of conserved regions not applicable for modifications.

Structural genomics, the structure-based analysis of multiple gene products, became a reality with the completion of the human and other genomes. Most

315

structural genomics programs have concentrated their efforts on soluble proteins because the requirements for their handling are less demanding than for membrane proteins. Typically, only some 50 high-resolution structures are available for membrane proteins, most of which are of bacterial origin.[2] Eukaryotic membrane proteins that have been crystallized successfully are generally present in exceptionally high concentrations in certain tissues, such as the nicotinic acetylcholine receptor in the electric organ of *Torpedo marmorata*.[3] Similarly, bovine rhodopsin is very abundantly expressed in the cow retina and that permits efficient purification and crystallization of this model GPCR.[4] Bovine rhodopsin remains the only representative of the family of GPCRs for which a high-resolution three-dimensional (3D) structure is available.

The situation is rather paradoxical as GPCRs are such important drug targets, as described in detail in other chapters. Knowledge of the protein structure has proven helpful in many cases for drug development. Structure-based drug design both facilitates the drug discovery process and also aids in generating more selective medicines with potentially fewer side effects. Additionally, it should shorten the time required for the whole drug development process.

The major problems of obtaining high-resolution structures of GPCRs arise from their topological structures. The presence of seven transmembrane spanning domains presents a much more difficult task for recombinant protein production than for soluble proteins. Various expression systems ranging from cell-free translation systems to bacterial, yeast, and animal cell-based systems have been applied as described in detail in Chapter 8. No expression system has been universally applied successfully to all GPCRs and in many cases modifications to the receptor gene sequences have been necessary.

Another factor that hampers success in determining the structure of GPCRs is the need to solubilize the receptors from the membranes with detergents prior to purification, which certainly increases the impacts on the quality and the quantity of the solubilized materials. Moreover, purification presents its own difficulties due to the biophysical and biochemical natures of the solubilized GPCRs. Finally, because crystallization is performed in the presence of detergents, special conditions must be developed. Taking into consideration all these special requirements, it is not that surprising that the success rate for obtaining high-resolution structures of GPCRs has been so low.

The clearest shortcoming at present is the dearth of appropriate technologies. In comparison, structure determination of soluble proteins some 20 years ago represented an immense struggle due to the lack of appropriate methods. Investment in strong technology development led to almost exponential increases in novel crystal structure determinations. Membrane proteins, and particularly GPCRs, are certainly demanding targets and it is hard to predict today whether similar success will be achieved despite major investments in technology. It seems, however, that the current trend certainly favors technology development, and parallel approaches should enhance success rates.

To make this approach feasible, large networks have been established either on national or international levels. These consortia have garnered general and specific interest from academia and industry. Figure 14.1 presents a model of the composition

Structural Genomics Network

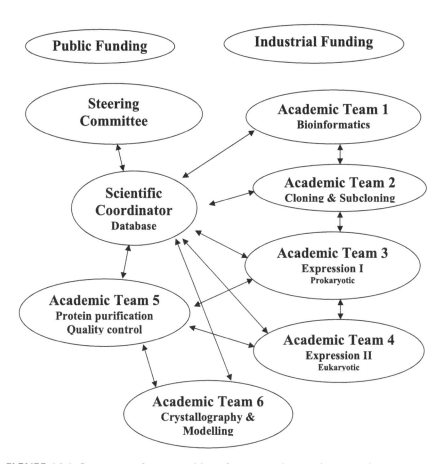

FIGURE 14.1 Organogram for composition of a structural genomics network.

and function of such a network. Table 14.1 is an overview of several existing consortia cited in this chapter. The networks are divided into three groups as discussed below. Section 14.2 covers networks concerned with both soluble and membrane protein targets but not with GPCRs. The second group discussed in Section 14.3 includes GPCRs and also other membrane proteins. The third group is represented by only one network known as MePNet and is discussed in Section 14.4.

14.2 STRUCTURAL GENOMICS PROGRAMS FOCUSING ON MEMBRANE PROTEINS

Several of the existing networks focusing on membrane proteins deal mainly with prokaryotic targets in agreement with the generally accepted view that the success rate for structure resolution on bacterial targets is much higher than for eukaryotic

TABLE 14.1
Structural Genomics Networks with Membrane Protein Targets

Network	Target Proteins	Expression Host	Website
American Networks			
Joint Center for Structural Genomics (JCSG)	Mouse genome 180 GPCRs	*E. coli* Baculovirus Adenovirus	www.jcsg.com
Southeastern Collaboratory on Structural Genomics (SECSG)	*C. elegans* MPs *P. furiousus* MPs Selected human cDNAs	*E. coli* *E. coli* *E. coli* COS, MEL	www.secsg.org/cel.htm www.secsg.org/pfu.htm www.secsg. org/hum.htm
Mycobacterium tuberculosis Structural Genomics Consortium	M. tuberculosis 1165 ORFs of MPs	*E. coli*	http://doe-mbi.ucla.edu/TB
Japanese Networks			
RIKEN Genomic Sciences Center's Structurome	Bacterial MPs *T. thermophilis* MPs	*E. coli* *E. coli,* cell-free	www.riken.go.jp www.thermus.org
Japan Biological Informatics Consortium and Biological Information Research Center	Bacteriorhodopsin Human GPCRs	*E. coli* Baculovirus	www.rcsb.org
European Networks			
European Membrane Proteins (E-MeP)	Prokaryotic MPs	*E. coli* *Lactococcus lactis*	www.e-mep.org
	GPCRs, ion channels, transporters	*S. cerevisiae* *P. pastoris* Baculovirus SFV	
Membrane Protein Network (MePNet)	101 GPCRs	*E. coli* *P. pastoris* SFV	www.mepnet.org
ProAMP–Protein-Wide Analysis of Membrane Proteins (ProAMP, Germany)	*S. typhimurium* MPs *H. pylori* MPs	*E. coli*	www.pst-ag.com
Membrane Protein Platform (SweGene MPP, Sweden)	Bacterial MPs Yeast MPs	*E. coli* Yeast	www.swegene.org

TABLE 14.1 (Continued)
Structural Genomics Networks with Membrane Protein Targets

Network	Target Proteins	Expression Host	Website
Swiss National Center of Competence Research (NCCR)	Membrane proteins	*E. coli*	www.structuralbiology.ethz.ch

MP = Membrane proteins. COS = African green monkey. MEL = Mouse erytholeukemia. ORF = Open reading frame. SFV = Semliki Forest Virus.

membrane proteins. The NIH-funded network on *Mycobacterium tuberculosis* aims to study all 3924 ORFs (open reading frames) of this bacterial genome.[5] It has been suggested that more than 1000 ORFs represent membrane proteins but no GPCRs. The overexpression of the targets is carried out in *E. coli* because this organism is anticipated to generate high success rates. One valuable aspect of *M. tuberculosis* targets is the presence of eukaryotic homologues among the serine and threonine kinases[6] and adenylate cyclases.[7] One goal of the program is to develop improved drugs for tuberculosis. Another structural genomics initiative dealing with a whole bacterial genome was initiated by the Southeast Collaboratory on Structural Genomics (SECSG) based in the southeast U.S.A. The relatively small genome of *Pyrococcus furiosus* contains some 2200 ORFs, of which soluble and membrane proteins will be expressed initially from conventional *E. coli* vectors.[8] Targets proving more difficult to express will be evaluated in improved prokaryotic and also in eukaryotic expression systems.

Among European networks, ProAMP (Protein-Wide Analysis of Membrane Proteins), was established in Germany with financial support from the German Research Ministry (BMBF). Membrane protein targets were selected from *Salmonella typhimurium* and *Helicobacter pylori*, with special emphasis on pathogenicity. The plan is to subject overexpressed and purified targets to x-ray crystallography, NMR spectroscopy, and electron microscopy. The structural information obtained within the network will be used for the development of novel anti-infective drugs.

The Postgenome Research and Technology Program named SweGene and located in Southwestern Sweden established a Membrane Protein Platform (MPP) with activities in the area of protein overproduction and structural biology. In general, bacterial and yeast expression vectors are employed for the overexpression of bacterial reaction centers, aquaporins, transhydrogenases, and glycerol facilitator proteins. Substantial efforts have been dedicated to the optimization of protein production featuring host cell engineering, fermentation technology improvement, and development of downstream processing. MPP also invests resources in crystallization technologies such as the lipidic cubic phase approach.

In Japan, structural genomics initiatives have been active through the RIKEN Genomics Sciences Center (GSC). In its Structurome Project, the small-size genome of *Thermus thermophilus* has been subjected to structural studies due to the thermostability and good crystallizability of its proteins.[9] More than 1100 ORFs have been overexpressed in *E. coli*, some 700 proteins purified, and 3D structures determined for more than 130 proteins. Incorporation of the Protein Folds Project and the Structurome Project into the RIKEN Structural Genomics Initiative made it possible to include programs involving other organisms such as the *Arabadopsis thaliana* plant. The target proteins are subjected to overexpression in a cell-free *E. coli* system and a modified wheat-germ cell-free translation system.[10]

14.3 STRUCTURAL GENOMICS INITIATIVES INCLUDING GPCRS

Another structural genomics program within SECSG that targets a whole genome focuses on *Caenerhabdhitis elegans*.[11] The goal is to solve as many structures as possible among the 19,092 expressed proteins. *C. elegans*, a representative of a eukaryotic worm, possesses several membrane proteins including a few GPCRs in its genome. Among the 250 *C. elegans* proteins expressed in *E. coli*, 27 have been crystallized and not one is a membrane protein. The SECSG studies also selected human targets, among them GPCRs that will be expressed from lentivirus vectors.

The Joint Center for Structural Genomics (JCSG) was established in San Diego by the Scripps Research Institute (TSRI), the Genomics Institute for the Novartis Research Foundation (GNF), the San Diego Supercomputer Center (SDSC) at University of California San Diego (UCSD), and the Stanford Synchrotron Radiation Laboratory (SSRL, a division of the Stanford Linear Accelerator Center, SLAC) at Stanford University. This network includes structural studies on the whole mouse genome in its program. Although most of the targets are soluble proteins, several mouse GPCRs are planned to be studied. Additionally, a large number of GPCRs will be studied by JCSG. The consortium will apply *E. coli*, baculovirus and adenovirus vectors for expression studies. In addition to the core members of the program, 29 scientific collaborators from other institutions such as the Salk Institute and Syrrx Inc. are included.

The National Center of Competence in Research (NCCR) was established in Switzerland as a network of the University of Zurich, the Swiss Federal Institute of Technology (ETH) in Zurich, the Paul Sherrer Institute (PSI) at Villigen, and the University of Basel. The main areas of interest of this consortium are the development of improved methods for recombinant protein expression, experimental structure determination, and computational biology. NMR spectroscopy and electron microscopy play particularly important parts in the program. The targets include membrane proteins from *Helicobacter pylori*, ABC transporters, and several mammalian GPCRs.

Within the Japan Biological Informatics Consortium (JBIC), a program was initiated to improve expression of GPCRs, 2D crystallization, and electron crystallography. Bacteriorhodopsin is used as a model receptor for the structural studies, but human GPCRs are also included. Baculovirus vectors are applied for overexpression of a mutant human endothelin B receptor (ETBR) in suspension cultures of Sf9 insect cells with yields of milligrams of active receptor.[12]

The European Union (EU)-funded E-MeP (European Membrane Proteins) network consisting of 20 laboratories was established recently. E-MeP has in its program both prokaryotic and eukaryotic membrane protein targets. In addition to some 100 eukaryotic ion channels and transporters, 100 GPCRs will be expressed from prokaryotic and eukaryotic vectors. Expression comparison will be initially carried out in *E. coli, Lactococcus lactis, P. pastoris, Saccharomyces cerevisiae*, baculovirus, and Semliki Forest virus (SFV). Large-scale production for purification and crystallization will be performed in the system providing the best expression patterns for selected targets.

14.4 MePNet APPROACH

MePNet, a network that uniquely focuses on GPCRs, was established in 2001. This consortium was originally composed of four academic research teams and a small biotech company (BioXtal). The program has been supported financially by more than 30 pharmaceutical and biotech companies. The major objectives of the network relate to technology innovation. One goal has been improvement of methods in all areas including expression, purification, and crystallography in a high throughput format.

The MePNet program selected 100 GPCRs based on availability of high affinity radioligands as a way to confirm functionalities of the recombinantly expressed receptors by binding assays. Nonradioactive ligands are also useful for the enhancement of expression levels and receptor stabilization during purification. The selection of targets with available ligands obviously excluded orphan GPCRs from the program. Other target selection criteria included representation of all GPCR families and a large selection of certain subtypes. Additionally, the project aimed at linkages to human diseases because many GPCRs are related to various diseases as described in other chapters.

Although several GPCR clones were received through generous donations, the majority of the GPCR cDNAs were cloned by applying reverse transcriptase polymerase chain reaction (RT-PCR) technologies to commercially available poly A+ RNA. Suitable restriction sites were introduced into the primer sequences before subcloning into the specifically engineered expression vectors, and various tags were introduced into the vectors to facilitate the purification of receptors by affinity chromatography.

Protease cleavage sites engineered between the GPCR and tag sequences allowed removal of tags potentially affecting the crystallization after purification. Three well-established robust systems were applied. Conventional *E. coli* pET15 vectors were chosen for GPCR expression in bacterial inclusion bodies.[13] Additionally, various *E. coli* Gateway fusion vectors were applied for a limited number of targets. Yeast expression was carried out in *Pichia pastoris*.[14] The third system applied is based on SFV vectors for expression in several mammalian cell lines.[15] Typically for expression in both yeast and mammalian cells, the GPCRs are expected to be localized in membranes.

A preliminary evaluation of the expression levels of the 100 GPCRs was carried out by dot blots and Western blots. Attention was also paid to optimization of expression by comparing expression levels of various *E. coli* strains using specially engineered bacterial strains with optimized codon usage for mammalian genes. Dramatic differences in expression levels were observed for different *E. coli* strains. Other factors of importance were the cell culture temperature and also the expression vector applied. Clearly, the C-terminally tagged pET vector generated better success rates for expression than those observed for the N-terminal constructs of the same GPCR. Overall, the Gateway vectors proved superior to the pET vectors.

Yeast expression was substantially improved by optimization. In the first phase, *P. pastoris* clones were subjected to selection and clones with suitable copy numbers were processed. Yeast cell culture conditions seemed to play an important role on expression levels. The growth temperature and additives to culture media improved expression levels by 20- to 100-fold. Because GPCRs are anticipated to be present in functional forms in yeast membranes, their binding activities can be verified by radioligand binding assays. Interestingly, binding activities could be increased significantly for several GPCRs by addition of ligands and other additives to the yeast cell cultures. A comparison of immunoblotting and binding data suggested that the receptor quantities in Western blots seemed unaffected by ligand addition, most likely affecting mainly the quality of the expressed material.

Expression of the MePNet GPCRs in mammalian cells from SFV vectors suggested that the choice of host cell line plays an important role in expression levels. Certain GPCRs showed higher expression levels in baby hamster kidney (BHK) cells whereas it was advantageous to express others in Chinese hamster ovary (CHO) or human embryonic kidney (HEK293) cells. Little attention has been paid to the optimization of SFV expression, but the expression of histamine receptors by SFV in the presence of ligands in the culture medium previously demonstrated significantly enhanced specific binding activity.[16]

In summary, the expression rate was relatively modest in *E. coli*, resulting in positive signals of expected size in Western blots for approximately 50% of the GPCR targets. The expression from *P. pastoris* and SFV vectors in yeast and mammalian cells, respectively, generated over 90% success rates. The comparison of the three systems presented a unique opportunity to evaluate the expression in relation to GPCR sequence, size, subtype and other parameters. Classification of the GPCRs according to their expression levels suggested that more than 60% are expressed at levels of 1 mg/L or higher — amounts that can be considered structural-biology compatible.

Selected well expressed GPCRs were subjected to large-scale production in *E. coli* flask and fermenter cultures generating yields of tens to hundreds of milligrams. Before GPCRs expressed in bacterial inclusion bodies can be subjected to crystallography, refolding is required. Although technological improvements have been achieved in this area,[17] refolding still represents a major obstacle for restoring receptor functionality. One of the objectives for MePNet is to develop improved refolding methods. Access to large quantities of numerous GPCRs provides a good basis for serious technology development in a high throughput format.

A limited number (15 to 20 targets) expressed from *P. pastoris* and SFV vectors were selected for further large-scale production. Yeast expression was carried out in fermenters known to generate large biomasses. The mammalian cells were grown either in suspension cultures in spinner flasks or in adherent cells in roller flasks. Before purification, the recombinant receptors must be subjected to solubilization in the presence of strong detergents. The large quantities of expressed receptors provide sufficient materials for testing a wide range of detergents under a multitude of conditions and concentrations. The GPCRs expressed in yeast and mammalian cells can be continuously monitored by ligand binding for functionality. This quality control is of great importance to ensure that the right protein is expressed, solubilized, and purified in its active state. Alternative methods for functional characterization of GPCRs, mainly based on fluorescence technologies, are currently under development.

Tags engineered into the MePNet vectors for *E. coli*, *P. pastoris*, and SFV vectors present the means for affinity column chromatography purification. Several targets have now been purified successfully and will be subjected to crystallography attempts, the ultimate challenge. Again, the more material available, the better the opportunities are to analyze as many parameters as possible. Furthermore, recent crystallography developments, especially in the area of nanodrop technology, have substantially reduced the material requirement needs. Another advantage of starting with a large number of GPCRs is the possibility of using back-up targets in case no further progress is realistically expected for a certain GPCR after evaluating a reasonable number of conditions and compounds. At the time of submission of this manuscript, several purified MePNet GPCRs have been subjected to crystallization attempts.

14.5 CONCLUSIONS

High-resolution structure determination of membrane proteins is still in its infancy compared to the achievements involving soluble proteins. It is obvious that major investments are required before significant progress can be made. Although some success has been obtained for individual targets, the progress is generally slow and the risk of failure high. Studies on a large number of targets in parallel can improve the understanding of sequence requirements, expressability of targets, refolding and solubilization conditions and, finally, crystallizability of receptors. The best approach clearly is the establishment of large networks that can provide a broad range of expertise including molecular biology, cell engineering, pharmacology, biochemistry, protein purification, and crystallography. Most importantly, the consortia described in this chapter include both academia and industry. Many national and international networks have been established in recent years. Obviously, the process of setting up a functional consortium requires materials, good coordination, and time. It is therefore expected that we will soon be able to harvest the fruits of conquering these difficult targets.

REFERENCES

1. Debe DA, Hambly A. Supporting your pipeline with structural knowledge. Curr Drug Disc March 2004; 15–18.

2. Abramson J, Smirnova I, Kasho V, Verner G, Kaback HR, Iwata S. Structure and mechanism of the lactose permease of *Escherichia coli*. Science 2003; 301: 610–615.

3. Unwin N. Structure and action of the nicotinic acetylcholine receptor explored by electron microscopy. FEBS Lett 2003; 555: 91–95.

4. Palczewski K, Kumasaka T, Hori T, Behnke CA, Motoshima H, Fox BA, Le Trong I, Teller DC, Okada T, Stenkamp RE et al. Crystal structure of rhodopsin: a G-protein-coupled receptor. Science 2000; 289: 739–745.

5. Goulding CW, Apostol M, Anderson DH, Gill HS, Smith CV, Kuo MR, Yang JK, Waldo GS, Suh SW, Chauhan R et al. The TB structural genomics consortium: providing a structural foundation for drug discovery. Curr Drug Targets Infect Disord 2002; 2: 121–141.

6. Cole ST, Brosch R, Parkhill J, Garnier T, Churcher C, Harris D, Gordon SV, Eiglmeier K, Gas S, Barry CE III et al. Deciphering the biology of *Mycobacterium tuberculosis* from the complete genome sequence. Nature 1998; 393: 537–544.

7. McCue LA, McDonough KA, Lawrence CE. Functional classification of cNMP-binding proteins and nucleotide cyclases with implications for novel regulatory pathways in *Mycobacterium tuberculosis*. Genome Res 2000; 10: 204–219.

8. Robb FT, Maeder DL, Brown JR, DiRuggiero J, Stump MD, Yeh RK, Weiss RB, Dunn DM. Glutamate dehydrogenases from hyperthermophiles. Methods Enzymol 2001; 330: 134–157.

9. Yokoyama S, Hirota H, Kigawa T, Yabuki T, Shirouzu M, Terada T, Ito Y, Matsuo Y, Kuroda Y, Nishimura Y et al. Structural genomics projects in Japan. Nat Struct Biol 2000; Suppl: 943–945.

10. Madin K, Sawasaki T, Ogasawara T, Endo Y. A highly efficient and robust cell-free protein synthesis system prepared from wheat embryos: plants apparently contain a suicide system directed at ribosomes. Proc Natl Acad Sci USA 2000; 97: 559–564.

11. Adams MW, Dailey HA, DeLucas LJ, Luo M, Prestegard JH, Rose JP, Wang B-C. The Southeast Collaboratory for Structural Genomics: a high-throughput gene to structure factory. Acc Chem Res 2003; 36: 191–198.

12. Doi T, Hiroaki Y, Arimono I, Fujiyoshi Y, Okamoto T, Satoh M, Furuichi Y. Characterization of human endothelin B receptor and mutant receptors expressed in insect cells. Eur J Biochem 1997; 248: 139–148.

13. Drew D, Fröderberg L, Baars L, De Gier J-WL. Assembly and overexpression of membrane proteins in *Escherichia coli*. Biochim Biophys Acta 2003; 1610: 3–10.

14. Weiss HM, Haase W, Michel H, Reilander H. Comparative biochemical and pharmacological characterization of the mouse 5-HT5A 5-hydroxytryptamine receptor and the human beta-2-adrenergic receptor produced in the methylotrophic yeast *Pichia pastoris*. Biochem J 1998; 330: 1137–1147.

15. Lundstrom K. Semliki Forest virus vectors for rapid high-level expression of integral membrane proteins. Biochim Biophys Acta 2003; 1610: 90–96.

16. Hoffmann M, Verzijl D, Lundstrom K, Simmen U, Alewijnse AE, Timmerman H, Leurs R. Recombinant Semliki Forest virus for over-expression and pharmacological characterization of the histamine H2 receptor in mammalian cells. Eur J Pharmacol 2001; 427: 105–114.

17. Lopez de Maturana R, Willshaw A, Kuntzsch A, Rudolph R, Donelly D. The isolated N-terminal domain of the glucagon-like peptide-1 (GLP-1) receptor binds exendin peptides with much higher affinity than GLP-1. J Biol Chem 2003; 278: 10195–10200.

15 Molecular Basis of Dimerization of Family A G Protein-Coupled Receptors

Graeme Milligan

CONTENTS

ABSTRACT

Analysis of the organization of rhodopsin in murine rod outer segment discs provided direct evidence of the dimeric and, indeed, higher-order quaternary structure of a family A G protein-coupled receptor (GPCR) in a physiological context. This builds on a wide range of observations of protein–protein interactions between both differentially epitope tagged and coexpressed forms of a single GPCR and between different GPCRs. Growing evidence suggests that interactions between many GPCRs occur early in the process of receptor synthesis and maturation and prior to membrane delivery, and that dimerization may be integral to function as structural models are consistent with a GPCR dimer providing an appropriate footprint for binding of a single G protein heterotrimer. Ongoing studies are exploring the interfaces between GPCRs that allow dimerization and the relative affinities of interaction between different GPCR pairings. No clear consensus on either of these issues has yet been reached, but the available information and the approaches employed are discussed in this chapter.

15.1 INTRODUCTION

G protein-coupled receptors (GPCRs) are encoded by one of the largest gene families in the human genome. About 400 genes encode GPCRs that respond to endogenously produced ligands[1,2] and some 350 others encode GPCRs that respond to exogenous ligands including odorants and tastes.[3] It has become increasingly clear that GPCRs can exist as dimers or higher-order oligomers and that this may be integral to their functions.[4–7]

The rhodopsin-like or family A GPCRs are by far the largest grouping. The metabotropic glutamate receptor-like family C receptors comprise the smallest group. Despite the small number of class C GPCRs, studies of their expression and function have been integral to appreciation of the importance of dimerization for function. For example, although cloning of a seven-transmembrane polypeptide family C GPCR that bound a high-affinity γ-aminobutyric acid B (GABA$_B$) receptor antagonist seemed initially to identify this receptor,[8] it was soon apparent that this polypeptide was not effectively transported to the surface of transfected cells. Furthermore, response of this receptor to GABA was either poor or non-existent.

The recognition that coexpression of a second, highly homologous family C GPCR known as GABA$_B$ r2 was required to allow effective cell surface delivery and function of the original polypeptide (now known as GABA$_B$ r1) helped determine that the functional GABA$_B$ receptor is an obligate heterodimer.[9,10] Although a special case in that only GABA$_B$ r1 can bind GABA and hence initiate a signal, the requirement for GABA$_B$ r2 to allow effective G protein coupling[11] provided an intellectual framework for related studies on homodimers and heterodimers of other GPCRs.

Taste cells that express one or pair-wise combinations of three GPCRs, T1R1, T1R2 and T1R3, allow the detection of sugars and amino acids that are mainly associated with positive taste sensations. Sweet and umami (the taste of monosodium glutamate) are the main positive taste sensations in humans. The heterodimer pairing of T1R1 + T1R3 generates a umami sensor whereas the T1R2 + T1R3 heterodimer defines a sweet receptor.[12,13] Other means exist to detect sweet and umami tastes in the absence of T1R3[14] but the molecular basis of this is currently unclear. The requirement for coexpression of specific pairs of class C GPCRs to produce particular pharmacologies of response is entirely consistent with key physiological roles requiring heterodimer formation. Such observations have influenced thinking about the contribution that heterodimerization between family A GPCRs may make to detailed pharmacology.

15.2 QUATERNARY STRUCTURES OF CLASS A GPCRs

Expression of a single gene or cDNA encoding a family A GPCR sequence in a heterologous cell line results in most cases in the production of a receptor able to recapitulate the pharmacology of a well-known pattern of physiological responses.

For example, following isolation of genomic DNA known initially as G21, expression generated a ligand-binding site characteristic of the pharmacologically defined 5-HT$_{1A}$ receptor[15] and responses to 5-HT and other relevant agonists that were blocked by antagonists with 5-HT$_{1A}$ receptor pharmacology.

Although it could be argued that the cells used for expression harbored other endogenously expressed polypeptides required to interact with the protein product of the G21 DNA to generate this receptor, the most obvious explanation is that the molecular structure of the 5-HT$_{1A}$ receptor was encoded entirely by the G21 DNA.[15] Of course, such studies did not address whether the 5-HT$_{1A}$ receptor functioned as a monomer. A wide range of approaches has been employed in recent years to indicate that family A GPCRs can exist as dimeric complexes. These include co-immunoprecipitation studies that employ differentially epitope-tagged forms of a single GPCR,[16,17] biophysical techniques utilizing a number of resonance energy transfer techniques,[18–20] and functional reconstitution following coexpression of pairs of nonfunctional constructs harboring distinct inactivating mutations.[21,22]

All these techniques have inherent limitations. For example, the resonance energy transfer approaches can only demonstrate formally that the appropriately labeled GPCRs are within proximity limits defined by the bases of the techniques. The co-immunoprecipitation studies require membrane fragments to be fully solubilized to provide convincing evidence for direct protein–protein interactions. The use of combinations of distinct approaches appropriate to each physical interaction increases the confidence in such conclusions.[23] The application of atomic force microscopy (AFM) to rod outer segments has indicated most directly that the rhodopsin photon receptor is organized in paracrystalline arrays of dimers.[24] It is clearly possible that the high local concentration of rhodopsin *in situ* imposes physical and organizational constraints that may not be generally applicable. Despite this caveat, these studies provide the best examples of the structural organization of family A GPCRs as dimers and higher order complexes in a native environment.

15.3 CONSEQUENCES OF DIMERIZATION OF FAMILY A GPCRs

Expression in *Escherichia coli* inclusion bodies followed by denaturation and refolding of the leukotriene B$_4$ BLT1 receptor results in the production of a homodimer that binds leukotriene B$_4$ with low affinity.[25] Addition of a GDP-loaded form of the heterotrimeric G protein G$\alpha_{i2}\beta_1\gamma_2$ leads to high-affinity ligand binding and the appearance of a pentameric (BLT1)$_2$G$\alpha_{i2}\beta_1\gamma_2$ complex.[25] This is indicative that a GPCR dimer may be required to generate a high-affinity binding site for a heterotrimeric G protein. If this is the case, efficient signal transduction may require GPCR dimers.[25]

It is clear that the βγ complex plays a key role in effective information transfer from GPCR to the G protein α subunit. For example, fusion of a form of Gα_{11} that binds βγ complex with low affinity to the C-terminal tail of the α_{1b}-adrenoceptor results in much lower levels of agonist-stimulated [^{35}S]GTPγS binding than an equivalent fusion protein containing wild-type Gα_{11} and this can be compensated for by overexpression of a βγ pairing.[26] Models based on the atomic level structures of bovine rhodopsin and heterotrimeric G proteins are consistent with a GPCR dimer providing an appropriate footprint for interaction with a single G protein heterotrimer.[24]

This combination of observations suggests that a GPCR dimer may be central to effective signal transduction. The literature on the effects of agonist ligands on the quaternary structures of GPCRs is large and complex and has recently been reviewed.[4,27] Various studies have indicated that agonists for different GPCRs may enhance or reduce the extent of dimerization or have no effect. Thus, agonist-induced dimerization may be anticipated to enhance G protein binding whereas agonist-mediated dissociation of GPCR dimers might be expected to reduce interaction affinity and result in the release of G protein. Both of these premises fit the models of agonist-mediated signal transduction, either by promoting G protein interaction and activation or via release of G protein from the receptor upon activation.

Interestingly, however, a GPCR dimer may also provide an appropriate footprint to bind a β-arrestin.[28] Many GPCRs bind β-arrestins, in an agonist-promoted manner with good affinity and become co-internalized with the β-arrestin, via clathrin-coated pits. If high-affinity interactions with β-arrestins require GPCR dimers, this would suggest that such GPCRs internalize as dimers and may even subsequently recycle to the cell surfaces as dimers following dissociation of the β-arrestin in the pool of acidic recycling endosomes.

15.4 MECHANISMS OF DIMERIZATION OF FAMILY A GPCRs

15.4.1 Organization of GPCR Dimers and Oligomers

The only direct visualization of GPCR organization *in situ* stems from the use of AFM to examine the distribution of rhodopsin in mouse rod outer segments.[24] A highly structured paracrystalline array of what appeared to be rows of dimers was observed. A number of calculations based on these observations were used to deduce that interdimer contacts were possibly provided by a region between transmembrane helices IV and V, and it was suggested that the rows of dimers might arise from contacts involving transmembrane regions I and II.

In addition to the interpretations of such images, a significant number of modelling and informatic studies have tried to predict the bases of GPCR dimerization. A number were based on correlated mutation[29–31] or evolutionary trace[32] analyses. These have suggested key interfaces on the outward facing regions of a number of the transmembrane helices. However, although interesting, clear linkage of such predictions and direct experimental analysis has only been reported for the CCR5 and CCR2 chemokine receptors, where prediction of sites of interaction in transmembrane helices I and IV were supported by mutations that resulted in disruption of dimerization[33] (see Section 15.4.3).

Modelling studies have also produced a series of conceptually possible structures that are based on contact interfaces in which the individual monomers remain essentially independent, self-contained units or in which transmembrane domains from each monomer are rearranged and exchanged within the dimer. Issues associated with the production of certain possible dimer conformations have been discussed in some detail by Dean et al.[32] One general approach that has provided evidence in favor of a "domain swapping" model is the use of pairs of receptors that are both unable to bind ligand for different reasons.

Such distinct GPCR mutants are sometimes able to reconstitute an effective binding site when they are coexpressed. A key example was derived from the coexpression of a pair of binding-deficient forms of the angiotensin II AT1 receptor.[34] Such a model is consistent with the chimeric receptor studies of Maggio and colleagues[35–37] that employed pairs of receptor chimeras in which the N-terminus and transmembrane regions I through V of receptor A linked to transmembrane regions VI and VII and the C-terminal tail of receptor B was coexpressed with the reverse chimera.

However, contact and domain-swap dimers are not inherently mutually exclusive and may coexist. Mutation of Asp[107] in transmembrane domain III or Phe[432] in transmembrane region VI to Ala results in two radioligand-binding deficient mutants of the histamine H1 receptor. Coexpression of these two mutants results in a reconstituted radioligand binding site that exhibits a pharmacological profile corresponding to the wild-type histamine H1 receptor.[38] Conceptually, this requires a domain swap to allow each element of the dimer to bind ligand. Although the [^{3}H]mepyramine histamine H1 receptor radioligand displayed much higher B_{max} than the [^{3}H]-(-)-*trans*-H2-PAT ligand following expression of the wild-type histamine H1 receptor, maximal binding of [^{3}H]mepyramine was substantially lower and was not greater than binding of [^{3}H]-(-)-*trans*-H2-PAT upon coexpression of the Asp[107] and Phe[432] mutant receptors.[38]

Although entirely speculative, such results would be consistent with [^{3}H]-(-)-*trans*-H2-PAT binding only to a domain swap dimer that would represent only a small fraction of the population of dimers of wild-type histamine H1 receptors. Detailed analysis of the binding of antagonist ligands reported to display substantially different B_{max} levels in the same preparation might be informative. Even in family A GPCR homodimers, it must be expected that allosteric or cooperative effects of the individual monomers should be detectable in ligand-binding studies. Whether allosteric or co-operative efforts can be observed has been analyzed for the binding of different ligands to the dopamine D2 receptor.[39]

Furthermore, heterodimerization between GPCRs offers the potential for variation in pharmacologies and responses in a number of ways. Cooperative effects of the receptor monomers on ligand binding and function are likely to be different than effects for homodimers. In domain swap heterodimers, the prospect of distinct ligand binding pharmacology seems virtually assured. Furthermore, differences in the identities of G proteins activated by heterodimers even when the two corresponding homodimers are believed to activate the same set of G proteins[40] have been reported.

Although many studies have reported distinct pharmacologies following coexpression of pairs of family A GPCRs that have the capacity to interact,[4] the requirement for ligands that selectively identify or activate specific GPCR heterodimer pairings is now clear. Development of such ligands represents the best opportunity to assess the prospective occurrences and roles of heterodimers in a physiological setting. Without such ligands, progress in this area will be as fraught with difficulty as attempts to understand the contribution of GPCR constitutive activity to physiology.[41] Considerable effort has been made to explain the roles of various elements of GPCRs in dimerization and how they might be modulated or controlled.

15.4.2 CONTRIBUTIONS OF INTRA- AND EXTRACELLULAR DOMAINS

Identification of protein–protein interactions between regions within the C-terminal tails of the $GABA_B$ r1 and $GABA_B$ r2 subunits of the $GABA_B$ receptor heterodimer[9] were consistent with a central function of these regions in dimerization. However, it appears that the key role of these interactions is based around the maturation and appropriate cell surface delivery of the heterodimeric receptor, and removal of the C-terminal region does not inherently prevent protein–protein interactions. The C-terminal region of $GABA_B$ r1 contains a well-defined endoplasmic reticulum retention motif that is the molecular basis for the lack of delivery of this polypeptide to the cell surface when expressed alone in heterologous cell systems. Interaction with the C-terminal region of $GABA_B$ r2 shields or masks this region and hence allows trafficking of the heterodimer to the cell surface.

A significant number of family A GPCRs also contain endoplasmic reticulum retention motifs within the sequences of their C-terminal tails.[42] However, the importance of these to maturation, cell surface delivery, and potentially to the selectivity of interactions in the generation of GPCR heterodimers remains to be studied in a systematic manner. Despite this, a number of reports have suggested important roles for the C-terminal tails of such GPCRs for homodimerization. The first report indicated that truncation of the C-terminal 15 amino acids of the DOP (delta opioid peptide) receptor prevented dimerization of this GPCR.[17]

Although this model has not been supported by subsequent studies on the DOP receptor, recent analysis of the dimerization of C-terminal tail swap chimeras between the somatostatin SSTR1 and SSTR5 receptor subtypes has suggested a role for the C-terminal tail in effective homodimerization and heterodimerization for the SSTR5 receptor.[43] The SSTR1 was reportedly unable to homodimerize until the C-terminal tail of this receptor was replaced with that from the SSTR5 receptor. Interestingly, in earlier studies the C-terminal tail of the SSTR5 receptor was also implicated in the effectiveness of signal transduction and thought to play a role in both receptor internalization and desensitization.[44]

Despite these examples, the C-terminal tails of a number of family A GPCRs can be removed or exchanged without altering dimerization. For example, truncation to Thr^{369} did not inhibit fluorescence resonance energy transfer signals corresponding to dimers or oligomers of the hamster α_{1b}-adrenoceptor.[45] Although a significant number of splice variants of the human α_{1a}-adrenoceptor vary only in the sequences in the C-terminal tails,[46] combinations of these can produce equivalent heterodimers.[23]

A number of studies have implicated roles for the N-terminal domain or possibly glycosylation in this region in dimerization.[47,48] The length of the N-terminal domain varies significantly among different members of family A GPCRs. For the bradykinin B2 receptor, addition of the endogenous bradykinin peptide agonist was reported to induce dimerization of the receptor. Truncation of the N-terminal region prevented this effect without altering binding of the ligand. A peptide corresponding to the N-terminal region of the receptor was also shown to limit receptor dimerization.[47] In a similar manner, mutation of the single site of N-glycosylation in the N-terminal region of the β_1-adrenoceptor has been reported to limit the ability of this receptor to homodimerize when analyzed by co-immunoprecipitation of differentially epitope-tagged forms.[48]

Interestingly, and by contrast, the glycosylation-deficient mutant of the β_1-adrenoceptor has been reported to show significantly enhanced heterodimerization with the α_{2A}-adrenoceptor.[49] Mutation to prevent glycosylation of the α_{2A}-adrenoceptor further increased these interactions. The implications of this are unclear and the increased heterodimerization observed may simply reflect the reduced homodimerization of the β_1-adrenoceptor that presumably also occurred in cells coexpressing the two mutated receptors. Again, however, this does not seem a general effect. A glycosylation-deficient form of the α_{1b}-adrenoceptor appears to have protein–protein interactions that are not different from the wild type.[45]

Studies of the dimerization and/or oligomerization of the *Saccharomyces cerevisiae* pheromone receptor that fragmented this GPCR and then analyzed interactions using fluorescence resonance energy transfer indicated that elimination of the N-terminal region along with transmembrane region I prevented protein–protein interactions and that a fragment consisting solely of the N-terminal domain plus transmembrane region I could self-associate.[50] These data imply a symmetrical interface produced from elements of the N-terminal region and/or TM1. Follow-up studies implicated the glycophorin A-like **GXXXG** dimerization sequence in transmembrane region I as playing a key role.[51] Although these authors pointed out that such motifs were present in either transmembrane region I or other transmembrane regions of a considerable number of mammalian GPCRs, the general applicability of this model remains far from clear, as discussed in Section 15.4.3.

15.4.3 CONTRIBUTIONS OF TRANSMEMBRANE DOMAINS

Despite evidence for roles of the intracellular and extracellular domains in the dimerization of certain family A GPCRs, the majority of recent studies have concentrated on contributions of the transmembrane helices. Whether considering contact or domain swap models of dimerization (see Section 15.4.1), contributions of the transmembrane helices seem likely. Initial studies of the mechanisms of dimerization of family A GPCRs were based on the ability of a peptide corresponding to transmembrane region VI of the β_2-adrenoceptor (amino acids 276 through 296) to limit the presence of immunologically detected dimers and interfere with agonist-mediated activation of adenylyl cyclase activity.[16]

A **GXXXGXXXL** sequence akin to that in transmembrane region I of the yeast pheromone is present in this region and when the above sequence within the peptide was converted to **AXXXAXXXA,** effectiveness of the peptide was lost.[16] Salahpour et al.[52] recently modified the [276]**GXXXGXXX**[284]**L** sequence of the β_2-adrenoceptor. Although single and combined alterations of the two Gly residues to Leu were without effect on dimerization and/or oligomerization, mutation of [284]Leu to Tyr, both with and without mutation of the glycine residues, reduced protein–protein interactions as monitored by bioluminescence resonance energy transfer. However, the lack of effect of mutation of the Gly residues within the **GXXXGXXXL** motif certainly does not suggest these results are directly akin to the effects noted for the pheromone receptor.

Further evidence for a role of transmembrane region VI in dimerization of family A GPCRs comes from analysis of the BLT1 leukotriene B_4 receptor. A fragment of this receptor containing the sequence of helix VI has also been shown to prevent or disrupt dimerization whereas fragments containing the sequences of the other transmembrane helices were without effect.[25] The BLT1 leukotriene B_4 receptor does not contain a glycophorin A-like dimerization motif within helix VI, indicating that this is not a generic mechanism. Although there are a significant number of Leu residues, this is hardly an uncommon amino acid in transmembrane helices.

The combination of the effect of the transmembrane helix VI-containing fragment and the lack of effects of the other transmembrane fragments is at least consistent with a single symmetrical TMVI–TMVI interaction defining the dimer interface. Mutational analysis of the **GXXXGXXXL** motif in transmembrane region I of the hamster α_{1b}-adrenoceptor does not suggest this to be a key interface. Mutation of either **GXXX[53]GXXXL** to [53]**L** (reference 45) or [49]**GXXX[53]GXXX[57]L** to [49]**LXXX[53]LXXX[57]A** (references 21 and 53) did not disrupt interaction. However, interaction mapping of fragments of the α_{1b}-adrenoceptor demonstrated symmetrical TM1–TM1 interactions and symmetrical TM4–TM4 interactions.

This may imply the capacity of the α_{1b}-adrenoceptor to form a homodimer in two distinct ways. However, a further interesting possibility reflects that such interactions can be accommodated easily if the α_{1b}-adrenoceptor is able to form an oligomeric chain.[53] Recent studies of amino acids required for dimerization of the chemokine CCR5 receptor focused on specific hydrophobic residues in transmembrane regions I and IV.[33] A double mutant, [52]I to V (in transmembrane region I) [150]Val to Ala (in transmembrane region IV) was reported not to dimerize or signal.[33]

In contrast to the view that GPCR dimerization occurs early in biogenesis and is integral to transport to the cell surface,[52] Hernanz-Falcon et al.[33] reported that the I [52]V, V [150]A CCR5 receptor was expressed equally well at the cell surface as the wild-type receptor as monitored by staining with an antireceptor antibody. Hence, although both these reports are of considerable interest, a series of key questions about the general validity of the models remain. It is also of interest to note that peptides encompassing the proposed dimerization interfaces of the CCR5 receptor were able to block fluorescence resonance energy transfer signals reflecting dimerization of the CCR5 receptor when the peptides were simply added to cells coexpressing CFP (cyan fluorescent protein) and YFP (yellow fluorescent protein)-tagged forms of the CCR5 receptor.[33] If these studies can be extended, this method may represent a generic means to disrupt dimerization and hence assess its importance for function.

Early peptide competition experiments suggested roles for transmembrane regions VI and VII in the dimerization of the dopamine D2 receptor.[54] However, cysteine cross-linking[55] and receptor fragmentation[56] approaches have indicated an important role of transmembrane helix IV in this receptor. These conclusions may not be incompatible. These apparent differences may simply reflect that multiple contact points are required to produce high affinity interactions. Indeed, direct evidence in favor of multiple contact points provided by interactions between transmembrane helices has been provided by atomic force microscopy studies on the *in situ* organization of rhodopsin.[24]

As noted earlier, intradimeric contacts appear likely to involve transmembrane helices IV and V. However, it has been suggested that the observed rows of rhodopsin dimers in rod outer segments may involve contacts provided by transmembrane helices I and II and possibly the third intracellular loop.[24] Cysteine cross-linking experiments on the complement C5a receptor[57] were interpreted as evidence for possible roles of transmembrane helices I, II and/or IV and indications that the receptor may exist as a higher order oligomer in membranes.

15.5 CONCLUSIONS

The diversity of data available at this point may suggest that no single set of interactions or common mechanism can explain the basis of homodimerization within the family of class A GPCRs. This is perhaps surprising given the overall homology of the family. Although more are clearly required, sufficient basic observational data are now available to employ informatic tools to search for common patterns and inform direct experimentation (see Section 15.4.1). It is likely that the basis of heterodimerization of family A GPCRs will still be more complex if simple, heavily repeated themes do not provide general explanations of the basis of homodimerization.

ACKNOWLEDGMENTS

Work on GPCR dimerization in the author's laboratory was supported by the Medical Research Council, the Biotechnology and Biosciences Research Council, and the Wellcome Trust.

REFERENCES

1. Fredriksson, R., Lagerstrom, M.C., Lundin, L.G., Schioth HB. The G-protein-coupled receptors in the human genome form five main families: phylogenetic analysis, paralogon groups, and fingerprints. Mol. Pharmacol. 2003, 63, 1256–1272.
2. Vassilatis, D.K., Hohmann, J.G., Zeng, H., Li, F., Ranchalis, J.E., Mortrud, M.T., Brown, A., Rodriguez, S.S., Weller, J.R., Wright, A.C., Bergmann, J.E., Gaitanaris, G.A. The G-protein-coupled receptor repertoires of human and mouse. Proc. Natl. Acad. Sci. USA 2003, 100, 4903–4908.
3. Malnic, B., Godfrey, P.A., Buck, L.B. The human olfactory receptor gene family. Proc. Natl. Acad. Sci. USA 2004, 101, 2584–2589.
4. Milligan, G. G-protein-coupled receptor dimerization: function and ligand pharmacology. Mol. Pharmacol. 2004, 66, 1–7.
5. Milligan, G., Ramsay, D., Pascal, G., Carrillo, J.J. GPCR dimerization. Life Sci. 2003, 74, 181–188.
6. Angers, S., Salahpour, A., Bouvier, M. Dimerization: an emerging concept for G-protein-coupled receptor ontogeny and function. Annu. Rev. Pharmacol. Toxicol. 2002, 42, 409–435.

7. George, S.R., O'Dowd, B.F., Lee, S.P. G-protein-coupled receptor oligomerization and its potential for drug discovery. Nat. Rev. Drug Disc. 2002, 1, 808–820.

8. Kaupmann, K., Huggel, K., Heid, J., Flor, P.J., Bischoff, S., Mickel, S.J., McMaster, G., Angst, C., Bittiger, H., Froestl, W., Bettler, B. Expression cloning of GABA(B) receptors uncovers similarity to metabotropic glutamate receptors. Nature 1997, 386, 239–246.

9. White, J.H., Wise, A., Main, M.J., Green, A., Fraser, N.J., Disney, G.H., Barnes, A.A., Emson, P., Foord, S.M., Marshall, F.H. Heterodimerization is required for the formation of a functional GABA(B) receptor. Nature 1998, 396, 679–682.

10. Jones, K.A., Borowsky, B., Tamm, J.A., Craig, D.A., Durkin, M.M., Dai, M., Yao, W.J., Johnson, M., Gunwaldsen, C., Huang, L.Y., Tang, C., Shen, Q., Salon, J.A., Morse, K., Laz, T., Smith, K.E., Nagarathnam, D., Noble, S.A., Branchek, T.A., Gerald, C. GABA(B) receptors function as a heteromeric assembly of the subunits GABA(B)R1 and GABA(B)R2. Nature 1998, 396, 674–679.

11. Blein, S., Hawrot, E., Barlow, P. The metabotropic GABA receptor: molecular insights and their functional consequences. Cell. Mol. Life Sci. 2000, 57, 635–650.

12. Li, X., Staszewski, L., Xu, H., Durick, K., Zoller, M., Adler, E. Human receptors for sweet and umami taste. Proc. Natl. Acad. Sci. USA 2002, 99, 4692–4696.

13. Zhao, G.Q., Zhang, Y., Hoon, M.A., Chandrashekar, J., Erlenbach, I., Ryba, N.J., Zuker, C.S. The receptors for mammalian sweet and umami taste. Cell 2003, 115, 255–266.

14. Damak, S., Rong, M., Yasumatsu, K., Kokrashvili, Z., Varadarajan, V., Zou, S., Jiang, P., Ninomiya, Y., Margolskee, R.F. Detection of sweet and umami taste in the absence of taste receptor T1r3. Science 2003, 301, 850–853.

15. Fargin, A., Raymond, J.R., Lohse, M.J., Kobilka, B.K., Caron, M.G., Lefkowitz, R.J. The genomic clone G-21 which resembles a beta-adrenergic receptor sequence encodes the 5-HT1A receptor. Nature 1988, 335, 358–360.

16. Herbert, T.E., Moffett, S., Morello, J-P., Loisel, T.P., Bichet, D.G., Barret, C., Bouvier, M. A peptide derived from a β2-adrenergic receptor transmembrane domain inhibits both receptor dimerization and activation. J. Biol. Chem. 1996, 271, 16384–16392.

17. Cvejic, S., Devi, L.A. Dimerization of the δ opioid receptor: implication for a role in receptor internalization. J. Biol. Chem. 1977, 273, 26959–26964.

18. Overton, M.C., Blumer, K.J. G-protein-coupled receptors function as oligomers *in vivo*. Curr. Biol. 2000, 10, 341–344.

19. Angers, S., Salahpour, A., Joly, E., Hilairet, S., Chelsky, D., Dennis, M., Bouvier, M. Detection of β2-adrenergic receptor dimerization in living cells using bioluminescene resonance energy transfer (BRET). Proc. Natl. Acad. Sci. USA 2000, 97, 3684–3689.

20. Rocheville, M., Lange, D.C., Kumar, U., Patel, S.C., Patel, R.C., Patel, Y.C. Receptors for dopamine and somatostatin: formation of hetero-oligomers with enhanced functional activity. Science 2000, 288, 154–157.

21. Carrillo, J.J., Pediani, J., Milligan, G. Dimers of class A G-protein-coupled receptors function via agonist-mediated trans-activation of associated G-proteins. J. Biol. Chem. 2003, 278, 42578–42587.

22. Lee, C., Ji, I., Ryu, K., Song, Y., Conn, P.M., Ji, T.H. Two defective heterozygous luteinizing hormone receptors can rescue hormone action. J. Biol. Chem. 2002, 277, 15795–15800.

23. Ramsay, D., Carr, I.C., Pediani, J., Lopez-Gimenez, J.F., Thurlow, R., Fidock, M., Milligan, G. High affinity interactions between human α_{1A}-adrenoceptor C-terminal splice variants produce homo- and heterodimers but do not generate the α_{1L}-adrenoceptor. Mol. Pharmacol. 2004, 66, 228–239.

24. Liang, Y., Fotiadis, D., Filipek, S., Saperstein, D.A., Palczewski, K., Engel, A. Organization of the G-protein-coupled receptors rhodopsin and opsin in native membranes. J. Biol. Chem. 2003, 278, 21655–21662.

25. Baneres, J.L., Parello, J. Structure-based analysis of GPCR function: evidence for a novel pentameric assembly between the dimeric leukotriene B4 receptor BLT1 and the G-protein. J. Mol. Biol. 2003, 329, 815–829.

26. Liu, S., Carrillo, J.J., Pediani, J. Milligan, G. Effective information transfer from the α_{1b}-adrenoceptor to $G_{11}\alpha$ requires both β/γ interactions and an aromatic group 4 amino acid from the C-terminus of the G-protein. J. Biol.Chem. 2002, 277, 25707–25714.

27. Devi, L.A. Heterodimerization of G-protein-coupled receptors: pharmacology, signaling and trafficking. Trends Pharmacol. Sci. 2001, 22, 532–537.

28. Han, M., Gurevich, V.V., Vishnivetskiy, S.A., Sigler, P.B. Schubert, C. Crystal structure of beta-arrestin at 1.9 Å: possible mechanism of receptor binding and membrane translocation. Structure 2001, 9, 869–880.

29. Filizola, M., Olmea, O., Weinstein, H. Prediction of heterodimerization interfaces of G-protein coupled receptors with a new subtractive correlated mutation method. Protein Eng. 2002, 15, 881–885.

30. Filizola, M., Weinstein, H. Structural models for dimerization of G-protein coupled receptors: the opioid receptor homodimers. Biopolymers 2002, 66, 317–325.

31. Gouldson, P.R., Dean, M.K., Snell C,R., Bywater, R.P., Gkoutos, G., Reynolds, C.A. Lipid-facing correlated mutations and dimerization in G-protein coupled receptors. Protein Eng. 2001, 14, 759–767.

32. Dean, M.K., Higgs, C., Smith, R.E., Bywater, R.P., Snell, C.R., Scott, P.D., Upton, G.J., Howe, T.J., Reynolds, C.A. Dimerization of G-protein-coupled receptors. J. Med. Chem. 2001, 44, 4595–4614.

33. Hernanz-Falcon, P., Rodriguez-Frade, J.M., Serrano, A., Juan, D., del Sol, A., Soriano, S.F., Roncal, F., Gomez, L., Valencia, A., Martinez-A, C., Mellado, M. Identification of amino acid residues crucial for chemokine receptor dimerization. Nat. Immunol. 2004, 5, 216–223.

34. Monnot, C., Bihoreau, C., Conchon, S., Curnow, K.M., Corvol, P., Clauser, E. Polar residues in the transmembrane domains of the type 1 angiotensin II receptor are required for binding and coupling: reconstitution of the binding site by co-expression of two deficient mutants. J. Biol. Chem. 1996, 271, 1507–1513.

35. Maggio, R., Vogel, Z., Wess, J. Coexpression studies with mutant muscarinic/adrenergic receptors provide evidence for intermolecular "cross-talk" between G-protein-linked receptors. Proc. Natl. Acad. Sci. USA. 1993, 90, 3103–3107.

36. Scarselli, M., Novi, F., Schallmach, E., Lin, R., Baragli, A., Colzi, A., Griffon, N., Corsini, G.U., Sokoloff, P., Levenson, R., Vogel, Z., Maggio, R. D2/D3 dopamine receptor heterodimers exhibit unique functional properties. J. Biol. Chem. 2001, 276, 30308–30314.

37. Maggio, R., Scarselli, M., Novi, F., Millan, M. J, Corsini, G.U. Potent activation of dopamine D3/D2 heterodimers by the antiparkinsonian agents, S32504, pramipexole and ropinirole. J. Neurochem. 2003, 87, 631–641.

38. Bakker, R.A., Dees, G., Carrillo, J.J., Booth, R.G., Lopez-Gimenez, J.F., Milligan, G., Strange, P.G., Leurs, R. Domain swapping in the human histamine H1 receptor. J. Pharmacol. Exp. Ther. 2004, 311, 131–138.

39. Armstrong, D., Strange, P.G. Dopamine D2 receptor dimer formation: evidence from ligand binding. J. Biol. Chem. 2001, 276, 22621–22629.

40. George, S.R., Fan, T., Xie, Z., Tse, R., Tam, V., Varghese, G., O'Dowd, B.F. Oligomerization of μ- and δ-opioid receptors. J. Biol. Chem. 2000, 275, 26128–26135.

41. Milligan, G. Constitutive activity and inverse agonists of G-protein-coupled receptors: a current perspective. Mol. Pharmacol. 2003, 64, 1271–1276.

42. Margeta-Mitrovic, M. Assembly-dependent trafficking assays in the detection of receptor–receptor interactions. Methods 2002, 27, 311–317.

43. Grant, M., Patel, R.C., Kumar, U. The role of subtype-specific ligand binding and the C-tail domain in dimer formation of human somatostatin receptors. J. Biol. Chem. 2004, 279, 38636–38643.

44. Hukovic, N., Panetta, R., Kumar, U., Rocheville, M., Patel, Y.C. The cytoplasmic tail of the human somatostatin receptor type 5 is crucial for interaction with adenylyl cyclase and in mediating desensitization and internalization. J. Biol. Chem. 1998, 273, 21416–21422.

45. Stanasila, L., Perez, J-B., Vogel, H., Cotecchia, S. Oligomerization of the α_{1a}- and α_{1b}-adrenergic receptor subtypes. J. Biol. Chem. 2003, 278, 40239–40251.

46. Coge, F., Guenin, S.P., Renouard-Try, A., Rique, H., Ouvry, C., Fabry, N., Beauverger, P., Nicolas, J.P., Galizzi, J.P., Boutin, J.A., Canet, E. Truncated isoforms inhibit [^{3}H]prazosin binding and cellular trafficking of native human alpha-1A-adrenoceptors. Biochem. J. 1999, 343, 231–239.

47. AbdAlla, S., Zaki, E., Lother, H., Quitterer, U. Involvement of the amino terminus of the B(2) receptor in agonist-induced receptor dimerization. J. Biol. Chem. 1999, 274, 26079–26084.

48. He, J., Xu, J., Castleberry, A.M., Lau, A.G., Hall, R.A. Glycosylation of beta(1)-adrenergic receptors regulates receptor surface expression and dimerization. Biochem. Biophys. Res. Commun. 2002, 297, 565–572.

49. Xu, J., He, J., Castleberry, A.M., Balasubramanian, S., Lau, A.G., Hall, R.A. Heterodimerization of alpha 2A- and beta 1-adrenergic receptors. J. Biol. Chem. 2003, 278, 10770–10777.

50. Overton, M.C., Blumer, K.J. The extracellular N-terminal domain and transmembrane domains 1 and 2 mediate oligomerization of a yeast G-protein-coupled receptor. J. Biol. Chem. 2002, 277, 41463–41472.

51. Overton, M.C., Chinault, S.L., Blumer, K.J. Oligomerization, biogenesis, and signaling is promoted by a glycophorin A-like dimerization motif in transmembrane domain 1 of a yeast G-protein-coupled receptor. J. Biol. Chem. 2003, 278, 49369–49377.

52. Salahpour, A., Angers, S., Mercier, J.F., Lagace, M., Marullo, S., Bouvier, M. Homodimerization of the beta 2-adrenergic receptor as a prerequisite for cell surface targeting. J. Biol. Chem. 2004, 279, 33390–33397.

53. Carrillo, J.J., López-Gimenez, J.F., Milligan, G. Multiple interactions between transmembrane helices generate the oligomeric α_{1b}-adrenoceptor. Mol. Pharmacol. 2004, 66, 1123–1137

54. Ng, G.Y., O'Dowd, B.F., Lee, S.P., Chung, H.T., Brann, M.R., Seeman, P., George, S.R. Dopamine D2 receptor dimers and receptor-blocking peptides. Biochem. Biophys. Res. Commun. 1996, 227, 200–204.

55. Guo, W., Shi, L., Javitch, J.A. The fourth transmembrane segment forms the interface of the dopamine d2 receptor homodimer. J. Biol. Chem. 2003, 278, 4385–4388.

56. Lee, S.P., O'Dowd, B.F., Rajaram, R.D., Nguyen, T., George, S.R. D2 dopamine receptor homodimerization is mediated by multiple sites of interaction, including an intermolecular interaction involving transmembrane domain 4. Biochemistry 2003, 42, 11023–11031.

57. Klco, J.M., Lassere, T.B., Baranski, T.J. C5a receptor oligomerization. I. Disulfide trapping reveals oligomers and potential contact surfaces in a G-protein-coupled receptor J. Biol. Chem. 2003, 278, 35345–35353.

16 Orphan Receptors: Promising Targets for Drug Discovery

Yumiko Saito, Zhiwei Wang, and Olivier Civelli

CONTENTS

ABSTRACT

The completion of the human genome project led to identification of approximately 720 genes belonging to the G protein-coupled receptor (GPCR) superfamily. Among these are a number of "orphan" GPCRs, receptors with no known endogenous ligands. Although orphan GPCRs are genes without functions, they offer the potential to discover new intercellular interactions and new insights for basic research and ultimately for drug discovery, but their endogenous ligands must be identified for this to happen. The deorphanization of GPCRs has been an ongoing effort since the first GPCRs were cloned by homology approaches. More recently, orphan GPCRs have been used as targets to identify novel endogenous transmitters. Since 1995, eight novel bioactive peptide families have been discovered and these have already greatly enriched our understanding of several brain-directed physiological responses. This chapter presents the strategies used in the identification of endogenous ligands of orphan GPCRs and discusses one deorphanized GPCR system, the melanin-concentrating hormone (MCH) system, that produced immediate impact on drug discovery.

16.1   INTRODUCTION

Guanine nucleotide-binding regulatory (G) protein-coupled receptors (GPCRs) constitute the largest superfamily of membrane-bound receptors. They all comprise seven membrane-spanning domains, each of which has an extracellular N-terminus and a cytoplasmic C-terminal tail. They bind neurotransmitters, neuropeptides, glycoprotein hormones, chemokines, odorants and a variety of other chemical transducers. They also regulate numerous physiological processes including neuronal excitability, hormone homeostasis, and behavior. The GPCR superfamily is of paramount importance to drug discovery, because it serves as a target for more than 60% of all the marketed drugs. GPCR-related drugs have demonstrated therapeutic benefits across a broad spectrum of diseases including pain, asthma, peptic ulcers, and hypertension.[1]

Of the 35,000 human genes, approximately 720 are GPCRs. Of these, 367 are viewed as "nonchemosensory" receptors because they are expected to be activated by ligands that are not involved in olfaction, gustation, or pheromone responses. These nonchemosensory GPCRs were the first to be considered potential drug targets. By the mid-1990s, some 200 of these GPCRs were shown to bind known transmitters,[2] leaving about 160 receptors whose natural ligands have not been identified — the so-called "orphan receptors." In view of their novelty and potential, the orphan GPCRs are exciting therapeutic targets,[3,4] but before they can be used as pharmaceutical targets, they must be "deorphanized" because drug discovery relies on receptor binding assays.

Progress in deorphanizing GPCRs has been impressive but more than 100 GPCRs remain orphans. Identification of their natural activators or transmitters is a prerequisite to understanding their functions and therapeutic potentials. This chapter summarizes the approaches used to identify the natural ligands for orphan GPCRs and describe the status of our knowledge about one deorphanized GPCR, the melanin-concentrating hormone (MCH) receptor.

16.2   SEARCH FOR ENDOGENOUS LIGANDS OF ORPHAN GPCRS

The overall strategy leading to the discovery of the natural ligands of orphan receptors followed a unique principle. An orphan GPCR is expressed by transfection into a suitable cell system (mammalian, yeast, *Xenopus laevis* cells) that is then exposed to a variety of naturally occurring molecules that may cause the activation of the GPCR. Orphan GPCR stimulation is monitored through changes in second messenger levels (Figure 16.1). The presence of the natural ligand is recognized by the change in second messenger levels. This approach has been further differentiated into a "reverse pharmacology" in which the orphan GPCR is exposed to known ligands and an "orphan receptor strategy" in which the orphan GPCR is exposed to tissue extracts and is intended to find new transmitters.[5]

The reverse pharmacology approach was the first to be applied in the late 1980s with the identification of GPCRs that were cloned by low-stringency homology screening: the dopamine D2 and the serotonin 5HT1A receptors.[5,6] At that time, orphan GPCRs were exposed to few neurotransmitters.[7] Now, with the application of high-throughput screening techniques, batteries of orphan GPCRs can be exposed to thousands of putative ligands (Figure 16.1A).

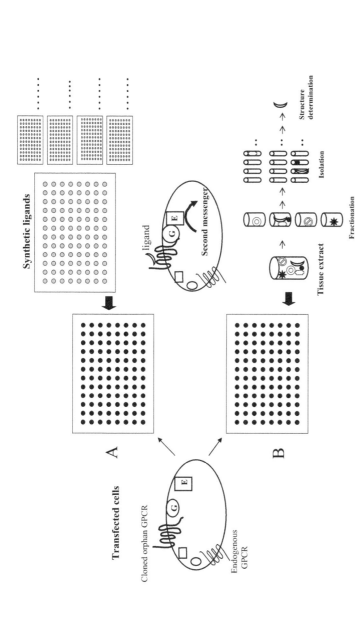

FIGURE 16.1 Approaches used to deorphanize orphan GPCRs. Orphan GPCRs are expressed in cell systems. The cells express both exogenous orphan GPCRs and endogenous GPCRs and are screened for changes in second messenger responses in high-throughput formats. A: Reverse pharmacology. Libraries of synthetic compounds are tested on the orphan GPCR-expressing cells. Because they are known compounds, this approach leads to the pairing of orphan GPCRs to known transmitters. B: Orphan receptor strategy. Tissue extracts are tested on the orphan GPCR-expressing cells. The extracts are prepared using classical biochemical approaches. If found positive, they are further fractionated until the active component is isolated. Its structure is then determined. This approach permits the discovery of novel transmitters.

Most often, because it is easily applicable to high-throughput screens, measurement of intracellular calcium release has been the parameter of choice in monitoring orphan GPCR reactivity. The application of the reverse pharmacology approach resulted in the pairing of several dozens of orphan GPCRs to natural ligands (Table 16.1 and 16.2). Essentially, it led to the pairing of known ligands to new receptors, but also led to the discovery of one unexpected ligand, UDP-glucose.[8] The discovery of a receptor for UDP-glucose was a surprise because UDP-glucose had been considered as a carrier of sugar moieties necessary for carbohydrate biosyntheses, but not as an extracellular signaling molecule.

It should also be noted that in some cases the reverse pharmacology approach led to discrepancies. For example, OGR-1 and GPR4 were reported to be putative receptors for sphingosylphosphorylcholine and lysophosphatidylcholine, respectively.[9,10] However, these receptors are in fact proton sensors, unrelated to lysophospholipid signals.[11] Bradykinin has been claimed as a ligand for GPR100/GPCR142,[12] but relaxin-3 has also been found to activate GPCR142/GPR100.[13] Furthermore, the orphan GPR99 (=GPR88) was found to be activated by both AMP/adenosine and α-ketoglutarate[14,15] (Table 16.2). The question of identifying the endogenous ligands of GPR99 and GPR100 remains to be answered and the application of the reverse pharmacology approach is winding down because some hundred GPCRs remain as orphans and few potential natural unpaired ligands exist.

With this in mind, the orphan receptor strategy was developed to discover novel transmitters in the mid-1990s.[3] The orphan receptor is exposed to fractionated tissue extracts that potentially contain the natural transmitter. Activation of the orphan receptor is again measured by changes in second messenger responses. Positive fractions are purified until the active component is isolated and characterized (Figure 16.1B). This can lead to the identification of novel transmitters.

The first success of this strategy was the discovery in 1995 of nociceptin/orphanin FQ, a neuropeptide that bears similarity to opioid peptides.[16,17] Since then, eight novel peptide families have been discovered[18–27] (Table 16.3).

The orphan receptor strategy is demanding because it faces two unknowns: the chemical nature of the transmitter and the type of second messenger response that the orphan GPCR will induce. These unknowns, added to the fact that any mammalian cell line contains a battery of endogenous active receptors, make application of the orphan receptor strategy a process requiring careful control and constant adjustment. The knowledge of the structures of GPCRs is of help but has not led to definitive rules in defining the chemical natures and signaling pathways of orphan GPCRs.

Translating information from the human genome into functional bioactive peptides has also been employed. Gene-based expression analyses alone, however, may not provide sufficient information because post-translational modifications play important roles in the structures and activities of ligands. For example, the serine 3 residue with a 28-amino acid peptide of ghrelin is modified by an n-octanoic acid and this unique post-translational modification is essential for the activity of ghrelin.[22] Therefore, the orphan receptor strategy is expected to lead to most of the GPCR deorphanizations in the future.

TABLE 16.1
Orphan GPCRs Paired to Peptides since 1995

Ligand and Reference	Orphan Receptor	Ligand Source	Read-Out	Major Functions	Year
MCH[30-34]	GPR24 (=SLC-1)	Brain, SL	cAMP, [Ca²⁺], GIRKs	Feeding	1999
UTII[65,66]	GPR14 (SENR in rats)	Brain, SL	[Ca²⁺]	Vasoconstriction	1999
Motilin[67]	GPR38	SL	[Ca²⁺]	Vasodilatation (indirect)	1999
NMU[68,69]	FM-3/4 (NMU1 and 2)	Brain, SL	[Ca²⁺]	Uterine contraction, unknown	2000
NPFF/NPAF[70,71]	HLWAR77 (hNPFF2)	SL	[Ca²⁺]	Pain	2000
MCH[45-47]	SLT	SL	[Ca²⁺]	Feeding	2001
PK-1/2[72,73]	GPR73	Bovine milk, SL	[Ca²⁺]	Angiogenesis, circadian rhythm	2002
BAM22[74]	SNSR3 and 4 (MrgX1)	SL	[Ca²⁺]	Unknown	2002
Relaxin[75]	LGR7/8	SL	cAMP	Pregnancy	2002
Cortistatin14[76]	MrgX2	SL	[Ca²⁺]	Unknown	2003
Bradykinin[12]	GPR100 (=GPCR142)	SL	[Ca²⁺]	Pain	2003
Relaxin-3[13,77]	GPCR135/142	Brain (for GPCR135)	cAMP	Pregnancy, unknown	2003
FLP-18/21[78*]	NRP-1	SL	GIRKs	Social feeding in *C. elegans*	2003
Stunted A/B[79*]	Methuselah	Drosophila	[Ca²⁺]	Aging in Drosophila	2004
hRFRP-1/3[80]	OT7T022 (GPR147, NPFF1R1)	*In silico* SL	cAMP	Unknown	2000
NPB[81]	GPR7	*In silico* SL	cAMP	Feeding, unknown	2002
RF-amide (QRFP)[82]	GPR103 (AQ27)	*In silico* SL	[Ca²⁺]	Unknown	2003

Note: Since 1995, CGRP,[83] GALP,[84] and TIP39[85] were also paired to GPCRs but these were already known receptors. SL = synthetic ligand. * = invertebrate. GIRK = G protein–gated inwardly rectifying potassium channels. CGRP = calcitonin gene–related peptide. MCH = melanin-concentrating hormone. UT II = urotensin II. NMU = neuromedin U. NPFF/NPAF = neuropeptide FF/AF. NPB = neuropeptide B. PK-1/2 = prokineticin-1/2.

TABLE 16.2
Orphan GPCRs Paired to Small Molecules and Chemokines since 1995

Ligand and Reference	Orphan Receptor	Ligand Source	Read-Out	Major Functions	Year
Anaphylatoxin C3a[86]	C3A receptor	SL	cAMP	Inflammation	1996
LTB4[87]	BLTR	SL	cAMP, [Ca^{2+}] etc.	Bronchial constriction	1997
S1P[88–92]	Edg1/3/5/6/8	SL	cAMP, [Ca^{2+}] etc.	Cell differentiation/growth	1998–2000
LPA[93–95]	Edg2/4/7	SL	Luciferase etc.	Cell differentiation/growth	1998–2000
LTD4[96]	HG55 (CysLT1)	SL	[Ca^{2+}]	Bronchial constriction	1999
Histamine[97]	Histamine H3	SL	cAMP	Analgesic action, dementia	1999
UDP-glucose[8]	GPR105 (KIAA0001)	SL	[Ca^{2+}], β-gal	Unknown	2000
SPC[9]	OGR-1	SL	[Ca^{2+}]	Cell proliferation	2000
LTB4[98]	JULF2	SL	[Ca^{2+}], cAMP	Bronchial constriction	2000
Histamine[99]	GPRv53 (=H4-R)	SL	cAMP	Immune modulator	2000
LTC4 and LTD4[100]	CysLT2	SL	[Ca^{2+}]	Bronchial constriction	2000
Eskine[101]	GPR2	SL	[Ca^{2+}]	Chemotactic function	2000
LPC[102]	G2A	SL	[Ca^{2+}]	Macrophage function	2001
SPC and LPC[10]	GPR4	SL	[Ca^{2+}]	Cell proliferation	2001
ADP[103,104]	P2Y12 GPR86(P2Y13)	SL	IP3, cAMP	Platelet aggregation	2001

Ligand	Receptor	Tissue/Assay	Readout	Function	Year
Psychosine[105]	TDAG8	SL	cAMP	Globoid cell formation	2001
Tyramine[106]	TA1, TA2	SL	cAMP	Vasoconstriction	2001
5-oxo-ETE[107]	TG1019	SL	GTPγS	Chemotactic function	2002
Bile acid[108]	BG37 (=TGR5)	SL	cAMP	Macrophage function	2002
Adenine[109]	MrgA10	Brain	$[Ca^{2+}]$	Unknown	2002
β-Alanine[110]	TGR7 (=MrgD)	SL	$[Ca^{2+}]$	Pain, unknown	2003
Medium and long fatty acid[111]	GPR40	SL	$[Ca^{2+}]$	Insulin regulation	2003
Short fatty acid[112,113]	GPR41/43	SL	β-gal	Leptin production	2003
Nicotinic acid[114]	HM74A/B	SL	GTPγS binding	Lipid lowering	2003
Proton[11]	OGR1, GPR4	pH change	cAMP	Cell proliferation	2003
Succinic acid[4]	GPR91	Kidney	$[Ca^{2+}]$	Blood pressure	2004
α-Ketoglutarate[14]	GPR99	SL	$[Ca^{2+}]$	Unknown	2004
AMP and adenosine[15]	GPR80 (=GPR99, P2Y15)	SL	$[Ca^{2+}]$	Unknown	2004

Note: SL = synthetic ligand. LTB4 = leukotriene B4. S1P = sphingosine 1-phosphate. LPA = lysophosphatidic acid. LTD4 = leukotriene D4. SPC = sphingosylphosphorylcholine. LPC = lysophosphatidylcholine. ADP = adenosine diphosphate. 5-oxo-ETE = 5-oxo-6*E*,8*Z*,11*Z*,14*Z*-eicosatetraenoic acid.

TABLE 16.3
Novel Peptides Discovered Using Orphan Receptor Strategy

Ligand and Reference	Orphan Receptor	Ligand Source	Read-Out	Major Functions	Year
Nociceptin/OFQ[16,17]	ORL-1	Brain	cAMP	Stress, pain	1995
Hypocretins/OXs[18,19]	HFGAN72	Brain	[Ca²⁺]	Feeding, sleep-wakefulness	1998
PrRP[20]	GPR10	Brain	Arachidonic acid	Sleep, absence seizure, blood pressure	1998
Apelin[21]	APJ	Stomach	Ext. pH	Blood pressure, unknown	1998
Ghrelin[22]	GHSR	Stomach	[Ca²⁺]	Feeding, GH secretion	1999
Metastin (KiSS/kisspeptin)[23]	GPR54	Brain	[Ca²⁺]	Cell proliferation, development	2001
NPB and NPW[24,25]	GPR7/8	Brain	cAMP, Xenopus melanophores	Feeding, unknown	2002
NPS[26,27]	NPS receptor	Brain	cAMP, [Ca²⁺], Arachidonic acid	Arousal, anxiety, asthma	2002, 2004

Note: OFQ – orphanin FQ. OX = orexin. PrRp = prolactin-releasing peptide. GALP = galanin-like peptide. NBW = neuropeptide W. NPS = neuropeptide S.

16.3 EXAMPLE OF DEORPHANIZATION: MCH SYSTEM AND ITS IMPACT ON DRUG DISCOVERY

The SLC-1 orphan GPCR was originally discovered as an expressed sequence tag exhibiting about 40% homology in its hydrophobic domains to the five somatostatin receptors.[28] A rat ortholog that shares 91% overall sequence identity to the human SLC-1 receptor but is 49 amino acids shorter in its N-terminal segment[29] was later identified. The existence of a shorter human form was also reported. In 1999, five independent groups, including our own, reported the identity of the natural ligand of SLC-1 by using the two different strategies described for the deorphanization of orphan GPCRs.[30–34] First, three groups described the successful application of the orphan receptor strategy. They used brain extracts as starting materials and although they reached the same result, they monitored SLC-1 activity via three different second messenger responses (calcium influx with chimeric Gα protein, cAMP inhibition assay, and G protein-gated potassium channels).[31–33] Two other groups used the reverse pharmacology approach and screened large libraries of synthetic molecules as potential activators of SLC-1[30,34] and monitored SLC-1 reactivity by calcium influx measurements. Each group reached the same conclusion: SLC-1 binds the known peptide MCH.

MCH is a 19-amino acid cyclic peptide that was originally discovered in teleost fish, where it acts as a skin-color-regulating hormone.[35] MCH in mammals has been implicated in a variety of physiological functions. Recently, however, its central role in the control of feeding has been the center of attention.[36] MCH injections into the lateral ventricles of rats led to increased food intake.[37,38] MCH knockout mice are hypophagic and exhibit increased metabolic rates, resulting in decreased body weight and body fat content.[39] Conversely, transgenic mice overexpressing MCH show a tendency to gain more weight than their wild-type counterparts, displaying hyper-leptinoma, glucose, and insulin.[40]

Another finding is that the MCH system is regulated by energy homeostasis and interacts with other energy metabolic modulators such as leptin. Prepro-MCH mRNA is upregulated both in fasted animals and in leptin-deficient *ob/ob* mice.[37] The MCH system also plays a key role as a downstream mediator of leptin action and is required for the obesity of leptin deficiency. Crossing MCH knockout mice with mice with obesity due to genetic leptin deficiency (*ob/ob*) attenuates phenotypic manifestations of leptin deficiency.[41] The marked reduction in weight in these double null mice is secondary to decreased total fat body fat. The mice display significantly increased locomotor activity and thermoregulation as compared to *ob/ob* mice but are surpris- ingly more hyperphagic than *ob/ob* mice. This indicates that the weight loss induced by the absence of MCH in *ob/ob* mice is resulted in increased energy expenditures. Thus, the importance of the MCH system as a potential candidate in obesity treatment has been well documented. However, the system could not be targeted in drug discovery due to the lack of adequate receptor binding assays. The deorphanization of SLC-1 changed this totally; the MCH system could be studied from both the ligand and receptor standpoints.

The exogenously expressed MCH-1 receptor (MCH-1R) coupled with various second messenger systems, including elevation of intracellular Ca^{2+} levels, inhibition of forskolin-stimulated cyclic AMP production, and activation of the mitogen-acti- vated protein kinase cascade.[30,31,34,42] Whereas it was known that MCH is prominently expressed only in two brain areas known to be involved in feeding behavior — the perikarya of the lateral hypothalamus and the zona incerta[43] — the MCH-1 receptor is widely distributed in the central nervous system and is found in particular in regions involved in rewarding behavior, feeding behavior, and metabolic regulation, for example, the arcuate nucleus, ventromedial hypothalamic nucleus, and nucleus accumbens in rats.[34,44]

A second high-affinity receptor for MCH was identified based on weak homol- ogy to MCH-1R.[45–47] This MCH-2 receptor appears to be coupled primarily through the $G\alpha_q$ signaling pathway.[46,47] Most importantly, the MCH-2 receptor is not expressed in a functional form in rodents but is present in dogs, ferrets, rhesus monkeys, and humans.[48] The distribution of the MCH-2 receptor in humans is a matter of controversy, as it was reported to be expressed mainly in the arcuate nucleus and the ventromedial hypothalamic nucleus,[46] although another study did not detect it in the human hypothalamus.[47] The fact that the MCH-2 receptor is absent in rodents dramatically impairs studies of its function and use in drug discovery.

MCH-1 receptor deficient (MCHR1–/–) mice that are lean and exhibit increased metabolic rates secondary to a hyperactive phenotype[49,50] have been developed. They are surprisingly hyperphagic, suggesting that the leanness observed in MCHR1–/– mice is due to stimulated energy expenditure. The discrepancy in the phenotypes of the MCH precursor knockout mice and the MCHR1–/– mice is a matter of debate that could be explained by the ablation of two additional unknown peptides, NEI and NGE,[36] encoded by the same precursor in MCH precursor knockout mice, but whose receptors are unknown.

Behavioral studies with agonists have been central in demonstrating a physiological role for the MCH system in regulating both food intake and energy homeostasis. Acute central administration of MCH leads to a rapid and significant increase in food intake,[38] and the potency at increasing food consumption is correlated with the affinity of the agonist for the MCH-1 receptor.[51] This suggests that the MCH-1 receptor directly mediates the orexigenic effects of MCH. Furthermore, chronic intracerebroventricular infusion of MCH or synthetic MCH-1 receptor agonists induced obesity in rodents.[52,53] The weight gain was accompanied by hyperphagia, reduced core temperature, and stimulated lipogenic activity in liver and white adipose tissue. This suggests that MCH may play a critical role in the development of obesity by modulating energy homeostasis.

Three MCH-1 receptor antagonists have confirmed the notion that pharmacological blockade of MCH-1 receptor is a promising therapeutic approach for obesity. T-226296, the first reported nonpeptide MCH-1 receptor-selective antagonist, effectively blocks food intake stimulated by high doses of intracerebroventricularly administrated MCH in rats,[54] proving its activity *in vivo*. A second nonpeptide antagonist, SNAP7941, provided the first evidence that chronic oral administration of an MCH-1 receptor antagonist can effect a sustained reduction in body weight and food intake greater than that elicited by fenfluamine, an effective anorectic agent.[55]

Furthermore, this antagonist exhibited unexpected evidence that blockade of MCH-1 receptor exerts anxiolytic and antidepressant activities. SNAP7941 was reported to increase swimming activity and decrease immobility in rats during a modified force swimming test, and have a profile similar to that of the selective serotonin reuptake inhibitor, fluoxetine. This result is in line with the MCH receptor localization studies[44] and a reciprocal interaction between serotoninergic and MCH pathways.[56]

The third antagonist reported is a MCH-modified peptide, and its chronic administration in rodents led to long-term reductions in appetite, caloric efficiency, body weight, and body fat gain without an effect on lean mass.[57] These reports indicate that the MCH system may represent a therapeutic indication for the treatment of obesity. Since the discovery of the MCH-1 receptor, an increased body of genetic and pharmacological evidence points at the MCH system as playing a central role in food intake and energy metabolism. The speed at which small molecule antagonists have been reported after the original description of the MCH-1 receptor exemplifies the importance of this system in drug discovery targeting obesity and abnormal food intake behaviors.[58]

16.4 CONCLUSIONS AND PERSPECTIVES

The search for the natural ligands of orphan GPCRs has opened new fields in basic research, in particular, neurobiology. Its impact may even be stronger on pharmaceutical research. Numerous drug discovery programs were initiated as soon as the natural ligands of orphan GPCRs were found. The OFQ/nociceptin system is sought for its activities related to pain and anxiety[17,59] and the orexin/hypocretin system is important for narcolepsy and food intake.[19,60] The ghrelin and MCH systems are the targets of drug discovery programs aimed at obesity,[61] whereas the metastin system is sought for its potential in oncology.[23] Moreover, because most of these deorphanization ventures have occurred recently, it is too early to evaluate the impact that the orphan GPCR research will ultimately have on drug discovery.

The next question is how much remains to be done; about 100 orphan GPCRs remain to be deorphanized. As discussed, the search for the natural ligands cannot hope to continue relying on the reverse pharmacology approach because few ligands remain to be matched. Much depends on the application of the orphan receptor strategy. Looking at the speed with which new ligands were found since the late 1990s (Table 16.3), it is clear that the discovery of novel ligands has become a difficult and demanding task.

Several factors probably caused this decline. First, the endogenous levels of the remaining ligands may be low or very low; they may even be detectable only under special circumstances, which would make their identification extremely difficult. Another possibility is that the remaining orphan GPCRs may need particular environments to become active. They either do not reach the plasma membrane or need additional molecules or heteropolymerization to be active.[62,63] In these cases, the search for the ligand may become impossible.

It is also possible that some of the orphan GPCRs do not couple to G proteins but activate G protein-independent signaling pathways,[64] in which case the quantitation of these pathways is necessary for monitoring the orphan GPCR activation. These issues make the application of the orphan receptor strategy a very risky venture and, consequently, few dare to attempt it, but the challenge must be balanced with the impact that the discovery of a novel neurotransmitter or neuropeptide will have on our understanding of brain function and new avenues in treating disease.

REFERENCES

1. Drews, J. Drug discovery: a historical perspective. Science 2000, 287, 1960–1964.
2. Vassilatis, D.K.; Hohmann, J.G.; Zeng, H.; Li, F.; Ranchalis, J.E.; Mortrud, M.T.; Brown, A.; Rodriguez, S.S.; Weller, J.R.; Wright, A.C.; Bergmann, J.E.; Gaitanaris, G.A. The G protein-coupled receptor repertoires of human and mouse. Proc. Natl. Acad. Sci. USA 2003, 100, 4903–4908.
3. Civelli, O.; Nothacker, H.P.; Saito, Y.; Wang, Z.; Lin, S.H.; Reinscheid, R.K. Novel neurotransmitters as natural ligands of orphan G-protein-coupled receptors. Trends Neurosci. 2001, 24, 230–237.
4. Douglas, S.A.; Ohlstein, E.H.; Johns, D.G. Techniques: cardiovascular pharmacology and drug discovery in the 21st century. Trends Pharmacol. Sci. 2004, 25, 225–233.
5. Bunzow, J.R.; Van Tol, H.H.; Grandy, D.K.; Albert, P.; Salon, J.; Christie, M.; C.A. Machida, C.A.; Neve, K.A.; Civelli, O. Cloning and expression of a rat D2 dopamine receptor cDNA, Nature 1988, 336, 783–787.
6. Fargin, A.; Raymond, J.R.; Lohse, M.J.; Kobilka, B.K.; Caron, M.J.; Lefkowitz, R.J. The genomic clone G-21 which resembles a beta-adrenergic receptor sequence encodes the 5-HT1A receptor. Nature 1988, 335, 358–360.

7. Libert, F.; Parmentier, M.; Lefort, A.; Dinsart, C.; Van Sand, J.; Maenhaut, C.; Simons, M.J.; Dumon, J.E.; Vassar, G. Selective amplification and cloning of four new members of the G protein-coupled receptor family. Science 1989, 244, 569–572.

8. Chambers, J.K.; Macdonald, L.E.; Sarau, H.M.; Ames, R.S.; Freeman, K.; Foley, J.J.; Zhu, Y.; McLaughlin, M.M.; Murdock.P.; McMillan, L.; Trill, J.; Swift, A.; Aiyar, N.; Taylor, P.; Vawter, L.; Naheed, S.; Szekeres, P.; Hervieu, G.; Scott, C.; Watson, J.M.; Murphy, A.J.; Duzic, E.; Klein, C.; Bergsma, D.J.; Wilson, S.; Livi, G.P. A G protein-coupled receptor for UDP-glucose. J. Biol. Chem. 2000, 275, 10767–10771.

9. Xu, Y.; Zhu, K.; Hong, G.; Wu, W.; Baudhuin, L.M.; Xiao, Y.; Damron, D.S. Sphingosylphosphorylcholine is a ligand for ovarian cancer G protein-coupled receptor 1. Nat. Cell Biol. 2000, 2, 261–267.

10. Zhu, K.; Baudhuin, L.M.; Hong, G.; Williams, F.S.; Cristina, K.L.; Kabarowski, J.H.; Witte, O.N.; Xu, Y. Sphingosylphosphorylcholine and lysophosphatidylcholine are ligands for the G protein-coupled receptor GPR4. J. Biol. Chem. 2001, 276, 41325–34135.

11. Ludwig, M.G.; Vanek, M.; Guerini, D.; Gasser, J.A.; Jones, C.E.; Junker, U.; Hofstetter, H.; Wolf, R.M.; Seuwen, K. Proton-sensing G protein-coupled receptors. Nature 2003, 425, 93–98.

12. Boels, K.; Schaller, H.C. Identification and characterisation of GPR100 as a novel human G-protein-coupled bradykinin receptor. Br. J. Pharmacol. 2003, 140, 932–938.

13. Liu, C.; Chen, J.; Sutton, S.; Roland, B.; Kuei, C.; Farmer, N.; Sillard, R.; Lovenberg, T.W. Identification of relaxin-3/INSL7 as a ligand for GPCR142. J. Biol. Chem. 2003, 278, 50765–50770.

14. He, W.; Miao, F.J.; Lin, D.C.; Schwandner, R.T.; Wang, Z.; Gao, J.; Chen, J.L.; Tian, H.; Ling, L. Citric acid cycle intermediates as ligands for orphan G protein-coupled receptors. Nature 2004, 429, 188–193.

15. Inbe, H.; Watanabe, S.; Miyawaki, M.; Tanabe, E.; Encinas, J.A. Identification and characterization of a cell-surface receptor, P2Y15, for AMP and adenosine. J. Biol. Chem. 2004, 279, 9790–9799.

16. Meunier, J.C.; Mollereau, C.; Toll, L.; Suaudeau, C.; Moisand, C.; Alvinerie, P.; Butour, J-L.; Guillemot, J-C.; Ferrara, P.; Monsarrat, B.; Mazargui, H.; Vassar, G.; Parmentier, M.; Costentin, J. Isolation and structure of the endogenous agonist of opioid receptor-like ORL_1 receptor. Nature 1995, 377, 532–535.

17. Reinscheid, R.K.; Nothacker, H.P.; Bourson, A.; Ardati, A.; Henningsen, R.A.; Bunzow, J.R.; Grandy, D.K.; Langen, H.; Monsma, F.J, Jr.; Civelli, O. Orphanin FQ: a neuropeptide that activates an opioid-like G protein-coupled receptor. Science 1995, 270, 792–794

18. de Lecea, L.; Kilduff, T.S.; Peyron, C.; Gao, X.; Foye, P.E.; Danielson, P.E.; Fukuhara, C.; Battenberg, E.L.; Gautvik, V.T.; F.S. Bartlett, F.S.; Frankel, W.N., 2nd; van den Pol, A.N.; Bloom, F.E.; Gautvik, K.M.; Sutcliffe, J.G. The hypocretins: hypothalamus-specific peptides with neuroexcitatory activity. Proc. Natl. Acad. Sci. USA 1998, 95, 322–327.

19. Sakurai, T.; Amemiya, A.; Ishii, M.; Matsuzaki, I.; Chemelli, R.M.; Tanaka, H.; Williams, S.C.; Richardson, J.A.; Kozlowski, G.P.; Wilson, S.; Arch, J.R.; Buckingham, R.E.; Haynes, A.C.; Carr, S.A.; Annan, R.S.; McNulty, D.E.; Liu, W.S.; Terrett, J.A.; Elshourbagy, N.A.; Bergsma, D.J.; Yanagisawa, M. Orexins and orexin receptors: a family of hypothalamic neuropeptides and G protein-coupled receptors that regulate feeding behavior. Cell 1998, 92, 573–585.

20. Hinuma, S.; Habata, Y.; Fujii, R.; Kawamata, Y.; Hosoya, M.; Fukusumi, S.; Kitada, C.; Masuo, Y.; Asano, T.; Matsumoto, H.; Sekiguchi, M.; Kurokawa, T.; Nishimura, O.; Onda, H.; Fujino, M. A prolactin-releasing peptide in the brain. Nature 1998, 393, 272–276.

21. Tatemoto, K.; Hosoya, M.; Habata, Y.; Fujii, R.; Kakegawa, T.; Zou, M.X.; Kawamata, Y.; Fukusumi, S.; Hinuma, S.; Kitada, C.; Kurokawa, T.; Onda, H.; Fujino, M. Isolation and characterization of a novel endogenous peptide ligand for the human APJ receptor. Biochem. Biophys. Res. Commun. 1998, 251, 471–476.

22. Kojima, M.; Hosoda, H.; Date, Y.; Nakazato, M.; Matsuo, H; Kangawa, K. Ghrelin is a growth-hormone-releasing acylated peptide from stomach. Nature 1999, 402, 656–660.

23. Ohtaki, T.; Shintani, Y.; Honda, S.; Matsumoto, H.; Hori, A.; Kanehashi, K.; Terao, S.; Kumano, S.; Takatsu, Y.; Masuda, Y.; Ishibashi, Y.; Watanabe, T.; Asada, M.; Yamada, T.; Suenaga, M.; Kitada, C.; Usuki, S.; Kurokawa, T.; Onda, H.; Nishimura, O.; Fujino, M. Metastasis suppressor gene KiSS-1 encodes peptide ligand of a G protein-coupled receptor. Nature 2001, 411, 613–617.

24. Shimomura, Y.; Harada, M.; Goto, M.; Sugo, T.; Matsumoto, Y.; Abe, M.; Watanabe, T.; Asami, T.; Kitada, C.; Mori, M.; Onda, H.; Fujino, M. Identification of neuropeptide W as the endogenous ligand for orphan G-protein-coupled receptors GPR7 and GPR8. J. Biol. Chem. 2002, 277, 35826–35832.

25. Tanaka, H.; Yoshida, T.; Miyamoto, N.; Motoike, T.; Kurosu, H.; Shibata, K.; Yamanaka, A.; Williams, S.C.; Richardson, J.A.; Tsujino, N.; Garry, M.G.; Lerner, M.R.; King, D.S.; O'Dowd, B.F.; Sakurai, T.; Yanagisawa, M. Characterization of a family of endogenous neuropeptide ligands for the G protein-coupled receptors GPR7 and GPR8. Proc. Natl. Acad. Sci. USA 2003, 100, 6251–6256.

26. Sato, S.; Shintani, Y.: Miyajima, N.; Yoshimura, K. Novel G protein-coupled receptor protein and DNA thereof. World patent application WO 02/31145 A1, April 18, 2002.

27. Xu, Y.L.; Reinscheid, R.K.; Huitron-Resendiz, S.; Clark, S.D.; Wang, Z.; Lin, S.H.; Brucher, F.A.; Zeng, J.; Ly, N.K.; Henriksen, S.J.; De Lecea, L.; Civelli, O. Neuropeptide S: A neuropeptide promoting arousal and anxiolytic-like effects. Neuron 2004, 43, 487–497.

28. Kolakowski, L.F. Jr.; Jung, B.P.; Nguyen, T.; Johnson, M.P.; Lynch, K.R.; Cheng, R.; Heng, H.H.; George, S.R.; O'Dowd, B.F. Characterization of a human gene related to genes encoding somatostatin receptors. FEBS Lett. 1996, 398, 253–258.

29. Lakaye, B.; Minet, A.; Zorzi, W.; Grisar, T. Cloning of the rat brain cDNA encoding for the SLC-1 G protein-coupled receptor reveals the presence of an intron in the gene. Biochim. Biophys. Acta 1998, 1401, 216–220.

30. Chambers, J.; Ames, R.S.; Bergsma, D.; Muir, A.; Fitzgerald, L.R.; Hervieu, G.; Dytko, G.M.; Foley, J.J.; Martin, J.; Liu, W.S.; Park, J.; Ellis, C.; Ganguly, S.; Konchar, S.; Cluderay, J.; Leslie, R.; Wilson, S.; Sarau, H.M. Melanin-concentrating hormone is the cognate ligand for the orphan G protein-coupled receptor SLC-1. Nature 1999, 400, 261–265.

31. Saito, Y.; Nothacker, H.P.; Wang, Z.; Lin, S.H.; Leslie, F.; Civelli, O. Molecular characterization of the melanin-concentrating-hormone receptor. Nature 1999, 400, 265–269.

32. Shimomura, Y.; Mori, M.; Sugo, T.; Ishibashi, Y.; Abe, M.; Kurokawa, T.; Onda, H.; Nishimura, O.; Sumino, Y.; Fujino, M. Isolation and identification of melanin-concentrating hormone as the endogenous ligand of the SLC-1 receptor. Biochem. Biophys. Res. Commun. 1999, 261, 622–626.

33. Bachner, D.; Kreienkamp, H.; Weise, C.; Buck, F.; Richter, D. Identification of melanin concentrating hormone (MCH) as the natural ligand for the orphan somatostatin-like receptor 1 (SLC-1). FEBS Lett. 1999, 457, 522–524.

34. Lembo, P.M.; Grazzini, E.; Cao, J.; Hubatsch, D.A.; Pelletier, M.; Hoffert, C.; St. Onge, S.; Pou, C.; Labrecque, J.; Groblewski, T.; O'Donnell, D.; Payza, K.; Ahmad, S.; Walker, P. The receptor for the orexigenic peptide melanin-concentrating hormone is a G-protein-coupled receptor. Nat. Cell. Biol. 1999, 1, 267–271.

35. Kawauchi, H.; Kawazoe, I.; Tsubokawa, M.; Kishida, M.; Baker, B.I. Characterization of melanin-concentrating hormone in chum salmon pituitaries. Nature 1983, 305, 321–323.

36. Pissios, P.; Maratos-Flier, E. Melanin-concentrating hormone: from fish skin to skinny mammals. Trends Endocr. Metabol. 2003, 14, 243–248.

37. Qu, D.; Ludwig, D.S.; Gammeltoft, S.; Piper, M.; Pelleymounter, M.A.; Cullen, M.J.; Mathes, W.F.; Przypek, R.; Kanarek, R.; Maratos-Flier, E. A role for melanin-concentrating hormone in the central regulation of feeding behaviour. Nature 1996, 380, 243–247.

38. Rossi, M.; Choi, S.J.; O'Shea, D.; Miyoshi, T.; Ghatei, M.A.; Bloom, S.R. Melanin-concentrating hormone acutely stimulates feeding, but chronic administration has no effect on body weight. Endocrinology 1997, 138, 351–355.

39. Shimada, M.; Tritos, N.A.; Lowell, B.B.; Flier, J.S.; Maratos-Flier, E. Mice lacking melanin-concentrating hormone are hypophagic and lean. Nature 1998, 396, 670–674.

40. Ludwig, D.S.; Tritos, N.A.; Mastaitis, J.W.; Kulkarni, R.; Kokkotou, E.; Elmquist, J.; Lowell, B.; Flier, J.S.; Maratos-Flier, E. Melanin-concentrating hormone overexpression in transgenic mice leads to obesity and insulin resistance. J. Clin. Invest. 2001, 107, 379–386.

41. Segal-Lieberman, G.; Bradley, R.L.; Kokkotou, E.; Carlson, M.; Trombly, D.J.; Wang, X.; Bates, S.; Myers, M.G.; Flier, J.S.; Maratos-Flier, E. Melanin-concentrating hormone is a critical mediator of the leptin-deficient phenotype. Proc. Natl. Acad. Sci. USA 2003, 100, 10085–10090.

42. Hawes, B.E.; Kil, E.; Green, B.; O'Neill, K.; Fried, S.; Graziano, M.P. The melanin-concentrating hormone couples to multiple G proteins to activate diverse intracellular signaling pathways. Endocrinology 2000, 141, 4524–4532.

43. Bittencourt, J.C.; Presse, F.; Arias, C.; Peto, C.; Vaughan, J.; Nahon, J.L.; Vale, W.; Sawchenko, P.E. The melanin-concentrating hormone system of the rat brain: an immuno- and hybridization histochemical characterization. J. Comp. Neurol. 1992, 319, 218–245.

44. Saito, Y.; Cheng, M.; Leslie, F.M.; Civelli, O. Expression of the melanin-concentrating hormone (MCH) receptor mRNA in the rat brain. J. Comp. Neurol. 2001, 435, 26–40.

45. Mori, M.; Harada, M.; Terao, Y.; Sugo, T.; Watanabe, T.; Shimomura, Y.; Abe, M.; Shintani, Y.; Onda, H.; Nishimura, O.; Fujino, M. Cloning of a novel G protein-coupled receptor, SLT, a subtype of the melanin-concentrating hormone receptor. Biochem. Biophys. Res. Commun. 2001, 283, 1013–1018.

46. Sailer, A.W.; Sano, H.; Zeng, Z.; McDonald, T.P.; Pan, J.; Pong, S.S.; Feighner, S.D.; Tan, C.P.; Fukami, T.; Iwaasa, H.; Hreniuk, D.L.; Morin, N.R.; Sadowski, S.J.; Ito, M.; Bansal, A.; Ky, B.; Figueroa, D.J.; Jiang, Q.; Austin, C.P.; MacNeil, D.J.; Ishihara, A.; Ihara, M.; Kanatani, A.; Van der Ploeg, L.H.; Howard, A.D.; Liu, Q. Identification and characterization of a second melanin-concentrating hormone receptor, MCH-2R. Proc. Natl. Acad. Sci. USA 2001, 98, 7564–7569.

47. Rodriguez, M.; Beauverger, P.; Naime, I.; Rique, H.; Ouvry, C.; Souchaud, S.; Dromaint, S.; Nagel, N.; Suply, S.; Audinot, V.; Boutin, J.A.; Galizzi, J.P. Cloning and molecular characterization of the novel human melanin-concentrating hormone receptor. Mol. Pharmacol. 2001, 60, 632–639.

48. Tan, C.P.; Sano, H.; Iwaasa, H.; Pan, J.; Sailer, A.W.; Hreniuk, D.L.; Feighner, S.D.; Palyha, O.C.; Pong, S.S.; Figueroa, D.J.; Austin, C.P.; Jiang, M.M.; Yu, H.; Ito, J.; Ito, M.; Guan, X.M.; MacNeil, D.J.; Kanatani, A.; Van der Ploeg, L.H.; Howard, A.D. Melanin-concentrating hormone receptor subtypes 1 and 2: species-specific gene expression. Genomics 2002, 79, 785–792.

49. Marsh, D.J.; Weingarth, D.T.; Novi, D.E.; Chen, H.Y.; Trumbauer, M.E.; Chen, A.S.; Guan, X.M.; Jiang, M.M.; Feng, Y.; Camacho, R.E.; Shen, Z.; Frazier, E.Z.; Yu, H.; Metzger, J.M.; Kuca, S.J.; Shearman, L.P.; Gopal-Truter, S.; MacNeil, D.J.; Strack, A.M.; MacIntyre, D.E.; Van der Ploeg, L.H.; Qian, S. Melanin-concentrating hormone 1 receptor-deficient mice are lean, hyperactive, and hyperphagic and have altered metabolism. Proc. Natl. Acad. Sci. USA 2002, 99, 3240–3245.

50. Chen, Y.; Hu, C.; Hsu, C.K.; Zhang, Q.; Bi, C.; Asnicar, M.; Hsiung, H.M.; Fox, N.; Slieker, L.J.; Yang, D.D.; Heiman, M.L.; Shi, Y. Targeted disruption of the melanin-concentrating hormone receptor-1 results in hyperphagia and resistance to diet-induced obesity. Endocrinology 2002, 143, 2469–2477.

51. Suply, T.; Della Zuana, O.; Audinot, A.; Rodriguez, M.; Beauverger, P.; Duhault, J.; Canet, E.; Galizzi, J.P.; Nahon, J.L.; Levens, N.; Boutin, J.A. SLC-1 receptor mediates effect of melanin-concentrating hormone on feeding behavior in rat: a structure-activity study, J. Pharmacol. Exp. Ther. 2001, 299, 137–146.

52. Della-Zuana, O.; Presse, F.; Ortola, C.; Duhault, J.; Nahon, J.L.; Levens, N. Acute and chronic administration of melanin-concentrating hormone enhances food intake and body weight in Wistar and Sprague-Dawley rats. Int. J. Obes. Relat. Metabol. Disord. 2002, 26, 1289–1295.

53. Ito, M.; Gomori, A.; Ishihara, A.; Oda, Z.; Mashiko, S.; Matsushita, H.; Yumoto, M.; Sano, S.; Tokita, H.; Moriya, M.; Iwaasa, H.; Kanatani, A. Characterization of MCH-mediated obesity in mice. Am. J. Physiol. Endocrinol. Metabol. 2003, 284, E940–E945.

54. Takekawa, S.; Asami, A.; Ishihara, Y.; Terauchi, J.; Kato, K.; Shimomura, Y.; Mori, M.; Murakoshi, H.; Kato, K.; Suzuki, N.; Nishimura, O.; Fujino, M. T-226296: a novel, orally active and selective melanin-concentrating hormone receptor antagonist. Eur. J. Pharmacol. 2002, 438, 129–135.

55. Borowsky, B.; Durkin, M.M.; Ogozalek, K.; Marzabadi, M.R.; DeLeon, J.; Lagu, B.; Heurich, R.; Lichtblau, H.; Shaposhnik, Z.; Daniewska, I.; Blackburn, T.P.; Branchek, T.A.; Gerald, C.; Vaysse, P.J.; Forray, C. Antidepressant, anxiolytic and anorectic effects of a melanin-concentrating hormone-1 receptor antagonist. Nat. Med. 2002, 8, 825–830.

56. Collin, M.; Backberg, M.; Onnestam, K.; Meister, B. 5-HA1A receptor immunoreactivity in hypothalamic neurons involved in body weight control. Neuroreport 2002, 13, 945–951.

57. Shearman, L.P.; Camacho, R.E.; Sloan Stribling, D.; Zhou, D.; Bednarek, M.A.; Hreniuk, D.L.; Feighner, S.D.; Tan, C.P.; Howard, A.D.; Van der Ploeg, L.H.; MacIntyre, D.E.; Hickey, G.J.; Strack, A.M. Chronic MCH-1 receptor modulation alters appetite, body weight and adiposity in rats. Eur. J. Pharmacol. 2003, 475, 37–47.

58. Collins, C.A.; Kym, P.R. Prospects for obesity treatment: MCH receptor antagonists. Curr. Opin. Invest. Drugs 2003, 4, 386–394.

59. Koster, A.; Montkowski, A.; Schulz, S.; Stube, E.M.; Knaudt, K.; Jenck, F.; Moreau, J.L.; Nothacker, H.P.; Civelli, O.; Reinscheid, R.K. Targeted disruption of the orphanin FQ/nociceptin gene increases stress susceptibility and impairs stress adaptation in mice. Proc. Natl. Acad. Sci. USA 1999, 96, 10444–10449.

60. Chemelli, R.M.; Willie, J.T.; Sinton, C.M.; Elmquist, J.K.; Scammell, T.; Lee, C.; Richardson, J.A.; Williams, S.C.; Xiong, Y.; Kisanuki, Y.; Fitch, T.E.; Nakazato, M.; Hammer, R.E.; Saper, C.B.; Yanagisawa, M. Narcolepsy in orexin knockout mice: molecular genetics of sleep regulation. Cell 1999, 98, 437–451.

61. Nakazato, M.; Murakami, N.; Date, Y.; Kojima, M.; Matsuo, H.; Kangawa, K.; Matsukura, S. A role for ghrelin in the central regulation of feeding. Nature 2001, 409, 194–198.

62. McLatchie, L.M.; Fraser, N.J.; Main, M.J.; Wise, A.; Brown, J.; Thompson, N.; Solari, R.; Lee, M.G.; Foord, S.M. RAMPs regulate the transport and ligand specificity of the calcitonin-receptor-like receptor. Nature 1998, 393, 333–339.

63. Kaupmann, K.; Malitschek, B.; Schuler, V.; Heid, J.; Froestl, W.; Beck, P.; Mosbacher, J.; Bischoff, S.; Kulik, A.; Shigemoto, R.; Karschin, A.; Bettler, B. GABAB-receptor subtypes assemble into functional heteromeric complexes. Nature 1998, 396, 683–687.

64. Yamauchi, T.; Kamon, J.; Ito, Y.; Tsuchida, A.; Yokomizo, T.; Kita, S.; Sugiyama, T.; Miyagishi, M.; Hara, K.; Tsunoda, M.; Murakami, K.; Ohteki, T.; Uchida, S.; Takekawa, S.; Waki, H.; Tsuno, N.H.; Shibata, Y.; Terauchi, Y.; Froguel, P.; Tobe, K.; Koyasu, S.; Taira, K.; Kitamura, T.; Shimizu, T.; Nagai, R.; Kadowaki, T. Cloning of adiponectin receptors that mediate antidiabetic metabolic effects. Nature 2003, 423, 762–769.

65. Nothacker, H.P.; Wang, Z.; McNeill, A.M.; Saito, Y.; Merten, S.; O'Dowd, B.; Duckles, S.P.; Civelli, O. Identification of the natural ligand of an orphan G protein-coupled receptor involved in the regulation of vasoconstriction. Nature Cell Biol. 1999, 1, 383–385.

66. Ames, R.S.; Sarau, H.M.; Chambers, J.K.; Willette, R.N.; Aiyar, N.V.; Romanic, A.M.; Louden, C.S.; Foley, J.J.; Sauermelch, C.F.; Coatney, R.W.; Ao, Z.; Disa, J.; Holmes, S.D.; Stadel, J.M.; Martin, J.D.; Liu, W.S.; Glover, G.I.; Wilson, S.; McNulty, D.E.; Ellis, C.E.; Elshourbagy, N.A.; Shabon, U.; Trill, J.J.; Hay, D.W.; Ohlstein, E.H.; Bergsma, D.J.; Douglas, S.A. Human urotensin-II is a potent vasoconstrictor and agonist for the orphan receptor GPR14. Nature 1999, 10, 282–286.

67. Feighner, S.D.; Tan, C.P.; McKee, K.K.; Palyha, O.C.; Hreniuk, D.L.; Pong, S.S.; Austin, C.P.; Figueroa, D.; MacNeil, D.; Cascieri, M.A.; Nargund, R.; Bakshi, R.; Abramovitz, M.; Stocco, R.; Kargman, S.; O'Neill, G.; Van Der Ploeg, L.H.; Evans, J.; Patchett, A.A.; Smith, R.G.; Howard, A.D. Receptor for motilin identified in the human gastrointestinal system. Science 1999, 284, 2184–2188.

68. Kojima, M.; Haruno, R.; Nakazato, M.; Date, Y.; Murakami, N.; Hanada, R.; Matsuo, H.; Kangawa, K. Purification and identification of neuromedin U as an endogenous ligand for an orphan receptor GPR66 (FM3). Biochem. Biophys. Res. Commun. 2000, 276, 435–438.

69. Howard, A.D.; Wang, R.; Pong, S.S.; Mellin, T.N.; Strack, A.; Guan, X.M.; Zeng, Z.; Williams, D.L. Jr.; Feighner, S.D.; Nunes, C.N.; Murphy, B.; Stair, J.N.; Yu, H.; Jiang, Q.; Clements, M.K.; Tan, C.P.; McKee, K.K.; Hreniuk, D.L.; McDonald, T.P.; Lynch, K.R.; Evans, J.F.; Austin, C.P.; Caskey, C.T.; Van der Ploeg, L.H.; Liu. Q. Identification of receptors for neuromedin U and its role in feeding. Nature 2000, 406, 70–74.

70. Elshourbagy, N.A.; Ames, R.S.; Fitzgerald, L.R.; Foley, J.J.; Chambers, J.K.; Szekeres, P.G.; Evans, N.A.; Schmidt, D.B.; Buckley, P.T.; Dytko, G.M.; Murdock, P.R.; Milligan, G.; Groarke, D.A.; Tan, K.B.; Shabon, U.; Nuthulaganti, P.; Wang, D.Y.; Wilson, S.; Bergsma, D.J.; Sarau, HM. Receptor for the pain modulatory neuropeptides FF and AF is an orphan G protein-coupled receptor. J Biol Chem. 2000, 275, 25965–25971.

71. Bonini, J.A.; Jones, K.A.; Adham, N.; Forray, C.; Artymyshyn, R.; Durkin, M.M.; Smith, K.E.; Tamm, J.A.; Boteju, L.W.; Lakhlani, P.P.; Raddatz, R.; Yao, W.J.; Ogozalek, K.L.; Boyle, N.; Kouranova, E.V.; Quan, Y.; Vaysse, P.J.; Wetzel, J.M.; Branchek, T.A.; Gerald, C.; Borowsky, B. Identification and characterization of two G protein-coupled receptors for neuropeptide FF. J. Biol. Chem. 2000, 275, 39324–39331.

72. Lin, D.C.; Bullock, C.M.; Ehlert, F.J.; Chen, J.L.; Tian, H.; Zhou, Q.Y. Identification and molecular characterization of two closely related G protein-coupled receptors activated by prokineticins/endocrine gland vascular endothelial growth factor. J. Biol. Chem. 2002, 277, 19276–19280.

73. Masuda, Y.; Takatsu, Y.; Terao, Y.; Kumano, S.; Ishibashi, Y.; Suenaga, M.; Abe, M.; Fukusumi, S.; Watanabe, T.; Shintani, Y.; Yamada, T.; Hinuma, S.; Inatomi, N.; Ohtaki, T.; Onda, H.; Fujino, M. Isolation and identification of EG-VEGF/prokineticins as cognate ligands for two orphan G-protein-coupled receptors. Biochem. Biophys. Res. Commun. 2002, 293, 396–402.

74. Lembo, P.M.; Grazzini, E.; Groblewski, T.; O'Donnell, D.; Roy, M.O.; Zhang, J.; Hoffert, C.; Cao, J.; Schmidt, R.; Pelletier, M.; Labarre, M.; Gosselin, M.; Fortin, Y.; Banville, D.; Shen, S.H.; Strom, P.; Payza, K.; Dray, A.; Walker, P.; Ahmad, S. Proenkephalin A gene products activate a new family of sensory neuron-specific GPCRs. Nat. Neurosci. 2002, 5, 201–209.

75. Hsu, S.Y.; Nakabayashi, K.; Nishi, S.; Kumagai, J.; Kudo, M.; Sherwood, O.D.; Hsueh, A.J. Activation of orphan receptors by the hormone relaxin. Science 2002, 295, 671–674.

76. Robas, N.; Mead, E.; Fidock, M. MrgX2 is a high potency cortistatin receptor expressed in dorsal root ganglion. J. Biol. Chem. 2003, 278, 44400–44404.

77. Liu, C.; Eriste, E.; Sutton, S.; Chen, J.; Roland, B.; Kuei, C.; Farmer, N.; Jornvall, H.; Sillard, R.; Lovenberg, T.W. Identification of relaxin-3/INSL7 as an endogenous ligand for the orphan G-protein-coupled receptor GPCR135. J. Biol. Chem. 2003, 278, 50754–50764.

78. Rogers, C.; Reale, V.; Kim, K.; Chatwin, H.; Li, C.; Evans, P.; de Bono M. Inhibition of *Caenorhabditis elegans* social feeding by FMRF amide-related peptide activation of NPR-1. Nat. Neurosci. 2003, 6, 1178–1185.

79. Cvejic S.; Zhu, Z.; Felice, S.J.; Berman, Y.; Huang, X.Y. The endogenous ligand Stunted of the GPCR Methuselah extends lifespan in Drosophila. Nat. Cell Biol. 2004, 6, 540–546.

80. Hinuma, S.; Shintani, Y.; Fukusumi, S.; Iijima, N.; Matsumoto, Y.; Hosoya, M.; Fujii, R.; Watanabe, T.; Kikuchi, K.; Terao, Y.; Yano, T.; Yamamoto, T.; Kawamata, Y.; Habata, Y.; Asada, M.; Kitada, C.; Kurokawa, T.; Onda, H.; Nishimura, O.; Tanaka, M.; Ibata, Y.; Fujino, M. New neuropeptides containing carboxy-terminal RFamide and their receptor in mammals. Nat. Cell Biol. 2000, 2, 703–708.

81. Fujii, R.; Yoshida, H.; Fukusumi, S; Habata, Y; Hosoya, M; Kawamata, Y.; Yano, T.; Hinuma, S.; Kitada, C.; Asami, T.; Mori, M.; Fujisawa, Y.; Fujino, M. Identification of a neuropeptide modified with bromine as an endogenous ligand for GPR7. J. Biol. Chem 2002, 277, 34010–34016.

82. Fukusumi, S.; Yoshida, H.; Fujii, R.; Maruyama, M.; Komatsu, H.; Habata, Y.; Shintani, Y.; Hinuma, S.; Fujino, M. A new peptidic ligand and its receptor regulating adrenal function in rats. J. Biol. Chem. 2003, 278, 46387–46395.

83. Aiyar, N.; Rand, K.; Elshourbagy, N.A.; Zeng, Z.; Adamou, J.E.; Bergsma, D.J.; Li, Y. A cDNA encoding the calcitonin gene-related peptide type 1 receptor. J. Biol. Chem. 1996, 271, 11325–11329.

84. Ohtaki, T.; Kumano, S.; Ishibashi, Y.; Ogi, K.; Matsui, H.; Harada, M.; Kitada, C.; Kurokawa, T.; Onda, H.; Fujino, M. 1999. Isolation and cDNA cloning of a novel galanin-like peptide (GALP) from porcine hypothalamus. J. Biol. Chem. 1999, 274, 37041–37045.

85. Usdin, T.B.; Hoare, S.R.; Wang, T.; Mezey, E.; Kowalak, J.A. TIP39: a new neuropeptide and PTH2-receptor agonist from hypothalamus. Nat. Neurosci. 1999, 2, 941–943.

86. Ames, R.S.; Li, Y.; Sarau, H.M.; Nuthulaganti, P.; Foley, J.J.; Ellis, C.; Zeng, Z.; Su, K.; Jurewicz, A.J.; Hertzberg, R.P.; Bergsma, D.J.; Kumar, C. Molecular cloning and characterization of the human anaphylatoxin C3a receptor. J. Biol. Chem. 1996, 271, 20231–20234.

87. Yokomizo, T.; Izumi, T.; Chang, K.; Takuwa, Y.; Shimizu, T. G protein-coupled receptor for leukotriene B4 that mediates chemotaxis. Nature 1997, 387, 620–624.

88. Lee, M.J.; Van Brocklyn, J.R.; Thangada, S.; Liu, C.H.; Hand, A.R.; Menzeleev, R.; Spiegel, S.; Hla, T. Sphingosine-1-phosphate as a ligand for the G protein-coupled receptor EDG-1. Science 1998, 279, 1552–1555.

89. An, S.; Bleu, T.; Huang, W.; Hallmark, O.G.; Coughlin, S.R.; Goetzl, E.J. Identification of cDNAs encoding two G protein-coupled receptors for lysosphingolipids. FEBS Lett. 1997, 417, 279–282.

90. An, S.; Zheng, Y.; Bleu, T. Sphingosine 1-phosphate-induced cell proliferation, survival, and related signaling events mediated by G protein-coupled receptors Edg3 and Edg5. J. Biol. Chem. 2000, 275, 288–296.

91. Graler, M.H.; Bernhardt, G.; Lipp, M. EDG6, a novel G protein-coupled receptor related to receptors for bioactive lysophospholipids, is specifically expressed in lymphoid tissue. Genomics 1998, 53, 164–169.

92. Takuwa, Y.; Takuwa, N.; Sugimoto, N. The Edg family G protein-coupled receptors for lysophospholipids: their signaling properties and biological activities. J. Biochem. (Tokyo) 2002, 131, 767–71.

93. Erickson J.R.; Wu, J.J.; Goddard, J.G.; Tigyi, G.; Kawanishi, K.; Tomei, L.D.; Kiefer, M.C. Edg-2/Vzg-1 couples to the yeast pheromone response pathway selectively in response to lysophosphatidic acid. J. Biol. Chem. 1999, 273, 1506–1510.

94. An, S.; Bleu, T.; Hallmark, O.G.; Goetzl, E.J. Characterization of a novel subtype of human G protein-coupled receptor for lysophosphatidic acid. J. Biol. Chem. 1998, 273, 7906–7910.

95. Bandoh, K.; Aoki, J.; Hosono, H.; Kobayashi, S.; Kobayashi, T.; Murakami-Murofushi, K.; Tsujimoto, M.; Arai, H.; Inoue, K. Molecular cloning and characterization of a novel human G-protein-coupled receptor, EDG7, for lysophosphatidic acid. J. Biol. Chem. 1999, 274, 27776–27785.

96. Lynch, K.R.; O'Neill, G.P.; Liu, Q.; Im, D.S.; Sawyer, N.; Metters, K.M.; Coulombe, N.; Abramovitz, M.; Figueroa, D.J.; Zeng, Z.; Connolly, B.M.; Bai, C.; Austin, C.P.; Chateauneuf, A.; Stocco, R.; Greig, G.M.; Kargman, S.; Hooks, S.B.; Hosfield, E.; Williams, D.L Jr.; Ford-Hutchinson, A.W.; Caskey, C.T.; Evans, J.F. Characterization of the human cysteinyl leukotriene CysLT1 receptor. Nature 1999, 399, 789–793.

97. Lovenberg, T.W.; Roland, B.L.; Wilson, S.J.; Jiang, X.; Pyati, J.; Huvar, A.; Jackson, M.R.; Erlander, M.G. Cloning and functional expression of the human histamine H3 receptor. Mol. Pharmacol. 1999, 55, 1101–1107.

98. Kamohara, M.; Takasaki, J.; Matsumoto, M.; Saito, T.; Ohishi, T.; Ishii, H.; Furuichi, K. Molecular cloning and characterization of another leukotriene B4 receptor. J. Biol. Chem. 2000, 275, 27000–27004.

99. Oda, T.; Morikawa, N.; Saito, Y.; Masuho, Y.; Matsumoto, S. Molecular cloning and characterization of a novel type of histamine receptor preferentially expressed in leukocytes. J. Biol. Chem. 2000, 275, 36781–36786.

100. Nothacker, H.P.; Wang, Z.; Zhu, Y.; Reinscheid, R.K.; Lin, S.H.; Civelli, O. Molecular cloning and characterization of a second human cysteinyl leukotriene receptor: discovery of a subtype selective agonist. Mol. Pharmacol. 2000, 58, 1601–1608.

101. Jarmin, D.I.; Rits, M.; Bota, D.; Gerard, N.P.; Graham, G.J.; Clark-Lewis, I.; Gerard, C. Cutting edge: identification of the orphan receptor G protein-coupled receptor 2 as CCR10, a specific receptor for the chemokine Eskine. J. Immunol. 2000, 164, 3460–3464.

102. Kabarowski, J.H.; Zhu, K.; Le, L.Q.; Witte, O.N.; Xu, Y. Lysophosphatidylcholine as a ligand for the immunoregulatory receptor G2A. Science 2000, 293, 702–705.

103. Hollopeter, G.; Jantzen, H.M.; Vincent, D.; Li, G.; England, L.; Ramakrishnan, V.; Yang, R.B.; Nurden, P.; Nurden, A.; Julius, D.; Conley, P.B. Identification of the platelet ADP receptor targeted by antithrombotic drugs. Nature 2001, 409, 202–207.

104. Zhang, F.L.; Luo, L.; Gustafson, E.; Lachowicz, J.; Smith, M.; Qiao, X.; Liu, Y.H.; Chen, G.; Pramanik, B.; Laz, T.M.; Palmer, K.; Bayne, M.; Monsma, F.J. Jr. ADP is the cognate ligand for the orphan G protein-coupled receptor SP1999. J. Biol. Chem. 2001, 276, 8608–8615.

105. Im, D.S.; Heise, C.E.; Nguyen, T.; O'Dowd, B.F.; Lynch, K.R. Identification of a molecular target of psychosine and its role in globoid cell formation. J. Cell Biol. 2001, 153, 429–434.

106. Borowsky, B.; Adham, N.; Jones, K.A.; Raddatz, R.; Artymyshyn, R.; Ogozalek, K.L.; Durkin, M.M.; Lakhlani, P.P.; Bonini, J.A.; Pathirana, S.; Boyle, N.; Pu, X.; Kouranova, E.; Lichtblau, H.; Ochoa, F.Y.; Branchek, T.A.; Gerald, C. Trace amines: identification of a family of mammalian G protein-coupled receptors. Proc. Natl. Acad. Sci. USA 2001, 98, 8966–8971.

107. Hosoi, T.; Koguchi, Y.; Sugikawa, E.; Chikada, A.; Ogawa, K.; Tsuda, N.; Suto, N.; Tsunoda, S.; Taniguchi, T.; Ohnuki, T. Identification of a novel human eicosanoid receptor coupled to G(i/o). J. Biol. Chem. 2002, 277, 31459–31465.

108. Maruyama, T.; Miyamoto, Y.; Nakamura, T.; Tamai, Y.; Okada, H.; Sugiyama, E.; Nakamura, T.; Itadani, H.; Tanaka, K. Identification of membrane-type receptor for bile acids (M-BAR). Biochem. Biophys. Res. Commun. 2002. 298, 714–719.

109. Bender, E.; Buist, A.; Jurzak, M.; Langlois, X.; Baggerman, G.; Verhasselt, P.; Ercken, M.; Guo, H.Q.; Wintmolders, C.; Van den Wyngaert, I.; Van Oers, I.; Schoofs, L.; Luyten, W. Characterization of an orphan G protein-coupled receptor localized in the dorsal root ganglia reveals adenine as a signaling molecule. Proc. Natl. Acad. Sci. USA 2002, 99, 8573–8378.

110. Shinohara, T.; Harada, M.; Ogi, K.; Maruyama, M.; Fujii, R.; Tanaka, H.; Fukusumi, S.; Komatsu, H.; Hosoya, M.; Noguchi, Y.; Watanabe, T.; Moriya, T.; Itoh, Y.; Hinuma, S. Identification of a G protein-coupled receptor specifically responsive to beta-alanine. J. Biol. Chem. 2004, 279, 23559–23564.

111. Briscoe, C.P.; Tadayyon, M.; Andrews, J.L.; Benson, W.G.; Chambers, J.K.; Eilert, M.M.; Ellis, C.; Elshourbagy, N.A.; Goetz, A.S.; Minnick, D.T.; Murdock, P.R.; Sauls, H.R. Jr.; Shabon, U.; Spinage, L.D.; Strum, J.C.; Szekeres, P.G.; Tan, K.B.; Way, J.M.; Ignar, D.M.; Wilson, S.; Muir, A.I. The orphan G protein-coupled receptor GPR40 is activated by medium and long chain fatty acids. J. Biol. Chem. 2003, 278, 11303–11311.

112. Brown, A.J.; Goldsworthy, S.M.; Barnes, A.A.; Eilert, M.M.; Tcheang, L.; Daniels, D.; Muir, A.I.; Wigglesworth, M.J.; Kinghorn, I.; Fraser, N.J.; Pike, N.B.; Strum, J.C.; Steplewski, K.M.; Murdock, P.R.; Holder, J.C.; Marshall, F.H.; Szekeres, P.G.; Wilson, S.; Ignar, D.M.; Foord, S.M.; Wise, A.; Dowell, S.J. The Orphan G protein-coupled receptors GPR41 and GPR43 are activated by propionate and other short chain carboxylic acids. J. Biol. Chem. 2003, 278, 11312–11319.

113. Xiong, Y.; Miyamoto, N.; Shibata, K.; Valasek, M.A.; Motoike, T.; Kedzierski, R.M.; Yanagisawa, M. Short-chain fatty acids stimulate leptin production in adipocytes through the G protein-coupled receptor GPR41. Proc. Natl. Acad. Sci. USA 2004, 101, 1045–1050.

114. Wise, A.; Foord, S.M.; Fraser, N.J.; Barnes, A.A.; Elshourbagy, N.; Eilert, M.; Ignar, D.M.; Murdock, P.R.; Steplewski, K.; Green, A.; Brown, A.J.; Dowell, S.J.; Szekeres, P.G.; Hassall, D.G.; Marshall, F.H.; Wilson, S.; Pike, N.B. Molecular identification of high and low affinity receptors for nicotinic acid. J. Biol. Chem. 2003, 278, 9869–9874.

Index